Theory of
MACHINES

Theory of
MACHINES

Vivek Kumar

MTech, LMISTE, LMISNT, LMAEE, FIE, PhD (pursuing)
Professor and Head
Department of Mechanical Engineering
Amity School of Engineering and Technology
Amity University, Uttar Pradesh, Noida

Chapters contributed by

Rahul Sindhwani
BTech (Mech), MTech (Mech), PhD (pursuing)
Assistant Professor, Department of Mechanical Engineering
Amity School of Engineering and Technology
Amity University, Uttar Pradesh, Noida

Edited by

Prerna Vivek

CBS

CBS Publishers & Distributors Pvt Ltd

New Delhi • Bengaluru • Chennai • Kochi • Kolkata • Mumbai
Hyderabad • Nagpur • Patna • Pune • Vijayawada

Theory of
Machines

ISBN: 978-93-86478-04-7

Copyright © Author and Publisher

First Edition: 2017

Published by Satish Kumar Jain and produced by Varun Jain for
CBS Publishers & Distributors Pvt Ltd
4819/XI Prahlad Street, 24 Ansari Road, Daryaganj, New Delhi 110 002, India.
Ph: 23289259, 23266861, 23266867   Website: www.cbspd.com
Fax: 011-23243014   e-mail: delhi@cbspd.com; cbspubs@airtelmail.in.

Corporate Office: 204 FIE, Industrial Area, Patparganj, Delhi-110092
Ph: 4934 4934   Fax: 4934 4935   e-mail: publishing@cbspd.com; publicity@cbspd.com

Branches

- **Bengaluru:** Seema House 2975, 17th Cross, K.R. Road,
  Banasankari 2nd Stage, Bengaluru 560 070, Karnataka
  Ph: +91-80-26771678/79   Fax: +91-80-26771680   e-mail: bangalore@cbspd.com
- **Chennai:** 7, Subbaraya Street, Shenoy Nagar, Chennai 600 030, Tamil Nadu
  Ph: +91-44-26680620, 26681266   Fax: +91-44-42032115   e-mail: chennai@cbspd.com
- **Kochi:** Ashana House, No. 39/1904, AM Thomas Road, Valanjambalam,
  Ernakulam 682 018, Kochi, Kerala
  Ph: +91-484-4059061-65   e-mail: kochi@cbspd.com
- **Kolkata:** 6/B, Ground Floor, Rameswar Shaw Road, Kolkata-700 014, West Bengal
  Ph: +91-33-22891126, 22891127, 22891128   e-mail: kolkata@cbspd.com
- **Mumbai:** 83-C, Dr E Moses Road, Worli, Mumbai-400018, Maharashtra
  Ph: +91-22-24902340/41   Fax: +91-22-24902342   e-mail: mumbai@cbspd.com

Representatives

- **Hyderabad** 0-9885175004   • **Nagpur**   0-9021734563   • **Patna** 0-9334159340
- **Pune**   0-9623451994   • **Vijayawada** 0-9000660880

Printed at:
Rashtriya Printers, Dilshad Garden, Delhi, India

to
my beloved father

Late Shri Balkishan Gupta

Foreword

It is with a great pleasure, I am writing the foreword to the book *Theory of Machines*. The author, Prof Vivek Kumar, has developed this volume with his long teaching experience in the field of theory of machines. The book can be used by undergraduate students of engineering as a textbook and for various competitive and entrance examinations. The book has been  written in a concise manner and use of language has been kept to minimum which makes the subject easier to understand and reflects the vast experience of the author.

The objective of this book is to educate students with basic concepts of mechanisms, force analysis, balancing of machines, friction, brakes and dynamometers, cams, gears, various types of governors, gyroscope and undesirable vibrations produced in machinery. Numerous illustrated examples and problems at the end of each chapter provide a good resource for the students to practise the use of basic principles given in the text. The content of this book covers the syllabi of most of the technical institutes/universities.

Salient features
- Theory in a concise form
- Vast coverage of the subject
- Numerous solved examples with illustrations
- Objective type questions and answers

I congratulate the author, Prof Vivek Kumar and chapters' contributor, Mr Rahul Sindhwani for their sincere efforts in bringing out this students friendly textbook. I expect that this book will certainly motivate a few students to specialize in key areas of the book for higher studies and projects based on these areas.

I strongly recommend this book to students and candidates preparing for various competitive and entrance examinations.

Prof (Dr) Ravi Prakash
PhD (Cranfield, UK), MSc (Salford, UK)
Advisor and Head
Amity Institute of Technology
Home Page: https://sites.google.com/site/profraviprakash/

Preface

Theory of Machines has been written keeping in mind the requirements of second and third year mechanical engineering students of various universities/engineering colleges. The content of this book covers the syllabi of most of the technical institutes/ universities. The objective of this book is to educate students with basic concepts of mechanisms, force analysis, balancing of machines, friction, brakes and dynamometers, cams, gears, various types of governors, gyroscope and undesirable vibrations produced in machinery. Theory has been written in a very concise form and simple language that makes the subject easy to understand.

Each chapter is supplemented with detailed illustrations, solved examples, problems for practice, and objective questions and answers to instill the basic underlying concepts.

Salient features
- Theory in a concise form
- Vast coverage of the subject
- Several solved examples with elaborate illustrations
- Objective type questions and answers

I am grateful to many of my colleagues for their comments and encouragement, namely, Prof (Dr) J K Rai (Amity University); Prof (Dr) Sudhir Kumar Singh (Skyline Institute); Prof (Dr) S S Chauhan (Skyline Institute); Prof (Dr) Ajay Chaudhury (NIET); Dr V K Dwivedi; Dr Asim Qadri (GCET); Dr Praveen Pachauri (NIET); Dr Sanjay Yadav (ITS); Dr Ashish Malik (ABES); Dr Bhumendra Kumar; Dr Devendra Vashistha (Manav Rachna University) and Mr Freedon Daniel (SRM University).

I extend my deep gratitude to Prof (Dr) Ravi Prakash, Head and Advisor, Engineering Design and Research, Amity University, Noida for his extraordinary support and writing the foreword to this book. I also extend deep gratitude to my mentor and guide Prof (Dr) L M Das (IIT, Delhi) for his constant help and encouragement.

I am thankful to Prof K M Soni, Deputy Dean, Engineering and Technology, Amity University; Prof (Dr) Abhay Bansal and Prof (Dr) M K Dutta, Joint Acting Head, Amity School of Engineering and Technology for being a constant support and source of encouragement to me. My colleague, Mr Rahul Sindhwani, Amity University has helped me in writing this book by contributing a few chapters. My colleagues, Dr G K Singh; Mr Shyamal Samant; Mr Priyank Srivastava; Mr K M Agrawal; Mr Sanjeev Sharma; Mr Sumit Sharma; Mr Vipin Kaushik; Mr Shubham Sharma; Mr Naveen Daniel and Mr Naveen Kumar, Amity University, Noida, have given their valuable input and helped me in various ways. I am thankful to all my faculty, colleagues and staff for this endeavor.

I wish to thank the senior faculty members from various institutions, Dr Anil Sethi (Galgotias); Dr U K Vates (Amity University); Dr D K Gupta (IPEC); Dr S N Satapathy (Galgotias University); Mr Rohan (Accurate); Mr Atishey Mittal (SRM University) and Mr Gagan Varshney (GNIT) for reviewing the manuscript. I am also thankful to Ms Anu Kamal; Ms Richa; Mr Leeladhar and Mr Bhupendra Sharma for helping me in writing and editing the manuscript of the book.

Last but not the least, I would like to thank my wife Smt Prerna, who not only took care of my needs during this period but helped me in writing the manuscript and the two children, Chaitanya and Richa for their invaluable cooperation.

I sincerely wish to express my appreciation to Mr S K Jain (CMD); Mr Y N Arjuna (Senior Vice President—Publishing, Editorial and Publicity); Mr Sumit Behl (Marketing Executive), CBS Publishers & Distributors for taking the initiative and encouraging me to write the book. I am thankful to Mr Kuldeep for the nice presentation of the book.

Utmost care has been taken to minimize the errors and typing mistakes. I will gratefully receive and acknowledge every comment and suggestions from the teachers and the students leading to improvements in the text as well as in the solved examples.

Vivek Kumar

Contents

1

Introduction

1.1 THEORY OF MACHINES

Theory of machines is defined as the branch of engineering science which deals with the study of relative motion between the various parts of a machine and forces acting on them (Fig. 1.1). Basic principles of mechanics are used in theory of machines for analysis of motion of machine elements.

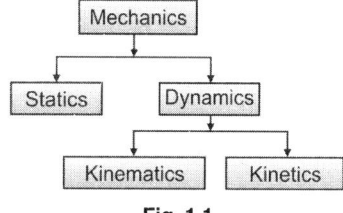

Fig. 1.1

The branch of scientific analysis that deals with motions, time and forces is called *mechanics*. It is divided into two parts: statics and dynamics.

Statics deals with analysis of machine parts at rest, i.e. forces and their effects on machine parts that are in equilibrium.

Dynamics deals with the forces and their effects on machine parts which are in motion.

Kinematics is the study of the geometrical aspect of motion without considering the forces involved, i.e. it is the study of position, displacement, rotation, velocity and acceleration.

Kinetics deals with the cause of motion, i.e. forces acting on machine parts and inertia forces due to mass and motion of machine parts.

Dynamics of machines is a part of theory of machines consisting of dynamics of various machine elements. Analysis of mechanisms is the study of motions and forces concerning different parts of an existing mechanism. Synthesis of mechanisms involves the design of the different parts.

1.2 MECHANISM AND MACHINE

1.2.1 Mechanism

The assembly of number of rigid bodies connected together in a way that the motion of one causes constrained and predictable motion to the other is known as mechanism, e.g. slider–crank mechanism, clock, watch, typewriter, etc.

1.2.2 Machine

A machine is a mechanism or a combination of mechanisms which imparts definite motion to parts, transmits and modifies the mechanical energy into useful work, e.g. slider–crank mechanism becomes a machine when used with valve mechanism, etc. in automobile engine.

1.2.3 Kinematic Link or Element

A kinematic link is also known as an element or simply a link. It is defined as a resistant body or a group of resistant bodies with rigid connections, preventing their relative motion, e.g. a slider–crank mechanism consists of four links: (1) frame (2) crank (3) connecting rod (4) slider

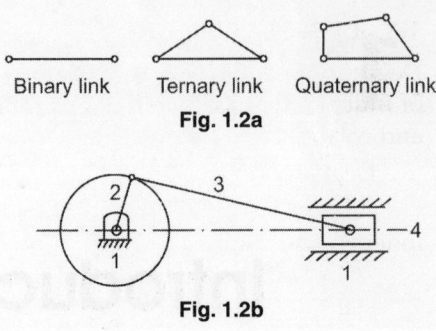

Fig. 1.2a

Fig. 1.2b

Links can be classified into binary, ternary, quaternary, etc. depending upon their ends (Fig. 1.2a).

1.2.4 Kinematic pair

A kinematic pair is a joint of two links having relative motion between them.

In slider–crank mechanism (Fig. 1.2b), link 2 and 1, link 2 and 3 and link 3 and 4 constitute turning pairs. Link 4 (slider) reciprocates relative to link 1 and is a sliding pair.

Classification of Kinetic Pair

A. According to the nature of contact, kinematic pairs are classified as:

  i. *Lower pair*: A kinematic pair is known as a lower pair if the two links have surface or area contact between them, e.g. a nut turning on a screw, shaft rotating in a bearing, universal joint, all pairs of slider–crank mechanism.

  ii. *Higher pair*: A pair of links having point or line contact between the links is known as a higher pair, e.g. cam and follower pair, tooth gears, ball and roller bearing, etc.

B. According to nature of mechanical constraint, kinetic pairs are classified as:

  i. *Closed pair*: When the two elements of a pair are held together mechanically, it is known as a closed pair. All the lower pairs and some of the higher pairs are closed pairs, e.g. a nut and screw pair.

  ii. *Unclosed pair*: When the two elements of a pair are in contact either due to force of gravity or spring action, it is known as a unclosed pair, e.g. a cam and follower pair.

C. According to the nature of relative motion, kinetic pairs are classified as:

  i. *Sliding pair*: When the two links have sliding motion relative to each other, they form a sliding pair, e.g. piston and cylinder, cross head and guides, ram and its guids in shaper, tail stock on the lathe bed.

  ii. *Turning pair*: A kinematic pair is known as a turning pair if one link has turning or revolving motion relative to the other, e.g. shaft revolving in a bearing, lathe spindle supported in head stock, etc.

  iii. *Rolling pair*: If one link has a rolling motion relative to the other, the pair is known as a rolling pair, e.g. ball and roller bearing, a wheel rolling on a flat surface, etc.

  iv. *Screw pair*: A kinematic pair is known as a screw pair if the two links have a turning as well as sliding motion between them, e.g. lead screw and the nut of a lathe, bolt with nut, etc. Screw pair is also known as helical pair.

  v. *Spherical pair*: If one link in the form of a sphere turns inside a fixed link, the kinematic pair is known as a spherical pair, e.g. ball and socket joint, pen stand, etc.

1.3 DEGREE OF FREEDOM

Degree of freedom of a pair is defined as the number of independent relative motions (both translational and rotational), a pair can have.

Degree of freedom = 6 – number of restraints

A body shown in Fig. 1.3 can describe following independent motions:

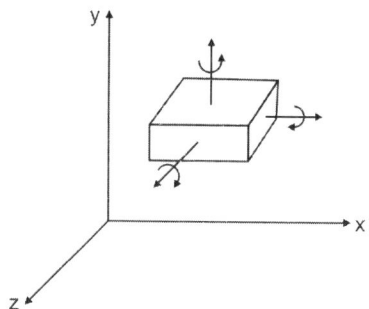

i. Translational motion along x-, y- and z-axes
ii. Rotational motion about x-, y- and z-axes.

Fig. 1.3

1.4 KINEMATIC CHAIN

When the kinematic pairs are coupled in a such a way that the last link is joined to the first link to transmit definite motion, it is called a *kinematic chain*. A redundant chain does not allow any motion of a link relative to the other, e.g. triangular chain.

For a four-link kinematic chain having lower pairs if p = no. of pairs, L = no. of links, j = no. of binary joints

then
$$L = 2p - 4 \qquad \qquad \text{... (1.1)}$$

$$j = \frac{3}{2}L - 2 \qquad \qquad \text{... (1.2)}$$

If LHS > RHS, then chain is locked
LHS = RHS, then chain is constrained
LHS < RHS, then chain is unconstrained.

For four-link kinematic chains having higher pairs, each higher pair is taken equivalent to two lower pairs and an additional link.

1.5 LINKAGE, MECHANISM AND STRUCTURE

A linkage is obtained if one of the links of a kinematic chain is fixed to the ground. A linkage is known as a mechanism if motion of one movable link results in definite motion of the others. If one of the links of a redundant chain is fixed, it is known as a structure. The structure with negative degree of freedom is known as a superstructure.

1.6 INVERSIONS OF SLIDER–CRANK CHAIN

The method of obtaining different mechanisms by fixing different links of a kinematic chain, is known as inversion of the mechanism. A slider–crank chain has the following inversions:

i. *First inversion*: When link 1 is fixed, link 2 is made crank and link 4 the slider (*see* Fig. 1.2b). The first inversion is used in reciprocating engine and reciprocating compressor.
ii. *Second inversion*: When link 2 (crank) of a slider–crank chain is fixed, then second inversion is obtained. This inversion is used in Whitworth quick-return mechanism and rotary engine.
iii. *Third inversion*: When link 3 (connecting rod) of the slider–crank chain is fixed, third inversion is obtained. This inversion is used in oscillating cylinder engine and crank and slotted lever mechanism.
iv. *Fourth inversion*: When link 4 of the slider–crank chain is fixed, the fourth inversion is obtained. This inversion is used in hand pumps.

Some of the topics discussed above have been described in detail in Chapter 2.

EXERCISE

1.1 Define the terms: mechanism, machine, link, kinematic pair, and kinematic chain.

1.2 Distinguish between analysis and synthesis of mechanism.

1.3 What do you understand by degree of freedom?

1.4 How are kinematic pairs classified? Explain with examples.

1.5 What is inversion of a mechanism? Describe various inversions of a slider–crank mechanism giving examples.

1.6 Differentiate between the following.
   i. Lower pair and higher pair
   ii. Turning pair and sliding pair
   iii. Screw pair and spherical pair
   iv. Closed pair and unclosed pair

OBJECTIVE TYPE QUESTIONS

1.1 A kinetic pair is a joint of
   (a) two links which are fixed
   (b) two links having same velocity
   (c) two links having relative motion between them
   (d) none of the above

1.2 Shaft with collars at both ends fitted into a circular hole forms a
   (a) turning pair
   (b) rolling pair
   (c) sliding pair
   (d) spherical pair

1.3 Ball-bearing and roller-bearing form a
   (a) turning pair
   (b) rolling pair
   (c) sliding pair
   (d) spherical pair

1.4 When two elements of a pair have a surface contact when in motion and the surface of one element slides over the surface of the other, the pair formed is called a
   (a) higher pair
   (b) lower pair
   (c) forced-closed pair
   (d) none of the above

1.5 The function of an element is to
   (a) transmit motion
   (b) to serve as a support
   (c) to guide other elements
   (d) all of the above
   (e) none

1.6 Whitworth quick return mechanism is an inversion of
   (a) double slider crank chain
   (b) single slider crank chain
   (c) four bar chain
   (d) none of the above

1.7 In a kinematic chain with four lower pairs, if one is sliding pair and three turning pairs, the mechanism is classified as
    (a) crossed slider–crank chain
    (b) four bar chain
    (c) slider crank chain
    (d) double slider–crank chain

1.8 If a kinematic chain has n links, then the number of mechanisms obtained is
    (a) $(n-1)$
    (b) $(n-2)$
    (c) $(n+1)$
    (d) n
    (e) none of these

1.9 A kinematic link or element is
    (a) any part of a machine
    (b) only the fixed part of a machine
    (c) any resistant body or assembly of resistant bodies which make a part of a machine connecting other pairs which have a motion relative to it
    (d) none of the above

1.10 In a reciprocating engine,
    (a) piston and gudgeon pin form two kinematic links
    (b) piston and gudgeon pin form one kinematic link
    (c) piston, gudgeon pin and connecting rod form one kinematic link
    (d) none of the above

1.11 A ball and socket joint forms a
    (a) turning pair
    (b) rolling pair
    (c) spherical pair
    (d) sliding pair

1.12 In a kinematic pair, if the elements have line contact or point contact when in motion, the pair is called a
    (a) higher pair
    (b) lower pair
    (c) closed pair
    (d) unclosed pair

1.13 Various kinematic pairs are given below. Choose the lower pair.
    (a) ball bearings
    (b) tooth gears in mesh
    (c) cam and follower
    (d) crankshaft and bearing

1.14 The motion of a rotating shaft in a footstep bearing constitutes between the elements of kinematic pair.
    (a) successfully constrained motion
    (b) completely constrained motion
    (c) incompletely constrained motion
    (d) none of the above. It is not a kinematic pair

1.15 The motion of a circular shaft with collars at each end rotating in a round hole constitutes between the elements of a kinematic pair
    (a) successfully constrained motion
    (b) completely constrained motion
    (c) incompletely constrained motion
    (d) none of the above. It is not a kinematic pair

1.16 Unconstrained rigid link in a plane has
  (a) one degree of freedom
  (b) two degrees of freedom
  (c) three degrees of freedom
  (d) zero degree of freedom

1.17 Piston and cylinder of a reciprocating steam engine form a
  (a) turning pair
  (b) rolling pair
  (c) sliding pair
  (d) none of the above

1.18 A bolt and a nut forms a
  (a) turning pair
  (b) rolling pair
  (c) screw pair
  (d) spherical pair

1.19 Oldham's coupling and elliptic trammels are the inversions of
  (a) double slider–crank chain
  (b) single slider–crank chain
  (c) four–bar chain
  (d) none of the above

1.20 In a kinematic chain with four lower pairs, if all the four lower pairs are turning pairs, the mechanism is classified as
  (a) four–bar chain
  (b) crossed slider–crank chain
  (c) slider–crank chain
  (d) double slider–crank chain

1.21 Choose the wrong statement:
  (a) a chain consisting of three links and three joints is known as a locked chain.
  (b) a chain consisting of four links and four joints is known as a kinematic chain.
  (c) quaternary joint is equivalent to three binary joints.
  (d) rectangular bar in a rectangular hole is an example of partially constrained motion.

1.22 'n' links are connected at the same joint, the joint is equivalent to
  (a) $(n - 1)$ binary joints
  (b) $(n - 2)$ binary joints
  (c) $(n - 3)$ binary joints
  (d) $(2n - 1)$ binary joints

1.23 In a reciprocating engine,
  (a) crankshaft and flywheel form two kinematic links
  (b) crankshaft and flywheel form one kinematic link
  (c) crankshaft and flywheel do not form kinematic link
  (d) flywheel and crankshaft separately form kinematic links

1.24 Joint of two elements that permits relative motion which is completely or successfully constrained is called a
  (a) mechanism
  (b) machine
  (c) structure
  (d) kinematic pair

1.25 The mass of the flywheel is concentrated in the rims because then it willl
  (a) store more energy
  (b) store less energy

(c) store zero energy

(d) make the flywheel stronger

1.26 In a kinematic pair, if the elements have surface contact when in motion, the pair is called a

(a) higher pair

(b) lower pair

(c) closed pair

(d) unclosed pair

1.27 Various kinematic pairs are given below. Choose the higher pair.

(a) roller bearing

(b) tooth gears in mesh

(c) cam and follower

(d) all of the above

1.28 Choose the correct statement

(a) tooth gears in mesh constitutes a higher pair

(b) belt on pulley drive constitutes a higher kinematic pair

(c) chain and sprocket drive constitute a higher kinematic pair

(d) all of the above

1.29 The motion of a circular shaft in a circular hole constitutes between the elements of a kinematic pair,

(a) successfully constrained motion

(b) completely constrained motion

(c) incompletely constrained motion

(d) none of the above

1.30 Kinematic chain

(a) comprises a chain of links in space

(b) comprises a chain of links with at least one link fixed and completely constrained motion

(c) comprises a chain of links with at least one link fixed and successfully constrained motion

(d) comprises a chain of links with incompletely constrained motion

1.31 Choose the correct statement

(a) mechanism transmits and modifies motion

(b) mechanism is the skeleton outline of machine to produce a definite motion between various links

(c) machine modifies mechanical work

(d) all of the above

1.32 Inversions are

(a) different mechanisms obtained by fixing different links in a kinematic chain with the object of changing relative motions of links with respect to one another

(b) different mechanisms obtained by fixing different links in a kinematic chain but keeping relative motions of links unchanged with respect to one another

(c) different mechanisms obtained by fixing different links in a kinematic chain to modify the mechanical advantage

(d) all of the above

1.33 The necessary condition for '*drag link quick return*' mechanism is that

(a) the shortest link is a fixed link. The sum of the shortest and the longest link is less than the sum of other two links

(b) the longest link is a fixed link. The sum of the shortest and the longest link is greater than the sum of other two links

(c) the shortest link is a fixed link. The sum of the shortest link and the longest link is greater than the sum of other two links

(d) the longest link is a fixed link. The sum of the shortest link and the longest link is less than the sum of other two links

---

ANSWERS

1.1 (c)	1.2 (a)	1.3 (b)	1.4 (b)	1.5 (d)	1.6 (b)
1.7 (c)	1.8 (d)	1.9 (c)	1.10 (b)	1.11 (c)	1.12 (a)
1.13 (d)	1.14 (a)	1.15 (b)	1.16 (c)	1.17 (c)	1.18 (c)
1.19 (a)	1.20 (a)	1.21 (d)	1.22 (a)	1.23 (b)	1.24 (d)
1.25 (a)	1.26 (b)	1.27 (d)	1.28 (d)	1.29 (c)	1.30 (a)
1.31 (d)	1.32 (b)	1.33 (a)			

2

Mechanisms

2.1 INTRODUCTION

2.1.1 Kinematics

It is that branch of theory of machines which deals with the relative motion between the various parts of a machines or it is the study of motion of bodies without considering external forces acting on them.

2.1.2 Dynamics

It is that branch of theory of machines which deals with the forces and their effects while acting upon the machine parts in motion.

2.1.3 Newton's Laws of Motion

Isaac Newton, the physicist gave three fundamental laws of motion. These are described below.

Newton's First Law of Motion or Law of Inertia

Every body continues to be in its state of rest or motion in a straight line, unless acted upon by some external force.

Inertia is that property of matter, by virtue of which a body cannot move by itself, nor change the motion imparted to it.

Newton's Second Law of Motion

The rate of change of momentum is directly proportional to the imposed force and takes place in the same direction in which the force acts.

Newton's Third Law of Motion

To every action there is an equal and opposite reaction.

2.2 KINEMATIC LINK OR ELEMENT

Each part of a machine, which moves relative to some other part is known as a kinematic link. A link may consists of several parts which are rigidly fastened together, so that they do not move relative to one another.

A link may or may not be in relative motion. But the link is said to be a kinematic link only when there is relative motion in the machine parts.

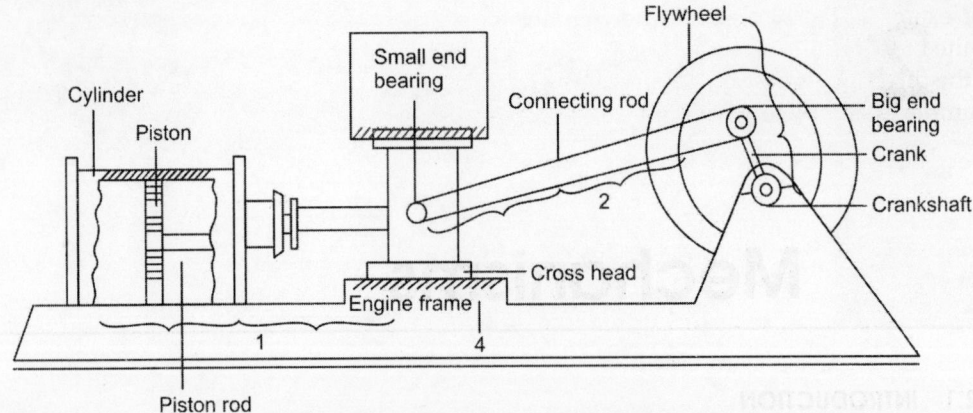

Fig. 2.1: Reciprocating steam engine

In reciprocating steam engine (Fig. 2.1), piston, piston rod and crosshead constitute one link; connecting rod with big and small end bearings constitute one link; crank, crankshaft and flywheel constitute one link, engine frame and main bearings constitute another link. So, a reciprocating steam engine consists of four links.

A link need not be a rigid body, but it must be a resistant body. A body is said to be resistant if it is capable of transmitting the required forces with negligible deformation. Thus, a link has two main characteristics:

 i. It should have relative motion.
 ii. It must be a resistant body.

Types of Links

In order to transmit motion, the driver and the follower may be connected by the following three types of link:

Rigid Link

A rigid link is one which does not undergo any deformation while transmitting motion. The connecting rod and crank of a reciprocating steam engine can be considered as a rigid link.

Flexible Link

A flexible link is one which is partly deformed in a manner not to affect the transmission of motion. For example, belts, ropes, chains and wires are flexible links and transmit tensile forces only.

Fluid Link

A fluid link is one which is formed by having a fluid in a receptacle and the motion is transmitted through the fluid by pressure or compression only, as in the case of hydraulic presses, jacks and brakes.

2.3 TYPES OF CONSTRAINED MOTION

2.3.1 Completely Constrained Motion

When the motion between a pair is limited to a definite direction irrespective of the direction of force applied, then the motion is said to be a completely constrained motion.

For example, the piston and the cylinder form a pair and the motion of the piston is limited to a definite direction (it will only reciprocate) relative to the cylinder irrespective of the direction of motion of the crank. The motion of a square bar in a square hole is an example of completely constrained motion (Fig. 2.2).

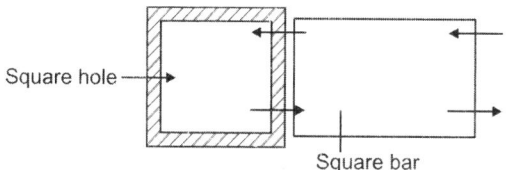

Fig. 2.2: Completely constrained motion

In constrained motion, the motion of the pair is limited to a particular direction only.

2.3.2 Incompletely Constrained Motion

When the motion between a pair can take place in more than one direction, then the motion is called an incompletely constrained motion. The change in the direction of impressed force may alter the direction of relative motion between the pair. A circular bar or shaft in a circular hole is an example of an incompletely constrained motion (Fig. 2.3).

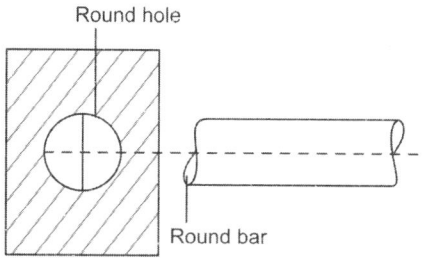

Fig. 2.3: Incompletely constrained motion

2.3.3 Successfully Constrained Motion

When the motion between the elements forming a pair, is such that the constrained motion is not completed by itself, but by some other means, then the motion is said to be successfully constrained motion. Consider a shaft in a foot–step bearing (Fig. 2.4). The shaft may rotate in bearing or it may move upwards. But if the load is placed on the shaft to prevent axial upward movement of the shaft, then the motion of the pair is said to be successfully constrained motion.

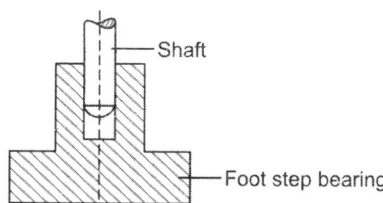

Fig. 2.4: Successfully constrained motion

2.4 KINEMATIC PAIR

The two links or elements of a machine when in contact with each other, are said to form a pair. If the relative motion between them is completely or successfully constrained, the pair is known as a kinematic pair.

Classification of Kinematic Pair

Kinematic pairs may be classified according to the following considerations:

According to the type of relative motion between the elements

(a) *Sliding pair*: When the two elements of a pair are connected in such a way that one can only slide relative to the other, the pair is known as sliding pair. The sliding pair has a completely constrained motion.
Example: The piston and cylinder, cross–head and guides of a reciprocating steam engine, ram and its guides in shaper, tail stock on the lathe bed.

(b) *Turning pair*: When the two elements of a pair are connected in such a way that one can only turn or revolve about a fixed axis of another link, the pair is known as a turning pair. A shaft with collars at both ends fitted into a circular hole, the crankshaft in a journal bearing in an engine, lathe spindle supported in head stock. A turning pair also has a completely constrained motion.

(c) *Rolling pair*: When the two elements of a pair are connected in such a way that one rolls over another fixed link, the pair is known as a rolling pair. Ball and roller bearings are examples.

(d) *Screw pair*: When the two elements of a pair are connected in such a way that one element can turn about the other by screw threads, the pair is a screw pair.
Example: The lead screw of lathe with nut, bolt with nut, etc.

(e) *Spherical pair*: When the two elements of a pair are connected in such a way that one element turns or swivels about the other fixed element, the pair formed is called a spherical pair.
Example: Car mirror attachment, pen stand, ball and socket joint, etc.

According to the type of contact between the elements

(a) *Lower pair*: If the two elements of a pair have a surface contact when relative motion takes place and the surface of one element slides over the surface of the other, the pair formed is known as a lower pair.
Example: Sliding pairs, turning pairs.

(b) *Higher pair*: If the two elements of a pair have a line or point contact when relative motion takes place and the motion between the two elements is partly turning and partly sliding, then the pair is known as a higher pair.
Example: Belt and rope drive, ball and roller bearings, cam and follower.

According to the type of closure

(a) *Self closed pair*: When the two elements of a pair are connected together mechanically in such a way that only required kind of relative motion occurs, it is known as a self closed pair. The lower pairs are self closed pair.

(b) *Force closed pair* or *unclosed pair*: When the two elements of a pair are not connected mechanically in such a way that they are kept in contact by the action of external forces, the pair is said to be a force closed pair.
Example: Cam and follower.

2.5 KINEMATIC CHAIN

When the kinematic pairs are coupled in such a way that the last link is joined to the first link to transmit definite motion (i.e. completely or successfully constrained motion), it is called a kinematic chain.

Example: The crankshaft of an engine forms a kinematic pair with the bearings which are fixed in a pair, the connecting rod with the crank forms a second kinematic pair, the piston with the connecting rod forms a third pair. The total combination of these links is a kinematic chain.

If each link is assumed to form two pairs with two adjacent links, then the relation between the number of pairs (p) forming a kinematic chain and the number of links (l) may be expressed in the form of an equation;

$$l = 2p - 4 \qquad \qquad ...(2.1)$$

Since in a kinematic chain each link forms a part of two pairs, therefore, there will be as many links as the number of pairs.

Another relation between the number of links (l) and the number of joints (j) which constitute a kinematic chain is given by

$$j = \frac{3}{2}l - 2 \qquad \qquad ...(2.2)$$

The equations are applicable only to kinematic chains, in which lower pairs are used.

There comes three outcomes from the above equations on the condition where the chain is kinematic or frame or unconstrained.

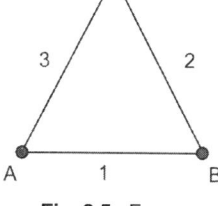

Fig. 2.5: Frame

Case 1: Consider the arrangement of three links *AB*, *BC* and *CA* with pin joints at *A*, *B*, *C* (Fig. 2.5).

No. of links, $d = 3$
No. of pairs, $p = 3$
No. of joints, $j = 3$

from, $\quad j = \frac{3}{2}l - 2$

$$3 = \frac{3}{2}(3) - 2$$

$3 > 2.5$, so a condition that arises is not a kinematic chain and hence no relative motion is possible. Such type of chain is called a frame or locked chain and forms a rigid structure. It is used in trusses.

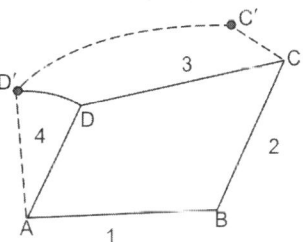

Fig. 2.6: Kinematic chain

Case 2: Consider the arrangement of four links *AB*, *BC*, *CD*, *DA* (Fig. 2.6). In this case

$$l = 4, p = 4, j = 4$$
$$l = 2p - 4$$

(or)

$$j = \frac{3}{2}l - 2$$

$$j = \frac{3}{2}(4) - 2$$

$$4 = 6 - 2$$
$$4 = 4$$
$$\therefore \qquad \text{LHS} = \text{RHS}$$

The condition is satisfied, hence, the chain is called kinematic chain.

Case 3: Consider an arrangement of five links *AB*, *BC*, *CD*, *DE*, *EA* (Fig. 2.7).

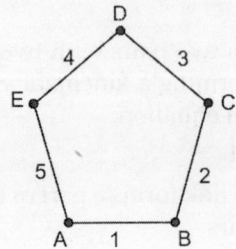

Fig. 2.7: Unconstrained chain

From
$$j = \frac{3}{2}l - 2$$

we get
$$5 = \frac{3}{2}(5) - 2$$
$$5 < 5.5$$
$$\text{LHS} < \text{RHS}$$

In this case, the condition is not satisfied, so it is not a kinematic chain. Such type of chain is called unconstrained chain, i.e. the relative motion is not completely constrained. *Note:* A chain having more than four links and satisfying the equation, is known as *compound kinematic link.*

2.6 TYPES OF JOINTS IN A CHAIN

2.6.1 Binary Joint

In Fig. 2.8, *A* is a binary joint where two links meet. The point of intersection of two links is known as a binary joint.

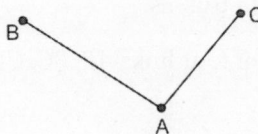

Fig. 2.8: Binary joint

2.6.2 Ternary joint

If three links meet at a point, then the joint is known as a ternary joint. The ternary joint is equivalent to two binary joints as one of the three links joined carry the pin for the other two links. In Fig. 2.9, A is a ternary joint.

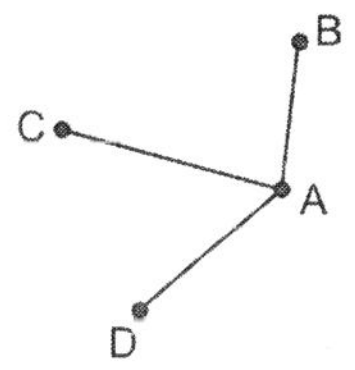

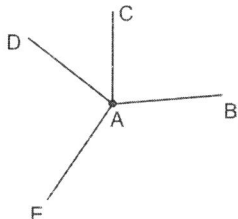

Fig. 2.9: Ternary joint

2.6.3 Quaternary Joint

If four links meet at a point, then the joint is known as a quaternary joint. A quaternary joint is equivalent to three binary joints. In general, when l number of links are joined at the same connection, the joint is equivalent to $(l - 1)$ binary joints. Figure 2.10 shows a quaternary joint A.

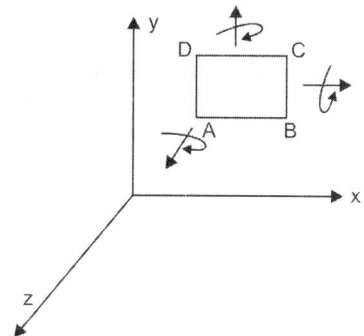

Fig. 2.10: Quaternary joint

2.7 DEGREE OF FREEDOM

An unconstrained rigid body moving in space can describe the following independent motions.

1. Translational motion along any three mutually perpendicular axes x, y and z.
2. Rotational motion about these axes.

Thus, a rigid body possesses six degrees of freedom. The connection of a link with another imposes certain constraints on their relative motion. The number of restraints can never be zero or six (Fig. 2.11).

Fig. 2.11: Degree of freedom

Degree of freedom of a pair is defined as the number of independent relative motions, both translational and rotational, a pair can have.

$$\boxed{\text{Degree of freedom} = 6 - \text{no. of restraints}}$$

2.7.1 Linkage

A linkage is obtained if one of the links of a kinematic chain is fixed to the ground.

2.7.2 Mechanism

If motion of any of the moveable link results in definite motion of the others, the linkage is known as mechanism.

2.7.3 Structure or Locked System

If one of the links of a redundant chain is fixed, it is known as a structure. The degree of freedom of a structure is zero. A structure with negative degree of freedom is known as a superstructure.

2.8 DEGREE OF FREEDOM EQUATION

A mechanism may consist of a number of pairs belonging to different classes having different number of restraints. It is also possible that some of the restraints imposed on the individual links are common or general to all the links of the mechanism.

A zero order mechanism will have no such general restraint. Of course, some of the pairs may have individual restraints. A first order mechanism has one general restraint; a second order mechanism has two general restraints, and so on up to the fifth order.

Expressing the number of degree of freedom of a linkage in terms of the number of links and the number of pair connections of different types is known as synthesis. Degree of freedom of a mechanism in space can be determined as follows.

Let

N = total number of links in a mechanism
F = degrees of freedom
P_1 = number of pairs having one degree of freedom and so on up to P_5.

Therefore,

number of movable links = $N - 1$ (as one link is fixed in Mechanism)
number of degrees of freedom of $(N - 1)$ movable links = $6(N - 1)$.

Each pair having one degree of freedom imposes 5 restraints, reducing its degrees of freedom by $5P_1$ and so on up to P_5.

So, the final equation is given as,

$$\boxed{F = 6(N - 1) - 5P_1 - 4P_2 - 3P_3 - 2P_4 - P_5}$$

But, in general, most of the mechanisms are of two-dimensional such as four-link or a slider–crank mechanism. So, there are three general restraints.

The equation is given by

$$\boxed{F = 3(N - 1) - 2P_1 - P_2}$$ (for only 2–D axes of rotation along x, y and z directions)

This is known as Gruebler's criterion for plane mechanisms. Each pair with one degree of freedom imposes two restraints on the mechanism.

Some authors mention the above equation as Kutzback's criterion and a simplified relation $[F = 3(N-1)-2P_1]$ for two restraints.

And also, the equation in terms of joints and no. of higher pairs is given as

$$F = 3(N-1)-2j-1h$$

All lower or binary joints have DOF as 1 and all higher joints have DOF as 2. The relative motion in mechanism was also described by DOF, as follows:

When DOF < 0, then it is a superstructure
              = 0, then it is a frame
              > 0, then there will be relative motion
              = 1, then the motion will be kinematic or constrained
              > 1, then the motion will be unconstrained.

Example: For the kinematic linkages, calculate the following:

• the no. of binary links (N_b) • the no. of ternary links (N_T) • the no. of quaternary links (N_q) • the no. of total links • the no. of loops • the no. of joints (p) • the no. of degrees of freedom.

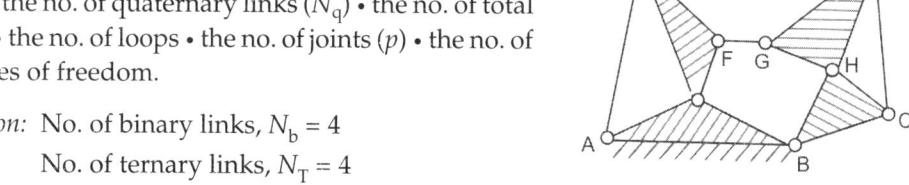

Fig. 2.12

Solution: No. of binary links, $N_b = 4$
            No. of ternary links, $N_T = 4$
            No. of quaternary links = 0
            No. of total links $(N) = 8$
            No. of loops $(L) = 4$
            No. of pairs = 11 by $P = (N+L-1) = (8+4-1) = 11$

$$\text{DOF or } F = 3(N-1) - 2P_1$$
$$= 3(8-1) - 2(11)$$
$$= 3(7) - 2(11)$$
$$= 21 - 22 = -1$$

$F = -1 < 0$

∴ The mechanism or structure has negative degree of freedom (F), so it is known as superstructure.

2.9 TYPES OF MECHANISMS

There are three types of mechanisms.

2.9.1 4R or Four-bar Chain

A four-bar chain is the most fundamental of the plane kinematic chains. It is a much prefered mechanical device for mechanisation and control of motion due to its simplicity. Basically, it consists of four rigid links which are connected in the form of a quadrilateral by four pin–joints. When one of the links is fixed, it is known as a linkage or mechanism. A link that makes complete revolution is called the crank, the link opposite to the fixed link is called the coupler, and the fourth link is called the lever or rocker if it oscillates or rotates.

It is possible to have a four-bar linkage, if the length of the one of link is greater than the sum of other three links. This includes the parts of the mechanism like frame, crank, etc.

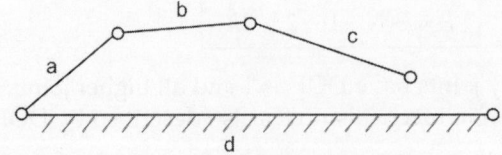

Fig. 2.13

Following are the necessary conditions for a link to act as a crank, coupler or rocker.

i. For the crank, the shortest link is fixed and the sum of the shortest and the longest links is less than the sum of the other two links.
ii. For the coupler or connecting rod, if any of the adjacent links are fixed, crank has a full revolution and link opposite to it oscillates due to the motion transferring through connecting rod and this motion is also known as crank–rocker motion.
iii. For the link opposite to the shortest link, it is said to be a rocker and the motion is oscillatory, then it is only due to the rocker–rocker motion.

These observations were made by Grashof. So, these are summarized as Grashof's law which states, "a four-bar mechanism has at least one revoluting link if the sum of the lengths of the largest and the shortest links is less than the sum of lengths of the other two links".

Applications of 4-bar chain or 4R-mechanism are described below.

Inversions of Four-bar Chain

Beam engine: A part of the mechanism of a beam engine which consists of four links is shown in Fig. 2.14. In this mechanism, when the crank rotates about the first centre A, the lever oscillates about a fixed centre D. The end E of the lever CDE is connected to a piston rod which reciprocates due to the rotation of the crank. In other words, the purpose of this mechanism is to convert rotary motion into reciprocating motion.

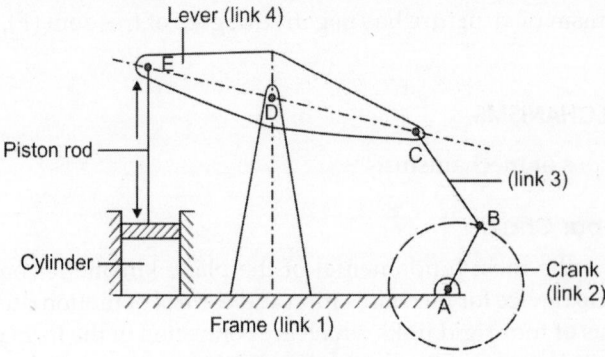

Fig. 2.14

The links are XA (frame), AB (crank), BC (connecting rod), CE (rocker). In this mechanism, the function is, as AB is pinned to the frame, so it can rotate (rotary motion) and is then pinned to the crank end B and it also can rotate (2R) and the end C is again pinned to the lever. So (3R) and the end E is pinned to the piston rod (4R). So, it is called 4R-bar chain or 4-bar chain mechanism. In this, the motion is crank–rocker motion. So, no. of links is 4.

Coupling Rod of a Locomotive

The mechanism of a coupling rod of a locomotive which consists of four links is shown in Fig. 2.15. In this, the links AD and BC (having equal length) act as cranks and are connected to the respective wheels. The link CD acts as a coupling rod and link AB is fixed in order to maintain a constant centre distance between them. This mechanism is meant for transmitting rotary motion from one wheel to the other wheel.

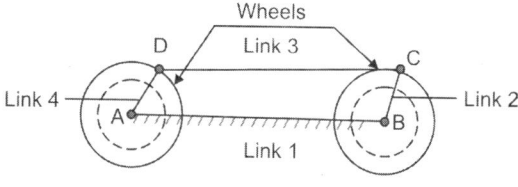

Fig. 2.15

This is based on the principle that if two opposite links are parallel and equal in length, then any of the links can be made fixed. The two links adjacent to the fixed link will always act as two cranks. The four links form a parallelogram in all the positions of the cranks, provided the cranks rotate in the same sense.

In this, there are four links and the motion is subjected to crank–crank motion.

Watt Indicator Mechanism

A Watt's indicator mechanism which consists of four links is shown in Fig. 2.16. The four links are, fixed link at A, link CE, link AC and link BFD. It may be noted that BF and FD form one link because these two parts have no relative motion between them. The links CE and BFD act as levers. The displacement of the link BFD is directly proportional to

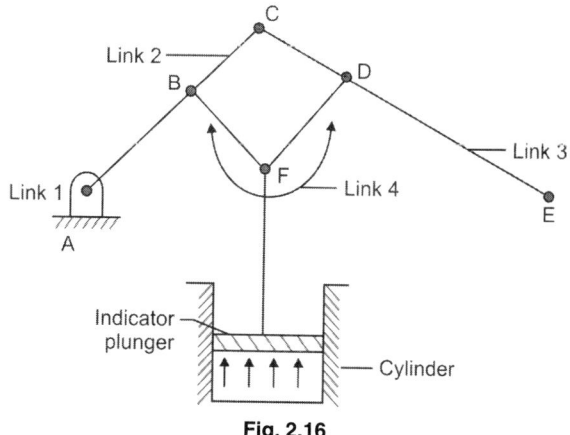

Fig. 2.16

the pressure of gas or steam which acts on the indicator plunger. For any small displacement of the mechanism, the tracing point E at the end of the link CE traces out as a straight line.

In this, the motion is due to rocker-rocker motion.

2.9.2 3R-1P Inversion

When one of the *turning pairs* of a four-bar chain is replaced by a *sliding* or a *prismatic pair*, it becomes a slider–crank chain or an 3R-1P inversion.

Turning Pair/Revolute Pair

When 2 elements are converted such that one element revolves around other element (link).

Sliding Pair/Prismatic Pair

When 2 elements or links have sliding motion.

There are 3 types of 3R-1P inversion, when it is :

   i. Crank fixed
  ii. Connecting rod fixed
 iii. Slider fixed

i. Firstly, lets talk about crank fixed. Crank fixed inversion is obtained, when crank link of the slider–crank chain is fixed.

Applications:

  (a) Rotary engine
  (b) Whitworth quick-return mechanism

  (a) *Rotary engine*: It is also known as *rotary internal combustion engine* is shown in Fig. 2.17. It can be observed that with the rotation of the link 3, the link 1 rotates about 0 and the slider 4 reciprocates on it. This also implies that if the slider is made to reciprocate on the link 1, crank 3 will rotate about *A* and the link 1 about 0. Moreover, instead of one cylinder, seven or nine cylinders symmetrically placed at a regular intervals in the same plane or in parallel planes, are used.

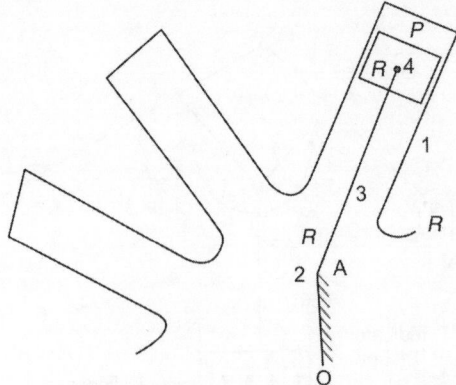

Fig. 2.17: 3R-1P slider makes prismatic pair with the bar and others are revolute pairs as mentioned in diagram (rotary engine)

(b) Whitworth quick-return mechanism: It is a mechanism used in workshop to cut metals. The forward stroke takes a little longer and cuts the metal, whereas the return stroke is idle and takes a shorter period and hence called quick-return.

Slider 4 rotates in a circle about A and slides on link 1. C is a point on the link 1 extended backwards where the link 5 is pivoted. The other end of the link 5 is pivoted to the tool, forward stroke of which cuts the metal. The axis of motion of slider 6 (tool) passes through O and is perpendicular to OA, the fixed link. The crank 3 rotates in the counter-clockwise direction.

Initially, let the slider 4 be at B so that C be at C'. Cutting tool 6 will have forward stroke. Finally, the slider B assumes the position B'' and the cutting tool 6 is in the extreme right position. The time taken for the forward stroke of the slider 6 is proportional to the obtuse angle $B''AB'$ at A. Similarly, the slider 4 completes the rest of the circle through the path $B''B'''B'$ and C passes through $C''C'''C'$. There is backward stroke of the tool 6. The time taken in this is proportional to the acute angle $B''AB'$ at A.

Let θ = obtuse angle $B'AB''$ at A

β = acute angle $B'AB''$ at A

Then, $\dfrac{\text{Time of cutting}}{\text{Time of return}} = \dfrac{\theta}{\beta}$

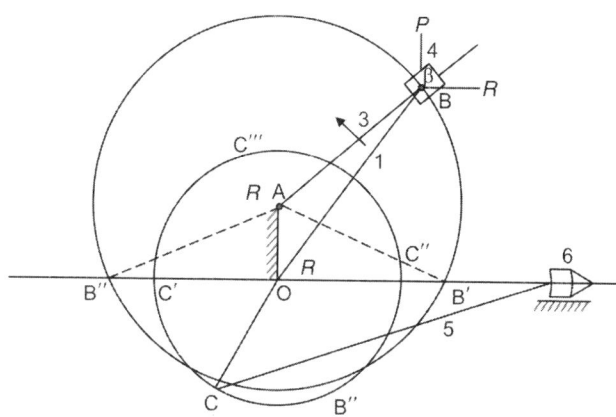

Fig. 2.18: Whitworth quick-return mechanism

There are 3 revolute pairs and 1 prismatic pair between the slider and bar link 1.

ii. Now, lets talk about the other type of inversion of 3R-1P, connecting rod fixed. In this, the connecting rod is fixed or taken as frame, and the other links are revolved around it.

The two applications are :

(a) Oscillating cylinder engine

(b) Crank or slotted lever mechanism

Oscillating cylinder engine, as shown in Fig. 2.19, the link 4 is made in the form of a cylinder and a piston is fixed to the end of the link 1. The piston reciprocates inside the cylinder pivoted to the fixed link 3. The arrangement is known as oscillating cylinder engine, in which as the piston reciprocates in the oscillating cylinder, the crank rotates.

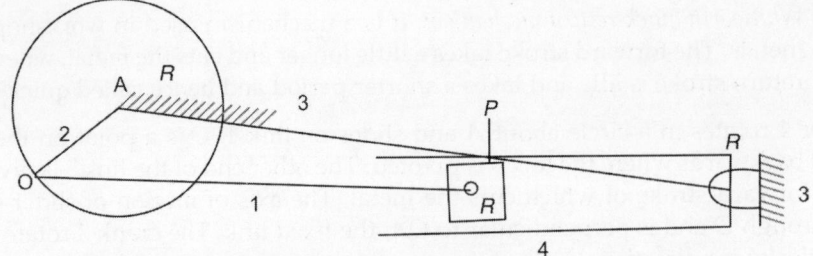

Fig. 2.19: One prismatic pair with the bar of slider and other revolute

(b) Crank and slotted lever mechanism: If the cylinder of an oscillating cylinder engine is made in the form of a guide and the piston in the form of a slider, the arrangment as shown in Fig. 2.20a, b is obtained. As the crank rotates about A, the guide 4 oscillates about B. At a point C on the guide, the link 5 is pivoted, the other end of which is connected to the cutting tool through a pivoted joint.

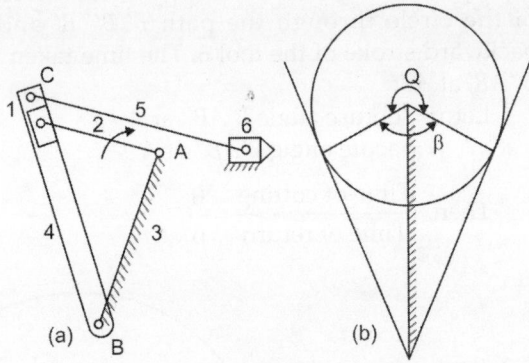

Figures 2.20a and b shows the extreme points of the oscillating guide 4. The time of the forward stroke is proportional to the angle θ whereas for the return stroke, it is proportional to angle b, provided the crank rotates clockwise.

Fig. 2.20

Slider Fixed

Application : Hand pump

Hand pump: Figure 2.21 shows a hand pump. Link 4 is made in the form of a cylinder and a plunger fixed to the link 1 reciprocates.

Example: The length of a fixed link of a crank and slotted-lever mechanism is 250 mm and that of the crank is 100 mm. Determine the

i. Inclination of the slotted lever with the vertical in the extreme position.

ii. Ratio of the time of cutting stroke to the time of return stroke, and

iii. Length of the stroke, if the length of the slotted lever is 450 mm and the line of stroke passes through the extreme positions of the free end of the lever.

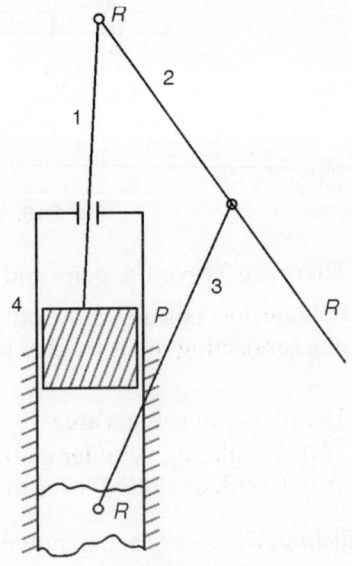

Fig. 2.21

Solution:

$OA = 250$ mm

$OP' = OP'' = 100$ mm

$AR' = AR'' = AR = 450$ mm

$$\cos\frac{\beta}{2} = \frac{OP'}{OA} = \frac{100}{250} = 0.4$$

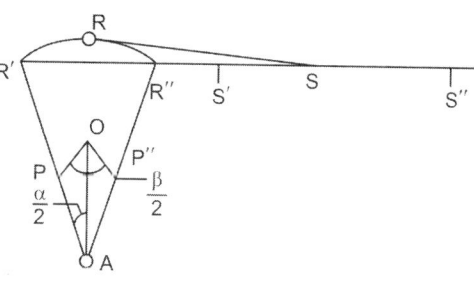

Fig. 2.22

or, $\dfrac{\beta}{2} = 66.4°,$ $\boxed{\beta = 132.8°}$

i. Angle of slotted lever with the vertical

$$\frac{\alpha}{2} = 90° - 66.4° = 23.6°$$

ii. $\dfrac{\text{Time of cutting stroke}}{\text{Time of return stroke}} = \dfrac{360° - \beta}{\beta} = \dfrac{360° - 132.8°}{132.8°} = \boxed{1.71}$

iii. Length of stroke $= S'S'' = R'R''$

$$= 2AR'.\sin\left(\frac{\alpha}{2}\right)$$

$$= \boxed{360.3 \text{ mm}}$$

2.9.3 Inversions of 2R-2P Kinematic Chain or Double Slider–Crank Chain

If in a four bar kinematic chain all links are free, motion will be unconstrained. When one link of kinematic chain is fixed, it works as a mechanism.

2R-2P is a four bar kinematic chain containing two revolute or turning pair, which allows only relative rotation between elements, and two prismatic or sliding pairs, which allows only a relative translation between elements.

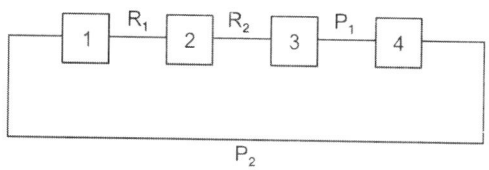

Fig. 2.23

where,

R_1 & R_2 are revolute pair

P_1 & P_2 are sliding pair

This chain gives three different mechanism:

i. *First Inversion:* Scotch yoke mechanism

ii. *Second Inversion:* Oldham's coupling

iii. *Third Inversion:* Elliptical trammer or scotted plate

First Inversion: The first inversion is obtained by fixing link 1, by doing so a mechanism called Scotch Yoke is obtained. The link 1 is a slider similar to link 3. Link 2 works as a crank. Link 4 is a slotted link. When link 2 rotates, link 4 has simple harmonic motion for angle 'θ' of link 2, the displacement of link 4 is given by

$$x = OA \cos \theta$$

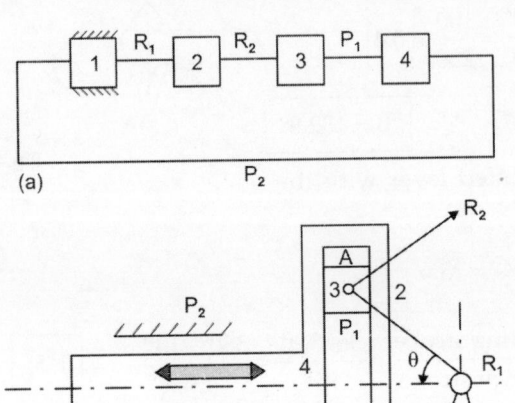

Fig. 2.24: Scotch yoke mechanism

Second Inversion: In this case, link 2 is fixed and a mechanism called Oldham's coupling is obtained. This coupling is used to connect two shafts which have eccentricity 'ε'. The axes of the two shafts are parallel but displaced by distance ε. The link 4 slides in the two slots provided in links 3 and 1. The centre of this link will move on a circle with diameter equal to eccentricity.

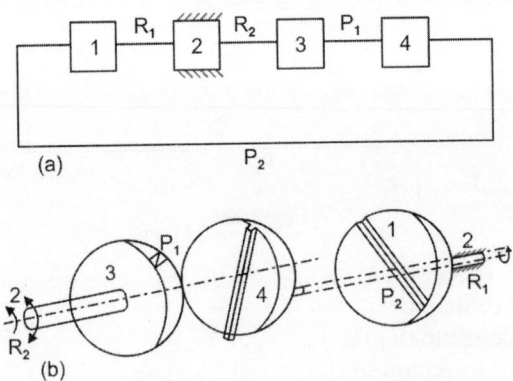

Fig. 2.25: Oldham's coupling

Third Inversion: This inversion is obtained by fixing link 4. The mechanism so obtained is called elliptical trammer. The mechanism is used to draw ellipse. The link 1, which is slider, moves in a horizontal slot of fixed link 4. The link 3 is also a slider and moves in a vertical slot. The point D on the extended portion of link 2 traces ellipse with system of axes. The position coordinates of point D are as follows:

$$x_D = AD \sin \theta$$

$$\because \sin^2 \theta + \cos^2 \theta = 1$$

$$y_D = CD \cos \theta$$

$$\boxed{\frac{x_D^2}{AD^2} + \frac{y_D^2}{CD^2} = 1}$$ equation of ellipse

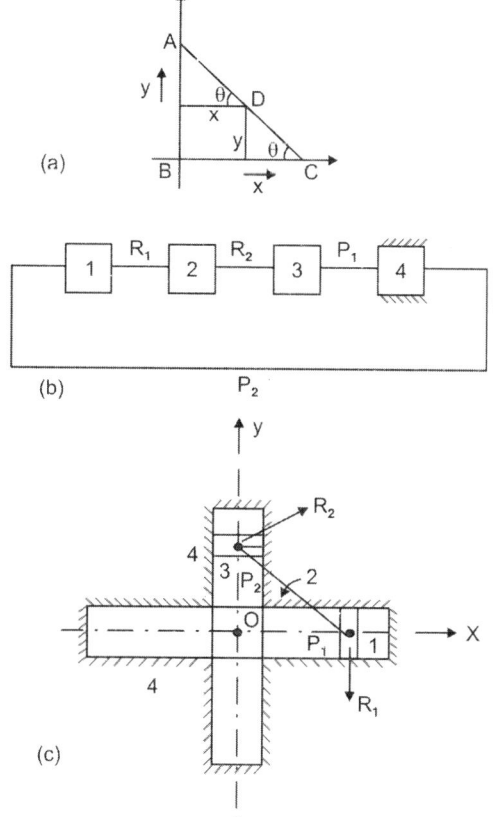

(a)

(b)

(c)

→ D is the rotating point
→ When D is at centre of AC, then

$$\boxed{x_D^2 + y_D^2 = 1}$$ equation of circle

$\because AD = DC$

Fig. 2.26: Elliptical trammer

EXAMPLES

Example 2.1: Calculate the degree of freedom of the linkage shown in Fig. 2.27. State whether it is a kinematic chain or not.

Solution: We know that degree of freedom is given by

$$F = 3 (n - 1) - 2p - h$$

where F = degrees of freedom
  p = number of lower pair (revolute)
  n = number of links
  h = number of higher pairs

Here $n = 5, p = 6, h = 0$
$F = 3(5 - 1) - 2 \times 6 = 0$

The linkage forms the structure.

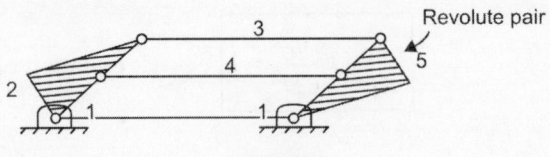

Fig. 2.27

Example 2.2: Find the degree of freedom of the mechanisms shown below in Fig. 2.28.

Solution: Given, Number of links = 10.

Number of binary joints.

There is rolling contact with sliding of pin in slot joint. It is equivalent to two binary joints. Total number of binary joints $p = 12$.

Applying Gruebler's equation

$$F = 3(n - 1) - 2p = 3 \times (10 - 1) - 2 \times 12 = 3$$

Thus, the degrees of freedom is 3.

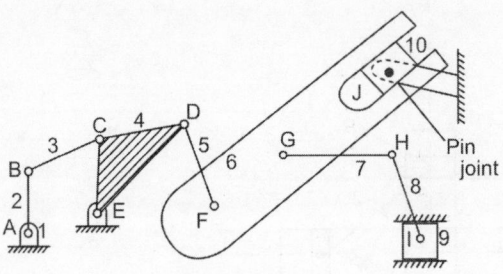

Fig. 2.28

Example 2.3: (i) State the Gruebler's criterion for ascertaining the degree of freedom of a planar mechanism having turning pairs only.

(ii) Extend Gruebler's criterion for planar mechanism to obtain the degree of freedom of a space mechanism as

$$F = 6(n - 1) - 5g - 4C - 4s$$

where g = total number of sliding pairs
C = total number of cylindrical pairs
s = total number of spherical pairs
n = numbers of links in the mechanism

(iii) Find the degree of freedom in each of the following cases given in Fig. 2.29.

(iv) Prove using Gruebler's criterion that the achieving constrained motion, the minimum number of binary links in a linkage = 4.

Solution: (i) If in a mechanism n is the number of links and p the number of binary joints or turning pairs or simple hinges, then $(n - 1)$ will be the number of movable links when obtaining the mechanism.

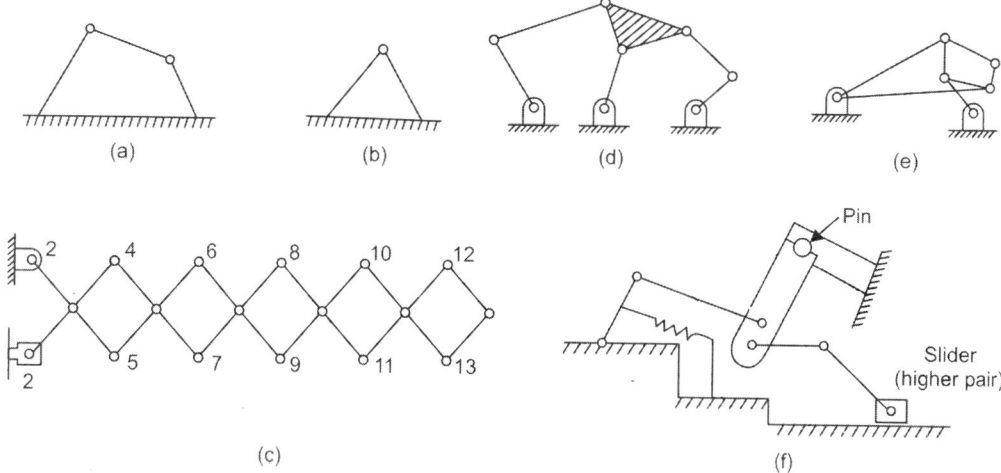

Fig. 2.29

The degree of freedom F is given by

$$F = 3(n - 1) - 2p$$

When the motion is constrained $F = 1$ and then Eq. (i) can be written as

$$1 = 3(n - 1) - 2p$$
$$3n - 2p = 4$$

Equation (i) is known as Gruebler's equation.

(ii) Suppose a mechanism has n links so the number of movable links $= n - 1$. In a space-motion, every link has six degrees of freedom.

The degrees of freedom are given by

$$F = 6(n - 1) - 5p1 - 4p2 - 3p3 - 2p4 - p5$$

where $p1$ = number of joints permitting 1 DOF   (where DOF = degree of freedom)
$p2$ = number of joints permitting 2 DOF
$p3$ = number of joints permitting 3 DOF
$p4$ = number of joints permitting 4 DOF

and so on.

We know that sliding joints permit 1 degree of freedom, cylindrical joints 2 and spherical joints permit 3 degrees of freedom.

Most usual joints in space mechanisms are the revolute ($F = 1$). The cylindrical joints ($F = 2$) and spherical joints ($F = 3$).

Thus terms $p4$ and $p5$ in Eq. (i) are eliminated and may be rewritten as

$$F = 6(n - 1)5g - 4c - 3s$$

where g = total number of sliding pairs
C = total number of cylindrical pairs
S = total number of spherical pairs

(iii) In Fig. 2.29 (a) $n = 4$, $p = 4$.

We know that DOF is given as

$$F = 3(n - 1) - 2p$$
$$= 3(3 - 1) - 2 \times 4 = 1$$

Constrained motion is obtained.

- In this case $n = 3, p = 3$.

We know that DOF is given as

$$F = 3(n-1) - 2p$$
$$= 3(3-1) - 2 \times 3 = 0$$

It is a structure.

- Here $n = 14, p = 18, h = 1$

$$F = 3(n-1) - 2p - h = 3(14-1) - 2 \times 18 - 1$$
$$= 39 - 36 - 1 = 2$$

- Here $n = 7, p = 8$

$$F = 3(n-1) - 2p$$
$$= 3(7-1) - 2 \times 8 = 2$$

- Here $n = 8, p = 10$ ($F = 1$ pair, $D = 1$ pair, but in points E, A, B and C, there are ternary joints, i.e. pairs at each point. Thus, total binary joints are 10.

$$F = 3(n-1) - 2p$$
$$= 3(8-1) - 2 \times 10 = 1$$

- In this case $n = 7, p = 7, h = 1$

$$F = 3(n-1) - 2p - h$$
$$= 3(7-1) - 2 \times 7 - 1 = 3$$

(iv) For constrained motion ($F = 1$) mechanism, Gruebler's criterion for degrees of freedom can be written as

$$F = 1 = 3(n-1) - 2p \qquad \qquad \text{...(ii)}$$

where p = number of simple hinges

       N = total number of links

which can be written as

$$N = n_2 + n_3 + n_4 + ...$$

where n_2 = number of binary links

      n_3 = number of ternary links

      n_4 = number of quaternary links in the mechanism and so on.

And the total member of elements (E) in a mechanism is given by

$$E = 2p = 2n_2 + 3n_3 + 4n_4 + 5n_5 \qquad \qquad \text{...(iii)}$$

Note: It is to be mentioned here that a simple hinge consists of two elements. A binary hinge has two elements, a ternary link will have three elements, a quaternary link has four elements and so on.

Substituting the value of n from Eq. (ii) and the value of $2p$ from Eq. (iii) into Eq. (i), one gets

$$\therefore \qquad 1 = 3(n_2 + n_3 + n_4 + ...1) - 2n_2 - 3n_3 - 4n_4 - 5n_5$$

or $\qquad \qquad n_2 = 4 + n_4 + 2n_5 + ...$

Since n_4 and n_5 are positive integers, so smallest possible value of n_2 is 4. So the minimum value of binary links is four.

EXERCISE

2.1 Define and explain the following terms.
   (a) Kinetic link, types of links
   (b) Mechanism
   (c) Mobility of mechanism
   (d) Kinematic chain

2.2 What do you understand by inversion of a mechanism? List various inversions of a four-bar chain.

2.3 Explain Gruebler's criterion for degrees of freedom for planar mechanisms.

2.4 Explain with sketches, inversions of single and double slider crank chain.

2.5 Define instantaneous centre and instantaneous axis. Describe with proof, the method of determining the velocity of a point on a moving link.

2.6 In a quick-return motion, mechanism of the oscillating link type, the distance between the fixed centres is 80 mm and the length of the driving crank is 20 mm. Determine the time ratio of the working stroke to the return stroke.

2.7 In an off-set slider crank mechanism, the eccentricity is 50 mm, length of crank is 300 mm, and length of connecting rod is 500 mm, determine the quick return ratio.

2.8 The distance between the axes of parallel shafts connected by Oldham's coupling is 25 mm. The speed of rotation of the shafts is 320 rpm. Determine the maximum velocity of sliding of each tongue in its slot.

2.9 In a quick-return motion, mechanism of the crank and slotted lever type, the ratio of maximum velocities is 2. If the length of stroke is 250 mm, find (a) the length of the slotted lever, (b) the ratio of times of cutting and return strokes, and (c) the maximum cutting velocity in m/s, if the crank rotates at 30 rpm.

2.10 In a crank and slotted lever mechanism. The length of crank is 560 mm and the ratio of time of working stroke to return stroke is 2.8. Determine (a) distance between the fixed centres, and (b) length of the slotted lever, if length of stroke is 250 mm.

2.11 Calculate the number of degrees of freedom of the mechanisms shown in Fig. 2.30.

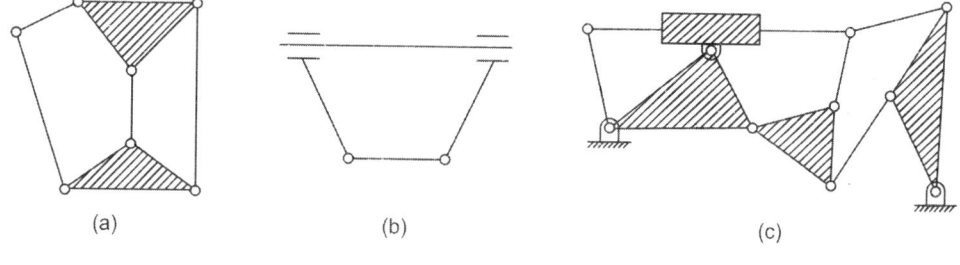

(a)   (b)   (c)

Fig. 2.30

2.12 Determine the number of degrees of freedom of all the devices shown in Fig. 2.31.

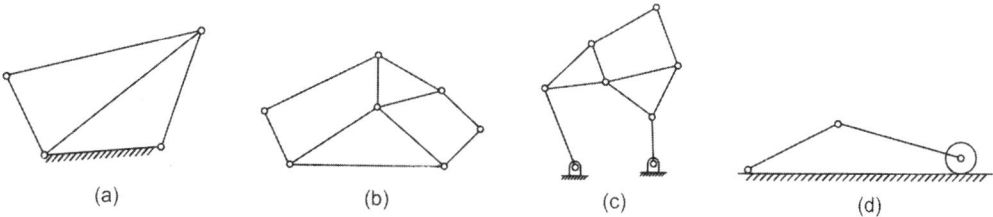

(a)   (b)   (c)   (d)

Fig. 2.31

2.13 Show that the chain shown in Fig. 2.32 is not a kinematic chain.

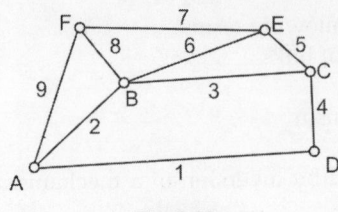

Fig. 2.32

2.14 Determine the type of chain in Fig. 2.33.

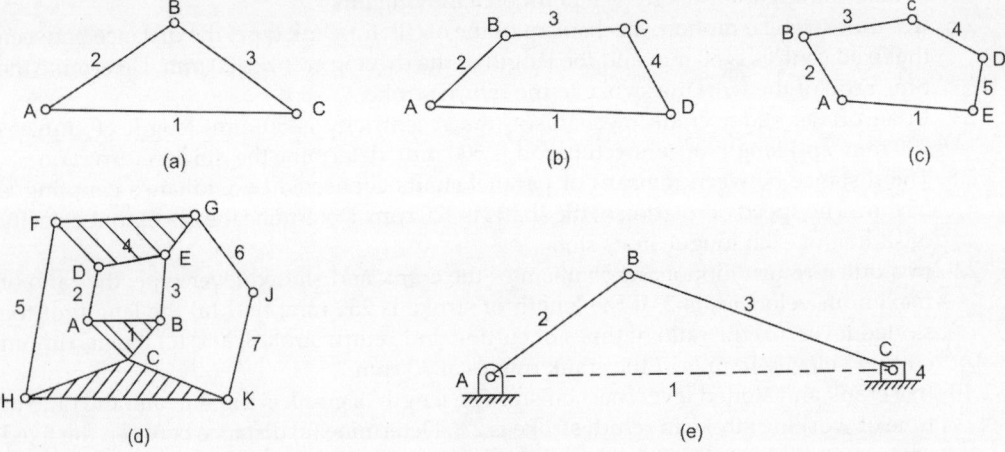

Fig. 2.33

2.15 Calculate the number of degrees of freedom of the linkages shown in Fig. 2.34

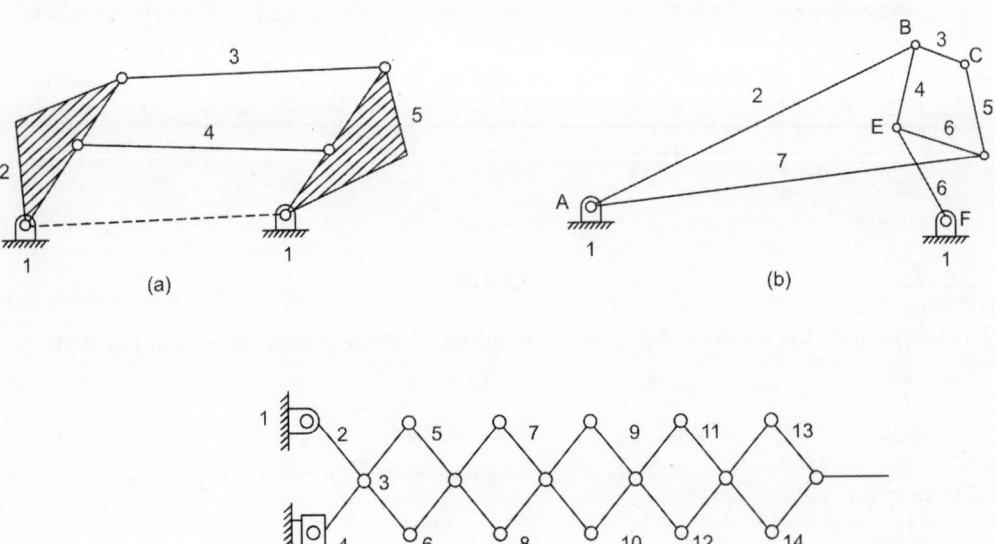

Fig. 2.34

OBJECTIVE TYPE QUESTIONS

2.1 In a reciprocating steam engine, which of the following forms a kinematic link?
  (a) cylinder and piston
  (b) piston road and connecting rod
  (c) crank shaft and flywheel
  (d) flywheel and engine frame

2.2 The motion of a piston in the cylinder of a steam engine is an example of
  (a) completely constrained motion
  (b) incompletely constrained motion
  (c) successfully constrained motion
  (d) none of these

2.3 The motion transmitted between the teeth of gears in a mesh is
  (a) sliding
  (b) rolling
  (c) may be rolling or sliding depending upon the shape of teeth
  (d) partly sliding and partly rolling

2.4 The motion
  (a) lower pair
  (b) higher pair
  (c) self closed pair
  (d) force closed pair

2.5 A ball and a socket joint forms a
  (a) turning pair
  (b) rolling pair
  (c) sliding pair
  (d) spherical pair

2.6 A kinetic chain is known as a mechanism when
  (a) none of the links is fixed
  (b) one of the links is fixed
  (c) two of the links are fixed
  (d) all of the links are fixed

2.7 The Grubler's criterion for determining the degrees of freedom (n) of a mechanism having plane motion is ...., where l = number of links, and j = number of binary joints.
  (a) $n = (l-1) - j$
  (a) $n = 2(l-1) - 2j$
  (c) $n = 3(l-1) - 2j$
  (d) $n = 4(l-1) - 3j$

2.8 The mechanism forms a structure, when the number of degrees of freedom (n) is equal to
  (a) 0
  (b) 1
  (c) 2
  (d) −1
  (a) $n = (l-1) - j$

2.9 In a four bar chain or quadric cycle chain
  (a) each of the four pairs is a turning pair
  (b) one if turning pair and three are sliding pairs
  (c) three are turning pairs and one is sliding pair
  (d) each of the four pairs is a sliding pair

2.10 Which of the following is an inversion of single slider–crank chain?
   (a) Beam engine
   (b) Watt's indicator mechanism
   (c) Elliptical trammels
   (d) Whitworth's quick-return motion mechanism
2.11 Which of the following in an inversion of double slider–crank chain?
   (a) Coupling rod of a locomotive
   (b) Pendulum pump
   (c) Elliptical trammels
   (d) Oscillating cylinder engine

ANSWERS

2.1 (c)	2.2 (a)	2.3 (d)	2.4 (c)	2.5 (d)	2.6 (b)
2.7 (c)	2.8 (a)	2.9 (a)	2.10 (d)	2.11 (c)	

3

Velocity and Acceleration Analysis

3.1 INTRODUCTION

After the study of mechanisms, the next step is to determine the velocities of various points on links of a mechanism. It is required to calculate the stored kinetic energy and determine the acceleration of links which are needed for dynamic force analysis. The velocities and accelerations of links can be determined mainly by two methods—graphical and analytical. The relative velocity method and instantaneous centre method are discussed in this chapter. The study of velocity and acceleration will be useful in the design of mechanisms of a machine.

Velocity

The rate of change of displacement with respect to time is velocity. Velocity can be linear or angular.

Linear velocity (v) is the rate of change of linear displacement with time = $\dfrac{dx}{dt}$

Angular velocity (ω) is rate of change of angular displacement with time = $\dfrac{d\theta}{dt}$

The relation between linear velocity and angular velocity is as follows:

$$x = r\theta$$
$$\frac{dx}{dt} = r\frac{d\theta}{dt}$$
$$v = r\omega$$
$$\omega = \frac{d\theta}{dt}$$

Acceleration

The rate of change of velocity with respect to time is termed acceleration. Acceleration can be linear or angular.

Linear acceleration is the rate of change of linear velocity with time,

$$f = \frac{dv}{dt} = \frac{d^2x}{dt^2}$$

Angular acceleration is the rate of change of angular velocity with time,

$$\alpha = \frac{d\omega}{dt} = \frac{d^2\theta}{dt^2}$$

Absolute Velocity

The velocity of a point with respect to a fixed point (zero velocity point) is the absolute velocity of that point (Fig. 3.1).

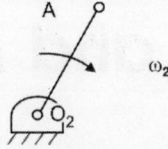

Fig. 3.1: Absolute velocity of a link

$$v_a = \omega_2 \times r$$
$$v_a = \omega_2 \times O_2A$$

Relative Velocity

The velocity of a point with respect to another point 'x' which is not fixed is the relative velocity of that point. For example, v_{ba}, the velocity of point B with respect to point A is relative velocity (Fig. 3.2).

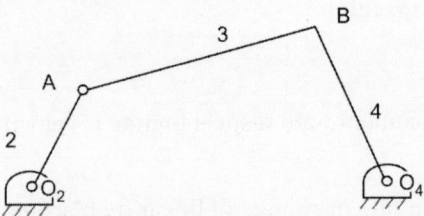

Fig. 3.2: Relative velocity of a link

Notation

Capital letters are used for configuration diagram and small letters for velocity vector diagram (Fig. 3.3).

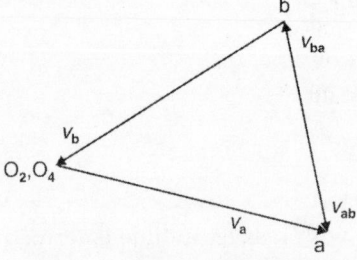

Fig. 3.3: Velocity vector diagram

$$\text{Vector } O_2 a = v_a = \text{Absolute velocity}$$
$$\text{Vector } ab = v_{ab}$$
$$ba = v_{ba}$$

v_{ab} is equal in magnitude with v_{ba} but opposite in direction.
Vector $O_4 b = v_b$ absolute velocity.

3.2 MOTION OF A LINK

Consider a rigid link OA, of length r, rotating about a fixed point O with a uniform angular velocity ω rad/s in the counterclockwise direction (Fig. 3.4a). In small time interval δt, link OA turns through a small angle $\delta\theta$, point A reaches A' in time δt (Fig. 3.4b).

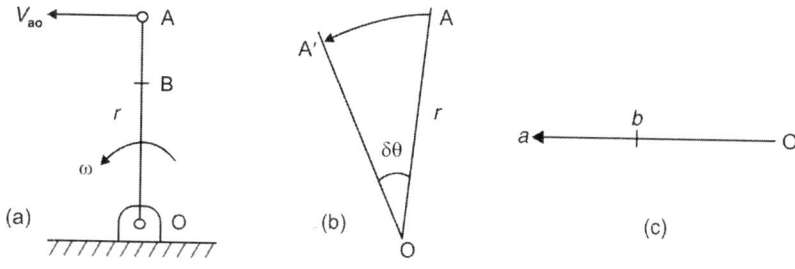

Fig. 3.4

$$\text{Velocity of A relative to } O = \frac{\text{arc } AA'}{\delta t}$$

$$v_{ao} = \frac{r\delta\theta}{\delta t} \quad (\text{arc } AA' = r\delta\theta)$$

Taking limit $\delta t \to 0$

$$v_{ao} = r\,\frac{d\theta}{dt} = r\omega \qquad \left(\frac{d\theta}{dt} = \omega\right)$$

$$v_{ao} = r\omega$$

The direction of v_{ao} is along the displacement of A and perpendicular to OA as $\delta t \to 0$. Now, consider a point B on the link OA.

Velocity of $\qquad B = \omega \times OB$

or $\qquad\qquad v_{bo} = \omega \times OB$

v_{ao} can be represented by vector $\overrightarrow{oa}$ and v_{bo} by vector $\overrightarrow{ob}$.

We find, $\qquad\qquad \dfrac{ob}{oa} = \dfrac{\omega OB}{\omega OA} = \dfrac{OB}{OA}$

Therefore, b divides the velocity vector in the same ratio as point B divides the link. This shows that the magnitude of the instantaneous linear velocity (v_{ao}) of a point A on a rotating body OA is proportional to its distance from the axis of rotation.

3.3 VELOCITIES IN FOUR BAR MECHANISM

Consider a four-bar mechanism $ABCD$ with AD as fixed link and BC as coupler. AB is the driver link rotating at an angular speed of ω rad/s in the clockwise direction (Fig. 3.5).

Velocity of A relative to A = Velocity of C relative to B + velocity of B relative to A

$$V_{ca} = V_{cb} + V_{ba}$$

Since both the points A and D lie on the fixed link AD, therefore, velocity of C relative to A is the same as velocity of C relative to D.

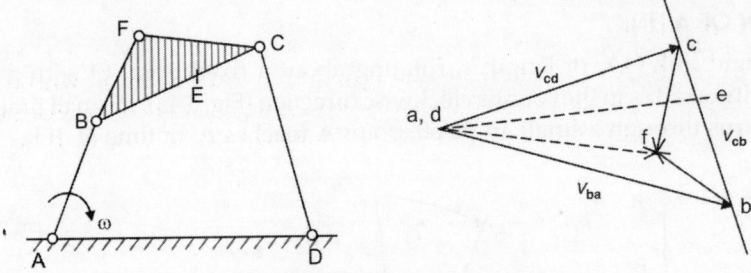

Fig. 3.5: Four bar mechanism

The related equation can be written as,

$$v_{cd} = v_{ba} + v_{cb}$$

or

$$\overrightarrow{dc} = \overrightarrow{ab} + \overrightarrow{bc}$$

In this,

$$v_{ba} = ab = \omega AB \qquad\qquad (\perp \text{ to } AB)$$
$$v_{cb} = bc = \text{unknown} \qquad\qquad (\perp \text{ to } BC)$$
$$v_{cd} = dc = \text{unknown} \qquad\qquad (\perp \text{ to } DC)$$

Velocity Diagram

The magnitude of unknown velocities can be determined by constructing the velocity diagram as discussed below:

i. From any point *a*, first draw vector *ab* of magnitude ωAB in the direction perpendicular to *AB*.

ii. Through *b*, draw a line perpendicular to *BC*, taken on both sides of *b*.

iii. From point *d (a)*, draw a line prependicular to *DC*. The line of vector *dc* and line of vector *bc* intersect at *C*.

iv. *dc* represents the magnitude of velocity of *C* relative to *A* (or *D*).

v. Intermediate point: The velocity of any point *E* on link *BC* can be found by dividing the velocity vector *bc* in the ratio as point *E* divides the link *BC*.

$$\frac{\overrightarrow{be}}{\overrightarrow{bc}} = \frac{BE}{BC}$$

Vector *ae* represents the absolute velocity of *E* with respect to *A*.

vi. For any offset point F, the equation can be writtten as:

$$v_{fb} + v_{ba} = v_{fc} + v_{cd}$$

or

$$v_{ba} + v_{fb} = v_{cd} + v_{fc}$$

$$\overrightarrow{ab} + \overrightarrow{bf} = \overrightarrow{dc} + \overrightarrow{cf}$$

To represent v_{fb}, draw a line $\perp BF$ through *b* and to represent v_{fc}, draw a line perpendicular to *CF* through *C*. The intersection of two lines gives the location of the point *f*.

Velocity Images

In Fig. 3.5 triangle *bfc* is similar to triangle *BFC* in which all the three sides *bc*, *cf* and *fb* are perpendicular to *BC*, *CF* and *FB* respectively. Such triangles like *bfc* are known as velocity images.

Velocity image of a link is scaled reproduction of configuration diagram rotated through 90° in the direction of the angular velocity.

3.4 ANGULAR VELOCITY OF LINKS

i. Angular Velocity of BC

Velocity of C relative to B, $V_{cb} = \overrightarrow{bc}$

The direction of $\overrightarrow{bc}$ is upwards from Fig. 3.5. Therefore, point C moves in counter-clockwise direction about B.

$$\omega_{cb} = \frac{v_{cb}}{CB}$$

Velocity of B relative to C, $V_{bc} = \overrightarrow{cb}$

The direction of $\overrightarrow{cb}$ is downwards from Fig. 3.5. Therefore, point B moves in counter clockwise direction about C.

$$\omega_{bc} = \frac{v_{bc}}{BC}$$

Thus, magnitude of $\omega_{cb} = \omega_{bc}$ and the direction of rotation is also same.

ii. Angular Velocity of CD

Velocity of C relative to D, $v_{cd} = \overrightarrow{dc}$

Looking at direction of $\overrightarrow{dc}$ at point C, C moves in clockwise direction about D.

$$\omega_{cd} = \frac{v_{cd}}{CD}$$

3.5 VELOCITY OF RUBBING

The links in a mechanism are generally connected by a pin, making a turning pair. The surfaces of links rub on the surface of pin. The velocity of rubbing of two surfaces depends on the angular velocity of a link relative to the other.

The velocity of rubbing is defined as the algebraic sum of the angular velocities of the two links which are connected by a pin joint, multiplied by the radius of the pin.

Let two links OA and OB are connected by a pin joint at O as shown in Fig. 3.6.

ω_{ao} = angular velocity of A with respect to O
ω_{bo} = angular velocity of B with respect to O
r = radius of the pin

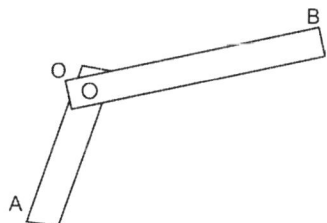

Fig. 3.6: Velocity of rubbing

Velocity of rubbing at O

$$= (\omega_{ao} - \omega_{bo})\, r, \text{ if the links move in the same direction}$$
$$= (\omega_{ao} + \omega_{bo})\, r, \text{ if the links move in the opposite direction.}$$

In Fig. 3.6 velocity of rubbing at $A = r_a\omega$, velocity of rubbing at $B = r_b(\omega_{ab} + \omega_{bc})$, since angular velocity of two links AB and BC are in opposite directions.

velocity of rubbing at $C = r_c (\omega_{bc} + \omega_{dc})$ ω_{bc} = counter-clockwise, ω_{dc} = clockwise velocity of rubbing at $D = r_d \times \omega_{cd}$.

3.6 VELOCITIES IN SLIDER–CRANK MECHANISM

Consider a slider–crank mechanism in which slider A is attached to the connecting rod AB. Let the radius of crank OB be r moving with uniform angular velocity ω (rad/s) in the clockwise direction. AB is the coupler joining A and B. The slider reciprocates along AO in fixed guide (Fig. 3.7). The velocity of slider A can be determined as explained below.

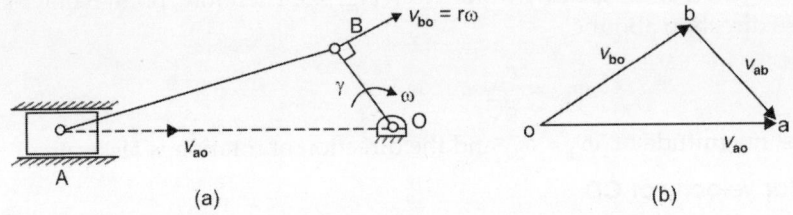

(a)   (b)

Fig. 3.7: (a) Slider–crank mechanism (b) Velocity diagram

Velocity of A relative to O = velocity of A relative to B + velocity of B relative to O.

$$v_{ao} = v_{ab} + v_{bo} = v_{bo} + v_{ab}$$

i. From any point O, draw vector $v_{bo} = r\omega$ perpendicular to OB.
ii. v_{ab} is perpendicular to AB. Draw a line perpendicular to AB through b.
iii. From point O, draw vector V_{ao} parallel to the motion of the slider A.

Intersection of two lines gives the location of point a. Now oa represents the magnitude of velocity of slider A.

The angular velocity of coupler AB,

$$\omega_{ab} = \frac{v_{ba}}{AB} = \frac{ab}{AB} \qquad\qquad \text{(counterclockwise about A)}$$

3.7 INSTANTANEOUS CENTRE METHOD

The instantaneous centre of rotation is a point common to two bodies in plane motion which has the same instantaneous velocity in each body.

To explain instantaneous centre, let us consider a plane body P having a nonlinear motion relative to another body Q. Consider two points A and B on body P having velocities as v_a and v_b respectively in the directions shown in Fig. 3.8.

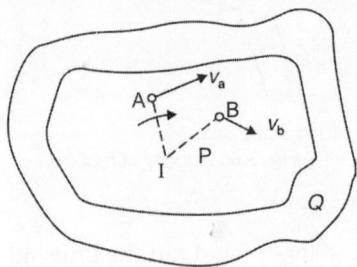

Fig. 3.8: Instantaneous centre method

If a line is drawn ⊥ to v_a at A, the body can be imagined to rotate about some point on the line. At the same time, centre of rotation of the body also lies on a line ⊥ to the direction of v_b at B. If I is the point of intersection of the two lines, the body P will be rotating about I at that instant. The point I is known as the *instantaneous centre of rotation* for the body P. The position of instantaneous centre changes with the motion of the body. The locus of all such instantaneous centres is known as *centrode*. A line drawn through an instantaneous centre and perpendicular to the plane of motion is called *instantaneous axis*. The locus of such an axis is known as *axode*.

In case ⊥ lines drawn from A and B meet outside the body P as shown in Fig. 3.9, I-centre will lie outside the body P (Fig. 3.9).

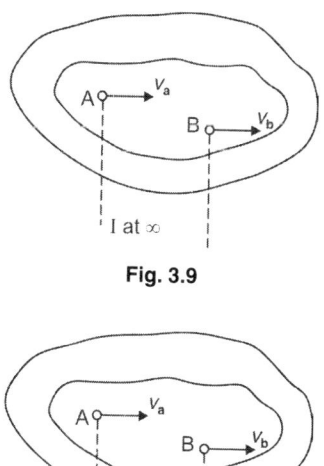

Fig. 3.9

Fig. 3.10

If the directions of v_a and v_b are parallel and the ⊥ at A and B meet at infinity, the I centre of the body lies at infinity (Fig. 3.10). This is the case when the body has linear motion.

3.7.1 Velocity of a Point on a Link by Instantaneous Centre Method

Consider two points A and B *on a rigid link*. Let v_A and v_B be the velocities of points A and B, at angles α and β as shown in Fig. 3.11. If v_A is known in magnitude and direction and v_B in direction only, then the magnitude of v_B may be determined by the instantaneous centre method as explained below.

Draw AI and BI perpendicular to the directions of v_A and v_B respectively. These lines intersect at I, which is instantaneous centre of the link. The complete rigid link can be considered to rotate about the centre I at that instant. Since A and B are the points on a rigid link, therefore there cannot be any relative motion between them along the line AB.

Now resolving the velocities along AB,

$$v_A \cos \alpha = v_B \cos \beta$$

or
$$\frac{v_A}{v_B} = \frac{\cos\beta}{\cos\alpha} = \frac{\sin(90° - \beta)}{\sin(90° - \alpha)} \qquad \qquad ...(3.1)$$

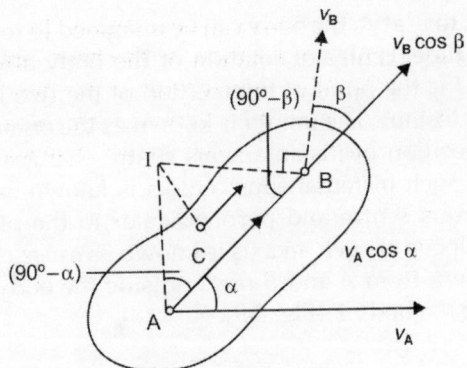

Fig. 3.11: Velocity of a link

Applying Lami's theorem to triangle ABI.

$$\frac{AI}{\sin(90°-\beta)} = \frac{BI}{\sin(90° - \alpha)}$$

or
$$\frac{AI}{BI} = \frac{\sin(90° - \beta)}{\sin(90° - \alpha)} \qquad ...(3.2)$$

From Eqs (3.1) and (3.2),

$$\frac{v_A}{v_B} = \frac{AI}{BI} \text{ or } \frac{v_A}{AI} = \frac{v_B}{BI} = \omega \qquad ...(3.2a)$$

where ω = angular velocity of the rigid link.

If C is any other point on the link, then

$$\frac{v_a}{AI} = \frac{v_B}{BI} = \frac{v_c}{CI}$$

From Eq. (3.2a), it can be observed that the magnitude of velocity of a point on a rigid link is inversely proportional to the distance of the point from the instantaneous centre and is perpendicular to the line joining the point to the instantaneous centre. The velocity of point I would be zero. This implies that there is no relative motion between two bodies at the instantaneous centre.

Notation

The I-centre of two bodies is named in ascending order of alphabets or digits, e.g. I-centre of two bodies 1 and 2 is named as 12.

3.7.2 Number of Instantaneous Centres

The number of instantaneous centres in a mechanism depends upon the number of links. If N is the number of instantaneous centres and n is the number of links, then

$$N = \frac{n(n-1)}{2}$$

A four bar linkage has 6 instantaneous centres, six bar has 15 and an eight bar has 28.

3.7.3 Types of Instantaneous Centres

There are three types of instantaneous centres namely (1) fixed instantaneous centre (2) permanent instantaneous centre and (3) neither fixed nor permanent instantaneous centre. The first two types, i.e. fixed and permanent instantaneous centres are together known as *primary instantaneous centres* and the third type is known as *secondary instantaneous centres*.

Example: Consider a four-bar mechanism (Fig. 3.12), i.e. $n = 4$.

The number of instantaneous centres (N) in a four-bar mechanism is given by

$$N = \frac{n(n-1)}{2} = \frac{4(4-1)}{2} = 6$$

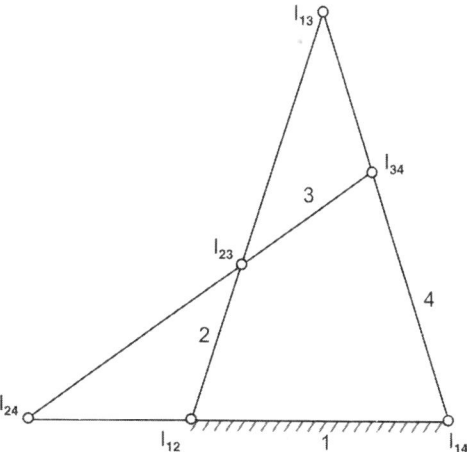

Fig. 3.12: Four bar mechanism

The instantaneous centre I_{12} and I_{14} are called the fixed instantaneous centres as they are fixed for all configurations of the mechanism. The instantaneous centres I_{23} and I_{34} are the permanent instantaneous centres as they move when the mechanism moves, but the joints are of permanent nature. The instantaneous centres I_{13} and I_{24} are neither fixed nor permanent instantaneous centres as they vary with the configuration of the mechanism.

3.7.4 Location of Instantaneous Centres

The centres in a mechanism can be located following certain rules described below.

i. When the two links, not fixed, are connected by a pin joint, the instantaneous centre lies on the centre of the pin as shown in Fig. 3.13a. Such an instantanesous centre is of permanent nature, but if one of the links is fixed, the instantaneous centre is of fixed type.

ii. When the two links have a pure rolling contact (*i.e.* link 2 rolls without slipping upon the fixed link 1 which may be straight or curved), the instantaneous centre lies on their point of contact, as shown in Fig. 3.13b.

iii. When the two links have a sliding contact, the instantaneous centre lies on the common normal at the point of contact which can be at infinity (Fig. 3.13c) or can lie on the centre of curvature of the curvilinear path (Fig. 3.13d).

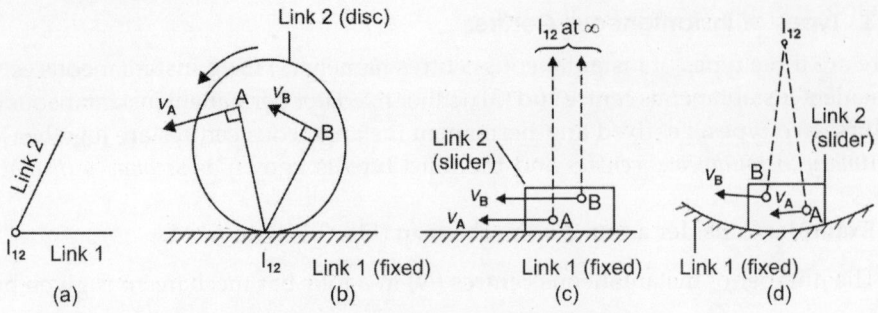

Fig. 3.13: Location of instantaneous centres

3.8 KENNEDY'S THEOREM OF THREE CENTRES

Arnold Kennedy's theorem states that any three bodies in plane motion will have exactly three instantaneous centres, and they will lie on the same straight line.

In other words, if three bodies have motion relative to each other, their instantaneous centres should lie on a straight line (Fig. 3.14).

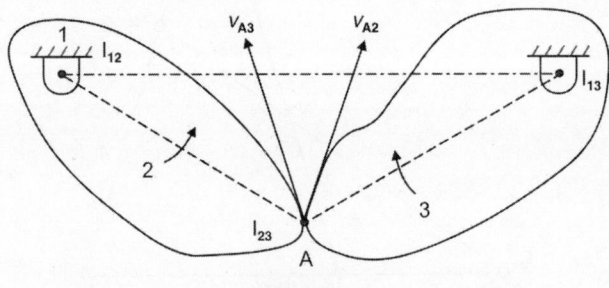

Fig. 3.14

Consider a three-link mechanism with link 1 being fixed, link 2 rotating about I_{12} and link 3 about I_{13}. Hence I_{12} and I_{13} are the instantaneous centres for links 1 and 2 and links 1 and 3 respectively. Let us assume that instantaneous centre of link 2 and 3 be at point A, i.e. I_{23}. Point A is a coincident point on link 2 and link 3.

Considering A on link 2, velocity of A with respect to I_{12} will be a vector $v_{A2} \perp$ to link A–I_{12}. Similarly, for point A on link 3, velocity of A with respect to I_{13} will be $\perp$ to A–I_{13}. It is seen that link velocity vectors of v_{A2} and v_{A3} are in different directions which is impossible. Hence, the instantaneous centre of the two links cannot be at the assumed position A.

It can be seen that the velocities v_{A2} and v_{A3} will be same in magnitude and direction only when I_{23} lies on the line joining I_{12} and I_{13}. Hence, if the three links have relative motion among themselves, all the three instantaneous centres must lie on a same straight line.

3.9 LOCATING INSTANTANEOUS CENTRES IN A MECHANISM

Following steps can be used to locate instantaneous centres in a mechanism:

Step 1: Draw the configuration diagram. Four bar mechanism has been taken as an example and shown in Fig. 3.15.

Step 2: Determine the number of instantaneous centres by using the relation

$$N = \frac{(n-1)n}{2}$$

In four bar mechanism, the number of instantaneous centres (N) is,

$$N = \frac{(n-1)n}{2} = \frac{4(4-1)}{2} = 6$$

Step 3: Make a list of all the instantaneous centres in a mechanism. For a four-bar mechanism, these centres are listed in the following table.

Link	1	2	3	4
Instantaneous Centres (6 in number)	12 13 14	23 24	34	–

Step 4: Locate the fixed and permanent instantaneous centres by inspection. In Fig. 3.15a, I_{12} and I_{14} are fixed instantaneous centres and I_{23} and I_{34} are permanent instantaneous centres. These are located at the centres of the pin joints.

Step 5: Locate the remaining two instantaneous centres by Kennedy's theorem. This is done by circle diagram as shown in Fig. 3.15b. Mark points on a circle equal to the number of links in the mechanism. For the present example, mark 1, 2, 3, and 4 on the circle. Join the points of which instantaneous centres have been found by inspection, by solid lines. In the circle diagram, these lines are 12, 23, 34 and 14 which indicate the centres I_{12}, I_{23}, I_{34} and I_{14}.

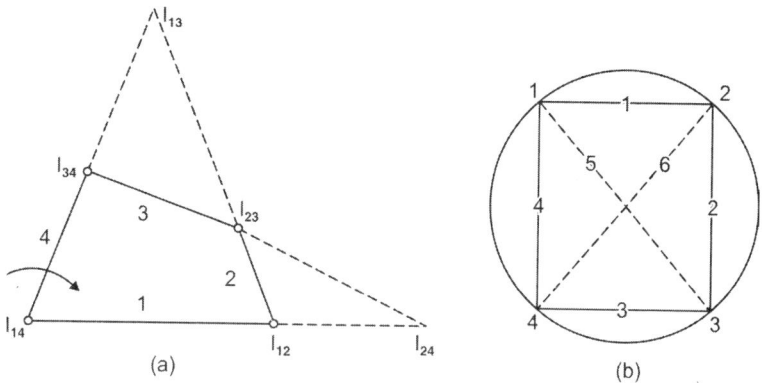

(a)   (b)

Fig. 3.15: Four bar mechanism

Step 6: To find the other two centres, join the points of which instantaneous centres are to be located. By dotted lines. In Fig. 3.15b, join 1 and 3 by dotted lines. The line 1–3 is common to both the similar triangles, therefore, the instantaneous centre I_{13} will lie on the intersection of $I_{12} I_{23}$ and $I_{14} I_{34}$ produced if necessary, on the mechanism. Thus, the instantaneous centre I_{13} is located at the intersection of two lines. Similarly, the instantaneous centre I_{24} can be found by joining points 2 and 4 and locating the point on the intersection of $I_{12} I_{14}$ and $I_{23} I_{34}$ produced if necessary, on the mechanism.

3.10 ANGULAR VELOCITY RATIO

The angular velocity ratio (m_v) is defined as the output angular velocity divided by the input angular velocity. For a four-bar mechanism, this is expressed as:

$$m_v = \frac{\omega_4}{\omega_2}$$

The angular velocity ratio is useful to find the angular velocity of second link (output link) when the angular velocity of first link (input link) is known. It has been found that the velocity of their common instantaneous centre relative to a fixed third link is same irrespective of whether the instantaneous centre is considered on first link or second link.

Example: In a four-bar mechanism shown in Fig. 3.15, it is required to find the angular velocity of link 4 when the angular velocity of link 2 is known. First, consider that, link 2 is a disc containing instantaneous centre I_{24} and revolving about I_{12},

$$v_{24} = (24 - 12)\omega_2$$

Now, consider link 4 to be containing instantaneous centre I_{24} and revolving about I_{14}.

$$v_{24} = (24 - 14)\omega_4$$

This expression is known as the angular velocity theorem which states that the angular velocity ratio of two links is inversely proportional to the distances of their common instantaneous centre from their respective centres of rotation.

3.11 ACCELERATION ANALYSIS

Once the velocity analysis is done, the next step is to determine the acceleration of various points of interest on links in the mechanism. The accelerations are required to calculate dynamic forces from $F = ma$ and these dynamic forces will contribute to the stresses in the links.

Acceleration is defined as the rate of change of velocity with respect to time, and it acts in the direction of the change of velocity.

3.11.1 Acceleration of a link

Consider a link OA, of length r moving in clockwise direction as shown in Fig. 3.16. Let point A moves with respect to O with an angular velocity ω (rad/s) and angular acceleration α (rad/s^2) at an instant.

When the velocity of a particle changes in magnitude and direction, it has two components of acceleration:

    i. Centripetal or radial acceleration which is perpendicular to the velocity of the particle at that instant,

$$f^c = \omega^2 r$$

    ii. Tangential acceleration which is parallel to the velocity of the particle at that instant,

$$f^t = \alpha \cdot r$$

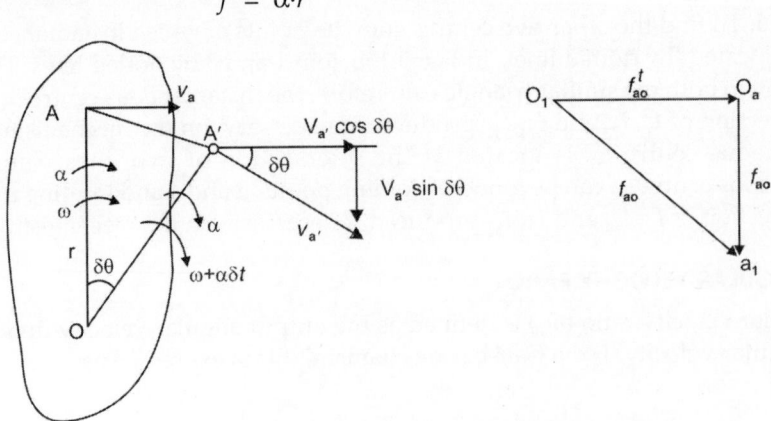

Fig. 3.16: Acceleration of a link

Tangential velocity of A, $v_a = \omega r$

Point A moves to A' after small time δt and link OA rotates through small angle $\delta\theta$.

Angular velocity of OA', $\omega_a' = \omega + \alpha\delta t$

Tangential velocity of A', $va' = \omega a'.r$

$$= (\omega + \alpha\delta t)r$$

3.11.2 Centripetal or radial acceleration

Change in velocity parallel to $OA = v_{a'}\sin\delta\theta - 0$

Acceleration of A parallel to $OA = \dfrac{(\omega + \alpha\,\delta t)r\sin\delta\theta}{\delta t}$

Taking limit $\delta t \to 0$, $\sin\delta\theta \to \delta\theta$

or centripetal acceleration,

$$f_{ao}^c = \omega r\,\frac{d\theta}{dt}$$

or

$$f_{ao}^c = \omega r\omega = \omega^2 r \qquad\qquad \left(\frac{d\theta}{dt} = \omega\right)$$

$$= \frac{v^2}{r} \qquad\qquad (v = \omega r)$$

3.11.3 Tangential Acceleration

Change in velocity of A perpendicular to OA

$$= v_{a'}\cos\delta\theta - v_a$$

Acceleration of A perpendiculer to OA,

$$= \frac{(\omega + \alpha\delta t)r\cos\delta\theta - \omega r}{\delta t}$$

Taking limit, $\delta t \to 0$, $\cos\delta\theta \to 1$

Tangential acceleration, $\quad f_{ao}^t = \alpha r$

or

$$f_{ao}^t = \frac{d\omega}{dt}\cdot r = \frac{dv}{dt}$$

When $\alpha = 0$, link OA rotates with uniform angular velocity $f_{ao}^t = \alpha r = 0$.

3.12 ACCELERATION IN SLIDER–CRANK MECHANISM

Consider a slider–crank mechanism as shown in Fig. 3.17.

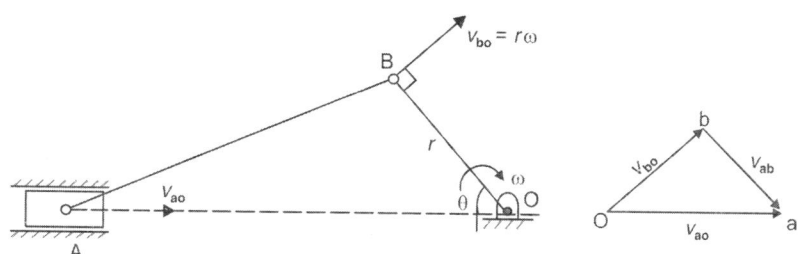

Fig. 3.17: Slider–crank mechanism

Acceleration of A relative to O = accelration of A reletive to B + acceleration of B relatvie to O.

$$f_{ao} = f_{ab} + f_{bo} = f_{bo} + f_{ab} = f_{bo} + f_{ab}{}^c + f^t{}_{ab}$$

or $\quad\quad\quad o_1 a_1 = o_1 b_1 + b_1 b_a + b_a a_1$

Since crank OB rotates at a uniform angular velocity ($\alpha = 0$), acceleration of B has only centripetal component (Fig. 3.18).

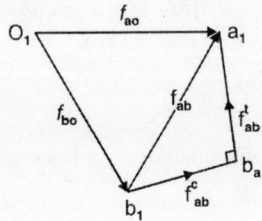

Fig. 3.18: Acceleration diagram

Acceleration of slider A is given by $o_1 a_1$

$$f_{ab}^c = v_{ab}^2 / BA \quad\quad\quad\quad\quad\quad\text{(Parallel to } BA)$$

$$f_{bo} = v_{bo}^2 / OB \quad\quad\quad\quad\quad\quad\text{(Parallel to } OB)$$

3.13 CORIOLIS ACCELERATION

The acceleration of a moving point relative to a fixed body can have two components of acceleration, viz. centripetal and tangential. But when a sliding point is present on a rotating link, an additional component of acceleration will be present, known as *coriolis acceleration*.

Consider a link OA and a slider B as shown in Fig. 3.19.

Let ω = angular velocity of link OA

$\quad v$ = linear velocity of slider B along OA

$\quad \alpha$ = angular acceleration of link OA

$\quad f$ = linear acceleration of slider on link

$\quad r$ = radial distance of point C on slider

$\quad \omega r$ = velocity of the slider B with respect to O, perpendicular to link OA.

Let $\delta\theta$ be the angular displacement of the link and δr be the radial displacement of the slider in outward direction in a short interval of time δt,

i.e. ω becomes $\quad\quad\quad \omega' = \omega + \delta\omega = \omega + \alpha\delta t$;

$\quad v$ becomes $\quad\quad\quad v' = v + \delta v = v + f.\delta t$;

$\quad r$ becomes $\quad\quad\quad r' = r + \delta r$ from velocity diagrams as shown in Fig. 3.19.

Total component of change of velocity along radial direction *(OA)* = component of change of velocity when linear velocity v is considered + component of change of velocity when change in $r\omega$ is considered.

$$= bx - yb_1 \quad\quad (yb_1 \text{ is radially inward, therefore taken negative})$$

$$= \{(v + \delta v) \cos \delta\theta - v\}\uparrow - \{(\omega + \delta\omega)(r + \delta r) \sin \delta\theta\}\downarrow \quad\quad ...(3.3)$$

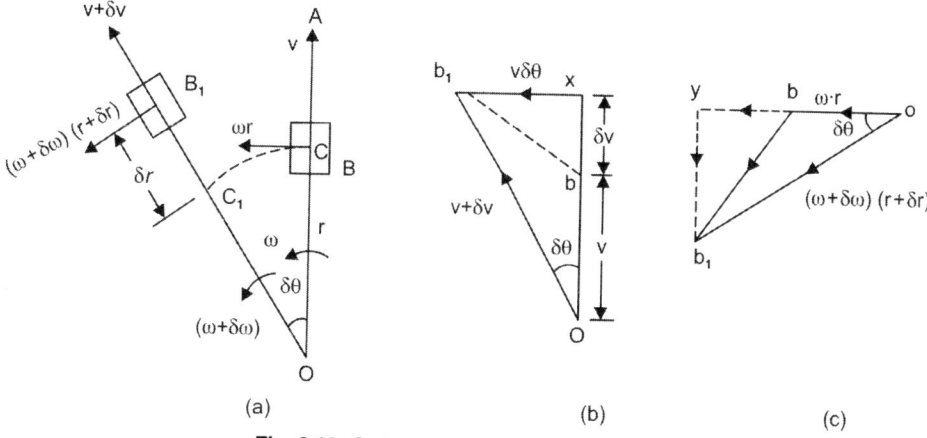

Fig. 3.19: Coriolis component of acceleration

Since $\delta\theta$ is very small, therefore substituting $\cos\delta\theta = 1$ and $\sin\delta\theta = \delta\theta$, Eq. (3.3) becomes

$$= \{v + \delta v - v\} - \{\omega r\delta\theta + \omega\delta r\delta\theta + \delta\omega \cdot r \cdot \delta\theta + \delta\omega\delta r\delta\theta\}$$
$$= (\delta v - \omega r\delta\theta)\uparrow \qquad \text{(neglecting small terms)}$$

Radial component of acceleration of slider *B* with respect to *O* acting radially outwards, taking the limit $\delta t \to 0$.

$$\alpha_{BO}^r = \lim_{\delta t \to 0} \frac{(\delta v - \omega r\delta\theta)}{\delta t} = \frac{dv}{dt} - \omega r\frac{d\theta}{dt}$$

$$= (f - \omega^2 r)\uparrow \qquad \left(\frac{d\theta}{dt} = \omega\right)$$

Similarly, total component of change in velocity along tangential direction,

$$= xb_1 + by$$
$$= \{(v + \delta v)\sin\delta\theta\} + \{(\omega + \delta\omega)(r + \delta r)\cos\delta\theta - \omega r\} \qquad \text{...(3.4)}$$

Since $\delta\theta$ is very small, $\cos\delta\theta = 1$, $\sin\delta\theta = \delta\theta$, Eq. (3.3) becomes

$$= v\delta\theta + (\omega\delta r + r\delta\omega) \qquad \text{(neglecting small terms)}$$

Tangential component of acceleration of slider *B* with respect to *O* acting perpendicular to *OA* ($\leftarrow$), is

$$\alpha_{BO}^t = \lim_{\delta t \to 0} \frac{v\delta\theta + (\omega\delta r + r\delta\omega)}{\delta t}$$

$$= v\frac{d\theta}{dt} + \omega\frac{dr}{dt} + r\frac{d\omega}{dt}$$

$$= v\omega + \omega v + r\alpha = 2\,v\omega + r\alpha$$

This tangential component of slider B, $2v\omega$ is known as coriolis acceleration component. It is always perpendicular to link and positive if both ω and v are either positive or negative.

The direction of coriolis acceleration component is obtained by rotating the radial velocity vector v through 90° in the direction of rotation of link. The direction of coriolis acceleration component ($2v\omega$) for all four possible cases is shown in Fig. 3.20.

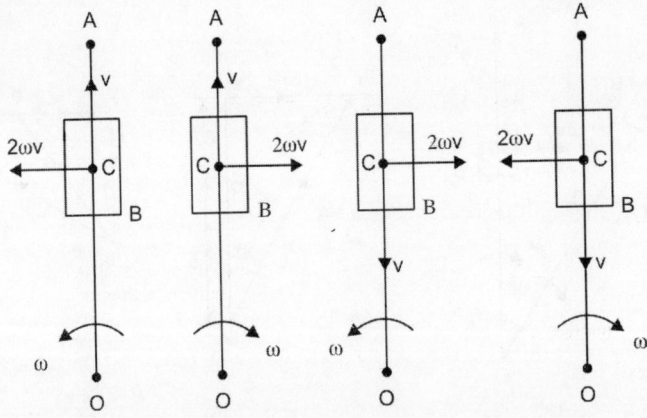

Fig. 3.20: Direction of coriolis components of acceleration

EXAMPLES

Example 3.1: The crank of a reciprocating engine revolves at a uniform speed of 310 rpm in clockwise direction. The crank and connecting rod are 15 cm and 65 cm long respectively. Find the velocity of piston for crank positions from 0° to 90° from inner dead centre at interval of 30°.

Solution: Crank radius r = 15 cm = 0.15 m

Connecting rod l = 65 cm = 0.65 m; N = 310 rpm

$$W = \frac{2\pi \times 310}{60} = 32.46 \text{ rad/s}; \quad n = \frac{l}{r} = \frac{0.65}{0.15} = 4.333$$

We know that approximate analytical method to find the velocity of the piston is given as

$$v = \omega \cdot r \{\sin \phi + \sin 2\phi/2n\} = 32.46 \times 0.15 \{\sin \phi + \sin \phi/2n\}$$

When $\phi = 30°$

$$v = 32.46 \times 0.15 \{\sin 30° + \sin 60°/2 \times 4.333\}$$

$$= 32.46 \times 0.15 (0.5999) = 2.92 \text{ m/s}$$

When $\phi = 60°$

$$v = 32.46 \times 0.15 \{\sin 60° + \sin 120°/2 \times 4.333\}$$

$$= 4.70 \text{ m/s}$$

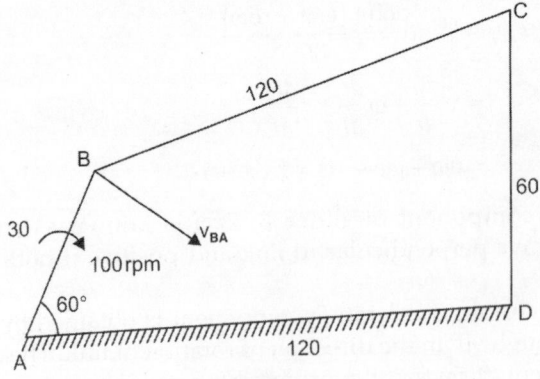

Fig. 3.21

When $\phi = 90°$

$$v = 32.46 \times 0.15 \{\sin 90° + \sin 180°/2 \times 4.333\}$$
$$= 4.8675 \text{ m/s}$$

Example 3.2: In a four bar chain ABCD, AD is fixed and 120 mm long. The crank AB is 30 mm long and rotates at 100 rpm clockwise, while the link CD = 60 mm oscillates about D, BC and AD are of equal length. Find the angular velocity of link CD when angle BAD = 60°.

Solution:

$N_{BA} = 100$ rpm

$$\omega_{BA} = 2\pi N_{ba}/60 = 2\pi \times 100/10 = 10.47 \text{ rad/s}$$

Configuration diagram is shown in Fig. 2.25(a).

Crank AB = 30 mm = 0.03 m

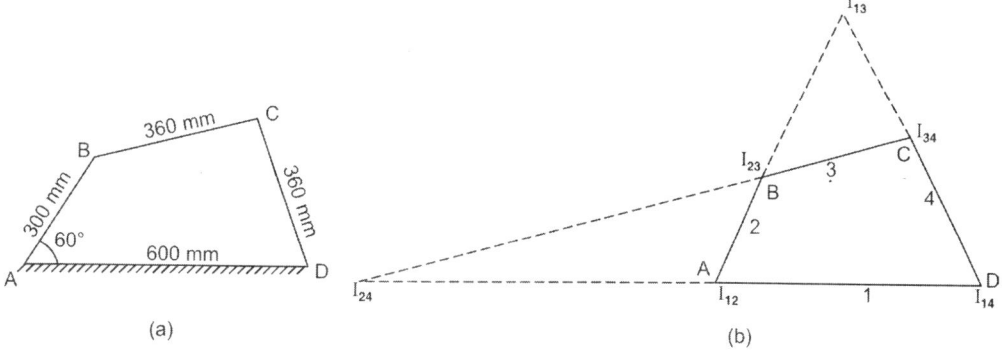

(a)　　　　　　　　　　　　　(b)

Fig. 3.22

Point A is fixed. The velocity of point B with respect to A,

$$v_{BA} = \omega_{BA} \cdot AB$$
$$= 10.47 \times 0.03 = 0.3141 \text{ m/s}$$

The velocity diagram shown in Fig. 3.22 is drawn as follows:

In the configuration diagram point A and D are fixed, so they may be taken as one point (a, d) in the velocity diagram. The velocity of point B with respect to A, v_{BA} is perpendicular to AB.

Vector AB represent v_{BA} is drawn perpendicular to AB with some suitable scale. $v_{BA} = 0.3141$ m/s.

The velocity of point C with respect to point B, v_{CB} is perpendicular to BC. Only the direction of v_{CB} is known.

From b draw a vector be representing v_{CB}. It contains point C.

The velocity of point C with respect to D, v_{CD} is perpendicular to CD.

From d (or a), draw vector dc representing v_{CD}. It intersects vector bc at point c. By measurement, we find

$$v_{CD} = dc = 0.24 \text{ m/s}$$

It is given that CD = 60 mm = 0.06 m

so angular velocity of link CD

$$\omega_{CD} = V_{CD}/CD = 0.24/0.06 = 4 \text{ rad/sec}$$

Its direction is clockwise about D.

Example 3.3: In a pin-jointed four bar mechanism as shown in Fig. 2.23, AB = 300 mm, BC = CD = 360 mm and AD = 600 mm, angle BAD = 60°, the crank AB rotates uniformly at 100 rpm, locate all the instantaneous centres and find the angular velocity of the link BC.

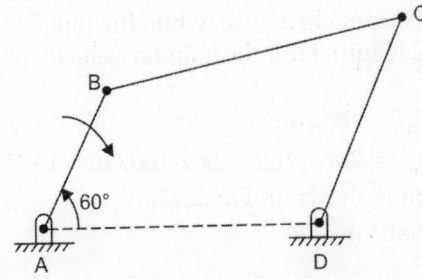

Fig. 3.23

Solution:
$$N_{AB} = 100 \text{ rpm}$$
$$\omega_{AB} = 2\pi n_{ab}/60 = 2\pi \times 100/60 = 1047 \text{ rad/s}$$
$$AB = 300 \text{ mm} = 0.30 \text{ m}$$
$$v_{BA} = \omega_{AB} \cdot AB = 10.47 \times 0.30 = 3.141 \text{ m/s}$$

Number of links $n = 4$, so the number of instantaneous centres is
$$N = n(n-1)/2 = 4(4-1)/2 = 6$$

The six instantaneous centres are: I12, I23, I34, I14, I13, I24.

Angular velocity of link BC:

B is point on link BC, so the velocity of point B on BC is
$$v_B = \omega_{BC} \cdot I13B \qquad (v_{BA} = v_B)$$

or
$$\omega_{BC} = v_B/I13B = 3.141/0.495 = 6.345 \text{ rad/s} \ (I13B = 0.495 \text{ m})$$

I13B is taken by measurement.

EXERCISE

3.1 What is configuration diagram? What is its use?

3.2 Describe the procedure to construct the diagram of a four-link mechanism.

3.3 What is a velocity image? State why it is known as a helpful device in the velocity analysis of complicated linkages.

3.4 Describe the procedure to draw velocity and acceleration diagrams of a four-link mechanism. In what way are the angular accelerations of the output link and the coupler found?

3.5 What is instantaneous centre of rotation? How do you find the number of instantaneous centres in a mechanism?

3.6 What is a velocity of rubbing? How is it found?

3.7 Define normal and tangential components of acceleration.

3.8 What is the coriolis component acceleration? In which cases does it occur? How is it determined?

3.9 How do you determine normal and tangential components of acceleration?

3.10 State and prove Kennedy's theorem as applicable to instantaneous centres of rotation of three bodies. How is it helpful in locating the various instantaneous centres of a mechanism?

3.11 What is acceleration image? How is it helpful in determining the acceleration of offset points in a mechanism?

3.12 State and explain angular velocity ratio theorem as applicable to mechanisms.

3.13 Define coriolis component of acceleration. When it occurs?

3.14 In a slider–crank mechanism, the stroke of the slider is one-half the length of the connecting rod. Draw a diagram to give the velocity of the slider at any instant assuming the crankshaft to run uniformly.

3.15 The dimension of a four-bar chain shown in Fig. 3.23 are: AD = BE = 120 mm, AB = 30 mm and CD = 600 mm. The crank AB rotates at 100 rpm. Determine the angular speed of link CD.

3.16 In a four-link mechanism, the crank AB rotates at 36 rad/s, the lengths of the links are AB = 200 mm, BC = 400 mm, CD = 450 mm and AD = 600 mm. AD is the fixed link. At the instant, when AB is at right angles to AD, determine the velocity of:
  i. The midpoint of link BC
  ii. A point on the link CD, 100 mm from the pin connecting the links CD and AD.

3.17 The crank AB of a four-bar mechanism shown in Fig. 3.24 rotates with an angular speed of 100 rad/s and an angular acceleration of 4400 rad/s² when the crank makes an angle of 53° to the horizontal. Determine (a) the angular acceleration of BC, and (b) linear acceleration of point E. Take AB = 75 mm, BC = 80 mm, AD = 125 mm and BE = 28 mm.

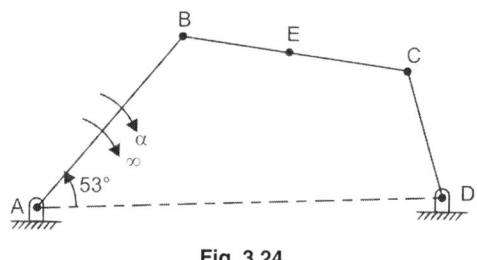

Fig. 3.24

3.18 The dimensions of the various links of the mechanism shown in Fig. 3.25 are: AD = DE = 150 mm, BC = CD = 450 mm, EF = 375 mm. The crank AB rotates at 120 rpm. The lever DC oscillates about the fixed D. Determine (a) velocity of slider F, and (b) angular speed of CD.

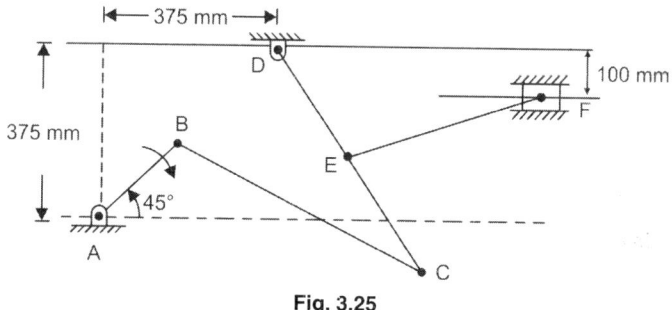

Fig. 3.25

3.19 In a toggle mechanism shown in Fig. 3.26, the crank OA rotates at 180 rpm and the slider is constrained to move on a horizontal path. OA = 180 mm, BC = 240 mm, AB = 360 mm and BD = 540 mm. Find (a) velocity of slider D, (b) angular speed of links AB, BC and BD, (c) velocity of rubbing on the pins of diameter 30 mm at A and D, and (b) torque applied to crank OA for a force of 2 kN at D.

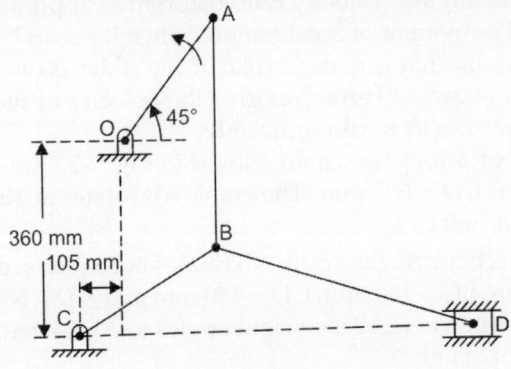

Fig. 3.26

3.20 Find the velocity of point C in the mechanism shown in Fig. 3.27 by using relative velocity method. Crank O_2A rotates at 20 rad/s clockwise.

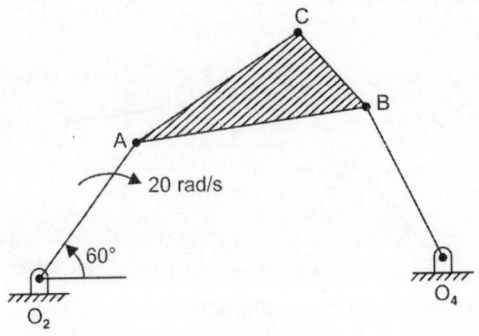

Fig. 3.27

3.21 Determine the angular acceleration of links 3 and 4 and the absolute acceleration of point C on link 3 in the mechanism shown in Fig. 3.28. $O_2A = 45$ mm, $AB = 130$ mm, $O_2O_4 = 90$ mm, $O_4B = 60$ mm, $AC = 55$ mm, $BC = 100$ mm.

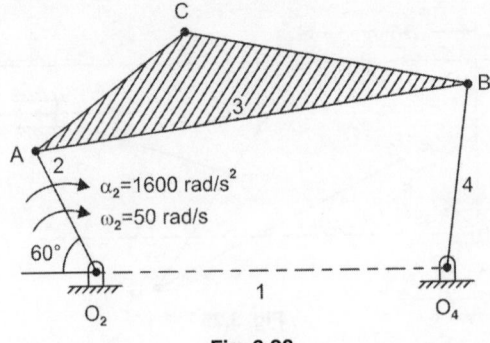

Fig. 3.28

3.22 In crank and slotted lever quick return mechanism, the distance between the fixed centres O and A is 250 mm. Other lengths are OP = 100 mm, AR = 400 mm, RS = 150 mm, and $\angle AOP = 120°$. Uniform speed of the crank is 602 rpm clockwise, line of stroke of the ram is perpendicular to OA and is 450 mm above A. Calculate the velocity and the acceleration of the ram S.

3.23 Figure 3.29 shows a mechanism in which the hydraulic actuator O_2A is expanding at a constant rate of 10 cm/s. Determine the directions and magnitudes of the angular velocity and acceleration of link O_4A.

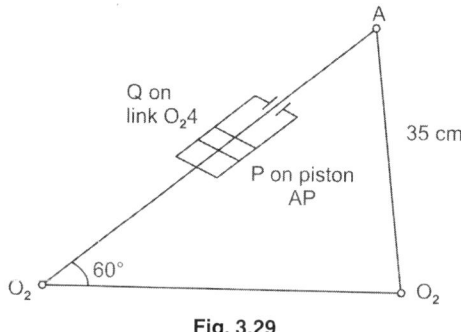

Fig. 3.29

3.24 In the mechanism shown in Fig. 3.30, $O_1O_2 = 210$ mm, $O_1B = 300$ mm and $O_2A = 60$ mm. The crank O_2A rotates at 300 rpm in the counterclockwise direction. Find (a) angular speed of link O_1A, and (b) velocity of slider.

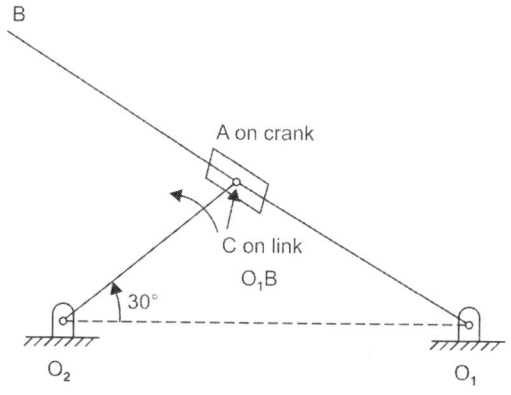

Fig. 3.30

3.25 For the mechanism shown in Fig. 3.31, find the acceleration of point B in link 3. Given: $\omega_2 = 30$ rad/s, $O_2A = 200$ mm, AB = BC = 175 mm, AC = 600 mm.

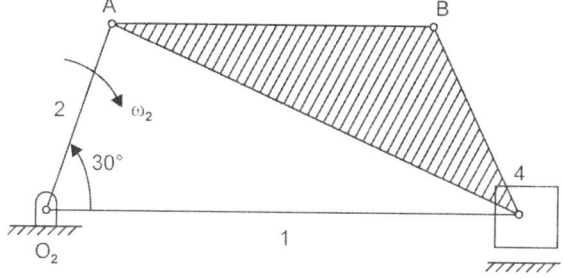

Fig. 3.31

3.26 The crank OP of a crank and slotted lever mechanism Fig. 3.32 rotates at 100 rpm in the counterclockwise direction. Various lengths of the links are OP = 60 mm,

AR = 480 mm and RS = 330 mm, the slider moves along an axis perpendicular to AO and is 120 mm from O. Determine the velocity of the slider when ∠ AOP is 135°. Also find the maximum velocity of the slider during cutting and return strokes.

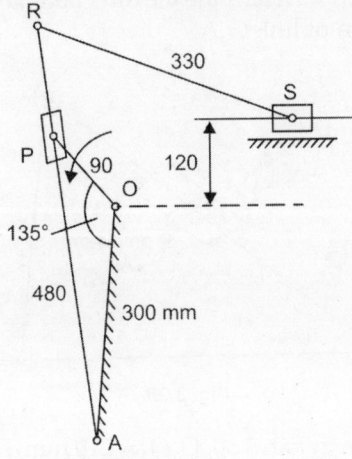

Fig. 3.32

3.27 In the steam mechanism shown in Fig. 3.33, at crank AB rotates at 200 rpm clockwise. Find the velocities of C, D, E, F and G, and the acceleration of slider at C. Given: AB = 120 mm, BC = 4780 mm, CD = 180 mm, DE = 360 mm, EF = 120 mm, and FG = 360 mm.

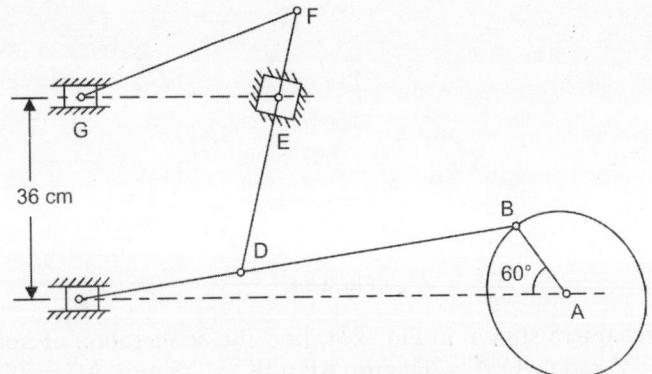

Fig. 3.33

3.28 For the mechanism shown in Fig. 3.34, determine the velocities of the points C, E and F and the angular velocities of the links BD, CDE and EF.

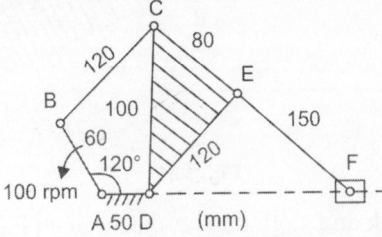

Fig. 3.34

3.29 A link AB of a four-bar mechanism ABCD revolves uniformly at 120 rpm in clockwise direction. Find the angular acceleration of links BC and CD and acceleration of point E in link BC. Given: AB = 75 mm, BC = 175 mm, EC = 50 mm, CD = 150 mm, AD = 100 mm and ∠BAD = 90°.

OBJECTIVE TYPE QUESTIONS

3.1 The total number of instantaneous centres for a mechanism consisting of *n* links are

(a) $\dfrac{n}{2}$

(b) *n*

(c) $\dfrac{n-1}{2}$

(d) $\dfrac{n(n-1)}{2}$

3.2 According to Aronhold Kennedy's theorem, if three bodies are moving relative to each other, their instantaneous centres will lie on a
(a) straight line
(b) parabolic curve
(c) ellipse
(d) none of these

3.3 The instantaneous centres which vary with the configuration of the mechanism are called
(a) permanent instantaneous centres
(b) fixed instantaneous centres
(c) neither fixed nor permanent instantaneous centres
(d) none of these

3.4 When a slider moves on a fixed link having curved surface, their instantaneous centre lies
(a) on their point of contact
(b) at the centre of curvature
(c) at the centre of circle
(d) at the pin joint

3.5 The direction of linear velocity of any point on a link with respect to another point on the same link is
(a) parallel to the link joining the points
(b) perpendicular to the link joining the points
(c) 45° to the link joining the points
(d) none of these

3.6 The magnitude of linear velocity of point B on a link AB relative to point A is .... where ω = angular velocity of the link AB.
(a) $\omega \cdot AB$
(b) $\omega (AB)^2$
(c) $\omega^2 \cdot AB$
(d) $(\omega \cdot AB)^2$

3.7 The two links OA and OB are connected by a pin joint at O. If the link OA turns with angular velocity ω_1 rad/s in the clockwise direction and the link OB turns with angular velocity ω_2 rad/s in the anticlockwise direction, then the rubbing velocity at the pin joint O is ...., where *r* = radius of the pin at O.

    (a) $\omega_1 \cdot \omega_2 \cdot r$
    (b) $(\omega_1 \cdot \omega_2)r$
    (c) $(\omega_1 + \omega_2)r$
    (d) $(\omega_1 \cdot \omega_2)2r$

3.8 In the above question, if both the links OA and OB turn in clockwise direction, then the rubbing velocity at the pin joint O is
    (a) $\omega_1 \cdot \omega_2 \cdot r$
    (b) $(\omega_1 \cdot \omega_2)r$
    (c) $(\omega_1 + \omega_2)r$
    (d) $(\omega_1 \cdot \omega_2)2r$

3.9 The component of the acceleration, parallel to the velocity of the particle, at a given instant is called
    (a) radial component
    (b) tangential component
    (c) coriolis component
    (d) none of these

3.10 A point B on a rigid link AB moves with respect to A with angular velocity ω (rad/s). The total acceleration of B with respect to A will be equal to
    (a) vector sum of radial component and coriolis component
    (b) vector sum of tangential component and coriolis component
    (c) vector sum of radial component and tangential component
    (d) vector difference of radial component and tangential component

3.11 The coriolis component of acceleration is taken into account for
    (a) slider–crank mechanism
    (b) four-bar chain mechanism
    (c) quick return motion mechanism
    (d) none of these

ANSWERS

3.1 (d)	3.2 (a)	3.3 (c)	3.4 (b)	3.5 (b)	3.6 (a)
3.7 (c)	3.8 (b)	3.9 (b)	3.10 (c)	3.11 (c)	

4

CAMS

4.1 INTRODUCTION

Cam is a rotating mechanical element which provides rotating, reciprocating or oscillating motion to the other element attached with it, termed follower. The cams may be flat-faced, roller type or knife-edged in shape. Complicated output motions which are otherwise difficult to obtain are possible by means of cams. The cams have applications in automatic mechanism, interval combustion engines for operating valves, machine tools, shoe making machines, etc. A cam and follower combination constitutes a higher pair. The combination as such has three elements namely cam, follower and frame. Frame acts as the support for the assembly. When cams are required to be produced on a large scale, punch press die-casting or milling methods are used as shown in Fig. 4.1.

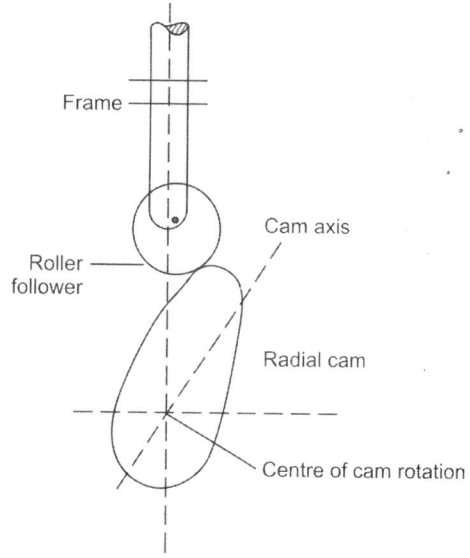

Fig. 4.1

4.2 TYPES OF CAM

Basically, the cams are of the following two types:
1. Radial cam
2. Cylindrical cam

The cams of different shapes are available such as conjugate cams having double discs which are in direct touch with the balls of the follower. This type of combination is suitable for high speed and high dynamic load. Since the grip of the follower is good, so there is less wear and tear. The operation is as shown in Fig. 4.2.

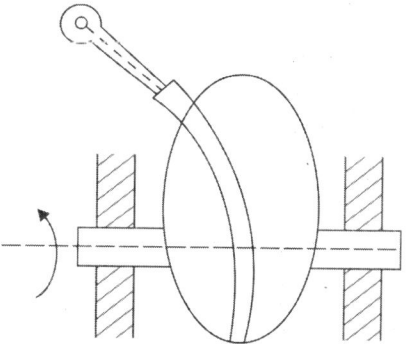

Fig. 4.2

Spherical cams are also found in use as shown in Fig. 4.3.

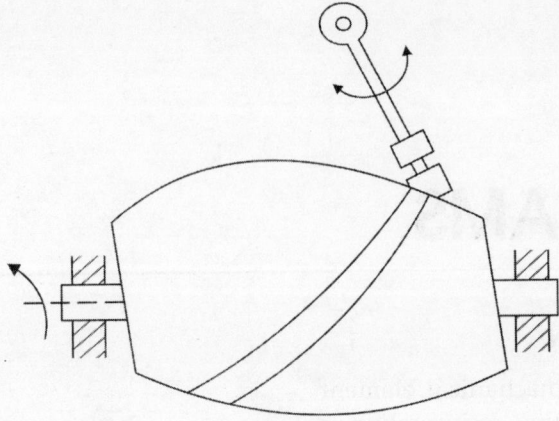

Fig. 4.3

4.2.1 Radial Cam

In this type of cam, follower moves in a plane perpendicular to the cam shaft. This type of cam is shown in Fig. 4.4. It is also known as disc or plate cam.

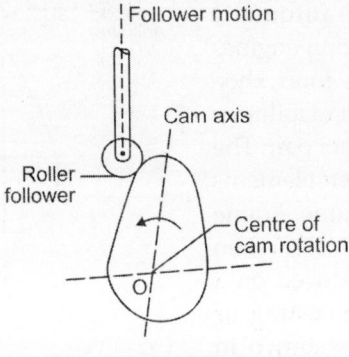

Fig. 4.4

4.2.2 Cylindrical Cam

In this type of cam, the follower moves in a plane parallel to the axis of rotation of the cam as shown in Fig. 4.5.

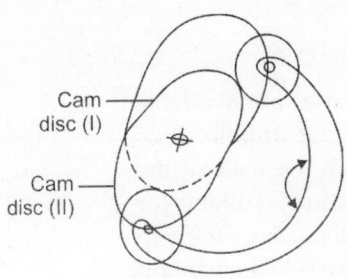

Fig. 4.5

4.3 TYPES OF FOLLOWER

A follower is guided by the movement of cam. It may be classified according to the following:

    i. Types of movement
    ii. Shape at the point of contact
    iii. Distance between cam shaft axis and follower axis.

4.3.1 Types of Movement

Oscillatory Follower

The follower shown in Fig. 4.6 is having oscillatory motion. The cam rotates with uniform speed.

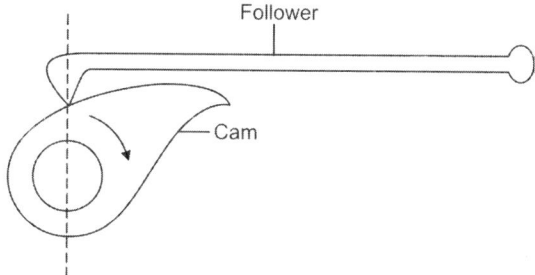

Fig. 4.6: Oscillatory follower

Translatory Follower

The follower shown in Fig. 4.7 has translatory movement while the cam rotates with uniform speed.

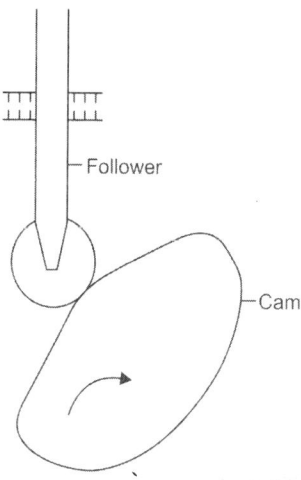

Fig. 4.7

4.3.2 Shape at the Point of Contact

Knife Edge Follower

Their use is not popular because of high wear rate. They are simple in construction and can be used with any type of cam (Fig. 4.8).

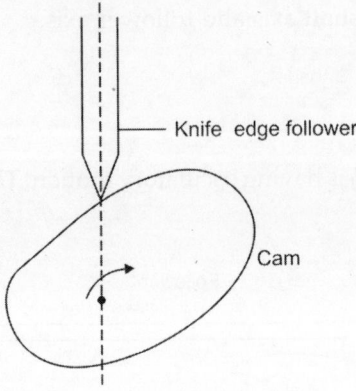

Fig. 4.8

Roller Follower

The rate of wear is considerably reduced as compared to the knife edge follower. The motion is rolling between the follower and cam surface. It has wide application in air craft engines. However, it is not used in automobiles because of space limitations and failure at pin points in the roller as shown in Fig. 4.9.

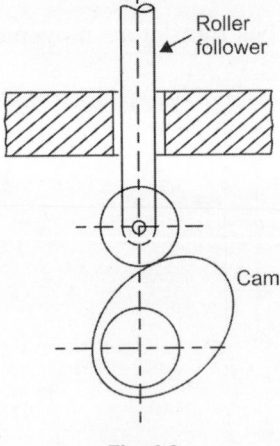

Fig. 4.9

Flat or Mushroom Follower

The contacting surface may be flat or spherical. Automobile engines use flat-faced mushroom followers with spherical curvature. They are used where the space is limited. They are best fitted with convex surface of cam. Spherical faced follower is preferred to flat faced as there is less surface stress and wear in the spherical follower as shown in Fig. 4.10.

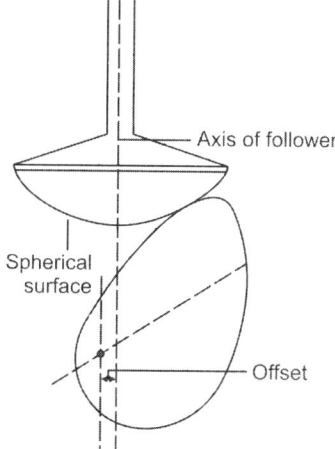

Fig. 4.10. Spherical faced follower

4.3.3 Distance between Cam Shaft Axis and Follower Axis

Sometimes there is a distance between cam shaft axis and follower axis. This distance is called offset. The follower is known as offset follower. In case of flat-faced follower wear can be considerably reduced by providing off setting. The off setting cause the follower to move about its own axis.

4.4 BASIC TERMINOLOGY OF CAM

With reference to having radial cam and roller follower, some important terms are explained below (Fig. 4.11).

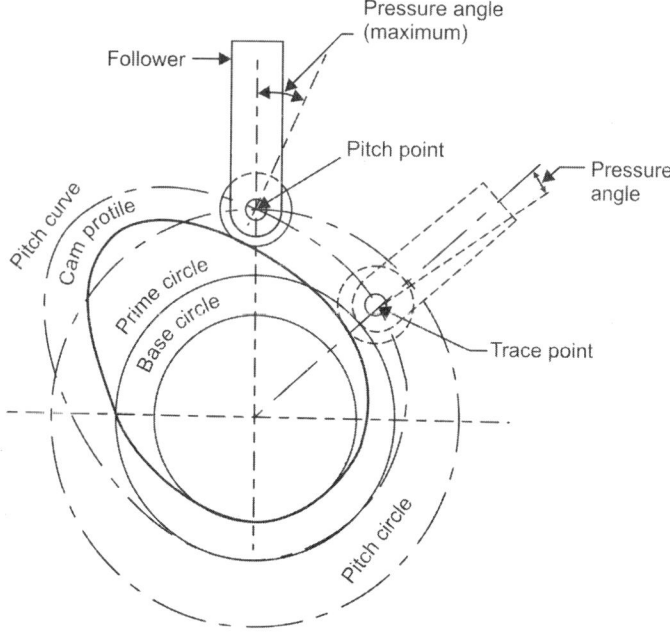

Fig. 4.11

4.4.1 Cam Profile

The surface of cam which comes into contact with the follower, is known as cam profile.

4.4.2 Base Circle

The smallest circle drawn from the centre of rotation of cam forming the part of cam profile, is known as base circle. The radius of this circle is called the least radius of the cam.

4.4.3 Trace Point

It is a reference point on the follower to make pitch curve. In case of knife edge follower, it is located at the knife edge itself, and in case of roller follower it is located at the roller centre.

4.4.4 Cam Angle

It is the angle of rotation of the cam for a definite displacement of the follower.

4.4.5 Pitch Curve

It is the path of trace point assuming that the cam is fixed and the follower moved around it.

4.4.6 Prime Circle

It is the smallest circle drawn to the pitch curve from the centre of rotation of cam.

4.4.7 Pressure Angle

It is the angle between the normal to the pitch curve and line of motion of the follower. This angle is very important in the design of cam. Its maximum value is about 30°. It is denoted by ϕ. If the value of ϕ exceeds 30°, a reciprocating type of follower will jam in the bearings. It depends upon the angle of ascent, lift of follower, follower motion offset, etc.

4.4.8 Lift of Follower

It is the maximum displacement of the follower from its lowest position to the top most position. It is also called the stroke of the follower.

4.5 MOTION OF FOLLOWER

4.5.1 Simple Harmonic Motion (SHM)

$s \rightarrow$ follower displacement
$h \rightarrow$ maximum follower displacement
$v \rightarrow$ velocity
$a \rightarrow$ acceleration of the follower
$\theta \rightarrow$ cam rotation angle
$\phi \rightarrow$ cam rotation angle for maximum follower displacement
$\beta \rightarrow$ angle on the harmonic circle.

The follower rises through a distance h while the cam turns through an angle ϕ (Fig. 4.12).

i. Draw a semicircle with cam rise as diameter.
ii. Divide the semicircle into n equal parts (even).

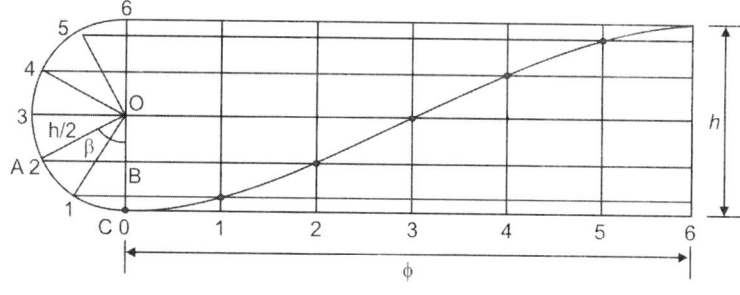

Fig. 4.12

iii. Divide the cam displacement interval into *n* equal divisions.

iv. Join the points which are drawn with a smooth curve. At any instant, displacement of the follower is given by

$$s = BC = OC - OB \qquad\qquad \left(\text{In } \Delta OAB, \frac{OB}{OA} = \cos \beta\right)$$

$$s = \frac{h}{2} - \frac{h}{2}\cos\beta$$

$$s = \frac{h}{2}[1 - \cos\beta]$$

For the rise of the follower displacement, the cam is rotated through an angle ϕ, whereas a point on the harmonic semicircle transverse an angle π (Figs 4.13 and 4.14)

$$\pi \rightarrow \phi$$

$$1 \rightarrow \frac{\phi}{\pi}$$

$$\beta \rightarrow \frac{\beta.\phi}{\pi} = \theta\beta = \frac{\pi\theta}{\phi}$$

$$s = \frac{h}{2}\left[1 - \cos\frac{\pi\theta}{\phi}\right]$$

ω = angular velocity of the cam

$$\theta = \omega t$$

$$s = \frac{h}{2}\left[1 - \cos\pi\frac{\omega t}{\phi}\right]$$

$$v = \frac{ds}{dt} = \frac{d}{dt}\left[\frac{h}{2}\left[1 - \cos\pi\frac{\omega t}{\phi}\right]\right]$$

$$v = \frac{h}{2}\frac{\pi\omega}{\phi}\sin\pi\frac{\omega t}{\phi} = \frac{h}{2}\frac{\pi\omega}{\phi}\sin\left(\frac{\pi\theta}{\phi}\right)$$

when $$\theta = \frac{\phi}{2}$$

$$v_{max} = \frac{h}{2}\frac{\pi\omega}{\phi}$$

$$a = \frac{dv}{dt} = \frac{d}{dt}\left[\frac{h}{2}\frac{\pi\omega}{\phi}\sin\frac{\pi\theta}{\phi}\right]$$

$$= \frac{h}{2}\frac{\pi\omega}{\phi}\frac{d}{dt}\left[\sin\frac{\pi\omega t}{\phi}\right]$$

$$a = \frac{h}{2}\left[\frac{\pi\omega}{\phi}\right]^2\cos\frac{\pi\theta}{\phi}$$

$$a_{max} = \frac{h}{2}\left[\frac{\pi\omega}{\phi}\right]^2$$

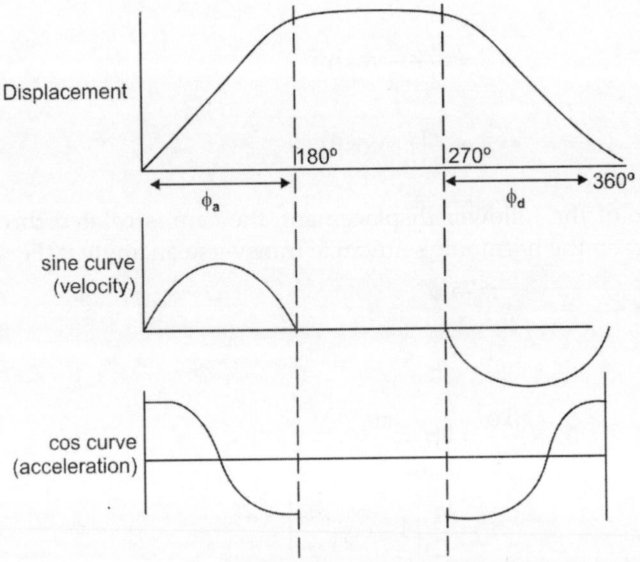

Fig. 4.13: Acceleration

4.5.2 Constant Acceleration and Deceleration

The equation for linear motion with constant acceleration

$$s = ut + \frac{1}{2}at^2$$

u is the initial velocity at the start of the motion $\therefore u \to 0$

$$s = \frac{1}{2}at^2$$

$$a = \frac{2s}{t^2} = \text{constant}$$

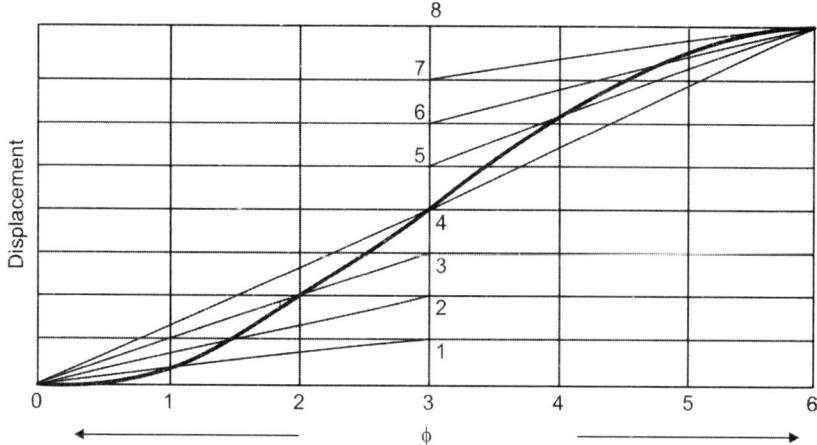

Fig. 4.14

a is constant, considering the follower at the middle, $s = \dfrac{h}{2}$

$$a = \frac{2h}{2t^2}$$

$$\theta = \omega t$$

$$t = \frac{\theta}{\omega} = \frac{\phi}{2\omega}$$

$$a = \frac{2h \times 4\omega^2}{2 \times \phi^2} = \frac{4h\omega^2}{\phi^2}$$

$$a = \frac{4h\omega^2}{\phi^2}$$

The velocity is linear during the period,

$$v = \frac{ds}{dt} = \frac{d}{dt}\left[\frac{1}{2}ft^2\right]$$

$$= \frac{1}{2} \times 2ft = ft$$

$$= \frac{4h\omega^2}{\phi^2} \times \frac{\theta}{\omega}$$

$$v = \frac{h\omega t}{\phi^2} \times \theta$$

Velocity is maximum when ϕ is maximum or follower is at midway

$$v_{max} = \frac{4h\omega}{\phi^2} \cdot \frac{\phi}{2}$$

$$\boxed{v_{max} = \frac{2h\omega}{\phi}}$$

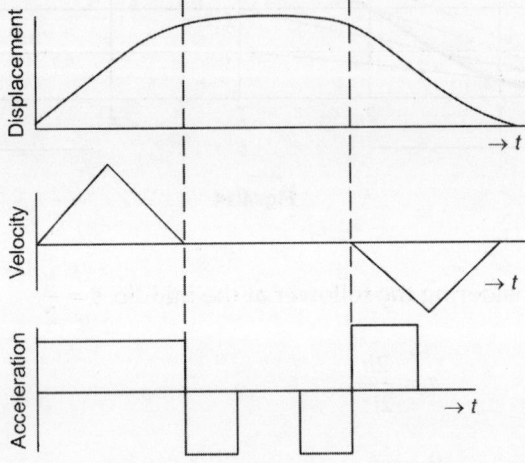

Fig. 4.15

4.5.3 Constant Velocity

Constant velocity of the follower implies that the displacement of the follower is proportional to the cam displacement.

$$h \rightarrow \phi \Rightarrow 1 = \frac{\phi}{h}$$

$$s \rightarrow s \frac{\phi}{h} = \theta$$

$$s = \frac{h\theta}{\phi} = \frac{h\omega t}{\phi}$$

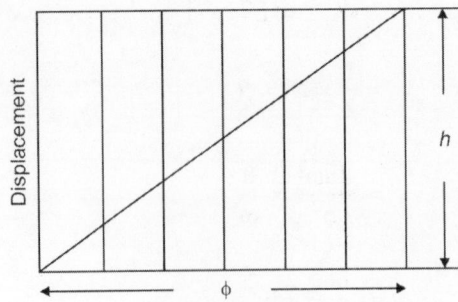

Fig. 4.16

$$v = \frac{ds}{dt} = \frac{d}{dt}\left[\frac{h\omega t}{\phi}\right] = \frac{h\omega}{\phi}$$

$$a = \frac{dv}{dt} = \frac{d}{dt}\left[\frac{h\omega}{\phi}\right] = 0$$

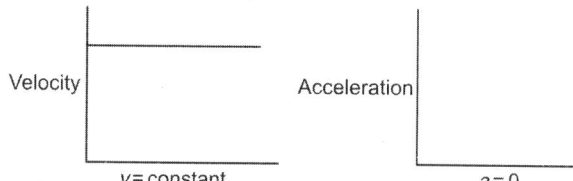

Velocity   Acceleration

$v =$ constant $a = 0$

Fig. 4.17

4.5.4 Cycloidal Motion

   i. Divide the cam displacement interval into eight equal parts.
   ii. Draw the diagonal of the diagram and extend it below.
 iii. Draw a circle with the centre anywhere on the lower portion of the diagram such that its circumference is equal to the follower displacement.

$$2\pi r = n$$

$\Rightarrow$ $$r = \frac{n}{2\pi}$$

  iv. Draw the circle and divide it into n equal arcs and do the numbering.
   v. Project the points on the circle to its vertical diagonal, then in a direction parallel to the diagonal of the diagram to the corresponding ordinates.

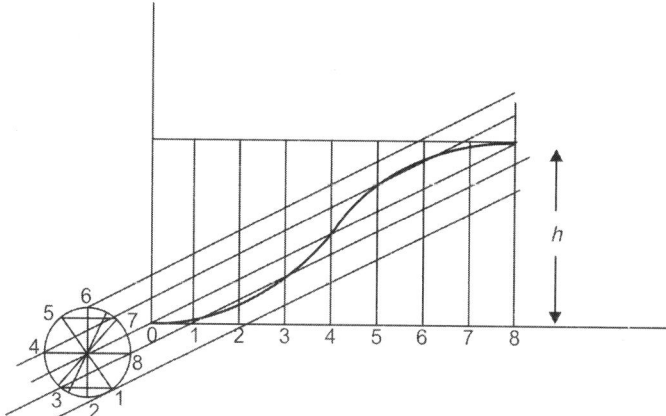

Fig. 4.18

$$s = \frac{h}{\pi}\left[\frac{\pi\theta}{\phi} - \frac{1}{2}\sin\frac{2\pi\theta}{\phi}\right]$$

$$\theta = \frac{ds}{dt} = \frac{ds}{d\theta}\times\frac{d\theta}{dt}$$

$$= \frac{d}{d\theta}\left[\frac{h}{\pi}\left(\frac{\pi\theta}{\phi} - \frac{1}{2}\sin\frac{2\pi\theta}{\phi}\right)\right]\frac{d\theta}{dt}$$

$$= \left[\frac{h}{\phi} - \frac{h}{2\pi}\times\frac{2\pi}{\phi}\times\cos\frac{2\pi\theta}{\phi}\right]\omega$$

$$\boxed{v_{max} = \frac{2h\omega}{\phi}} \quad \text{at } \theta = \frac{\phi}{2}$$

$$a = \frac{dv}{dt} = \frac{dv}{d\theta}\times\frac{d\theta}{dt}$$

$$= \left[\frac{h\omega}{\phi}\cdot\frac{2\pi}{\phi}\cdot\sin\frac{2\pi\theta}{\phi}\right]\omega$$

$$a = \frac{2h\pi\omega^2}{\phi^2}\sin\frac{2\pi\theta}{\phi}$$

$$\boxed{a_{max} = \frac{2\pi h\omega^2}{\phi^2}} \quad \text{at } \theta = \frac{\phi}{4}$$

EXAMPLES

Example 4.1: Draw the profile of a cam operating a knife edge follower having a lift of 30 mm. The cam raises the follower with SHM for 150° rotation followed by a period of dwell for 60°. The follower descends for the next 100° rotation of the cam with uniform velocity, again followed by a dwell period. The cam rotates at a uniform velocity of 120 rpm and has a least radius of 20 mm. What will be the max. velocity and acceleration of follower during the lift and the return?

Solution:

$$h = 30 \text{ mm}$$
$$\Psi_a = 150°$$
$$N = 120 \text{ rpm}$$
$$\delta_1 = 60°$$
$$r_c = 20 \text{ mm}$$
$$\psi_d = 100°$$
$$\delta_2 = (360° - 150° - 100° - 60°)$$
$$= 50°$$

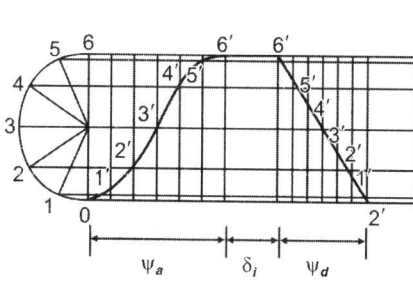
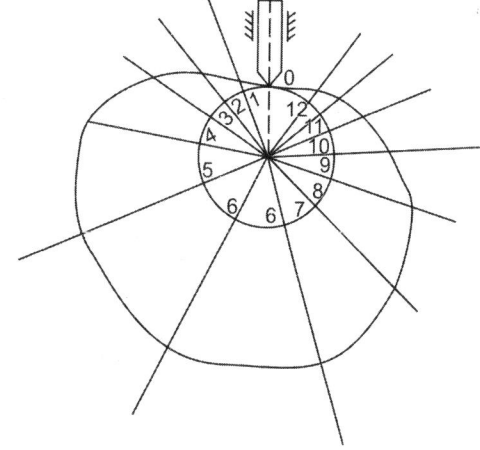

Fig. 4.19

During ascent,

$$\omega = \frac{2\pi N}{60} = \frac{2\pi \times 120}{60} = 12.57 \text{ rad/s}$$

$$v_{max} = \frac{h}{2} \times \frac{\pi\omega}{\psi_a}$$

$$= \frac{30}{2} \times \frac{\pi \times 12.57}{150 \times \dfrac{\pi}{180}} = 226.3 \text{ mm/s}$$

$$a_{max} = \frac{h}{2}\left(\frac{\pi\omega}{\psi_a}\right)^2$$

$$= \frac{30}{2} \times \left(\frac{\pi \times 12.57}{150 \times \dfrac{\pi}{180}}\right)$$

$$= 3413 \text{ mm/s}^2$$
$$= 3.413 \text{ m/s}^2$$

During desent,

$$v_{max} = \frac{h\omega}{\psi_d}$$

$$= 30 \times \frac{12.57}{100 \times \dfrac{\pi}{180}}$$

$$= 216 \text{ mm/s}$$
$$f_{max} = f = 0$$

Example 4.2: A cam with a minimum radius of 25 mm is to be designed for a knife edge follower with the following data:

- To raise the follower through 35 mm during 60° rotation.
- Dwell for next 40°.
- Descending during the next 90°.
- Dwell for the rest.

Draw the profile of a cam if the ascending and descending of the cam is with SHM and the line of stroke of the follower is offset common from the axis of the cam shaft.

Solution:

$$h = 35 \text{ mm}$$
$$N = 150 \text{ rpm}$$
$$r_c = 24 \text{ mm}$$
$$x = 10 \text{ mm}$$
$$\psi_a = 60°$$
$$\delta_1 = 40°$$
$$\psi_d = 90°$$

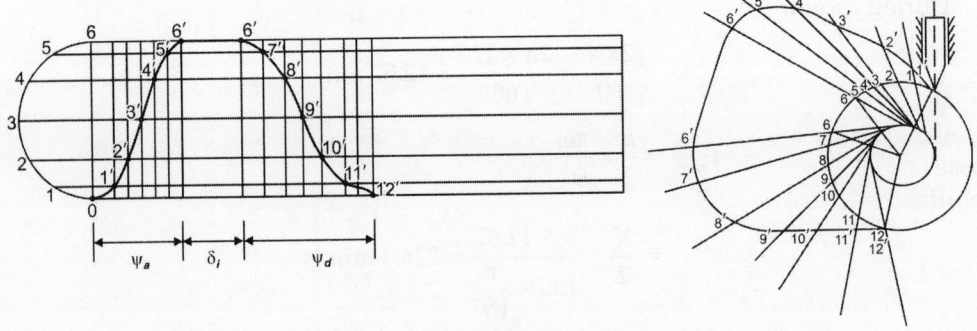

Fig. 4.20

During ascent,

$$\omega = \frac{2\pi \times 150}{60}$$
$$= 5\pi \text{ rad /s}$$

$$v_{max} = \frac{h}{2} \times \frac{\pi\omega}{\psi_a}$$

$$v_{max} = \frac{35}{2} \times \frac{\pi \times 5\pi}{60 \times \frac{\pi}{180}}$$

$$= 824.7 \text{ mm/s}$$

$$a_{max} = \frac{h}{2}\left(\frac{\pi\omega}{\psi_a}\right)^2$$

$$= \frac{35}{2}\left(\frac{\pi \times 5\pi}{60 \times \frac{\pi}{180}}\right)^2$$

$$= 38862 \text{ mm/s}^2$$
$$= 38.882 \text{ m/s}^2$$

During descent,

$$v_{max} = \frac{35}{2} \times \frac{\pi \times 5\pi}{90 \times \dfrac{\pi}{180}}$$

$$= 549.8 \text{ mm/s}$$

$$a_{max} = \frac{35}{2} \times \left(\frac{\pi \times 5\pi}{90 \times \dfrac{\pi}{180}} \right)^2$$

$$= 17272 \text{ mm/s}^2$$
$$= 17.272 \text{ m/s}^2$$

Example 4.3: A cam is to give the following motion to a knife edged follower:
- To raise the follower through 30 mm with uniform acceleration and deceleration during 120° rotation of the cam.
- Dwell for next 30°.
- To lower the follower with SHM during the next 90°.
- Dwell for rest of the cam rotation.

The cam has a minimum radius of 30 mm and rotates counter-clockwise at a uniform speed of 800 rpm. Draw the profile of the cam if the line of stroke of follower passes through the cam shaft.

Solution:

$$h = 30 \text{ mm}$$
$$\delta_1 = 30°$$
$$r_c = 30 \text{ mm}$$
$$\psi_a = 120°$$
$$\delta_1 = 30°$$
$$\psi_d = 90°$$
$$\delta_2 = 360° - 120° - 30° - 90° = 120°$$
$$N = 800 \text{ rpm}$$

$$\omega = \frac{2\pi \times 840}{60} = 88 \text{ rad/s}$$

During ascent,

$$v = \frac{4h\omega}{\psi_a^2} \times \theta$$

$$\therefore \qquad v_{max} = 2h \times \frac{\omega}{\psi_a}$$

$$= 2 \times 0.03 \times \frac{88}{\dfrac{120\pi}{180}}$$

$$= 2.52 \text{ m/s}$$

$$f_{uniform} = \frac{4h\omega^2}{\psi_a^2}$$

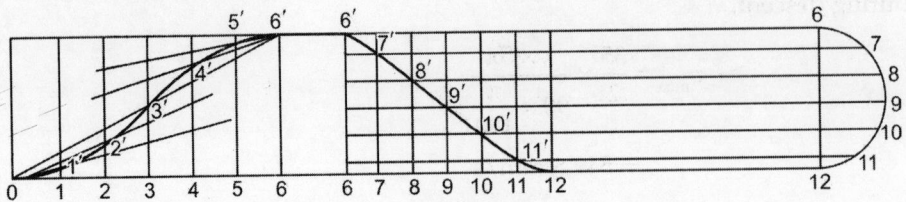

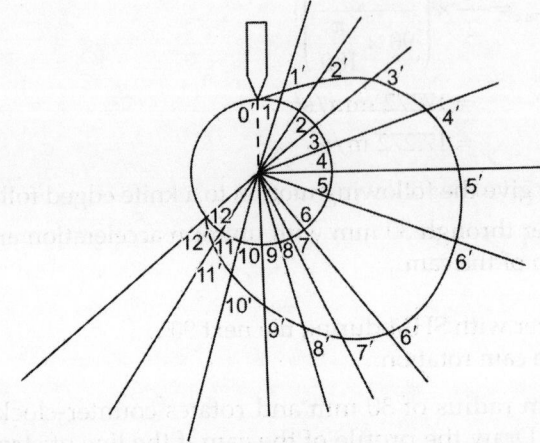

Fig. 4.21

$$= \frac{4 \times 0.03 \times 88^2}{\left(\dfrac{120\pi}{180}\right)^2}$$

$$= 211.9 \ \text{m/s}^2$$

During descent,

$$v = \frac{h}{2} \frac{\pi\omega}{\psi_d} \sin \frac{\pi\theta}{\psi_d}$$

$$\theta = \frac{\psi_d}{2}$$

$$v_{max} = \frac{h}{2} \times \frac{\pi\omega}{\psi_d}$$

$$= \frac{0.03}{2} \times \frac{\pi \times 88}{\dfrac{90\pi}{180}}$$

$$= 2.64 \ \text{mm/s}$$

The acceleration variation is given by

$$f = \frac{h}{2} \left(\frac{\pi\omega}{\psi}\right)^2 \cos \frac{\pi\theta}{\psi}$$

At $\qquad \theta = 0,$

$$f_{max} = \frac{h}{2} \left(\frac{\pi\omega}{\psi_d} \right)^2$$

$$= \frac{0.03}{2} \times \left(\frac{\pi \times 88}{\frac{90\pi}{180}} \right)$$

$$= 464.6 \text{ m/s}^2$$

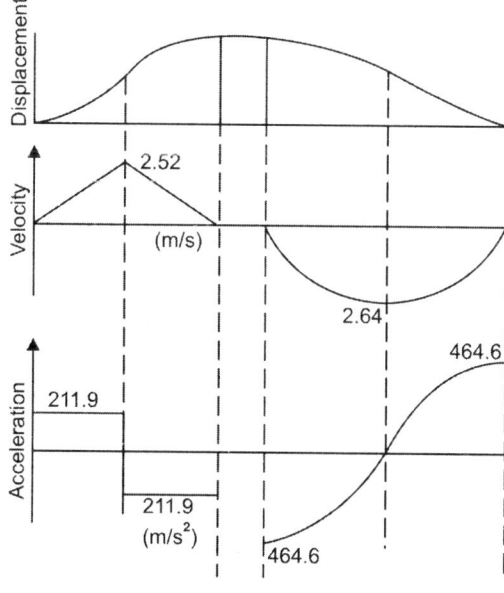

Fig. 4.22

Example 4.4: Draw the profile of a cam operating a roller reciprocating follower.

    Min. radius of cam = 25 mm

    Lift = 30 mm

    Roller diameter = 15 mm

The cam lifts the follower for 120° with SHM and dwell period of 30°, then the follower lowers down during 150° of the cam rotation with uniform acceleration and deceleration followed by a dwell period.

Solution:

$$h = 30 \text{ mm}$$
$$\psi_a = 120°$$
$$N = 150 \text{ mm}$$
$$\delta_1 = 30°$$
$$r_c = 2.5 \text{ mm}$$
$$r_r = 7.5 \text{ mm}$$
$$\psi_d = 150°$$
$$\delta_2 = 60°$$

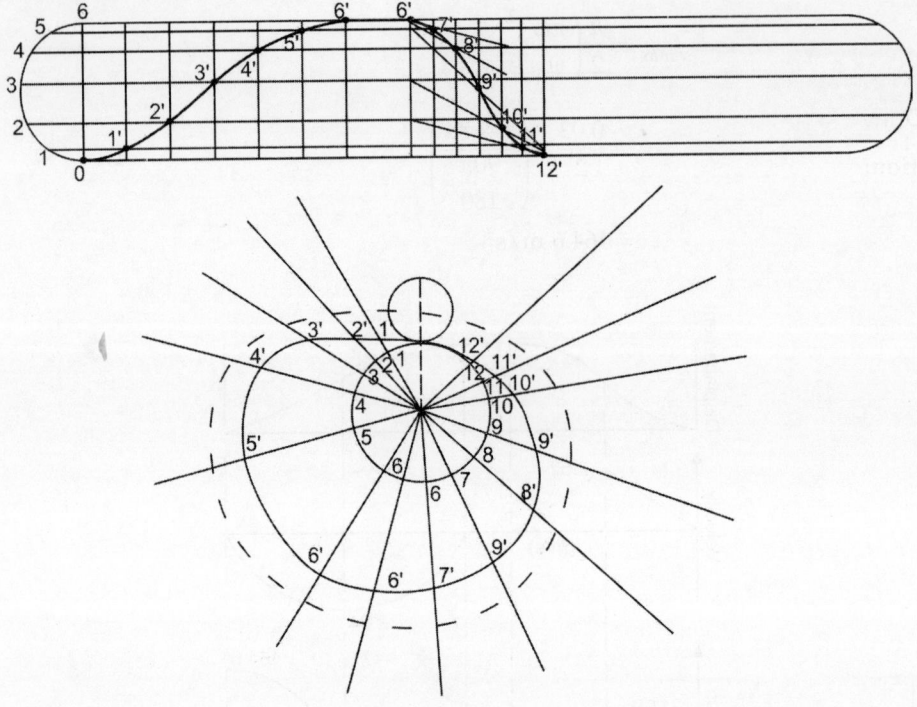

Fig. 4.23

$$v_{max} = \frac{2h\omega}{\psi_d}$$

$$= 2 \times 30 \times \frac{\dfrac{2\pi \times 150}{60}}{150 \times \dfrac{\pi}{180}}$$

$$= 360 \text{ mm/s}$$

$$f_{max} = f_{uniform} = \frac{4h\omega^2}{\psi_d^2}$$

$$= \frac{4 \times 30 \times \left(2\pi \times \dfrac{150}{60}\right)^2}{150 \times \dfrac{\pi}{180}}$$

$$= 4320 \text{ mm/s}^2$$

Example 4.5: The following data relate to a cam profile in which the follower moves with uniform acceleration and deceleration during ascent and descent.

$$\text{Min. rad} = 25 \text{ mm}$$
$$\text{Roller diameter} = 7.5 \text{ mm}$$
$$\text{Lift} = 28 \text{ mm}$$
$$\text{Offset of follower axis} = 12 \text{ mm}$$

angle of ascent = 60°
angle of descent = 90°
Speed of the cam = 200 rpm

Draw the profile of the cam.

Solution:

$$h = 28 \text{ mm}$$
$$r_c = 25 \text{ mm}$$
$$r_r = 7.5 \text{ mm}$$
$$x = 12 \text{ mm}$$
$$N = 120 \text{ mm}$$
$$\delta_2 = 165°$$
$$\psi_a = 60°$$
$$\delta_1 = 45°$$
$$\psi_d = 90°$$

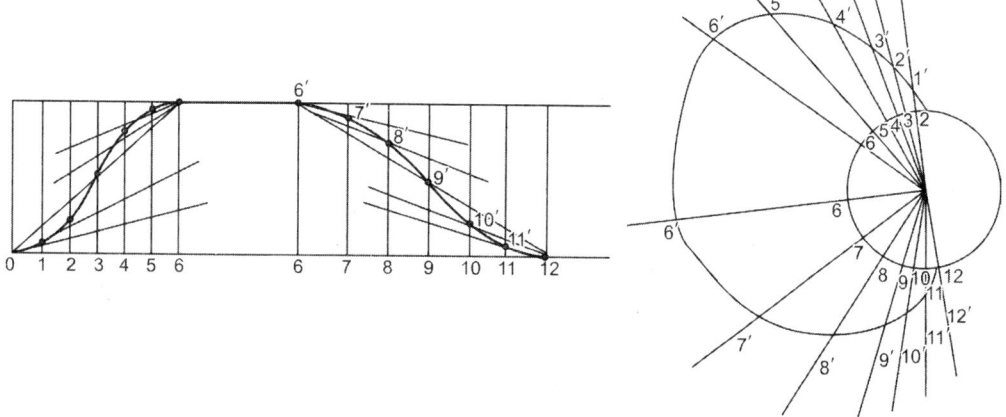

Fig. 4.24

During outstroke,

$$\omega = \frac{2\pi \times 200}{60} = 20.94 \text{ rad/s}$$

$$v_{max} = \frac{2h\omega}{\psi_a}$$

$$= \frac{2 \times 28 \times 20.94}{60 \times \dfrac{\pi}{180}}$$

$$= 1120 \text{ mm/s}$$
$$= 1.12 \text{ m/s}$$

$$f_{uniform} = \frac{4h\omega^2}{\psi_a^2}$$

$$= \frac{4 \times 28 \times (20.94)^2}{\left(60 \times \dfrac{\pi}{180}\right)^2}$$

$$= 44800$$

$$= 44.8 \text{ m/s}^2$$

During return stroke,

$$v_{max} = \frac{2h\omega}{\psi_d}$$

$$= \frac{2 \times 28 \times 20.94}{90 \times \dfrac{\pi}{180}}$$

$$= 747 \text{ mm/s}$$

$$= 0.747 \text{ m/s}$$

$$f_{uniform} = \frac{4 \times 28 \times (20.94)^2}{\left(90 \times \dfrac{\pi}{180}\right)^2}$$

$$= 19900 \text{ mm/s}^2$$

$$= 19.9 \text{ m/s}^2$$

Example 4.6: A flat-faced mashroom follower is operated by a uniformly rotating cam. The follower is raised through a distance of 25 mm in 120° rotation of the cam, remains at rest for the next 30° and is lowered during further 120° rotation of the cam. The raising of the follower takes place with cycloidal motion and the lowering with uniform acceleration and deceleration. However, the uniform acceleration is 2/3 of the uniform deceleration. The least radius of the cam is 25 mm which rotates at 300 rpm.

Draw the cam profile and determine the values of the max. velocity and max. acceleration during rising.

Solution:

$$h = 25 \text{ mm}$$
$$r_c = 25 \text{ mm}$$
$$N = 300 \text{ mm}$$
$$\psi_a = 120°$$
$$\delta_1 = 30°$$
$$\psi_d = 120°$$
$$\delta_2 = 90°$$

Time of acceleration

$$v = u + at = at \ (\because u = 0)$$

$$t = \frac{v}{a}$$

Time of deceleration

$$o = v - (3/2) \, at'$$

$$t' = \frac{2}{3} \times \frac{v}{a}$$

Displacement,

$$s = ut + \frac{1}{2} ft^2$$

$$= \frac{1}{2} ft^2$$

During deceleration,

$$= ft\left(\frac{2}{3}t\right) - \frac{1}{2}\left(\frac{2}{3}f\right)\left(\frac{2}{3}t\right)^2 = \frac{2}{3}ft^2 - \frac{1}{3}ft^2$$

$$= \frac{2}{3}\left(\frac{1}{2}ft^2\right)$$

During ascent,

$$v_{max} = \frac{2h\omega}{\psi_a}$$

$$\omega = \frac{2\pi \times 300}{60}$$

$$= 31.4 \text{ rad/s}$$

$$v_{max} = \frac{2 \times 25 \times 31.4}{\dfrac{120 \times \pi}{180}}$$

$$= 750 \text{ mm/s}$$
$$= 0.75 \text{ m/s}$$

$$f_{max} = \frac{2h\pi\omega^2}{\psi_a^2}$$

$$= \frac{2 \times 25 \times \pi \times (31.4)^2}{\left(\dfrac{120 \times \pi}{180}\right)^2}$$

$$= 35310 \text{ mm/s}^2$$
$$= 35.3 \text{ m/s}^2$$

During descent,

$$v = ft$$

$$v = \frac{2s}{t^2}$$

$$s = \frac{vt^2}{2}$$

$$s = \frac{3}{5} \times 25 = 15 \text{ mm}$$

Time for 300 rev = 60 s

$$1 \text{ rev} = \frac{60}{300}$$

$$= 0.25$$

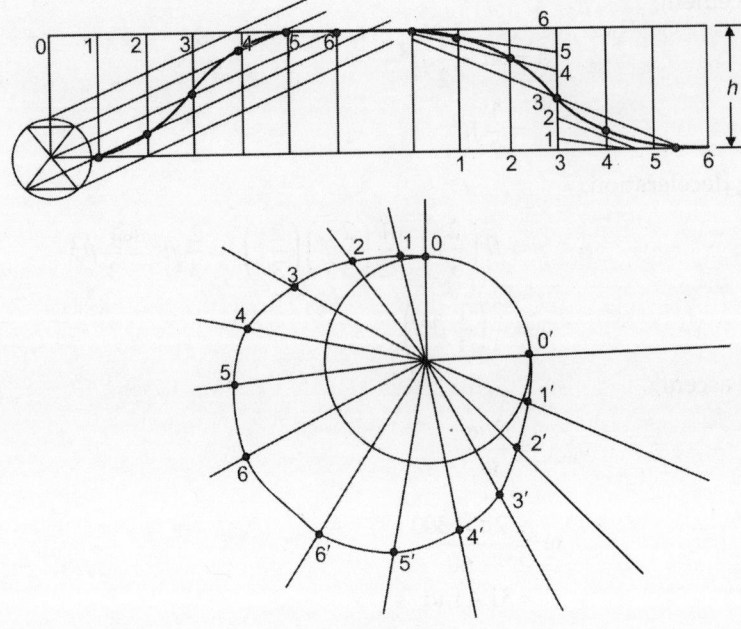

Fig. 4.25

$$\text{Time for } \left(\frac{3}{5} \times 120\right) = \frac{0.2}{360} \times 72$$

$$= 0.045$$

$$v_{\max} = \frac{2 \times 15}{0.04}$$

$$= 18.75 \text{ m/s}^2$$

$$\text{Uniform deceleration} = 18.75 \times \frac{3}{2}$$

$$= 28.13 \text{ m/s}^2$$

EXERCISE

4.1 With the help of neat sketches, explain the types of cam and follower. Give the specific applications of each type of cam.

4.2 Sketch displacement, velocity and acceleration vs time curves for SHM of a follower.

4.3 Give a neat sketch of a cam with offset roller follower.

4.4 Give the classification of cam followers.

4.5 Classify cams and followers.

4.6 What is a cam? What type of motion can be transmitted with a cam and follower combination? What are its elements?

4.7 Define pitch circle, trace point, pitch point, pitch curve and pressure angle.

4.8 Deduce an expression for the velocity and acceleration of a follower when it moves with SHM.

4.9 Draw the profile of a cam operating a knife edge follower when the axis of the follower passes through the axis of the cam shaft. The following data is given: Lift = 40 mm, angle of ascent = 60°, dwell = 45°, angle of descent = 90° and dwell for the remaining period of cam rotation.

The motion of the cam is simple harmonic during both ascent and descent. The least radius of cam is 50 mm. If the cam rotates at 300 rpm, determine the maximum velocity and acceleration of the follower during ascent and descent.

4.10 A cam with 30 mm as minimum diameter is rotating clockwise at a uniform speed of 1200 rpm and operates a roller follower of 10 mm diameter as given below:
   (i) Outward stroke of 30 mm during 120° of cam rotation with equal uniform acceleration and retardation.
   (ii) Follower is to dwell for 50° of cam rotation.
   (iii) Inward stroke during 90° of cam rotation with equal uniform acceleration and retardation.
   (iv) Follower is to dwell for the remaining period of cam rotation.
   Draw the cam profile if the axis of follower passes through the axis of the cam. Determine the maximum velocity and acceleration during outward and inward strokes.

   (*Ans.* 3.6 m/s, 4.8 m/s, 432 m/s^2, 768 m/s^2)

4.11 Draw the cam profile from the following data if the radial follower moves with simple harmonic motion during ascent and uniform acceleration and deceleration during descent: Lift = 40 mm, least radius of cam = 60 mm, angle of ascent = 54°, dwell = 40°, angle of descent = 72°, roller diameter = 20 mm.

4.12 The following data refers to a circular arc cam operating a flat-faced reciprocating follower. Minimum radius of cam = 30 mm, total angle of cam = 120°, radius of circular arc = 100 mm, nose radius = 10 mm, angular velocity of cam = 10 rad/s.

   Determine the velocity and acceleration of the follower when the cam has turned through 20°.

   (*Ans.* 239.4 mm/s, 6578 mm/s^2)

4.13 Draw the profile of a cam to give reciprocating motion to a flat-faced follower for the angle of following data: Lift = 25 mm, ascent = 120°, dwell = 30°, descent = 120°, dwell = 90°, minimum radius of cam = 25 mm.

   The ascent and descent is to take place with SHM. The line of movement of follower passes through the cam centre.

   The following data refers to a symmetrical circular arc cam operating a flat-faced follower: least radius of cam = 30 mm, lift = 12.5 mm, angle of lift = 55°, nose radius = 3 mm, speed of cam = 600 rpm. Calculate: (a) distance between cam and nose centres, (b) radius of circular flank, and (c) angle of contact on the circular flank.

   (*Ans.* (a) 39.5 mm, (b) 125.6 mm, (c) 15.3°)

4.14 The following data refers to a tangent cam operating a radial roller follower: minimum radius of cam = 45 mm, lift = 15 mm, nose radius = 18 mm, radius of roller = 20 mm, semi angle of cam action = 70°, angular velocity of cam = 10 rad/s. Draw the displacement, velocity and acceleration diagrams for one rotation of cam.

4.15 A symmetrical tangent cam with least radius of 30 mm operates a roller follower of 10 mm radius. The angle of ascent is 60° and lift is 20 mm. The speed of cam is 450 rpm. Calculate: (a) distance between cam and nose centres, (b) nose radius, (c) angle of contact of cam with straight flank, and (d) acceleration of follower in the following cases:
   (i) at the beginning of lift
   (ii) where the roller just touches the nose
   (iii) at the apex of circular nose.

   (*Ans.* (a) 40 mm, (b) 10 mm; (i) 88.33 m/s^2, (ii) 293.7 m/s^2, (iii) −266.5 m/s^2)

OBJECTIVE TYPE QUESTIONS

4.1 The size of cam depends upon
   (a) base circle
   (b) pitch circle
   (c) prime circle
   (d) pitch curve
4.2 The angle between the direction of the follower motion and a normal to the pitch curve is called
   (a) pitch angle
   (b) prime angle
   (c) base angle
   (d) pressure angle
4.3 A circle drawn with centre as the cam centre and radius equal to the distance between the cam centre and the point on the pitch curve at which the pressure angle is maximum, is called
   (a) base circle
   (b) pitch circle
   (c) prime circle
   (d) none of these
4.4 The cam follower generally used in automobile engines is
   (a) knife edge follower
   (b) flat-faced follower
   (c) sphericdal faced follower
   (d) roller follower
4.5 The cam follower extensively used in aircraft engines is
   (a) knife edge follower
   (b) flat-faced follower
   (c) spherical-faced follower
   (d) roller follower
4.6 In a radial cam, the follower moves
   (a) in a direction perpendicular to the cam axis
   (b) in a direction parallel to the cam axis
   (c) in any direction irrespective of the cam axis
   (d) along the cam axis
4.7 A radial follower is one
   (a) that reciprocates in the guides
   (b) that oscillates
   (c) in which the follower translates along an axis passing through the cam centre of rotation
   (d) none of the above
4.8 Offset is provided to a cam–follower mechanism to
   (a) minimize the side thrust
   (b) accelerate
   (c) avoid jerk
   (d) none of these
4.9 For the low and moderate speed engines, the cam–follower should move with
   (a) uniform velocity
   (b) simple harmonic motion
   (c) uniform acceleration and retardation
   (d) cycloidal motion

4.10 For high speed engines, the cam–follower should move with
  (a) uniform velocity
  (b) simple harmonic motion
  (c) uniform acceleration and retardation
  (d) cycloidal motion

ANSWERS

4.1 (a)	4.2 (d)	4.3 (b)	4.4 (c)	4.5 (d)	4.6 (a)
4.7 (a)	4.8 (a)	4.9 (b)	4.10 (d)		

5

Friction

5.1 INTRODUCITON

When a body slides or rolls over another, an opposing force is exerted at the contact surface acting tangentially. This force is known as *force of friction* or *friction*, where projections on the surface of two bodies get interlocked and oppose relative motion between the two and external force is required to break these interlockings (Fig. 5.1).

Friction (f) is both useful as well as harmful. It is useful where power is transmitted by belt drive and friction clutches, etc. It is harmful in case of bearings where power is lost due to friction.

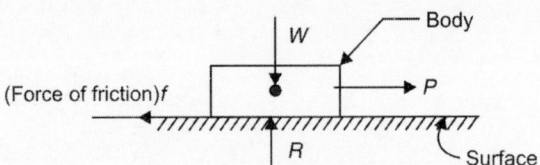

Fig. 5.1: Force of friction

5.2 TYPES OF FRICTION

Depending upon the condition of the surface, friction is of three types.

5.2.1 Dry Friction

The friction between two dry (completely unlubricated) surfaces in contact is known as dry friction. Dry friction is further of two types:

1. **Solid friction:** When two surfaces have a sliding motion relative to each other, it is called solid friction.
2. **Rolling friction:** When one surface rolls over another, then friction between two surfaces is known as rolling friction, e.g. ball and roller bearing.

5.2.2 Greasy Friction

When the two surfaces in contact have a very thin layer of lubricant between them, friction is known as greasy friction.

5.2.3 Film Friction

When the two surfaces in contact have a thick layer of lubricant between them, the friction between the surfaces is known as film or viscous friction. In this, friction occurs due to shearing of different layers of lubricant.

5.3   LAWS OF DRY FRICTION

i. Force of friction:
- depends upon the material of the two surfaces.
- independent of the area of contact.
- does not depend upon the velocity of sliding of one surface over the other.
- acts tangentially at the contact surface.
- acts in a direction opposite to the tendency of motion of the body.
- equal to the applied force so long as the body is at rest.

ii. When the motion of the body is impending, force of friction is maximum. This maximum frictional force is known as limiting friction.

iii. The limiting frictional force bears a constant ratio to the normal reaction, i.e.

$$\frac{f}{R_n} = \text{constant} = \mu$$

5.4   COEFFICIENT OF FRICTION

The ratio of the limiting force of friction (f) to the normal reaction (R_n) between two bodies is known as coefficient of friction. It is denoted by μ.

$$\mu = \frac{\text{limiting force of friction}}{\text{normal reaction}} = \frac{f}{R_n} \qquad ...(5.1)$$

5.5   ANGLE OF FRICTION

It is the angle made by the resultant of the normal reaction and the limiting force of friction with the normal reaction. It is denoted by ϕ. From Fig. 5.2.

$$\tan \phi = \frac{f}{R_n} = \frac{\mu R_n}{R_n} = \mu$$

$$\tan \phi = \mu \qquad ...(5.2)$$

Hence, tangent of angle of limiting friction is equal to the coefficient of friction.

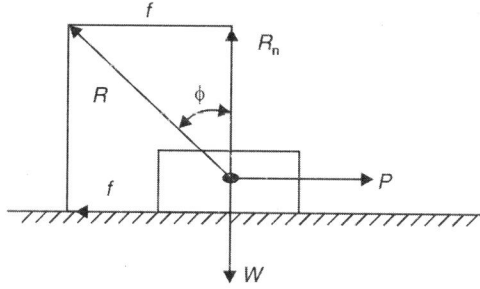

Fig. 5.2: Angle of friction

5.6   ANGLE OF REPOSE

The angle of repose is defined as the maximum angle of inclination of a plane at which a body remains in equilibrium over the inclined plane due to friction only. It is denoted by α.

The angle of inclination of plane is gradually increased till the body just starts sliding down. This angle of inclination is angle of repose (Fig. 5.3).

Resolving the forces along the plane,

$$f = W \sin \alpha \qquad \qquad ...(5.3)$$

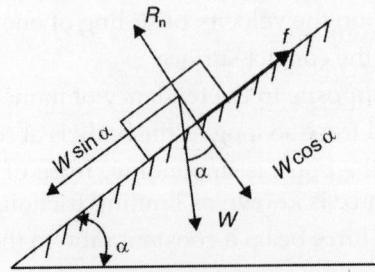

Fig. 5.3: Angle of repose

Resolving the forces normal to the plane,

$$R_n = W \cos \alpha \qquad \qquad ...(5.4)$$

Dividing Eq. (5.3) by Eq. (5.4), we have

$$\frac{f}{R_n} = \frac{W \sin \alpha}{W \cos \alpha} = \tan \alpha$$

But

$$\tan \phi = \frac{f}{R_n}$$

∴

$$\tan \alpha = \tan \phi = \frac{f}{R_n}$$

or

$$\alpha = \phi$$

i.e. angle of repose = angle of limiting friction

5.7 CONE OF FRICTION

Cone of friction is the right circular cone with vertex at the point of contact of the two bodies, axis in the direction of normal reaction (R_n) and semi-cone angle equal to the angle of friction (ϕ) (Fig. 5.4).

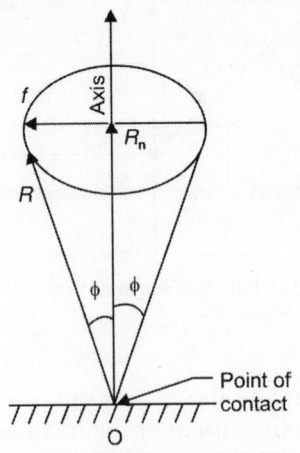

Fig. 5.4: Cone of friction

Example 5.1: A body of weight 70 N is placed on a rough horizontal plane. To just move the body on the horizontal plane, a push of 20 N inclined at 20° to the horizontal plane is required. Find the coefficient of friction.

Solution:

Weight of body $W = 70$ N
Force applied $P = 20$ N
Inclination of P $\theta = 20°$.
Let μ = coefficient of friction
              R_n = normal reaction
              f = force of friction = μR_n

When a push of 20 N at an angle 20° to the horizontal is applied to the body, the body just moves towards left.

Resolving forces along the plane

$$\mu R_n = 20 \cos 20° \qquad \text{(i)}$$

Resolving forces normal to the plane

$$R_n = 70 + 20 \sin 20°$$
$$= 70 + 20 \times 0.342$$
$$= 70 + 6.84 = 76.84$$

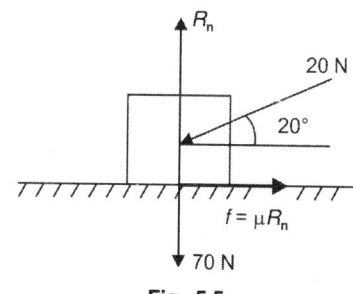

Fig. 5.5

Substituting the value of R_n in Eq. (i)

$$\mu \times 76.84 = 20 \cos 20°$$

$$\mu = \frac{20 \cos 20°}{76.84} = \frac{20 \times 0.9397}{76.84} = 0.244$$

Example 5.2 A body of weight 400 N is pulled up an inclined plane, by a force of 250 N. The inclination of the plane is 30° to the horizontal and the force is applied parallel to the plane. Determine the coefficient of friction.

Solution:

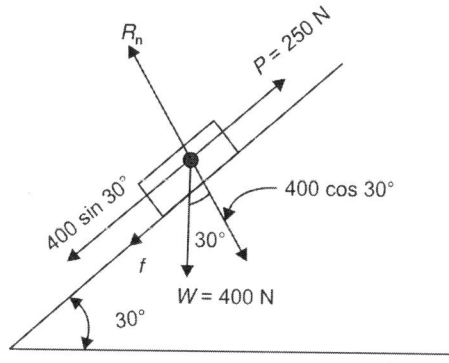

Fig. 5.6

Resolving the forces along the plane,

$$400 \sin 30° + f = 250$$

or

$$400 \times \frac{1}{2} + \mu R_n = 250$$

or

$$\mu R_n = 250 - 200 = 50 \qquad \qquad \text{... (i)}$$

Resolving the forces normal to the plane,

$$R_n = 400 \cos 30° = 400 \times 0.866$$

or

$$R_n = 346.4 \text{ N} \qquad \qquad \dots \text{(ii)}$$

Substituting the value of R_n in Eq. (i)

$$\mu \times 346.4 = 50$$

or

$$\mu = \frac{50}{346.4}$$

$$\mu = 0.144$$

5.8 PIVOT AND COLLAR FRICTION

The rotating shafts are subjected to axial thrust producing lateral motion of the shaft along its axis that is not desirable. To prevent lateral motion and to keep the shaft in correct position, thrust is taken either by a pivot or a collar.

5.8.1 Collar Bearing

A collar is provided at any position along the shaft to take the axial load on a mating surface. The surface of the collar may be flated (Fig. 5.7) or conical shape (Fig. 5.8). Collar bearings are also known as thrust bearings.

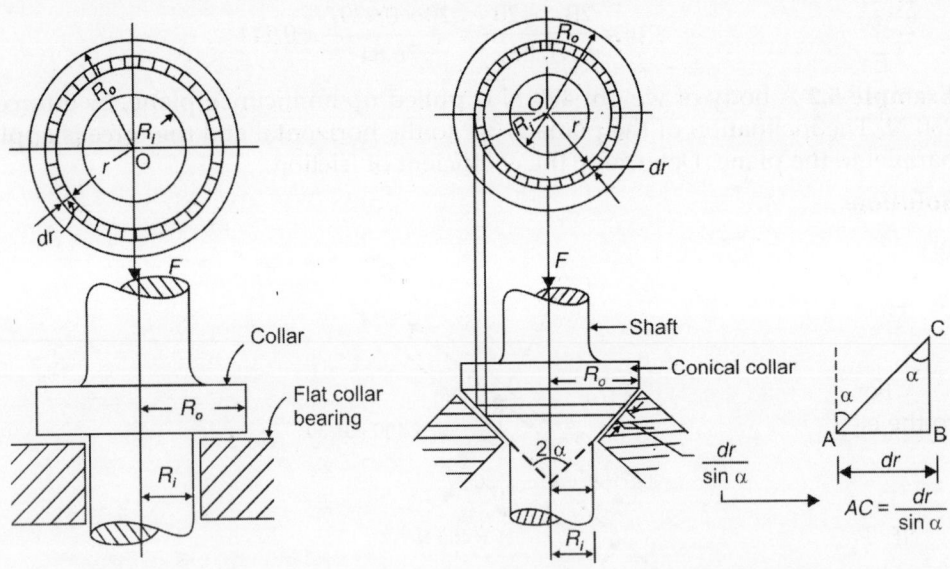

Fig. 5.7: Flat collar   Fig. 5.8: Conical collar

5.8.2 Pivot Bearing

When the axial load is taken by the end of the shaft which is inserted in a recess to bear the thrust, it is known as *pivot bearing*. Pivot bearing is also known as foot-step bearing. The pivot can have a flat surface (Fig. 5.9a) or a conical surface (Fig. 5.9b).

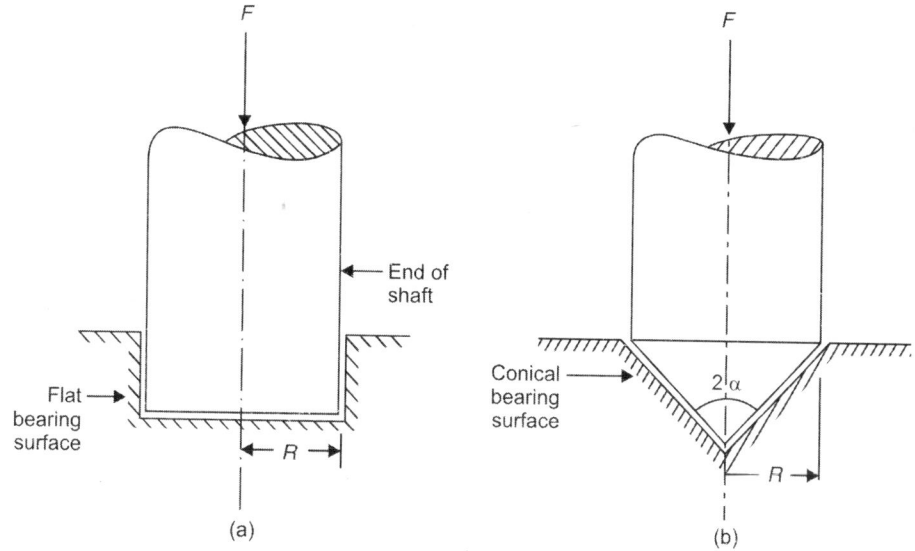

Fig. 5.9: Pivot bearing

5.8.3 Uniform Pressure and Uniform Wear

Frictional torque of bearing is determined based on the following two assumptions:

1. The pressure is uniformly distributed over the bearing surface.
2. Wear is uniform over the bearing surface.

Each assumption gives a different value of torque.

In uniform pressure, intensity of pressure is given as

$$\text{Pressure} = \frac{\text{axial force}}{\text{cross-sectional area}}$$

or
$$p = \frac{F}{\pi(R_o^2 - R_i^2)} \qquad \qquad \text{... (5.5)}$$

Here,
R_o = outer radius of collar

R_i = inner radius of collar

For uniform wear, the intensity of pressure should be inversely proportional to the elementary area, i.e.

$$P \propto \frac{1}{a} \text{ or } pa = \text{constant}$$

If P_1 = intensity of pressure at radius r_1

P_2 = intensity of pressure at radius r_2

b = width of the surface at radii r_1 and r_2

Then
$$P_1 a_1 = P_2 a_2$$
$$P_1 \times (2\pi r_1 \times b) = P_2 \times (2\pi r_2 \times b)$$

or
$$P_1 \times r_1 = P_2 \times r_2$$

or
$$Pr = C \text{ (constant) or } P = \frac{C}{r} \qquad \qquad \text{... (5.6)}$$

Therefore, for uniform wear, product of the normal pressure and the corresponding radius must be constant.

Axial force
$$F = \int_{R_i}^{R_o} \text{axial force on the elemental area}$$

$$= \int_{R_i}^{R_o} \text{pressure on the element} \times \text{elemental area}$$

$$= \int_{R_i}^{R_o} P \times 2\pi r dr \qquad (dr = \text{width of surface})$$

$$= \int_{R_i}^{R_o} \frac{C}{r} \times 2\pi r dr \qquad \left(P = \frac{C}{r}\right)$$

$$= \int_{R_i}^{R_o} 2\pi C dr$$

or
$$F = 2\pi C [r]_{R_i}^{R_o} = 2\pi \times Pr \times [R_o - R_i]$$

$\therefore$ Intensity of pressure $P = \dfrac{F}{2\pi r (R_o - R_i)}$ ... (5.7)

5.8.4 Collars

i. Flat Collar

A collar with flat bearing surface is known as a flat collar.

Let F = axial thrust

μ = coefficient of friction between the two surfaces

N = speed of the shaft

P = uniform intensity of normal pressure over an area

T = total frictional torque

R_o = outer radius of the collor

R_i = inner radius of the collor

Consider a circular element of the collar of width dr at radius r (Fig. 5.7)

Load on the element = pressure × area of element
$$= P \times 2\pi r dr$$

Frictional force on the element = μ × load

or
$$df = \mu \times P \times 2\pi r dr \qquad ...(5.8)$$

Frictional torque about the axis of the shaft,
$$dT = df \times r$$
$$dT = \mu \times P \times 2\pi r dr \times r$$
$$= 2\mu P \pi r^2 dr \qquad ... (5.9)$$

Total frictional torque $T = \int_{R_i}^{R_o} 2\mu P \pi r^2 dr$ ... (5.10)

a. Uniform Pressure Theory

Uniform pressure over whole area, $P = \dfrac{F}{\pi(R_o^2 - R_i^2)}$

Total frictional torque, $T = \int_{R_i}^{R_o} 2\mu \times \dfrac{F}{\pi r(R_o^2 - R_i^2)} \times \pi r^2 dr$ ...(5.11)

or

$$T = \frac{2\mu F}{(R_o^2 - R_i^2)} \int_{R_i}^{R_o} r^2 dr = \frac{2\mu F}{(R_o^2 - R_i^2)} \left[\frac{r^3}{3}\right]_{R_i}^{R_o}$$

$$= \frac{2\mu F}{(R_o^2 - R_i^2)} \times \frac{R_o^3 - R_i^3}{3}$$

$$= \frac{2}{3}\mu F \frac{(R_o^3 - R_i^3)}{(R_o^2 - R_i^2)}$$

Power lost in friction, $P = \dfrac{2\pi NT}{60}$

b. Uniform Wear Theory

From Eq. (5.7), pressure P at a radius r of the collar

$$= \frac{F}{2\pi r(R_o - R_i)}$$

Therefore, total frictional torque,

$$T = \int_{R_i}^{R_o} 2\mu \times \frac{F}{2\pi r(R_o - R_i)} \pi r^2 dr \qquad ...(5.12)$$

$$= \frac{\mu F}{(R_o - R_i)} \int_{R_i}^{R_o} r dr = \frac{\mu F}{(R_o - R_i)} \left[\frac{r^2}{2}\right]_{R_i}^{R_o}$$

$$= \frac{\mu F(R_o^2 - R_i^2)}{2(R_o - R_i)} = \frac{\mu F}{2}(R_o + R_i)$$

$$T = \mu F \times \left(\frac{R_o + R_i}{2}\right)$$

or $T = \mu F \times$ mean radius of the collar

ii. Conical Collar

Consider a circular element of width dr at radius r (Fig. 5.8).

Normal force on the element $= \dfrac{\text{axial force}}{\sin \alpha}$

Normal pressure on the element $= \dfrac{\text{normal force}}{\text{surface area}}$

or

$$= \frac{\text{axial force}}{\sin \alpha} \times \frac{1}{2\pi r \dfrac{dr}{\sin \alpha}}$$

$$= \frac{\text{axial force}}{2\pi r \, dr} = \text{axial pressure } (p)$$

Frictional force on the element,

$$df = \mu \times (P \times \text{area of the element})$$

$$= \mu \times P \times 2\pi r \frac{dr}{\sin \alpha} = 2\mu P \frac{\pi r \, dr}{\sin \alpha}$$

Frictional torque about the axis of the shaft,

$$dT = df \times r = \frac{2\mu P \pi r^2}{\sin \alpha} dr$$

$\therefore$ Total frictional torque $T = \displaystyle\int_{R_i}^{R_o} \frac{2\mu P \pi r^2}{\sin \alpha} dr$ ... (5.13)

a. Uniform Pressure Theory

Using Eq. (5.13), total frictional torque,

$$T = \int_{R_i}^{R_o} \frac{2\mu \pi r^2}{\sin \alpha} \times \frac{F}{\pi(R_o^2 - R_i^2)} dr \quad\quad\quad ...(5.14)$$

$$= \frac{2\mu F}{\sin \alpha (R_o^2 - R_i^2)} \int_{R_i}^{R_o} r^2 dr$$

$$= \frac{2\mu F}{\sin \alpha (R_o^2 - R_i^2)} \times \left[\frac{r^3}{3} \right]_{R_i}^{R_o}$$

$$= \frac{2\mu F}{3\sin \alpha} \times \frac{(R_o^3 - R_i^3)}{R_o^2 - R_i^2}$$

Torque (T) is increased by $\dfrac{1}{\sin \alpha}$ times from that for flat collar.

b. Uniform Wear Theory

Substituting value of P for uniform wear in Eq. (5.14), total frictional torque,

$$T = \int_{R_i}^{R_o} \frac{2\mu \pi r^2}{\sin \alpha} \times \frac{F}{2\pi r(R_o - R_i)} dr \quad\quad\quad ... (5.15)$$

$$= \frac{\mu F}{\sin \alpha (R_o - R_i)} \int_{R_i}^{R_o} r \, dr = \frac{\mu F}{\sin \alpha (R_o - R_i)} \left[\frac{r^2}{2} \right]_{R_i}^{R_o}$$

$$= \frac{\mu F}{2\sin \alpha} \times \frac{(R_o^2 - R_i^2)}{(R_o - R_i)} = \frac{\mu F}{2\sin \alpha} \times (R_o + R_i)$$

$$T = \frac{\mu F}{\sin \alpha} \times \frac{(R_o + R_i)}{2}$$

or
$$T = \frac{\mu F}{\sin \alpha} \times \text{mean radius of the collar}$$

So, torque is again increased by $\dfrac{1}{\sin \alpha}$ times from that for flat collar.

5.8.5 Pivots

i. Flat Pivots

Flat pivot bearing is also known as footstep bearing. Relations of torque for pivot can be obtained by integrating corresponding relations for collars from $R_i = 0$ to $R_o = R$.

a. Uniform pressure theory

Integrating Eq. (5.11) from $R_i = 0$ to $R_o = R$,

$$T = \int_0^R 2\mu \frac{F}{\pi R^2} \times \pi r^2 dr = \frac{2\mu F}{R^2} \int_0^R r^2 dr$$

$$= \frac{2\mu F}{R^2} \times \left[\frac{r^3}{3} \right]_0^R = \frac{2}{3} \frac{\mu F}{R^2} \times R^3$$

or
$$T = \frac{2}{3} \mu F R$$

b. Uniform wear theory

Integrating Eq. (5.12) from 0 to R

$$T = \int_0^R 2\mu \frac{F}{2\pi r R} \pi r^2 dr$$

$$= \frac{\mu F}{R} \int_0^R r dr = \frac{\mu F}{R} \left[\frac{r^2}{2} \right]_0^R = \frac{1}{2} \frac{\mu F}{R} \times R^2$$

or
$$T = \frac{1}{2} \mu F R$$

ii. Conical Pivot

a. Uniform pressure theory

Integrating Eq. (5.14) from 0 to R, total frictional

Torque,
$$T = \int_0^R \frac{2\mu \pi r^2}{\sin \alpha} \times \frac{F}{\pi R^2} dr$$

$$= \frac{2\mu F}{R^2 \sin \alpha} \times \int_0^R r^2 dr = \frac{2\mu F}{R^2 \sin \alpha} \times \left[\frac{r^3}{3} \right]_0^R$$

$$= \frac{2}{3} \frac{\mu F}{\sin \alpha \times R^2} R^3$$

or

$$T = \frac{2\mu FR}{3 \sin \alpha}$$

b. Uniform wear theory

Integrating Eq. (5.15) from 0 to R, total frictional torque,

$$T = \int_0^R \frac{2\mu\pi r^2}{\sin \alpha} \times \frac{F}{2\pi rR} dr$$

$$= \frac{\mu F}{\sin \alpha \times R} \times \int_0^R r dr = \frac{\mu F}{\sin \alpha \times R} \left[\frac{r^2}{2} \right]_0^R$$

$$= \frac{1}{2} \frac{\mu F}{\sin \alpha \times R} R^2$$

or

$$T = \frac{1}{2} \frac{\mu FR}{\sin \alpha}$$

It is clear from above expressions that the value of frictional torque is more when uniform pressure theory is used. In practice, the value of frictional torque lies between that given by two theories. A particular theory is selected for a specific application.

Example 5.3: In a thrust bearing, the external and the internal diameters of the contacting surface are 320 mm and 200 mm respectively. The total axial load is 80 kN and the intensity of pressure is 350 kN/m^2. The shaft rotates at 400 rpm. Taking the coefficient of friction as 0.06, calculate the power lost in overcoming the friction. Also, find the number of collars required for the bearing.

Solution:

$R_o = 0.16$ m, $F = 80 \times 10^3$ N, $R_i = 0.1$ m, $\mu = 0.06$, $N = 400$ rpm, $p = 350 \times 10^3$ N/m^2.

Using uniform pressure theory,

$$T = \frac{2}{3} \mu F \left(\frac{R_o^3 - R_i^3}{R_o^2 - R_i^2} \right)$$

$$= \frac{2}{3} \times 0.06 \times 80 \times 10^3 \left[\frac{(0.16)^3 - (0.10)^3}{(0.16)^2 - (0.10)^2} \right]$$

$$= 3200 \times 0.1985 = 635.12 \text{ N·m}$$

$$P = T\omega = T \frac{2\pi N}{60} = 635.12 \times \frac{2\pi \times 400}{60}$$

$$= 26602 \text{ W} \quad \text{or} \quad 26.602 \text{ kW}$$

$$\text{Number of collars} = \frac{\text{total load}}{\text{load per collar}} = \frac{F}{P \times \pi (R_o^2 - R_i^2)}$$

$$= \frac{80 \times 10^3}{350 \times 10^3 \times \pi[(0.16)^2 - (0.10)^2]}$$

$$= 4.66 \cong 5 \text{ collars}$$

Example 5.4: Find the power lost in friction assuming (i) uniform pressure and (ii) uniform wear when a vertical shaft of 100 mm diametr rotating at 150 rpm rests on a flat end footstep bearing. The coefficient of friction is 0.05 and shaft carries a vertical load of 15 kN.

Solution:

Diameter $\qquad D = 100 \text{ mm} = 0.1 \text{ m}, \quad \therefore R = \frac{0.1}{2} = 0.05 \text{ m}$

$$\text{Speed } N = 150 \text{ rpm}, \mu = 0.05$$
$$\text{Load} = 15 \text{ kN} = 15 \times 10^3 \text{ N}$$

(i) *Power lost in friction assuming uniform pressure:*
  For uniform pressure, the frictional torque is given as

$$T = \frac{2}{3}\mu FR$$

$$= \frac{2}{3} \times 0.05 \times 15 \times 10^3 \times 0.05 \text{ N·m} = 25 \text{ N·m}$$

$$\therefore \quad \text{Power lost in friction} = \frac{2\pi NT}{60}$$

$$= \frac{2\pi \times 150 \times 25}{60} \text{ W} = 392.7 \text{ W}$$

(ii) *Power lost in friction assuming uniform wear:*
  For uniform wear, the frictional torque is

$$T = \frac{1}{2}\mu FR$$

$$= \frac{1}{2} \times 0.05 \times 15 \times 10^3 \times 0.05 \text{ N·m} = 18.75 \text{ N·m}$$

$$\therefore \text{ Power lost in friction} = \frac{2\pi NT}{60}$$

$$= \frac{2\pi \times 150 \times 18.75}{60} \text{ W} = 0.294.5 \text{ W}$$

Example 5.5: A conical pivot with angle of cone as 120°, supports a vertical shaft of diameter 300 mm. It is subjected to a load of 20 kN. The coefficient of friction is 0.05 and the speed of shaft is 210 rpm. Calculate the power lost in friction assuming (i) uniform pressure and (ii) uniform wear.

Solution: $\qquad\qquad 2\alpha = 120° \qquad\qquad\qquad \therefore \alpha = 60°$

$$D = 300 \text{ mm} = 0.3 \text{ m} \qquad\qquad \therefore R = 0.15 \text{ m}$$
$$F = 20 \text{ kN} = 20 \times 10^3 \text{ N}; \qquad\qquad \mu = 0.05$$
$$N = 210 \text{ rpm}$$

(i) *Power lost in friction for uniform pressure*:

The frictional torque is

$$T = \frac{2}{3} \times \frac{\mu F R}{\sin \alpha}$$

$$= \frac{2}{3} \times \frac{0.05 \times 20 \times 10^3 \times 0.15}{\sin 60°} = 115.33 \text{ N·m}$$

∴　　　　Power lost $= \dfrac{2\pi N T}{60}$

$$= \frac{2\pi \times 210 \times 115.53}{60} = 2540.6 \text{ W} = 2.54 \text{ kW}$$

(ii) *Power lost in friction for uniform wear*:

The friction torque is given as

$$T = \frac{1}{2} \times \frac{\mu F R}{\sin \alpha}$$

$$= \frac{1}{2} \times \frac{0.05 \times 20 \times 10^3 \times 0.15}{\sin 60°} = 86.6 \text{ N·m}$$

∴　　　　Power lost $= \dfrac{2\pi N T}{60}$

$$= \frac{2\pi \times 210 \times 86.6}{60} = 1903.47 \text{ W} = 1.9 \text{ kW}$$

5.9 GREASY FRICTION AT A JOURNAL

The friction between two surfaces in contact having an extreme thin layer of lubricant between them and metal to metal contact can take place between high spots is known as greasy friction.

A journal bearing is shown in Fig. 5.10. When the shaft is at rest in the bearing, the weight of the shaft W acts through its centre of gravity at O. The shaft rests at the bottom of the bearing at A, known as the seat of pressure and metal to metal contact exists between the two. The reaction of the bearing acts at A in the vertically upward direction in line with W.

When a torque is applied to the shaft, it rotates (say in anticlockwise direction) and the seat of pressure climbs up the bearing in a direction opposite to that of rotation (i.e. in clockwise direction here) at B. Greasy friction condition is applicable at B as metal to metal contact and a thin layer of lubricant exists.

Normal reaction R_n at B and the force of friction μR_n can be combined into a resultant reaction force R which is inclined at an angle ϕ with R_n.

Therefore, the shaft is in equilibrium under the following two forces:

　i. Weight of shaft W acting vertically downward
　ii. Resultant force R

For equilibrium, R must be equal to W and must act vertically upwards. W and R form a couple known as friction couple

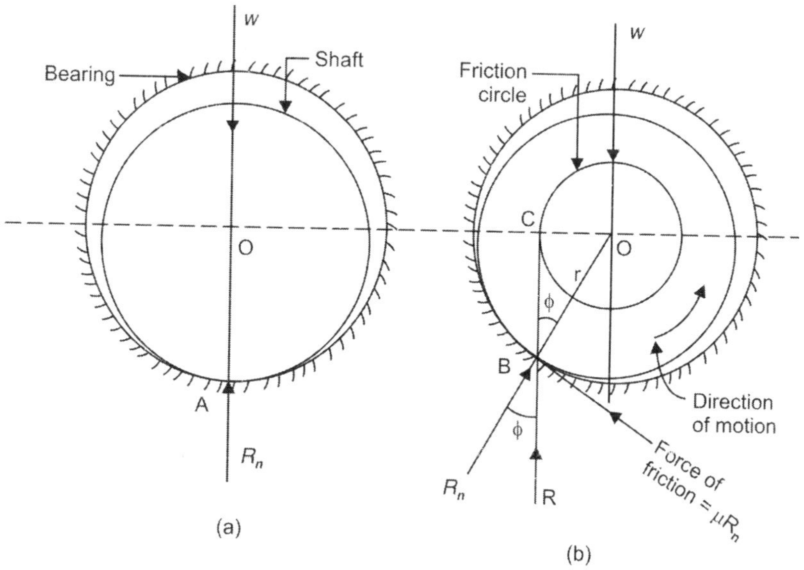

Fig. 5.10

Moment of friction couple $= W \times OC = W \times r \sin \phi$

$\approx W \times r \tan \phi \quad (\phi \text{ is small})$

$\approx Wr\mu \quad (\mu = \tan \phi) \qquad \qquad ...(5.16)$

This friction couple acts in a direction opposite to the direction of rotation.

A circle drawn with radius equal to $OC = r \sin \phi \approx r \tan \phi \approx r \times \mu$ is known as the friction circle of the journal. The radius of friction circle ($r\mu$) is independent of the weight of the shaft. It is constant.

The effect of friction is that the reaction is displaced through a distance equal to $r \sin \phi$ or such that it is tangential to the friction circle.

Power loss in friction $= T \times \omega \qquad \qquad (\omega = \text{angular speed})$

$= Wr\mu \times \dfrac{V}{r}$

$= W\mu V \text{ watt}$

5.10 MITCHELL THRUST BEARING

A Mitchell thrust bearing, shown in Fig. 5.11, consists of a series of metallic pads arranged around a rotating collar fixed to the shaft. Each pad is held by the housing of the bearing. So, pad cannot rotate but is able to tilt. When the thrust is transmitted to the pads, they get tilted slightly on edges. The oil is dragged by moving collar around the pad and a wedge-shaped oil film is formed that carries the axial load.

Mitchell thrust bearing is useful in transmitting very heavy thrusts, e.g. thrust transmitted by propeller of ship to the hull.

5.11 FRICTION CLUTCHES

A clutch is a device used to transmit the rotary motion of one shaft to another, the axes of which are coincident. It is used to transmit power from one shaft to another by engaging or disengaging them as per requirement. In automobiles, friction clutch is used to connect or disconnect gear box shaft from engine shaft.

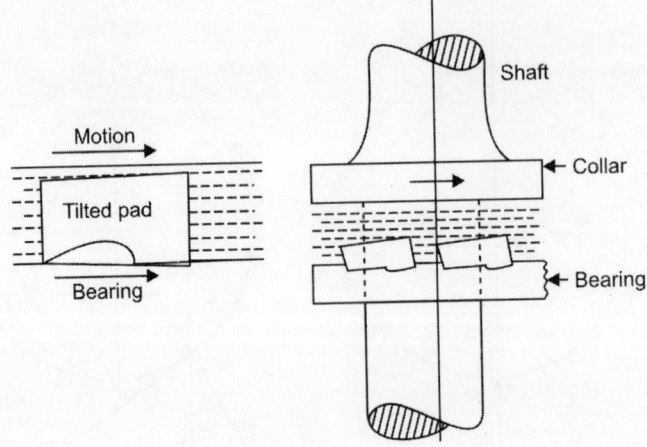

Fig. 5.11: Mitchell thrust bearing

Three types of friction clutches generally used:

i. Disc clutch or single plate clutch

ii. Multiplate clutch

iii. Cone clutch

5.11.1 Disc Clutch or Single Plate Clutch

A disc clutch consists of a single clutch plate, friction lined on both sides, which is free to slide axially on splines cut on the driven shaft.

A spring loaded pressure plate inside the clutch body presses the clutch plate against the flywheel when the clutch is engaged (Fig. 5.12a). The flywheel and pressure plate rotate with the driving shaft. The movement of the clutch pedal is transferred to the pressure plate through a thrust bearing.

When the clutch pedal is pressed down by foot, clutch is disengaged (Fig. 5.12b). Springs press against a cover attached to the flywheel and pressure plate moves away, thus removing pressure from the clutch plate. Now, the flywheel will rotate without driving the clutch plate and driven shaft.

If the torque due to frictional force is more than the torque to be transmitted, there will be no slip between driving and driven shafts.

Torque transmitted

Let the two friction surfaces be held together by an axial thrust W.

Theory of single plate clutch is same as that of flat collar bearing. In a single plate clutch, there are two friction surfaces, one on each side of friction plate. Therefore, total torque transmitted by the clutch is twice that for flat collar.

Total torque transmitted,

$$T = 2 \times \frac{2}{3} \mu W \left[\frac{R_o^3 - R_0^3}{R_0^2 - R_i^3} \right] \qquad \text{(for uniform pressure)}$$

$$T = 2 \times \frac{1}{2} \mu W (R_o + R_i) \qquad \text{(for uniform wear)}$$

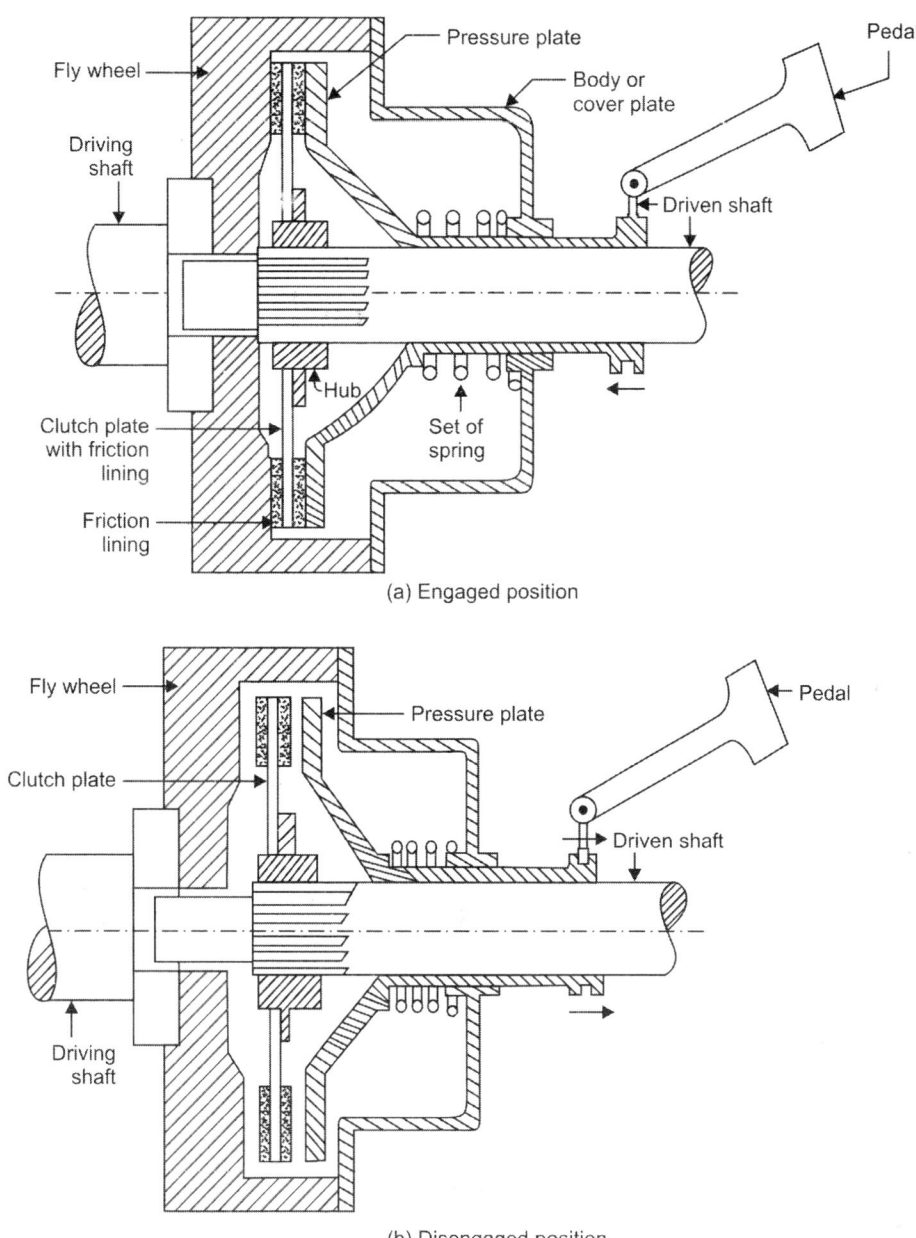

Fig. 5.12: Single plate clutch

For power transmission by friction through a clutch, uniform wear theory gives safer result.

5.11.2 Multiplate Clutch

In multiplate clutch, the number of metal plates and friction linings are increased so that the clutch can transmit large torque. Multiplate clutch is used in motor cars and machine tools. Friction plates are connected on the top of the flywheel. Hence, the friction plates rotate with the flywheel and the engine shaft. They can slide axially.

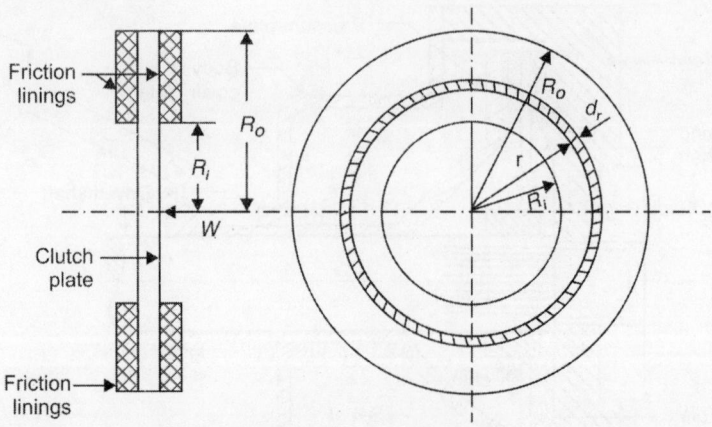

Fig. 5.13: Single plate

The discs or plates are also supported on splines of the driven shaft. Hence, these plates rotate with the driven shaft. They are situated in between the friction plates and can also slide axially.

In the engaged position when the foot is taken off from the clutch pedal, the set of springs will press the discs into contact with the friction plates. Hence, the power will be transmitted from the driving to the driven shaft.

Let n_1 = no. of friction plates on driving shaft

n_2 = no. of discs on the driven shaft

No. of active surfaces, $n = n_1 + n_2 - 1$

Total torque transmitted,

$$T = n \times \frac{2}{3}\mu W \left[\frac{R_o^3 - R_i^3}{R_o^2 - R_i^2} \right] \qquad \text{(for uniform pressure)}$$

$$T = n \times \frac{1}{2}\mu W (R_o + R_i) \qquad \text{(for uniform wear)}$$

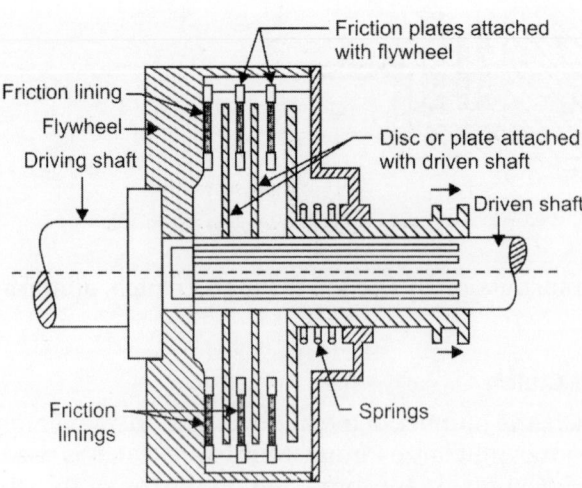

Fig. 5.14: Multiplate clutch in disengaged position

5.11.3 Cone clutch

In cone clutch (Fig. 5.15), the contact surfaces are in the form of cones. In the engaged position, the friction surfaces of the two cones are in complete contact due to spring pressure. In this position, torque is transmitted from the driving shaft to driven shaft through the flywheel and friction cones. For disengaging the clutch, the driven cone is pulled back through a lever system against the force of spring.

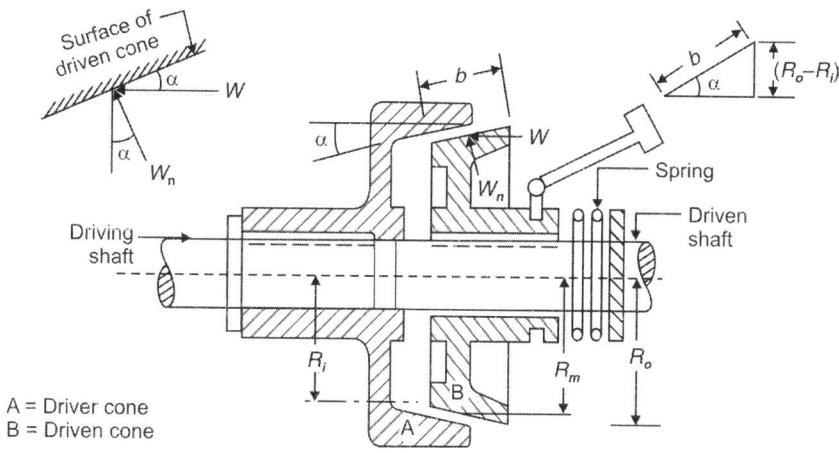

Fig. 5.15: Cone clutch

Let α = semi cone angle

W = total axial load

W_n = normal force $= \dfrac{W}{\sin \alpha}$

b = width of cone face $= \dfrac{(R_o - R_i)}{\sin \alpha}$

Analysis of cone clutch is same as that of truncated conical pivot.

For uniform pressure:

$$P = \frac{W}{\pi(R_o^2 - R_i^2)}$$

Torque transmitted, $\quad T = \dfrac{2}{3} \dfrac{\mu W}{\sin \alpha} \left[\dfrac{R_o^3 - R_i^3}{R_o^2 - R_i^2} \right]$

For uniform wear:

$$P \times r = c \text{ (constant)}$$
$$W = 2\pi p r (R_o - R_i)$$

Torque transmitted $\quad T = \dfrac{1}{2} \dfrac{\mu W}{\sin \alpha} (R_o + R_i)$

$$= \mu \times \frac{W}{\sin \alpha} \times \frac{R_o + R_i}{2}$$

$$= \mu W_n \times R_m \quad R_m = \text{mean radius}$$

Also
$$W_n = \frac{W}{\sin\alpha} = \frac{2\pi pr(R_o - R_i)}{\sin\alpha} = 2\pi Fr \times \frac{(R_o - R_i)}{\sin\alpha}$$

$$W_n = 2\pi Pr \times b \qquad \left(\because \frac{R_o - R_i}{\sin\alpha} = b\right)$$

The advantage of cone clutch is that the normal force on the contact surface $\left(W_n = \frac{W}{\sin\alpha}\right)$ is increased. However, cone clutches have become obsolete these days as it is difficult to disengage them due to small cone angles and dust and dirt binding the two cones.

5.11.4 Centrifugal clutch

A centrifugal clutch has a driving shaft carrying the spider, springs and four sliding shoe blocks. The driven shaft is connected to the pulley. As the speed of the shaft increases, the centrifugal force on radially mounted shoes increases. When the centrifugal force exceeds the resisting force of the springs, the shoes move outwards and press against the inside of the rim and in this way clutch is engaged and torque is transmitted to the rim. The outer surfaces of the shoes are lined with some friction material.

Let m = mass of each shoe

R = inner radius of the pulley rim

r = distance of centre of mass of each shoe from the shaft axis

n = number of shoes

w = normal speed of the shaft in rad/s

w' = speed at which the shoe moves forward

μ = coefficient of friction between the shoe and the rim.

Centrifugal force exerted by each shoe at the time of engagement with the rim = mrw^2

This will be equal to the resisting force of the spring $F = kx$

Centrifugal force exerted by each shoe at normal speed = mrw^2

Net normal force exerted by each shoe on the rim = $mrw^2 - mrw'^2$

Frictional force acting tangentially on each shoe = $\mu mr(w^2 - w'^2)$

Frictional torque acting on each shoe = $\mu mr(w^2 - w'^2)R$

Total frictional torque acting = $\mu mr(w^2 - w'^2)Rn$

If P is the maximum pressure intensity exerted on the shoe, then

$$mr(w^2 - w'^2) = Plb$$

where l and b are the contact length and width of each shoe.

Usually, the clearance between the shoe and the rim is very small and is neglected. However, it can be taken into account if required.

Example 5.6: A single-plate clutch transmits 25 kW at 900 rpm. The maximum pressure intensity between the plates is 85 kN/m². The outer diameter of the plate is 360 mm. Both the sides of the plate are effective and the coefficient of friction is 0.25. Determine

i. the inner diameter of the plate.

ii. the axial force to engage the clutch.

Solution:

Now,
$$P = 25 \text{ kW}, \mu = 0.25, N = 900 \text{ rpm}, R_o = 0.18 \text{ m}, p_i = 85 \text{ kN/m}^2,$$
$$P = T\omega$$

$$25000 = T \times \frac{2\pi \times 900}{60}$$

$$T = 265.26 \text{ N/m}$$

i.
$$T = \frac{\mu F}{2}(R_o + R_i) \times n \qquad (n = \text{number of surfaces})$$

$$= \frac{\mu}{2}[2\pi \times p_i R_i (R_o - R_i)(R_o + R_i)] \times n$$

$$265.26 = 0.25 \times \pi \times 85000 \times R_i (0.18 - R_i)(0.18 + R_i) \times 2$$

$$R_i = [(0.18)^2 - R_i^2] = 0.001987 \text{ or } 0.0324 R_i - R_i^3 = 0.001987$$

Solving the equation by trial and error method:

$$R_i = 0.1315 \text{ m or } 131.5 \text{ mm}$$

ii. Axial force
$$F = 2\pi p_i R_i (R_o - R_i) \times n$$
$$= 2\pi \times 85000 \times 0.1315(0.18 - 0.1315) \times 2$$
$$= 6812 \text{ N or } 6.812 \text{ kN}$$

Inner diameter of the plate $= 2 \times R_i$
$$= 2 \times 0.1315$$
$$= 0.263 \text{ or } 263 \text{ mm}$$

Example 5.7: If the capacity of a single plate clutch decreases by 13% during the initial wear period, determine the minimum value of the ratio of internal diameter to external diameter for the same axial load. Consider both the sides of the clutch plate to be effective.

Solution: A new clutch has a uniform pressure distribution, but after the initial wear the clutch exhibits the characteristics of uniform wear. Capacity of a clutch means the maximum torque transmitted. Thus, according to the given condition.

$$T_{wear} = 0.87 T_{pressure}$$

$$\frac{\mu F}{2}(R_o + R_i) \times n = 0.87 \frac{2}{3}\pi F\left[\frac{R_o^3 - R_i^3}{R_o^2 - R_i^2}\right] \times n$$

or
$$(R_o + R_i) = 1.16\left[\frac{R_o^3 - R_i^3}{R_o^2 - R_i^2}\right]$$

$$(R_o + R_i)(R_o^2 - R_i^2) = 1.16(R_o^3 - R_i^3)$$

$$(R_o + R_i)(R_o + R_i)(R_o - R_i) = 1.16(R_o - R_i)(R_o^2 + R_o R_i + R_i^2)$$

$$(R_o + R_i)^2 = 1.16(R_o^2 + R_o R_i + R_i^2)$$

$$(R_o^2 + 2R_o R_i + R_i^2) = 1.16(R_o^2 + R_o R_i + R_i^2)$$

$$0.16R_o^2 + 0.16R_i^2 - 0.84R_o R_i = 0$$

Dividing throughout by $0.16R_o^2$

$$1 + \left(\frac{R_i}{R_o}\right)^2 - 5.25\left(\frac{R_i}{R_o}\right) = 0$$

or

$$\left(\frac{R_i}{R_o}\right)^2 - 5.25\left(\frac{R_i}{R_o}\right) + 1 = 0$$

Taking

$$\frac{R_i}{R_o} = r \Rightarrow r^2 - 5.25r + 1 = 0$$

or

$$r = \frac{5.25 \pm \sqrt{5.25^2 - 4}}{2} = \frac{5.25 \pm 4.854}{2}$$

Positive value is not possible as ratio r cannot be more than 1

∴

$$r = \frac{5.25 - 4.854}{2} = 0.198 \text{ or } \frac{R_i}{R_o} = 0.198$$

Example 5.8: A single plate clutch, having two active surfaces, transmits 10 kW of power and the maximum torque developed is 120 N·m. Axial pressure is not to exceed 100 kN/m². The outer diameter of the friction plate is 1.3 times the inner diameter. Determine these diameters and the axial force exerted by the springs. Assume uniform wear and take coefficient of friction as 0.25.

Solution : No. of active surfaces, $n = 2$ and $T = 120$ N·m

$$P = 10 \text{ kW} = 10000 \text{ W}$$
$$p_{max} = 100 \text{ kN/m}^2$$
$$R_o = 1.3R_i, \mu = 0.25$$

For uniform wear, $P \times r = $ constant.

∴ Pressure will be maximum at inner radius

or

$$P_{max} \times R_i = \text{constant} = C \text{ or } 100000 \times R_i = C$$

$$\text{Axial thrust } W = 2\pi C (R_o - R_i)$$
$$= 2\pi \times 100000 R_i \times (1.3 R_i - R_i)$$

$$W = 188400R_i^2$$

Frictional torque

$$T = 2 \times \frac{\mu W}{2} \times (R_o + R_i)$$

or

$$120 = \frac{2 \times 0.25 \times 188400R_i^2}{2} \times (1.3R_i + R_i)$$

$$120 = 108330R_i^3$$

⇒

$$R_i = 103.5 \text{ mm}$$

Inner diameter, $D_i = 2 \times 103.5 = 207$ mm

Outer diameter, $D_o = 1.3D_i = 1.3 \times 207 = 269$ mm

Axial force, $W = 188400R_i^2 = 188400 \times (0.1035)^2$

$$W = 2018.2 \text{ N}$$

Example 5.9: A multiplate clutch transmits 55 kW of power at 1800 rpm. Coefficient of friction for the friction surfaces is 0.1. Axial intensity of pressure is not to exceed 160 kN/m². The internal radius is 80 mm and 0.7 times the external radius. Find the number of plates needed to transmit the required torque.

Solution: $p_i = 160 \times 10^3 \text{ N/m}^2$, $R_i = 0.08$ m, $R_o = \dfrac{0.08}{0.7} = 0.1143$ m, $\mu = 0.1$, $N = 1800$ rpm, $P = 55$ kW.

Assuming uniform wear conditions,

$$F = 2\pi p_i r_i (R_o - R_i)$$
$$= 2\pi \times 160 \times 10^3 \times 0.08\,(0.1143 - 0.08) = 2759 \text{ N}$$

$$T = \frac{1}{2}\mu F\left(R_o + R_i\right)$$

$$= \frac{1}{2} \times 0.1 \times 2759 \times (0.1143 + 0.08) = 26.78 \text{ N/m/surface}$$

Total torque transmitted $= \dfrac{P}{\omega} = \dfrac{55000}{2\pi \times 1800} = 291.8$ N/m

Number of friction surfaces required $= \dfrac{291.8}{26.78} = 10.9$ or 11 surfaces

So, there will be 12 plates. Six plates (rings) revolve with the driving or engine shaft and the other 6 with the driven shaft.

Example 5.10: A multiplate clutch transmits 25 kW of power at 1600 rpm. It has three discs on the driving shaft and two on the driven shaft. Coefficient of friction for the friction surfaces is 0.25. The external and internal radii of friction surfaces are 100 mm and 50 mm respectively. Find the maximum intensity of pressure between the discs. Assume uniform wear.

Solution:

$$P = 25 \text{ kW} = 25 \times 10^3 \text{ W}; N = 1600 \text{ rpm}.$$
$$\mu = 0.25; R_o = 100 \text{ mm} = 0.1 \text{ m}; R_i = 50 \text{ mm} = 0.05 \text{ m}$$

No. of discs on driving shaft $n_1 = 3$; No. of discs on driven shaft $n_2 = 2$.

∴ No. of friction (or active) surfaces $n = n_1 + n_2 - 1 = 3 + 2 - 1 = 4$

For uniform wear:

Let $\qquad p_{max} = $ max intensity of pressure

$$P = \frac{2\pi NT}{60}$$

or $\qquad 25 \times 10^3 = \dfrac{2\pi \times 1600 \times T}{60}$

$$T = \frac{25 \times 10^3 \times 60}{2\pi \times 1600} = 149.207 \text{ N} \cdot \text{m}$$

Total torque transmitted,

$$T = n \times \mu \times W \times R_m \qquad\qquad\qquad (i)$$

where $\qquad\qquad R_m = $ mean radius of friction surface

$$= \frac{R_o + R_i}{2} \qquad \text{(for uniform wear)}$$

$$= \frac{0.1 + 0.05}{2} = 0.075 \text{ m}$$

$$n = 4 \text{ and } \mu = 0.25$$

Substituting the values of T, n, μ and R_m in Eq. (i),

$$149.207 = 4 \times 0.25 \times W \times 0.075$$

$$\therefore \qquad W = \frac{149.207}{4 \times .25 \times 0.075} = 1989.426 \text{ N}$$

The expresssion for axial load (W) for uniform wear is also given as

$$W = 2\pi C \, (R_o - R_i)$$

Substituting the values of W, R_o and R_i

$$1989.426 = 2\pi \times C \times (0.1 - 0.05)$$

or

$$C = \frac{1989.426}{2\pi \times 0.05} = 6332.54$$

For uniform wear, $p \times R =$ (constant)

The pressure is maximum at internal radius,

$$p_{max} \times R_i = C$$

or

$$p_{max} \times 0.05 = 6332.54 \qquad [\because C = 6332.54 \text{ and } R_i = 0.05]$$

$$p_{max} = \frac{6332.54}{0.05} = 1266.50 \text{ N/m}^2 = 0.12665 \text{ N/mm}^2$$

Example 5.11: Determine the axial force required to engage a cone clutch transmitting 25 kW of power at 750 rpm. Average friction diameter of the cone is 400 mm, semicone angle 10°, $p_m = 60 \text{ kN/m}^2$ and coefficient of friction 0.25. Also, find the width of the friction cone.

Solution:

Let axial force = W

Power P = 25 kW = 25000 W

N = 750 rpm

$$\omega = \frac{2\pi \times 750}{60} = 78.5 \text{ rad/s}$$

$$\alpha = 10°, \mu = 0.25, R_m = 200 \text{ mm} = 0.2 \text{ m}$$

Power P = $T\omega$ or $T = \dfrac{P}{\omega} = \dfrac{25000}{78.5} = 318.47 \text{ N} \cdot \text{m}$

For uniform wear theory,

$$T = \frac{\mu W}{\sin \alpha} \times R_m$$

or

$$318.47 = \frac{0.25 \times W}{\sin 10°} \times 0.2$$

or

$$W = 1106 \text{ N}$$

Now $\qquad W_n = 2\pi p_m R_m \times b$ or $\dfrac{W}{\sin \alpha} = 2\pi p_m R_m \times b$

or $\qquad b = \dfrac{W}{\sin \alpha \times 2\pi p_m R_m}$

$$= \dfrac{1106}{\sin 10° \times 2\pi \times 60 \times 1000 \times 0.2} = 0.84 \text{ m}$$

or width of cone $\qquad b = 84.5 \text{ mm}.$

Example 5.12: The shaft of a collar thrust bearing rotates at 200 rpm and carries an end thrust of 10 tones. The outer and the inner diameters of the bearing are 480 mm and 280 mm respectively. If the power lost in friction is not to exceed 8 kW, determine the coefficient of friction of the lubricant of the bearing.

Solution: $\qquad N = 200 \text{ rpm}, W = \dfrac{2\pi N}{60} = \dfrac{2\pi \times 200}{60} = 20.94 \text{ rad/s}$

End thrust $\qquad F = 10 \text{ ton} = 10 \times 1000 \times 9.81 \text{ N} = 98100 \text{ N}$

$\qquad R_o = \dfrac{480}{2} = 240 \text{ mm} = 0.2 \text{ m}, R_i = \dfrac{280}{2} = 140 \text{ mm} = 0.14 \text{ m}$

Power lost $\qquad P = 8 \text{ kW} = 8000 \text{ W}$

Let coefficient of friction $= \mu$

$$P = Tw \Rightarrow T = \dfrac{P}{\omega} = \dfrac{8000}{20.94} = 382.04 \text{ N} \cdot \text{m}$$

For uniform pressure, torque transmitted,

$$T = \dfrac{2}{3}\mu F \left(\dfrac{R_o^3 - R_i^3}{R_o^2 - R_i^2} \right)$$

or $\qquad 382.04 = \dfrac{2}{3}\mu \times 98100 \times \dfrac{(0.24)^3 - (0.14)^3}{(0.24)^2 - (0.14)^2}$

or $\qquad \mu = 0.02$

Example 5.13: A conical pivot with angle of cone as 100° supports a load of 18 kN. The external radius is 2.5 times the internal radius. The shaft rotates at 150 rpm. If the intensity of pressure is to be 300 kN/m² and coefficient of friction 0.05, what will be the power lost in working against the friction?

Solution:

$F = 18 \text{ kN}, R_o = 2.5 R_i, p = 300 \text{ kN/m}^2, N = 150 \text{ rpm}, \mu = 0.05, \alpha = 50°$

In case of uniform pressure, normal pressure $p = \dfrac{F}{\pi(R_o^2 - R_1^2)}$

or $\qquad 300 \times 10^3 = \dfrac{18 \times 10^3}{\pi[(2.5 R_i)^2 - R_i^2]}$

or $\qquad (2.5 R_i)^2 - R_i^2 = \dfrac{18}{300 \times \pi}$

$\qquad R_i = 0.0603 \text{ m and } R_o = 0.0603 \times 2.5 = 0.1508 \text{ m}$

Torque transmitted $\quad T = \dfrac{2}{3}\dfrac{\mu F}{\sin\alpha}\left[\dfrac{R_o^3 - R_i^3}{R_o^2 - R_i^2}\right]$

or $\qquad T = \dfrac{2}{3}\times\dfrac{0.05\times18000}{\sin 50^\circ}\left[\dfrac{(0.1508)^3 - (0.0603)^3}{(0.1508)^2 - (0.0603)^2}\right] = 131.6\ \text{N}$

$$P = T\omega = T\dfrac{2\pi N}{60} = 131.6\times\dfrac{2\pi\times150}{60}$$

$$= 2067\ \text{W or } 2.067\ \text{kW.}$$

EXERCISE

5.1 Derive the frictional torque relation considering uniform pressure for conical pivot bearing.

5.2 Derive the frictional torque relation considering uniform wear for conical clutch.

5.3 Derive the frictional torque relation considering uniform wear for a flat pivot bearing.

5.4 Explain the working of a multiplate friction clutch with the help of a neat sketch.

5.5 What are uniform pressure and uniform wear theories? Derive expressions for the frictional torque considering both the theories for a flat collar.

5.6 What is friction circle? Derive an expression for the radius of friction circle in terms of angle of friction.

5.7 Which theory do you recommend, the uniform pressure theory or uniform wear theory for the frictional torque of a bearing?

5.8 What is a clutch? Describe with a neat sketch the working of a single plate friction clutch.

5.9 Describe the working of a Mitchell thrust bearing.

5.10 Derive the expression for frictional torque of a conical pivot considering uniform wear theory.

5.11 Derive the expression for frictional torque of a flat pivot considering uniform pressure theory.

5.12 Derive the relation for the torque transmitted by a cone clutch considering uniform pressure theory.

5.13 Cone clutches have become obsolete, although they provide high frictional torque. Why?

5.14 Describe with the help of a neat sketch a centrifugal clutch and derive a relation for the frictional torque.

5.15 A pull of 20 N, inclined at 25° to the horizontal plane, is required just to move a body placed on a rough horizontal plane. But the push required to move the body is 25 N. If the push is inclined at 25° to the horizontal, find the weight of the body and coefficient of friction. (*Ans.* 84.547 kg, 0.238)

5.16 In a collar thrust bearing, the external and internal radii are 250 mm and 150 mm respectively. The total axial load is 50 kN and shaft is rotating at 150 rpm. The coefficient of friction is equal to 0.05. Find the power lost in friction assuming uniform pressure. (*Ans.* 8.0176 kW)

5.17 A total load of 25 kN is supported by a conical pivot with angle of cone as 120°. The intensity of pressure is not to exceed 350 kN/m². The external radius is 2 times the internal radius. The shaft is rotating at 180 rpm and coefficient of friction is 0.05. Find the power absorbed in friction assuming uniform pressure. (*Ans.* 3.6826 kW)

5.18 In a thrust bearing, the external and internal radii of the contact surfaces are 210 mm and 160 mm respectively. The total axial load is 60 kN and coefficient of friction is 0.05. The shaft is rotating at 380 rpm. The intensity of presure is not to exceed 350 kN/m². Calculate:

i. Power lost in overcoming the friction

ii. Number of collars required for the thrust bearing. (*Ans.* 22.22 kW, 3)

5.19 A conical pivot bearing 15 cm in diameter has a cone angle of 120°. If the shaft supports an axial load of 2 tons and coefficient of friction is 0.03, find the power (in hp) lost in friction when shaft rotates at 200 rpm: assuming (i) uniform pressure and (ii) uniform wear. (*Ans.* 0.87 hp; 0.65 hp)

5.20 The inner and outer radii of a single plate clutch are 40 mm and 80 mm respectively. Determine the maximum, minimum and average pressure when the axial force is 3 kN. (*Ans.* 298.4 kN/m², 149.2 kN/m², 198.9 kN/m²)

5.21 Calculate the power transmitted by a single plate clutch at a speed of 2000 rpm, if the outer and inner radii of friction surfaces are 150 mm and 100 mm respectively. The maximum intensity of presure at any point of contact surface should not exceed 0.8×10^5 N/m². Take both sides of the plate as effective and coefficient of friction as 0.3. Assume uniform wear.

(*Ans.* 39.477 kW)

5.22 A power of 60 kW is transmitted by a multiplate clutch at 1500 rpm. Axial intensity of pressure is not to exceed 0.15 N/mm². The coefficient of friction for the friction surfaces is 0.15. The external radius of friction surface is 120 mm. Also, the external radius is equal to 1.25 times the internal radius. Find the number of plates needed to transmit the required power. Assume uniform wear. (*Ans.* 12)

5.23 A cone clutch with cone angle 30° is used to transmit a power of 10 kW at 800 rpm. the intensity of pressure between the contact surfaces is not to exceed 85 kN/m². The width of the conical friction surface is half of the mean radius. If coeffcient of friction is 0.15, then find the dimensions of the contact surfaces. Assume uniform wear. Also find the axial load or force required to hold the clutch while transmitting the power. What is the width of the friction surface? (*Ans.* 138 mm, 157 mm, 1400. 3N, 73.4 mm)

OBJECTIVE TYPE QUESTIONS

5.1 The efficiency of a screw jack depends on
(a) pitch of threads
(b) load
(c) both pitch and load
(d) none

5.2 The efficiency of a screw jack increases with
(a) decrease in load
(b) increase in load
(c) decrease in pitch
(d) increase in pitch

5.3 The efficiency of a screw jack is

(a) $\eta = \dfrac{\tan\alpha}{\tan(\alpha - \phi)}$

(b) $\eta = \dfrac{\tan(\alpha + \phi)}{\tan\alpha}$

(c) $\eta = \dfrac{\tan\alpha}{\tan(\alpha + \phi)}$

(d) $\eta = \dfrac{\tan(\alpha - \phi)}{\tan\alpha}$

5.4 The efficiency of a screw jack is maximum when

(a) $\alpha = 45° - \dfrac{\phi}{4}$

(b) $\alpha = 45° + \dfrac{\phi}{2}$

(c) $\alpha = 45° + \dfrac{\phi}{4}$

(d) $\alpha = 45° - \dfrac{\phi}{2}$

5.5 No force is required for downward motion of a load on screw jack if
   (a) $\alpha < \phi$
   (b) $\alpha > \phi$
   (c) $\alpha > 2\phi$
   (d) $\alpha < 2\phi$

5.6 Maximum efficiency of a screw jack is

(a) $\dfrac{1 - \sin\phi}{1 + \sin\phi}$

(b) $\dfrac{1 + \sin\phi}{1 - \sin\phi}$

(c) $\dfrac{\sin\phi}{1 - \cos\phi}$

(d) $\dfrac{\sin\phi}{1 + \cos\phi}$

5.7 Frictional force is more in
   (a) dry sliding friction
   (b) dry rolling friction
   (c) nonviscous friction
   (d) viscous friction

5.8 Viscous friction is also known as
   (a) film friction or boundary friction
   (b) perfect friction or nonviscous
   (c) film friction or greasy friction
   (d) film friction or perfect friction

5.9 Which of the following is correct?
   (a) kinetic friction is less than maximum static friction
   (b) kinetic friction is more than maximum static friction

(c) kinetic friction is equal to maximum static friction

(d) impending sliding friction is more than the limiting friction

5.10 The angle of friction is the maximum between

(a) the normal reaction and the frictional force

(b) the frictional force and the resultant reaction

(c) the normal reaction and the resultant reaction

(d) the frictional force and the normal reaction

5.11 If ϕ is the angle of friction, then

(a) $\tan\phi = \dfrac{\text{maximum static friction}}{\text{normal reaction}}$

(b) $\mu = \tan^{-1}\phi$

(c) $\tan\phi = \dfrac{\text{normal reaction}}{\text{frictional force}}$

(d) $\tan\phi = \dfrac{\text{frictional force}}{\text{normal reaction}}$

5.12 A screw and a nut will be self-locking if efficiency is

(a) more than 50%

(b) equal to 50%

(c) less than 50%

(d) equal to that for overhauling

5.13 In a journal bearing, the load acting on the shaft is

(a) axial

(b) coaxial

(c) transverse to its axis

(d) inclined to axis

5.14 Pivot and collar bearings are meant to carry

(a) vertical and axial loads respectively

(b) axial and vertical loads respectively

(c) axial load only

(d) vertical load only

5.15 Footstep bearing is provided

(a) at the end of a vertical shaft

(b) at the end of horizontal shaft

(c) along the length of the shaft

(d) to carry transverse load

5.16 Footstep bearing is

(a) a pivot

(b) a collar

(c) a journal bearing

(d) a clutch

5.17 In respect of flat pivot bearing, the ratio between the frictional torque at uniform rate of wear and uniform intensity of pressure is

(a) 1

(b) $\dfrac{2}{3}$

(c) $\dfrac{4}{3}$

(d) $\dfrac{3}{4}$

5.18 If r is the radius of a shaft in a journal bearing, the radius of friction circle is

    (a) μ

    (b) r

    (c) $\dfrac{\mu}{r}$

    (d) μr

5.19 In a square thread, if p is the pitch is D outside diameter, the mean diameter d_m

    (a) $D + p$

    (b) $\dfrac{D - p}{2}$

    (c) $\dfrac{D + p}{2}$

    (d) $D - \dfrac{p}{2}$

5.20 Collar bearings are provided on shafts so that

    (a) the intensity of pressure is decreased

    (b) the intensity of pressure is increased

    (c) the frictional torque is decreased

    (d) the frictional torque is increased

5.21 The moment on a pulley which produces rotation is called

    (a) work

    (b) energy

    (c) momentum

    (d) torque

5.22 If T_1 and T_2 are the tensions on tight and slack sides of a belt in newtons and v velocity in m/s, the power transmitted is given in watt as

    (a) $(T_1 - T_2)\, v$

    (b) $\dfrac{(T_1 - T_2)v}{4500}$

    (c) $\dfrac{(T_1 - T_2)v}{75}$

    (d) $\dfrac{(T_1 - T_2)v}{60}$

5.23 In the relation $\dfrac{T_1}{T_2} = e^{\mu\theta}$, the angle of contact θ is the angle made by the belt on the

    (a) smaller pulley

    (b) larger pulley

    (c) both the pulleys

    (d) none of the above

5.24 A Jockey pulley is fitted on

    (a) the slack side

    (b) the right side

    (c) may be fitted on any side

    (d) none of the above

5.25 The relation for centrifugal tension is given by

(a) $T_c = mv^2$

(b) $T_c = \dfrac{mv^2}{2}$

(c) $\dfrac{mv^2}{2g}$

(d) m^2v

5.26 Due to creep of belts, there is

(a) loss of motion

(b) loss of power to be transmitted

(c) some length of the belt passes off the follower without any contract with it

(d) all of the above take place

5.27 Crowning of pulley is done

(a) to make them more sturdy

(b) to avoid the slipping of the belt

(c) to make pulley look more pleasant in appearance

(d) to enable pulley rigidly fixed to the shaft

5.28 To avoid slipping of the belt

(a) the pulleys are provided with flanges

(b) the rim of the pulley is made convex

(c) lateral stiffness in the belt is provided

(d) All of the above

5.29 For power transmission by belt it is assumed that

(a) the belt material is elastic

(b) the length of the belt remains unchanged

(c) Both of the above

(d) None of the above

5.30 The V-belt sheaves of pulleys normally have groove angle of

(a) 50° to 65°

(b) 20° to 30°

(c) 35° to 40°

(d) 15° to 20°

5.31 If the initial tension in the belt is increased

(a) the power transmitted bythe belt decreases

(b) the power transmitted by belt increases

(c) no change in power

(d) the power may increase or decrease

5.32 For constant velocity ratio, positive drive with large centre distance between driver and driven shaft

(a) chain drive is used

(b) flat belt drive is used

(c) gear drive is used

(d) V-belt drive is used

5.33 Considering centrifugal tension in belt, the maximum permissible velocity is

(a) inversely proportional to maximum tension

(b) proportional to maximum tension

(c) proportional to cube root of maximum tension

(d) proportional to square root of maximum tension

5.34 In a full conical of radius R, cone angle 2α and carrying load W, the frictional torque is

(a) $\dfrac{\mu W}{\sin \alpha} \times \dfrac{2}{3} R$ at uniform rate of wear

(b) $\dfrac{\mu W}{\sin \alpha} \times \dfrac{R}{2}$ at uniform pressure

(c) $\dfrac{\mu W}{\sin \alpha} \times R$ at uniform rate of wear

(d) $\dfrac{\mu W}{\sin \alpha} \times \dfrac{2}{3} R$ at uniform pressure

5.35 Unless otherwise specifically stated
  (a) uniform intensity of pressure is assumed in pivots
  (b) uniform rate of wear is assumed in clutches
  (c) uniform intensity of pressure is assumed in pivots
  (d) uniform rate of wear is assumed in clutches

5.36 The virtual coefficient of friction for V-threads, with respect to actual coefficient of friction is
  (a) more
  (b) less
  (c) equal
  (d) None of the above

5.37 In pivot bearing, the wear at the contact area is
  (a) zero at the centre
  (b) uniform throughout
  (c) maximum at the centre
  (d) maximum at the outer radius

5.38 The efficiency of a wedge is

(a) $\eta = \dfrac{\tan \alpha}{\tan (\alpha - 2\phi)}$

(b) $\eta = \dfrac{\tan (\alpha + 2\phi)}{\tan \alpha}$

(c) $\eta = \dfrac{\tan \alpha}{\tan (\alpha + 2\phi)}$

(d) $\eta = \dfrac{\tan (\alpha - 2\phi)}{\tan \alpha}$

5.39 For flat and conical pivots, ratio of friction torque with uniform wear to friction torque with uniform pressure is
  (a) 2/3
  (b) 3/2
  (c) 4/3
  (d) 3/4

5.40 The frictional torque for the same diameter in a conical bearing is .... than in a flat bearing.
  (a) more
  (b) less
  (c) equal

5.41 For a safe design, a friction clutch is designed assuming
  (a) uniform pressure theory
  (b) uniform wear theory
  (c) any one of the two
  (d) none
5.42 In a multiple friction clutch, the number of active friction surface is
  (a) $2n$
  (b) n
  (c) $2(n-1)$
  (d) $(n-1)$
5.43 In a screw jack, the effort required to lift the load w is given by
  (a) $p = W \tan (\alpha - \phi)$
  (b) $p = W \tan (\alpha + \phi)$
  (c) $p = W \cos (\alpha - \phi)$
  (d) $p = W \cos (\alpha - \phi)$
5.44 The radius of a friction circle for a shaft of radius 'r' rotating inside a bearing is
  (a) $r \sin \phi$
  (b) $r \cos \phi$
  (c) $r \tan \phi$
  (d) $r \cot \phi$
5.45 The frictional torque transmitted in a flat pivot bearing, considering uniform pressure, is
  (a) $\dfrac{1}{2} \mu WR$

  (b) $\dfrac{2}{3} \mu WR$

  (c) $\dfrac{3}{4} \mu WR$

  (d) μWR
5.46 The frictional torque transmitted in a conical pivot bearing, considering uniform wear is
  (a) $\dfrac{1}{2} \mu WR \operatorname{cosec} \alpha$

  (b) $\dfrac{2}{3} \mu WR \operatorname{cosec} \alpha$

  (c) $\dfrac{3}{4} \mu WR \operatorname{cosec} \alpha$

  (d) $\mu WR \operatorname{cosec} \alpha$
5.47 The frictional torque transmitted by a disc or plate clutch is same that of
  (a) flat pivot bearing
  (b) flat collar bearing
  (c) conical pivot bearing
  (d) trapezoidal pivot bearing
5.48 The frictional torque transmitted by a cone clutch is same as that of
  (a) flat pivot bearing
  (b) flat collar bearing
  (c) conical pivot bearing
  (d) trapezoidal pivot bearing

ANSWERS

5.1 (a)	5.2 (d)	5.3 (c)	5.4 (d)	5.5 (b)	5.6 (a)
5.7 (d)	5.8 (d)	5.9 (a)	5.10 (c)	5.11 (a)	5.12 (c)
5.13 (c)	5.14 (c)	5.15 (a)	5.16 (a)	5.17 (d)	5.18 (d)
5.19 (d)	5.20 (a)	5.21 (d)	5.22 (a)	5.23 (a)	5.24 (a)
5.25 (a)	5.26 (d)	5.27 (b)	5.28 (d)	5.29 (c)	5.30 (c)
5.31 (b)	5.32 (a)	5.33 (d)	5.34 (b)	5.35 (b)	5.36 (a)
5.37 (a)	5.38 (c)	5.39 (d)	5.40 (a)	5.41 (b)	5.42 (d)
5.43 (b)	5.44 (a)	5.45 (b)	5.46 (a)	5.47 (b)	5.48 (d)

6 Gears

6.1 INTRODUCTION

Gears are used to transmit motion from one shaft to another or between a shaft and a slide. This is accomplished by successively engaging teeth. Gears use no intermediate link or connector and transmist the motion by direct contact. In this method, the surfaces of two bodies make a tangential contact. The two bodies have either a rolling or a sliding motion along the tangent at the point of contact. No motion is possible along the common normal as that will either break the contact or no body will tend to penetrate into the other.

If power transmitted between two shafts is small, motion between them may be obtained by using two plain cylinders or discs 1 and 2 as shown in Fig. 6.1. If there is no slip of one surface relative to the other, a definite motion of 1 can be transmitted to 2 and *vice versa*. Such wheels termed *friction wheels*. However, as the power transmitted increases, slip occurs between the discs and the motion no longer remains definite.

Assuming no slipping between the two surfaces, the following kinematic relationship exists for their linear velocity;

$$v_p = \omega_1 r_1 = \omega_2 r_2$$
$$= 2\pi N_1 r_1 = 2\pi N_2 r_2$$

or

$$\frac{\omega_1}{\omega_2} = \frac{N_1}{N_2} = \frac{r_2}{r_1}$$

where N = speed of disc (rpm)

ω = angular velocity of disc (rad/s)

r = radius of disc

6.2 CLASSIFICATION OF GEARS

Gears may be classified according to the position of axis of the shafts. The axis of the two shafts between which the motion is to be transmitted, may be parallel (spur gears, helical gears), intersecting, nonintersecting and nonparallel.

6.2.1 Spur Gears

The two parallel and coplanar shafts connected by the gears are called spur gears. These gears have teeth parallel to the axis of the wheel or shaft.

6.2.2 Helical Gears

The gears having teeth inclined to the axis of the shaft are known as helical gears.

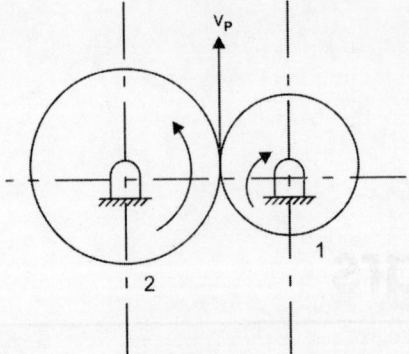

Fig. 6.1

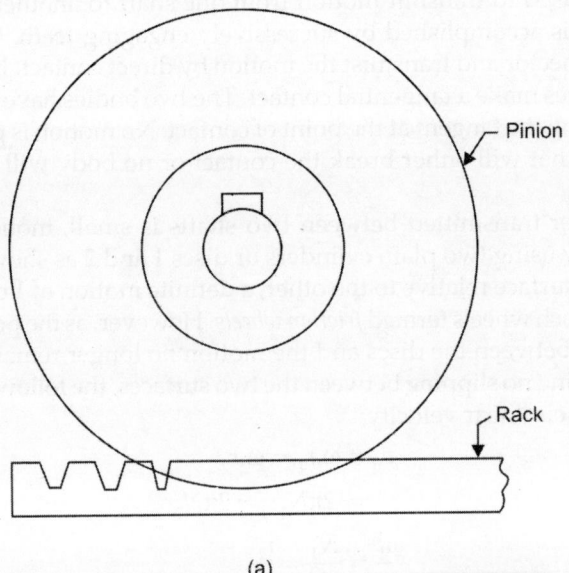

(a)

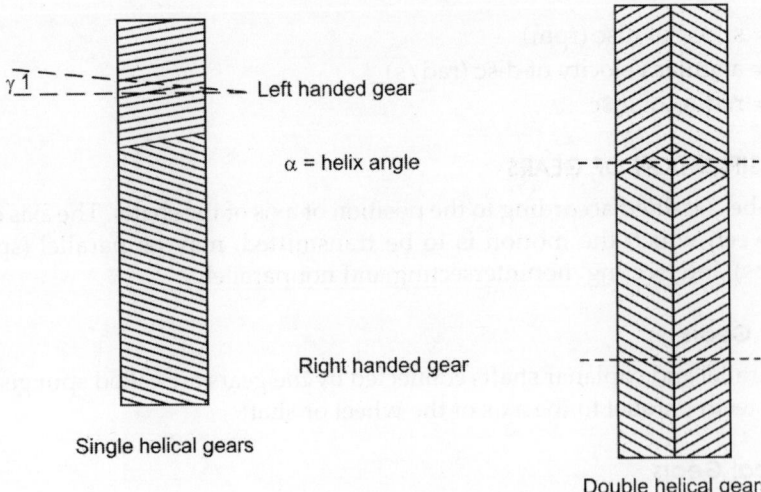

(b)

Fig. 6.2

6.2.3 Bevel Gear

Bevel gears are used to transmit motion from one shaft to another where their axes intersect. The shaft may be inclined at any angle from 0° to 180°. In case when this angle is 90° and giving equal speed, the gearing is called mitre. They are used in differentials of automobiles. Refer to Fig. 6.2. Here θ is the shaft angle. They are out on conical surfaces.

6.2.4 Worm and Worm Gear

Worm is a cylindrical body having one or more threads cut on it in the form of helix. Worm with one thread is called as single threaded worm. Worm having two, three or four threads cut on it is called double, triple or quardruple threaded worm and so on. Worm and worm gear drive are used for high speed ratio up to 500:1. The system is used to transfer motion from input shaft to the follower shaft. The worm may be conical in shape. The diameter of worm is small as compared to worm gear.

6.3 GEAR TERMINOLOGY

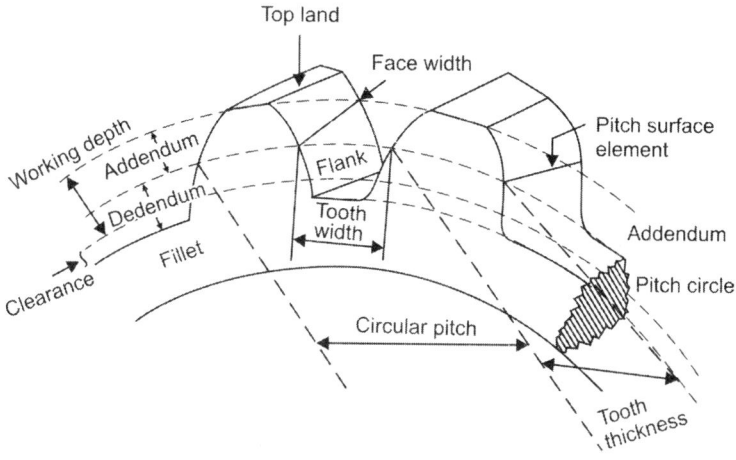

Fig. 6.3

6.3.1 Pitch Circle

It is an imaginary circle which by pure rolling action, would give the same motion as the actual gear.

6.3.2 Pitch Circle Diameter (PCD)

It is the diameter pitch circle. The size of the gear is usually specified by the pitch circle diameter. It is also known as pitch diameter.

6.3.3 Pitch Point

It is a common point of contact between two pitch circles.

6.3.4 Pitch Surface

It is the surface of the rolling disc which the meshing gears have replaced at the pitch circle.

6.3.5 Addendum

It is the radial distance of a tooth from the pitch circle to the top of the tooth.

6.3.6 Dedendum

It is the radial distance of a tooth from the pitch circle to the bottom of the tooth.

6.3.7 Addendum Circle

It is the circle drawn through the top of the teeth and concentric with the pitch circle.

6.3.8 Dedendum Circle

It is the circle drawn through the bottom of the teeth. It is also called root circle.

Root circle diameter = Pitch circle diameter × cos ϕ, where ϕ is the pressure angle.

6.3.9 Pressure Angle or Angle of Obliquity

It is the angle between the common normal to two gear teeth at the point of contact and the common normal at the pitch point. It is usually denoted by ϕ. The standard pressure angles are 14½° and 20°.

6.3.10 Circular Pitch (P_c)

It is the distance measured on the circumference of the pitch circle from a point of one tooth to the corresponding point on the next tooth. It is the usually denoted by P_c

Mathematically, $\qquad P_c = \dfrac{\pi D}{T}$

where D = pitch circle diameter

$\qquad T$ = number of teeth

A little consideration will show that the two gears will mesh together correctly, if the two wheels have the same circular pitch.

Note: If D_1 and D_2 are the diameters of the two meshing gears having teeth T_1 and T_2 respectively, then for them to mesh correctly,

$$P_c = \frac{\pi D_1}{T_1} = \frac{\pi D_2}{T_2} \text{ or } \frac{D_1}{D_2} = \frac{T_1}{T_2}$$

6.3.11 Diameteral Pitch (P_d)

It is the ratio of number of teeth to the pitch circle diameter (in millimeters). It is denoted by P_d. Mathematically

$$P_d = \frac{T}{D} = \frac{\pi}{P_c}$$

where, T = number of teeth, D = pitch circle diameter.

6.3.12 Module

It is the ratio of pitch circle diameter to the number of teeth. It is usually denoted by m, mathematically,

Module $(m) = \dfrac{D}{T}$

6.3.13 Clearance

Radial difference between the addendum and dedendum of a tooth.

6.3.14 Total Depth

It is the radial distance between the addendum and the dedendum circles of a gear. It is equal to the sum of the addendum and dedendum.

6.3.15 Working Depth of Teeth

The maximum depth to which a tooth penetrates into the tooth space of the mating gear is the working depth of a tooth.

6.3.16 Tooth Space

It is the width of space between the two adjacent teeth mesasured along the pitch circle.

6.3.17 Tooth Thickness

It is the thickness of the tooth measured along the pitch circle.

6.3.18 Backlash

It is the difference between the tooth space and tooth thickness as measured along the pitch circle, i.e.

$$\text{backlash} = \text{tooth space} - \text{tooth thickness.}$$

6.3.19 Face Width

It is the width of the gear tooth measured parallel to its axis.

6.4 LAW OF GEARING

The law of gearing states that the condition which must be fulfiled by the gear tooth profiles is to maintain a constant angular velocity ratio between two gears, Fig. 6.4 shows two bodies 1 and 2 representing a portion of two gears in mesh A point C on the tooth profile of the gear 1 is in constant with a point D on the tooth profile of gear 2. Let it be n–n.

Let ω_1 = instantaneous angular velocity of gear 1
ω_2 = instantaneous angular velocity of gear 2
v_c = linear velocity of C
v_d = linear velocity of D

Then $v_c = \omega_1 \cdot AC$ in a direction perpendicular to AC
$v_c = \omega_2 \cdot BD$ in a direction perpendicular to BD

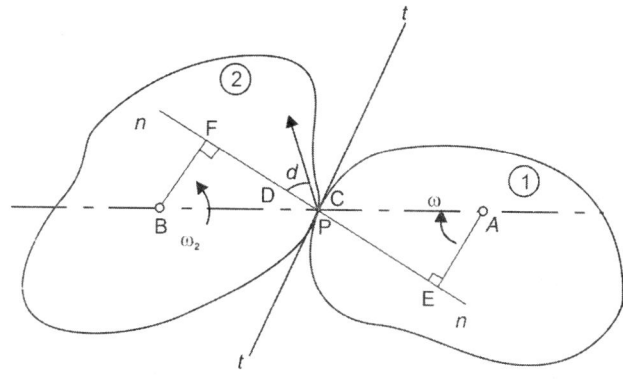

Fig. 6.4

Component of v_c along $n-n = v_c \cos \alpha$

Component of v_d along $n-n = v_d \cos \beta$

Relative motion along $n-n = v_c \cos \alpha - v_d \cos \beta$

For proper constant,

$$v_c \cos \alpha - v_d \cos \beta = 0$$

$$\omega_1 AC \cos \alpha - \omega_2 BD \cos \beta = 0$$

$$\omega_1 AC \frac{AE}{AC} - \omega_2 BD \frac{BF}{BD} = 0$$

$$\omega_1 AE - \omega_2 BF = 0$$

$$\frac{\omega_1}{\omega_2} = \frac{BF}{AE} = \frac{BP}{AP}$$

Thus, it is seen that the centre line AB is divided at P by the common normal in the inverse ratio of the angular velocities of the two gears.

Also as the $\triangle AEP$ and $\triangle BFP$ are similar, we have

$$\frac{BP}{AP} = \frac{FP}{EP} \Rightarrow \omega_1 EP = \omega_2 FP$$

6.5 VELOCITY OF SLIDES

If the curved surfaces of the two teeth of the gears 1 and 2 are to remain in contact, one can have a sliding motion relative to the other along the common tangent $t-t$ at $C-D$.

Component of v_c along $t-t = v_c \sin \alpha$

Component of v_d along $t-t = v_d \sin \alpha$

Velocity of sliding $= v_c \sin \alpha - v_d \sin \beta$

$$= \omega_1 AC \frac{EC}{AC} - \omega_2 BD \frac{FD}{BD}$$

$$= \omega_1 EC - \omega_2 FD$$

$$= \omega_1 (EP + PC) - \omega_2 (FP - PD)$$

$$= (\omega_1 + \omega_2) PC + \omega_1 EP - \omega_2 FP$$

$$= (\omega_1 + \omega_2) PC$$

= sum of angular velocities × distance between the pitch point and the point of contact

Gears are classified on the basis of the velocity of the drive also such as:

Low speed gearing — up to 3 m/sec

Medium speed gearing — 3 to 15 m/sec

High speed gearing — more than 15 m/sec

6.6 FORMS OF TEETH

In two meshing teeth, the shape of one of the teeth may be chosen arbitrarily and the other tooth is designed in such a way that they both follow the law of correct gearing. Such teeth are called conjugate teeth. Gears having conjugate teeth can be successfully used for transmitting motion but they are diffcult to manufacture as special devices are used for this purpose which are costly. So conjugate teeth are not much common in use. The common forms of teeth used widely are worm gear. In this there is surface contact.

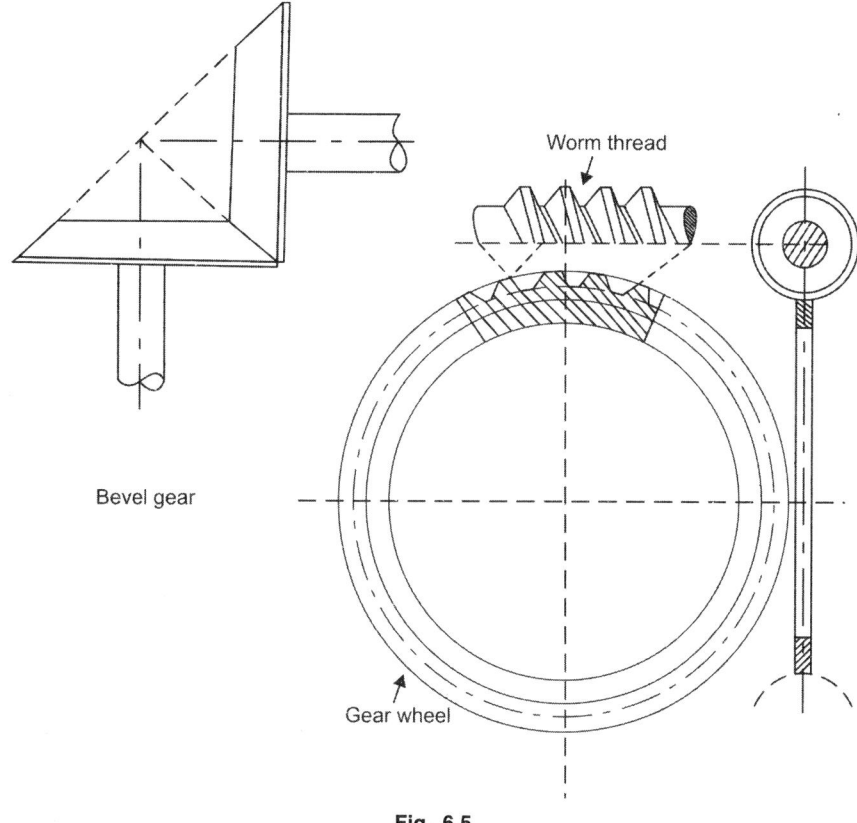

Worm thread

Bevel gear

Gear wheel

Fig. 6.5

The worm rotates with high speed and there is sliding between the worm threads and the wheel teeth. So large amount of heat is generated in the direction tangential to the thread surface. The spiral angle of worm thread for maximum efficiency is slighlty more than 45° as shown in Fig. 6.2.

   i. Cycloidal tooth profile
   ii. Involute tooth profile

6.7 CYCLOIDAL TOOTH PROFILE

A cycloidal is the locus of a point on the circumference of a circle which rolls without slipping on a fixed straight line as shown in Fig. 6.3.

   Two circles 1 and 2 are touching at point P. Circle 2 is moving to the right without slipping on fixed straight line which is the pitch line. The point P traces a curve PA on circle as it rolls. PA represents the face of the cycloidal tooth profile. When circle 1 rolls to the left without slipping on fixed straight line P traces a curve BPA. Similarly CPD represents the other side of the tooth.

Epicycloid

When a circle rolls outside the circumference of a fixed circle without slipping, a point on the circle forms a curve which is known as epicycloid.

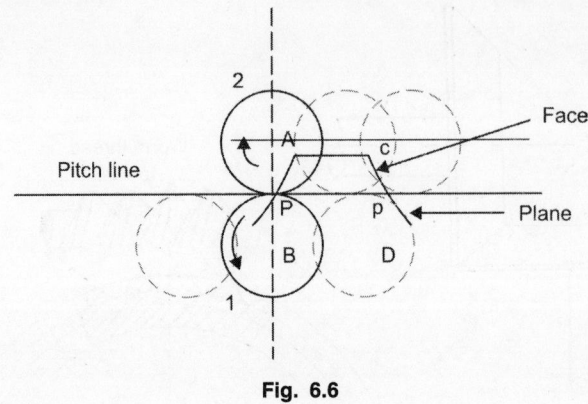

Fig. 6.6

Hypocycloid

When a circle rolls inside the circumference of a fixed circle without slipping, a point on the circle forms a curve which is called as hypocycloid. Figure 6.7 shows curve ABL and A'B'N as the epicycloid and hypocycloid respectively.

The curves PA and PB form the face and flank of the cycloidal tooth respectively. In a similar way, the other side of the tooth A'A'B can be constructed.

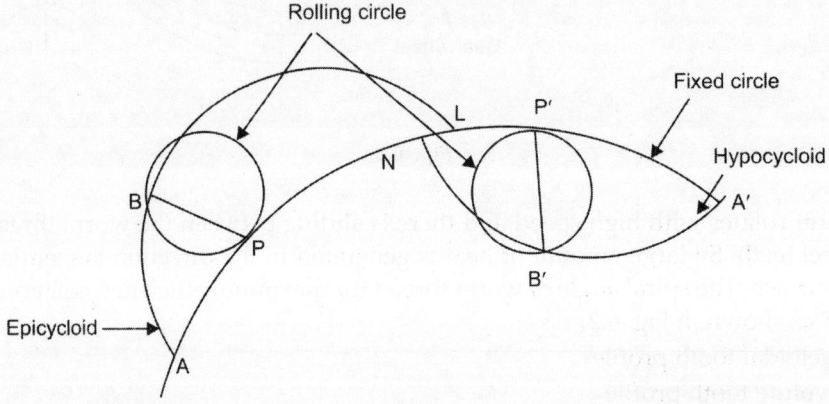

Fig. 6.7

6.8 INVOLUTE TOOTH PROFILE

From Fig. 6.7, a string AB (arc) is wrapped on the cirumference of a circle. When the string is unwound from the circle it takes the straight form AB'. AB' is perpendicular to OA. The point B finally reaches to point B' passing through point B_1 B_2...B_5. The curve BB' is known as the involute.

So, involute is the locus of a point located at the end of a (straight) thread when it is unwound from a circle. The base circle is divided into equal no. of parts (say six) A-5, 5-4, 4-3, etc. Tangents at points 5, 4, 3, etc. are drawn and lengths of tangents at these points are set off equal to arcs B-5, B-4, B-3, etc. to form the involute. Thus curve. BB_1B_2... B^1 is the involute.

Note: When the rolling circle touches the fixed circle at point P. The point A which is the generating point reaches at point B and the line BP is found normal to the epicycloid ABL similarly B'P' is normal to the hypocycloid A'B'N and arc AB = arc AP.

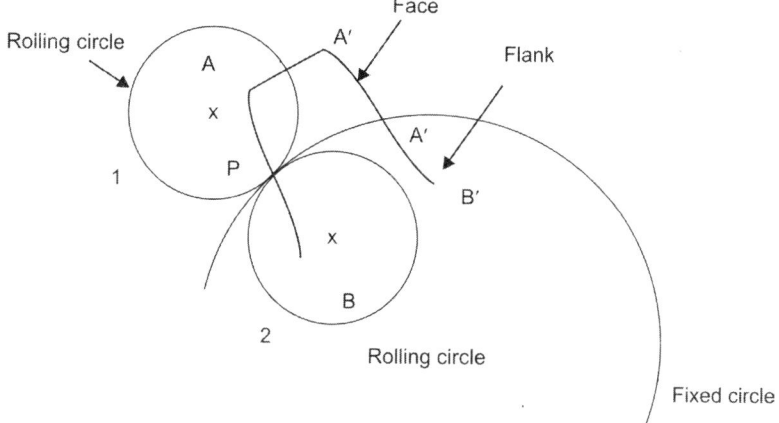

Fig. 6.8

The construction of epicycloid and hypocycloid curves is shown in Fig. 6.7. Two rolling circles 1 and 2 touch each other at point *P*. Circles 1 and 2 roll outside and inside of a fixed circle respectively. The point *P* on circle 1 traces a curve *PA* which is known as epicycloid. Similarly, when circle 2 rolls to the left inside the fixed circle the point *P* traces a curve *PB* known as hypocycloid.

Thus, the shape of involute will be as shown in Fig. 6.8, where *A* is the straight point and *AE* is the initial short profile (length) of the tooth which when extended forms one side of the tooth, i.e. *AM*. The other side of tooth will be generated in the same way starting from point *O* in the reverse direction; arc *AQD* = arc *DB*.

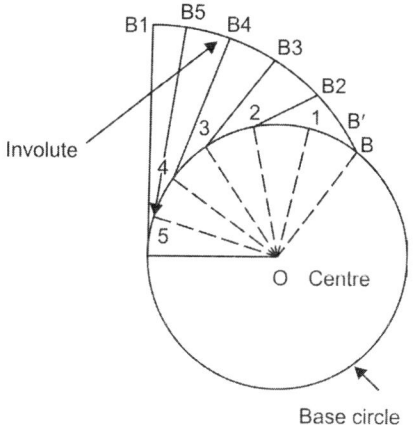

Fig. 6.9

6.9 PATH OF CONTACT

Pinion 1 is the driver and rotating clockwise, wheel 2 is driven in the counter-clockwise direction, *EF* is their common tangent to the base circle.

Contact of the two is made where the addendum circle of the wheel meets the line of action *EF*.

Let   *r* = pitch circle radius of pinion
                                         R = pitch circle of wheel

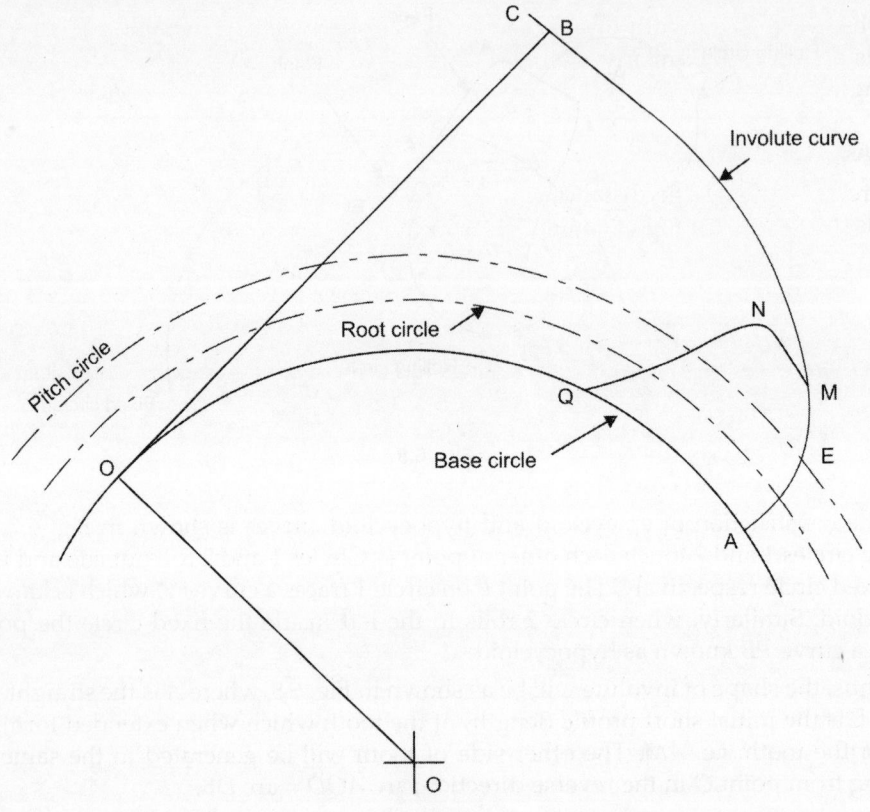

Fig. 6.10

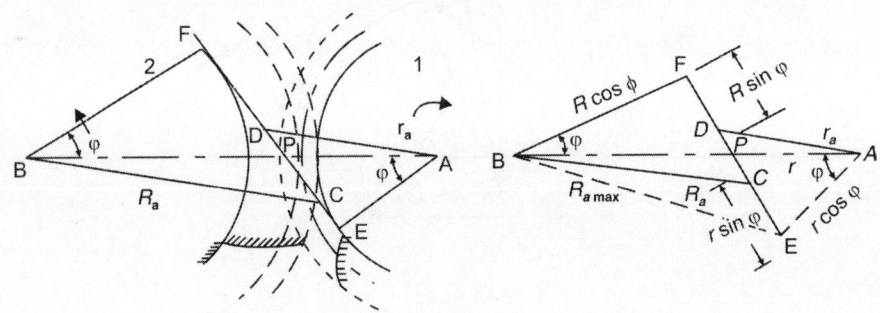

Fig. 6.11

R_a = addendum circle radius of wheel

r_a = addendum circle of pinion

Path of contact = path of approach + path of recess

$$CD = CP + PD$$

$$= (CF - PF) + (DE - PE)$$

$$= \left[\sqrt{R_a^2 - R^2 \cos^2 \phi} - R \sin \phi\right] + \left[\sqrt{r_a^2 - r^2 \cos^2 \phi} - r \sin \phi\right]$$

$$= \left[\sqrt{R_a^2 - R^2 \cos^2 \phi}\right] + \left[\sqrt{r_a^2 - r^2 \cos^2 \phi}\right] - (R + r)\sin \phi$$

Observe that the path of approach can be found if the dimensions of the driven wheels are known. Similarly, the path of recess is known from the dimensions of the driving wheel.

6.10 ARC OF CONTACT

The arc of contact is the distance travelled by a point on either pitch circle of the two wheels during the period of contact of pair of teeth.

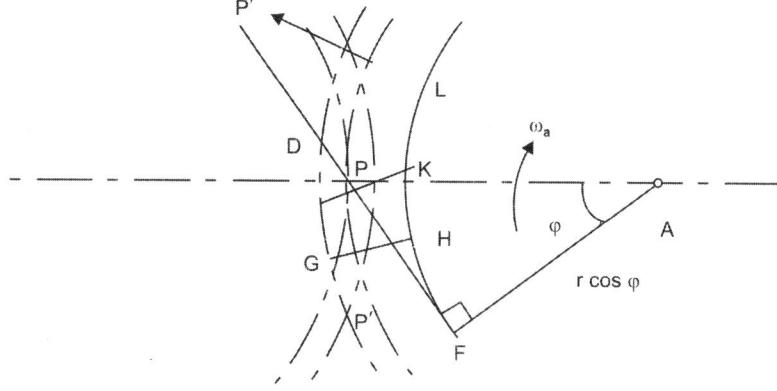

Fig. 6.12

At the beginning of engagement, the driver involute is shown as *GH*; when the point of contact is at *P*, it is shown as *JK* and when at the end of engagement, it is *DL*.

Let the time of transverse the arc of approach be t_a, then

arc of approach = $P'P$ = tangential velocity of P' × time of approach.

$$= \omega_a r \times t_a$$

$$= \omega_a (r \cos \phi) \frac{1}{\cos \phi} t_a$$

$$= (\text{tangential velocity of } H) \; t_a . \frac{1}{\cos \phi}$$

$$= \frac{\text{arc } HK}{\cos \phi} = \frac{\text{arc } FK - \text{arc } FH}{\cos \phi}$$

$$= \frac{FP - FC}{\cos \phi} = \frac{CP}{\cos \phi}$$

Arc *FK* is equal to the path *FP* as the point *P* is on the generator *FP* that rolls on the base circle *FHK* to generate involute *PK*. Similarly

$$\text{arc } FH = \text{path } FC$$

arc of recess = PP'' = tangential velocity of P × time of recess

$$= \omega_a \times t_r$$

$$= \omega_a (r \cos \phi) \frac{1}{\cos \phi} t_r$$

$$= (\text{tangential velociy of } K) \; t_r \frac{1}{\cos \phi}$$

$$= \frac{\text{arc } KL}{\cos \phi} = \frac{\text{arc } FL - \text{arc } FK}{\cos \phi}$$

$$PP'' = \frac{FD - FP}{\cos \phi} = \frac{PD}{\cos \phi}$$

$$\text{arc of contact} = \frac{CP}{\cos \phi} + \frac{PD}{\cos \phi} = \frac{CP + PD}{\cos \phi} = \frac{CD}{\cos \phi}$$

$$\text{arc of contact} = \frac{\text{path of contact}}{\cos \phi}$$

6.11 NUMBER OF PAIRS OF TEETH IN CONTACT: CONTACT RATIO

The arc of contact is the length of the pitch circle traversed by a point on it during the mating of a pair of teeth.

Thus, all the teeth lying in between the arc of contact will be meshing with the teeth on the other wheel.

Therefore, the number of teeth in contact is

$$n = \frac{\text{arc of contact}}{\text{circular pitch}} = \frac{CD}{\cos \beta} \cdot \frac{1}{P}$$

As the ratio of arc of contact to the circular pitch is also the contact ratio, the number of teeth is also expressed in terms of contact ratio.

If n lies between 1 and 2, the no. of teeth in contact at any time will not be less than one and never more than two.

Example 6.1: Each of two gears in a mesh has 48 teeth and module of 8 mm. The teeth are of 20° involute profile. The arc of contact is 2.25 times the circular pitch. Determine the addendum.

Solution. Given $\phi = 20°, t = T = 48, m = 8$ mm

$$R = r = \frac{mT}{2} = \frac{8 \times 48}{2} = 192 \text{ mm}; R_a = r_a$$

arc of contact = 2.25 × circular pitch = 2.25 πm

$$= 2.25 \times \pi \times 8 = 56.55 \text{ mm}$$

Path of contact = 56.55 × cos 20° = 53.14 mm

or

$$= \left[\sqrt{R_a^2 - R^2 \cos^2 \phi} - R \sin \phi \right] + \left[\sqrt{r_a^2 - r^2 \cos^2 \phi} - r \sin \phi \right]$$

$$= 53.14$$

or

$$= 2\sqrt{R_a^2 - 192^2 \cos^2 20° } - 192 \sin 20° = 53.14 \quad [\text{as } R_a = r_a, R = r]$$

$$\Rightarrow R_a = 202.6 \text{ mm}$$

$$\text{Addendum} = R_a - R = 202.6 - 192 \text{ mm}$$

$$= 10.6 \text{ mm}$$

Example 6.2: Two involute gears in mesh have 20° pressure angle. The gear ratio is 3 and the number of teeth on the pinion is 24. The teeth have a module of 6 mm. The pitch line velocity is 1.5 m/s and the addendum equal to one module. Determine the angle of action of the pinion and the maximum velocity of sliding.

Solution: Given $\phi = 20°, t = 24, m = 6$ mm

$$T = 24 \times 3 = 72$$

$$\left[r = \frac{mr}{2} = \frac{6 \times 24}{2} = 72 \text{ mm} \right.$$

$$R = 72 \times 3 = 216 \text{ mm}$$

$$\left. r_a = 72 + 6 = 78 \text{ mm} \right]$$

$$R_a = 216 + 6 = 222 \text{ mm}$$

$$\text{Path of contact} = \left[\sqrt{R_a^2 - R^2 \cos^2 \phi} - R \sin \phi \right] + \left[\sqrt{r_a^2 - r^2 \cos^2 \phi} - r \sin \phi \right]$$

$$= \left[\sqrt{(222)^2 - 216^2 \cos^2 20° } - 216 \sin 20° \right]$$

$$= + \left[\sqrt{78^2 - 72^2 \cos^2 20°} - 72 \sin 20° \right]$$

$$= 16.04 + 14.18 = 30.22 \text{ mm}$$

$$\text{Arc of contact} = \frac{\text{path of contact}}{\cos \phi} = \frac{30.22}{\cos 20°} = 32.16$$

$$\text{Angle of action} = \frac{\text{arc of contact}}{r} = \frac{32.16}{72} = 0.4467 \text{ rad}$$

$$= 0.4467 \times \frac{180}{\pi} = 25.59°$$

$$\text{Velocity of sliding} = (\omega_p + \omega_g) \times \text{path of approach}$$

$$= \left[\frac{v}{r} + \frac{v}{R} \right] \times \text{path of approach}$$

$$= \left[\frac{1500}{72} + \frac{1500}{216} \right] \times 16.04 = 445.6 \text{ mm/s}$$

Example 6.3: Two involute gears in a mesh have a module of 8 mm and a pressure angle of 20°. The larger gear has 57 teeth while the pinion has 23. If the addendum on pinion and gear wheels are equal to one module, find:
   i. contact ratio
  ii. angle of action of the pinion and the gear wheel
 iii. ratio of the sliding to rolling velocity at the
     a. beginning of contact
     b. pitch point
     c. end of contact

Solution. Given

$$\phi = 20°; T = 57 ; t = 23 ; m = 8 \text{ mm}$$

$$R = \frac{mT}{2} = \frac{8 \times 57}{2} = 228 \text{ mm}$$

$$R_a = R + m = 228 + 8 = 236 \text{ mm}$$

$$r = \frac{mt}{2} = \frac{8 \times 57}{2} = 228 \text{ mm}$$

$$r_a = r + m = 100 \text{ mm}$$

$$n = \frac{\text{arc of contact}}{\text{circular pitch}} = \frac{\text{path of contact}}{\cos\phi} \times \frac{1}{\pi m}$$

$$= \frac{\text{path of approach} + \text{path of recess}}{\cos\phi \times \pi m}$$

$$= \left[\frac{\sqrt{R_a^2 - R^2 \cos^2\phi} - R\sin\phi + \sqrt{r_a^2 - r^2 \cos^2\phi} - r\sin\phi}{\cos\phi + \pi m}\right]$$

$$= \left[\frac{\left[\sqrt{236^2 - 228^2 \cos^2 20°} - 228\sin 20°\right] + \sqrt{100^2 - 92^2 \cos^2 20°} - 92\sin 20°}{\cos 20° \times \pi \times 8}\right]$$

$$= \frac{20.97 + 18.79}{\cos 20° \times \pi \times 8} = 42.31 \times \frac{1}{\pi \times 8} = 1.68$$

ii. Angle of action $\quad \delta_p = \dfrac{\text{arc of contact}}{r} = \dfrac{42.31}{92}$

$$= 0.46 \text{ rad or } 0.46 \times \frac{180°}{\pi} = 26.3°$$

$$\delta_g = \frac{\text{arc of contact}}{r} = \frac{42.31}{228} = 0.1856 \text{ rad}$$

or $\quad\quad\quad 0.1856 \times \dfrac{180°}{\pi} = 10.63°$

iii. (a) $\dfrac{\text{Sliding velocity}}{\text{Rolling velocity}} = \dfrac{\left(\omega_p + \omega_g\right) \times \text{path of approach}}{\text{Pitch line velocity } (\omega_p r)}$

$$= \frac{\left[\omega_p + \dfrac{23}{57}\omega_p\right] \times 20.97}{\omega_p \times 92} = 0.32$$

(b) $\dfrac{\text{Sliding velocity}}{\text{Rolling velocity}} = \dfrac{\left(\omega_p + \omega_g\right) \times 0}{\text{Pitch line velocity}} = 0$

(c) $\dfrac{\text{Sliding velocity}}{\text{Rolling velocity}} = \dfrac{\left[\omega_p \times \dfrac{23}{57}\omega_p\right] \text{path of recess}}{\omega_p \times r}$

$$= \frac{\left[1 + \dfrac{23}{57}\right] \times 18.79}{92} = 0.287$$

6.12 MINIMUM NUMBER OF TEETH

The maximum value of the addendum radius of the wheel to avoid interference can be up to *BE* (Fig. 6.8).

$$(BE)^2 = (BF)^2 + (FE)^2$$
$$= (BF)^2 + (FP + PE)^2$$
$$= (R\cos\phi)^2 + (R\sin\phi + r\sin\phi)^2$$
$$= R^2\cos^2\phi + R^2\sin^2\phi + r^2\sin^2\phi + 2Rr\sin^2\phi$$
$$= R^2(\cos^2\phi + \sin^2\phi) + \sin^2\phi\,(r^2 + 2rR)$$
$$= R^2 + (r^2 + 2rR)\sin^2\phi$$
$$= R^2\left[1 + \frac{1}{R^2}\left(r^2 + 2rR\right)\sin^2\phi\right]$$

$$BE = R\sqrt{1 + \frac{r}{R}\left[\frac{r}{R} + 2\right]\sin^2\phi}$$

Therefore, the maximum value of the addendum of the wheel can be equal to BE pitch circle radius.

$$a_{w\,max} = \left[\sqrt{1 + \frac{r}{R}\left[\frac{r}{R} + 2\right]\sin^2\phi}\right] - R$$

$$= R\left[\sqrt{1 + \frac{r}{R}\left[\frac{r}{R} + 2\right]\sin^2\phi} - 1\right]$$

Let t = number of teeth on the pinion
T = number of teeth on the wheel

$$R = \frac{mT}{2},\ r = \frac{mt}{2},\ G = \frac{T}{t} = \text{gear ratio}$$

Hence $a_{w\,max} = \dfrac{mT}{2}\left[\sqrt{1 + \dfrac{t}{T}\left[\dfrac{t}{T} + 2\right]\sin^2\phi} - 1\right]$

$$= \frac{mT}{2}\left[\sqrt{1 + \frac{1}{G}\left[\frac{1}{G} + 2\right]\sin^2\phi} - 1\right]$$

Let the adopted value of the addendum in some case be a_w, times the module of teeth. Then, the adopted value of the addendum must be less than the maximum value of the addendum to avoid interference.

$$\frac{mT}{2}\left[\sqrt{1 + \frac{1}{G}\left[\frac{1}{G} + 2\right]\sin^2\phi} - 1\right] \geq a_w m$$

or

$$T \geq \frac{2a_w}{\sqrt{1 + \dfrac{1}{G}\left[\dfrac{1}{G} + 2\right]\sin^2\phi} - 1}$$

In the limits,

$$T = \frac{2a_w}{\sqrt{1 + \dfrac{1}{G}\left[\dfrac{1}{G} + 2\right]\sin^2\phi} - 1}$$

This gives the minimum no. of teeth on the wheel for the given value of the gear ratio, the pressure angle and addendum coefficient a_w.

The minimum number of teeth on the pinion

$$t = \frac{T}{G}$$

For $a_w = 1$, when the addendum is equal to one module

$$T \geq \frac{2}{\sqrt{1 + \dfrac{1}{G}\left[\dfrac{1}{G} + 2\right]\sin^2\phi} - 1}$$

For equal no. of teeth on pinion and the wheel, $G = 1$

$$T_{min} = \frac{2}{\sqrt{1 + 3\sin^2\phi} - 1}$$

In case of pinion, the maximum value of addendum radius to avoid interference is AF

$$(AF)^2 = (r\cos\phi)^2 + [R\sin\phi + r\sin\phi]^2$$

and it can be shown that the maximum value of the addendum of the pinion is

$$a_p\max = r\sqrt{1 + \frac{R}{r}\left[\frac{R}{r} + 2\right]\sin^2\phi} - r$$

or

$$a_p\max = \frac{mt}{2}\left[\sqrt{1 + G(G+2)\sin^2\phi} - 1\right]$$

Example 6.4: Two involute gears in mesh have a velocity ratio of 3. The arc of approach is not to be less than the circular pitch when the pinion is the driver. The pressure angle of the involute teeth is 20°. Determine the least no. of teeth on each gear. Also, find the addendum of the wheel in terms of module.

Solution: Since $\phi = 20°$, $VR = 3$

Arc of approach= circular pitch = πm

$\therefore$ Path of approach= $\pi m \cos 20° = 2.952$ m

Maximum length of path of approach = $r \sin\phi$

$$= \frac{mt}{2}\sin 20° = 0.171mt$$

$\therefore \qquad 0.171mt = 2.952m$ or $t = 17.26 \approx 18$

and $\qquad\qquad T = 18 \times 3 = 54$

Maximum addendum of the wheel,

$$a_{w\max} = \frac{mT}{2}\left[\sqrt{1 + \frac{1}{G}\left(\frac{1}{G} + 2\right)\sin^2\phi} - 1\right]$$

$$= \frac{m \times 54}{2}\left[\sqrt{1 + \frac{1}{3}\left(\frac{1}{3} + 2\right)\sin^2 20°} - 1\right]$$

$$= 1.2 \text{ m}$$

Example 6.5: Two 20° involute spur gears mesh externally and give a velocity ratio of 3. The module is 3 mm and addendum is equal to 1.1 times the module. If the pinion rotates at 120 rpm, determine:

i. minimum number of teeth on each wheel to avoid interference
ii. contact ratio

Solution: Given

$$\phi = 20°; N_p = 120 \text{ rpm}$$
$$VR = 3; \text{Addendum} = 1.1m$$
$$m = 3 \text{ mm}; a_w = 1.1$$

(i)

$$T = \frac{2a_w}{\sqrt{1 + \dfrac{1}{G}\left[\dfrac{1}{G} + 2\right]\sin^2 \phi} - 1}$$

$$= \frac{2 \times 1.1}{\sqrt{1 + \dfrac{1}{3}\left[\dfrac{1}{3} + 2\right]\sin^2 \phi} - 1} = 49.44$$

Taking the highest whole number divisible by the velocity ratio

$$T = 51, t = \frac{51}{3} = 17$$

ii. Contact ratio or number of pair of teeth in contact

$$n = \frac{\text{arc of contact}}{\text{circular pitch}} = \left[\frac{\text{path of contact}}{\cos \phi}\right] \times \frac{1}{\pi m}$$

$$n = \frac{\sqrt{R_a^2 - R^2 \cos^2 \phi} - R \sin \phi + \sqrt{r_a^2 - r^2 \cos^2 \phi} - r \sin \phi}{\cos \phi \times \pi m}$$

We have

$$R = \frac{mT}{2} = \frac{3 \times 51}{2} = 76.5 \text{ mm}$$

$$R_a = R + 1.1m$$
$$= 76.5 + 1.1 \times 3 = 79.8 \text{ mm}$$

$$r = \frac{mT}{2} = \frac{3 \times 17}{2} = 25.5 \text{ mm}$$

$$r_a = 25.5 + 1.1 \times 3 = 28.8 \text{ mm}$$

$$n = \frac{\left[\sqrt{(79.8)^2 - (76.5 \cos 20°)^2} - 76.5 \sin 20° + \sqrt{(28.8)^2 - (25.5 \cos 20°)^2} - 25.5 \sin 20°\right]}{\cos 20° \times \pi \times 3}$$

$$\Rightarrow n = 1.78$$

Thus, one pair of teeth will always remain in contact whereas for 78% of the time, two pairs of teeth will be in contact.

6.13 GEAR TRAINS

6.13.1 Introduction

A gear train is a combination of gears used to transmit motion from one shaft to another. It becomes neccessary when if is required to obtain large speed reduction within a small space.

Types of gear trains:
  i. Simple gear train
  ii. Compound gear train
  iii. Reverted gear train
  iv. Planetary or epicyclic gear train.

Simple Gear Train

A series of gears, capable of recieving and transmitting motion from one gear to another is called simple gear train. In it all the gear axes remain fixed relative to the frame and each gear is on a separate shaft.

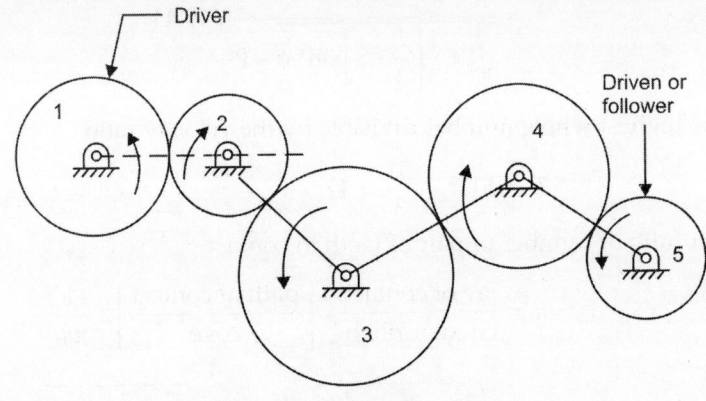

Fig. 6.13

1. Two external gears of a pair always move in opposite directions.
2. All odd number gears move in one direction and all even number gears move in the opposite direction.
3. *Speed ratio*: The ratio of the driving to that of the driven shaft, is negative when the input and output gears rotate in the opposite directions and positive when the two rotate in the same direction. The reverse of the speed ratio is known as the train value of the gear train.
4. All the gears can be in straight line or arranged in a zig-zag manner. A simple gear train can also have level gears.

$$\text{Let } T = \text{no. of teeth on a gear}$$
$$N = \text{speed of a gear in rpm}$$

from gear 1 and gear 2

$$\frac{N_2}{N_1} = \frac{T_1}{T_2} \left(\because \frac{\omega_2}{\omega_1} = \frac{2\pi N_2}{2\pi N_1} = \frac{N_2}{N_1} \right)$$

Similary

$$\frac{N_3}{N_2} = \frac{T_2}{T_1}, \frac{N_4}{N_3} = \frac{T_3}{T_4} \text{ and } \frac{N_5}{N_4} = \frac{T_4}{T_5}$$

Multiplying

$$\frac{N_2}{N_1} \times \frac{N_3}{N_3} \times \frac{N_4}{N_3} \times \frac{N_5}{N_4} = \frac{T_1}{T_2} \times \frac{T_2}{T_3} \times \frac{T_3}{T_4} \times \frac{T_4}{T_5}$$

Train value

$$\left(\frac{N_5}{N_1} = \frac{T_1}{T_5} \right) = \frac{\text{No. of teeth on driving gear}}{\text{No. of teeth on driven gear}}$$

$$\text{Speed ratio} = \frac{1}{\text{train value}}$$

i.e. $\qquad N_1/N_5 = T_5/T_1.$

Compound Gear Train

When a series of gears are connected in such a way that two or more gears rotate about an axis with the same angular velocity, it is known as compound gear train. In this type, some of the intermediate shafts, i.e other than the input and output shafts, carry more than one gear as shown in Fig. 6.14.

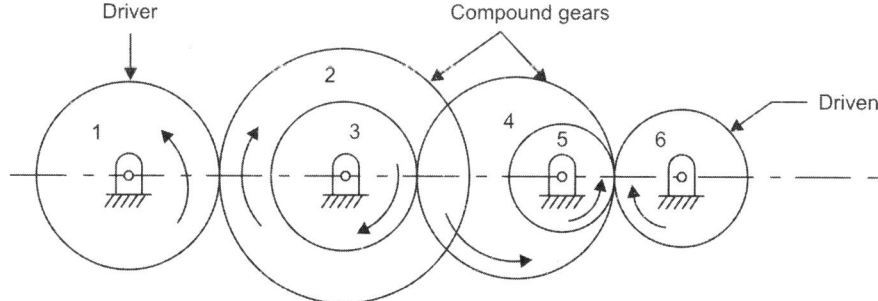

Fig. 6.14

If gear 1 is the driver then

$$\frac{N_2}{N_1} = \frac{T_1}{T_2}, \frac{N_4}{N_3} = \frac{T_3}{T_4} \text{ and } \frac{N_6}{N_5} = \frac{T_5}{T_6}$$

Multiplying $\qquad \dfrac{N_2}{N_1} \times \dfrac{N_4}{N_3} \times \dfrac{N_6}{N_5} = \dfrac{T_1}{T_2} \times \dfrac{T_3}{T_4} \times \dfrac{T_5}{T_6} \quad (\because N_3 = N_2 \text{ and } N_5 = N_4)$

$$\frac{N_6}{N_1} = \frac{T_1}{T_2} \times \frac{T_3}{T_4} \times \frac{T_5}{T_6}$$

$$\text{Train value} = \frac{\text{product of no. of teeth on driving gears}}{\text{product of no. of teeth on driven gears}}$$

or Speed ratio $\qquad \dfrac{N_1}{N_6} = \dfrac{T_2 \times T_4 \times T_6}{T_1 \times T_3 \times T_5}$

Reverted Gear Train

If the axes of the first and the last wheels of a compound gear coincide, it is called a reverted gear train. Such an arrangement is used in clocks and simple latter where back gear is used to give a slow speed to the chuck.

Since the distance between the centres of the shafts of gear 1 and gear 2 as well as gears 3 and 4 are same.

$\therefore \qquad\qquad r_1 + r_2 = r_3 + r_4$ ...(6.1)

Also circular pitch or module of all the gears is assumed to be same, therefore no. of teeth on each gear is directly proportional to its circumference or radius.

$$\left(\because r = \frac{mT}{2} \right)$$

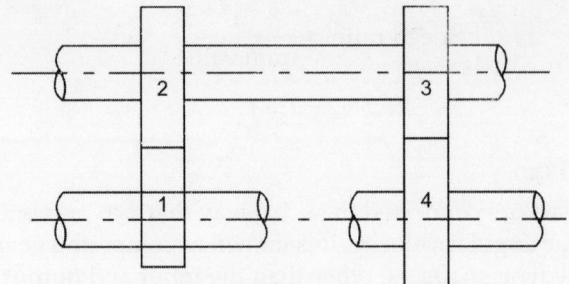

Fig. 6.15

$$m\frac{T_1}{2} + m\frac{T_2}{2} = m\frac{T_3}{2} + m\frac{T_4}{2}$$

$$T_1 + T_2 = T_3 + T_4 \qquad \qquad ...(6.2)$$

$$\text{Speed ratio} = \frac{\text{product of no. of teeth on driven gears}}{\text{product of no. of teeth on driving gears}}$$

$$\therefore \quad \frac{N_1}{N_4} = \frac{T_2 \times T_4}{T_1 \times T_3} \qquad \qquad ...(6.3)$$

Planetary or Epicyclic Gear Train

A gear train having a relative motion of axes is called a planetary or epicyclic gear train. In epicyclic gear train, the axis of at least one of the gears also moves relative to the frame.

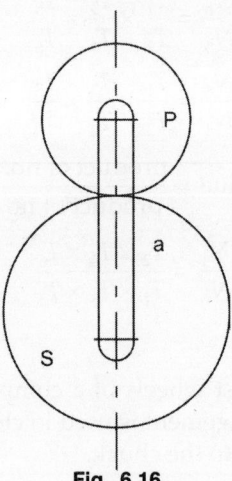

Fig. 6.16

Consider two gear wheels S and P, the the axes of which are connected by an arm a. If the arm a is fixed, the wheels S and P constitute a simple gear train. If the wheel S is fixed so that the arm can rotate about the axis of S, the wheel P would also move around S. Therefore, it is an epicyclic gear train.

Velocity Ratio of Epicyclic Gear Train

The following two methods may be used for finding out the velocity ratio of an epicyclic gear train.

Tabular Method

Consider an epicyclic gear train as shown in Fig. 6.16.

Let T_A = no. of teeth on gear A

T_B = no. of teeth on gear B

First of all, let us suppose that the arm is fixed. Therefore, the axes of both the gears are also fixed relative to each other. When gear A makes one revolution anticlockwise, the gear B will make (T_A/T_B) revolutions clockwise.

Assuming anticlockwise rotation as positive and clockwise rotation as negative, we may say that when gear A makes + 1 revolution, then gear B makes $\left(-\dfrac{T_A}{T_B}\right)$ revolutions. This statement of relative motion is entered in the first row of the Table 6.1.

Secondly, if the gear A makes $+x$ revolutions, then the gear B will make $\left(-x\,\dfrac{T_A}{T_B}\right)$ revolution. This statement is entered in second row of Table 6.1. In other words, multiply each motion by x in the gear train. But the two conditions are usually supplied in any epicyclic train. Some element is fixed and the other has specified motion. These two conditions are sufficient to solve all the equations, and hence to determine the motion of any element in the epicyclic gear train. Let the arm C be fixed in an epicyclic gear train as shown in Fig. 6.17. Therefore, speed of the gear A relative to the arm C

$$= N_A - N_C$$

and speed of the gear B relative to arm C.

$$= N_B - N_C$$

Since gear A and B are meshing directly, therefore they will revolve in opposite direction

$$\frac{N_B - N_C}{N_A - N_C} = -\frac{T_A}{T_B}$$

Since the arm C is fixed, therefore its speed $N_C = 0$.

$$\frac{N_B}{N_A} = \frac{-T_A}{T_B}$$

If gear A is fixed, then $N_A = 0$.

$$\frac{N_B - N_C}{0 - N_C} = \frac{-T_A}{T_B}$$

or $\qquad N_B/N_C = 1 + T_A/T_B$

Thirdly, each element of an epicyclic is given $+y$ revolutions and entered in the third row. Finally, the motion of each of the gear train is added up and entered in the fourth row.

Table 6.1

Step no.	Conditions of motion	Revolutions of elements Arm C	Gear A	Gear B
1.	Arm is fixed gear A rotates through + 1 revolution i.e. 1 revolution anticlockwise	0	+ 1	$-\dfrac{T_A}{T_B}$
2.	Arm fixed and gear A rotates through +x revolutions.	0	+ x	$-x\dfrac{T_A}{T_B}$
3.	Add +y revolutions to all elements	+ y	+ y	+ y
4.	Total motion	+ y	x + y	$y - x\dfrac{T_A}{T_B}$

Note: When two conditions about the motion of rotation of any two elements are known, then the unknown speed of third element may be obtained by substituting the given data in the third column and fourth row.

Algebraic Method

In this method, the motion of each element of epicyclic train relative to the arm is set in the form of equations. The number of equations depends upon the number of elements.

If gear B is fixed, then $N_B = 0$

$$\frac{0 - N_C}{N_A - N_C} = -\frac{T_A}{T_B} \Rightarrow \frac{N_A - N_C}{N_C} = \frac{T_B}{T_A}$$

$$\therefore \qquad \frac{N_A}{N_C} = 1 + \frac{T_B}{T_A}$$

Example 6.6: In an epicyclic gear train, an arm carries two gears A and B having 36 and 45 teeth respectively. If the arm rotates at 150 rpm in the anticlockwise direction about the centre of the gear A which is fixed, determine the speed of gear B. If the gear A instead of being fixed, makes 300 rpm in the clockwise direction, what will be the speed of gear B?

Solution: Given
$$T_A = 36$$
$$T_B = 45$$
$$N_C = 150 \text{ rpm (anticlockwise)}$$

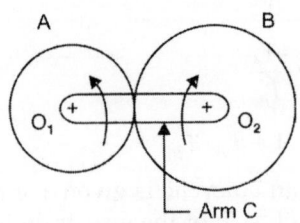

Fig. 6.17

We shall solve this problem by two methods:

   i. Tabular method

   ii. Algebric method

i. *Tabular method*:

		Revolutions of elements		
Step no.	Conditions of motion	Arm C	Gear A	Gear B
1.	Arm is fixed, gear A rotates through +1 revolution (i.e. +1 revolution anticlockwise)	0	+1	$-\dfrac{36}{45}$
2.	Arm fixed, gear A rotates + x revolutions	0	$+x$	$-\dfrac{36}{45}x$
3.	Add +y revolutions to all elements	$+y$	$+y$	$+y$
4.	Total motion	$+y$	$x+y$	$y-\dfrac{36}{45}x$

Table 6.2

Speed of gear B when gear A is fixed: Since the speed of arm is 150 rpm anticlockwise, therefore from the fourth row of Table 6.2.

$$y = + 150 \text{ rpm}$$

Also gear A is fixed, therefore $x + y = 0$

or $$x = -y = -150 \text{ rpm}$$

$\therefore$ Speed of gear B (N_B) $= y - \dfrac{36}{45}x$

$$N_B = 150 - \frac{36}{45}(-150)$$

$$N_B = 270 \text{ rpm (anticlockwise)}$$

Speed of gear B when gear A makes 300 rpm clockwise: Since gear A makes 300 rpm clockwise, therefore from the fourth row of Table 6.2.

$$x + y = -300$$

or

$$x = -300 - y$$

$$x = -300 - 150$$

$$x = -450 \text{ rpm}$$

$\therefore$ Speed of the gear B,

$$N_B = y - \frac{36}{45}x$$

$$= 150 - \frac{36}{45} \times (-450)$$

$$= 510 \text{ rpm (anticlockwise)}$$

ii. *Algebraic method*:

$$\text{Let } N_A = \text{speed of gear } A$$

$$N_B = \text{speed of gear } B$$

$$N_C = \text{speed of gear } C$$

Assuming the arm C to be fixed, speed of gear A relative to arm $C = N_A - N_C$ and speed of gear B relative to arm $C = N_B - N_C$. Since both the gears A and B revolve in opposite directions $\dfrac{N_B - N_C}{N_A - N_C} = \dfrac{-T_A}{T_B}$.

Speed of the gear B when gear A is fixed: $N_A = 0$, $N_C = +150$ rpm

$$\frac{N_B - N_C}{N_A - N_C} = \frac{-T_A}{T_B}$$

$$\frac{N_B - 150}{0 - 150} = \frac{-36}{45}$$

$$N_B = 120 + 150$$

$$N_B = 270 \text{ rpm}$$

Speed of gear B when gear A makes 300 rpm clockwise: $N_A = -300$ rpm

$$\frac{N_B - 150}{-300 - 150} = \frac{-36}{45}$$

$$N_B = -450 \times -0.8 + 150$$

or

$$= 360 + 150$$

$$= 510 \text{ rpm}$$

Example 6.7: In a reverted epicyclic gear train, the arm A carries two gears B and C and a compound gear D-E. The gear B meshes with gear E and the gear C meshes with gear D. The no. of teeth on gears B, C and D are 75, 30 and 90 respectively. Find the speed and direction of gear C when gear B is fixed and the arm A makes 100 rpm clockwise.

Solution: Given

$$T_B = 75$$
$$T_C = 30$$
$$T_D = 90$$
$$N_A = 100 \text{ rpm (clockwise)}$$

Let d_B, d_C, d_D and d_E be the pitch circle diameters of gears B, C, D and E respectively. From the geometry of Fig. 6.18.

$$d_B + d_E = d_C + d_D$$

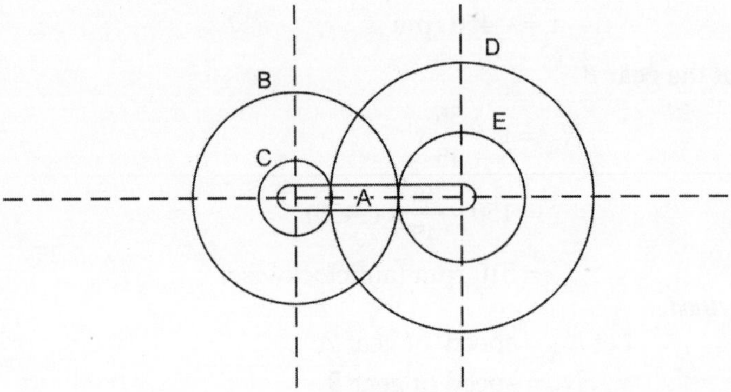

Fig. 6.18

Since the no. of teeth on each gear, for the same module, are proportional to their pitch circle diameters, therefore

$$T_B + T_E = T_C + T_D$$

$\therefore$
$$T_E = T_C + T_D - T_B$$
$$= 30 + 90 - 75$$
$$= 45$$

Step No.	Condition of motion	Revolutions of elements			
		Arm A	Compound Gear D–E	Gear B	Gear C
1.	Arm fixed - compound gear D - E rotate through + 1 revolution	0	+ 1	$\dfrac{-T_E}{T_B}$	$\dfrac{-T_D}{T_C}$
2.	Arm fixed - compound gear D - E rotated through + x revolution	0	+ x	$-x \times \dfrac{T_E}{T_B}$	$-x \times \dfrac{T_D}{T_C}$
3.	Add + y revolutions to all elements	+ y	+ y	+ y	+ y
4.	Total motion	+ y	x + y	$y - x \times \dfrac{T_E}{T_B}$	$y - x \times \dfrac{T_D}{T_C}$

Since the gear B is fixed, therefore from the fourth row of the table,

$$y - x \times \frac{T_E}{T_B} = 0 \text{ or } y - x \times \frac{45}{75} = 0$$

$$y - 0.6x = 0 \qquad \qquad ...(1)$$

Also arm A makes 100 rpm clockwise, therefore

$$y = -100$$

Substituting $y = -100$ in Eq. (1), we get

$$-100 - 0.6x = 0$$
$$x = -166.67$$

$$N_C = y - x \times \frac{T_D}{T_C}$$

$$= -100 + 166.67 \times \frac{90}{30}$$

$$= +400 \text{ rpm}$$
$$= 400 \text{ (anticlockwise)}$$

Compound Epicycle Gear Train: Sun and Planet Gear

It consists of two coaxial shafts s_1 and s_2, an annulus gear A which is fixed, the compound gear B-C, the sun gear D and the arm H. The sun gear is coaxial with the annulus gear and the arm but independent of them.

The annulus gear A meshes with the gear B and the sun gear D meshes with the gear C. It may be noted that when the annulus gear is fixed, the sun gear provides the drive and when the sun gear is fixed, the annulus gear provides the drive. In both the cases, the arm acts as a follower.

Note: The gear at the centre is called the *sun gear* and the gears whose axes move are called *planet gears*.

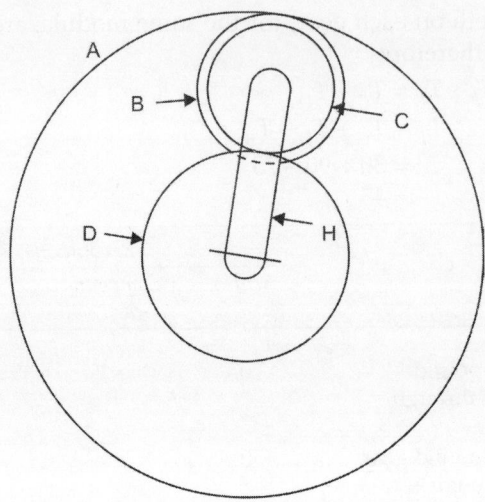

Fig. 6.19

Let T_A, T_B, T_C and T_D be the no. of teeth and N_A, N_B, N_C and N_D be the speeds of the gears A, B, C and D respectively. A little consideration will show that when the arm is fixed and the sun gear D is turned anticlockwise, then the compound gear B-C and the annulus gear A will rotate to the clockwise direction.

Step no.	Conditions of motion	Revolutions of elements			
		Arm	Gear D	Compound gear B–C	Gear A
1.	Arm is fixed, gear D rotates through +1 revolution	0	+1	$\dfrac{-T_D}{T_C}$	$\dfrac{-T_D}{T_C} \times \dfrac{T_B}{T_A}$
2.	Arm is fixed, gear D rot;ates through +x revolution	0	+x	$-x \times \dfrac{T_D}{T_C}$	$-x \times \dfrac{T_D}{T_C} \times \dfrac{T_B}{T_A}$
3.	Add +y revolutions to all elements	+y	+y	+y	+y
4.	Total motion	+y	$x+y$	$y - x \times \dfrac{T_D}{T_C}$	$y - x \times \dfrac{T_D}{T_C} \times \dfrac{T_B}{T_A}$

Example 6.8: An epicycle gear consists of three gears A, B and C as shown in Fig. 6.20. The gear A has 72 internal teeth and gear C has 32 external teeth. The gear B meshes with both A and C and is carried on an arm EF which rotates about the centre of A at 18 rpm. If the gear A is fixed, determine the speed of gears B and C.

Solution: Given $\qquad T_A = 72$

$\qquad\qquad\qquad\qquad T_C = 32$

Speed of arm $\qquad EF = 18$ rpm

Step no.	Conditions of motion	Revolutions of elements			
		Arm EF	Gear C	Gear B	Gear A
1.	Arm is fixed, gear C rotate through + 1 revolution	0	+ 1	$\dfrac{-T_C}{T_B}$	$\dfrac{-T_C}{T_A}$
2.	Arm is fixed, gear C rotates through + x revolutions	0	+x	$-x \times \dfrac{T_C}{T_B}$	$-x \times \dfrac{T_C}{T_A}$
3.	Add + y revolutions to all elements	+y	+y	+ y	+ y
4.	Total motion	+y	x + y	$y - x \times \dfrac{T_C}{T_B}$	$y - x \times \dfrac{T_C}{T_A}$

Speed of gear C:

$$y = 18 \text{ rpm}$$

As *A* is fixed,

$$y - x \times \frac{T_C}{T_A} = 0$$

$$18 - x \times \frac{32}{72} = 0$$

$$x = 40.5$$

∴ Speed of gear

$$C = x + y$$
$$= 40.5 + 18$$
$$= + 58.5 \text{ (in the direction of arm)}$$

Speed of gear B:

Let d_A, d_B and d_C be the pitch circle diameter

$$d_B + \frac{d_C}{2} = \frac{d_A}{2}$$

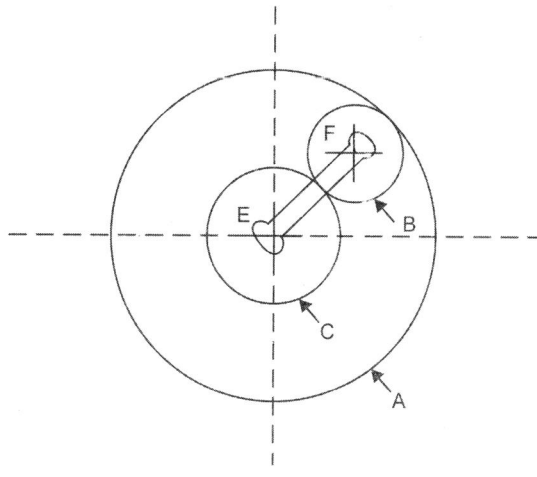

Fig. 6.20

$$2d_B + d_C = d_A$$

$$\Rightarrow \qquad 2T_B + T_C = T_A$$

or $\qquad 2T_B + 32 = 72$

$$T_B = 20$$

$\therefore$ Speed of gear $\qquad B = y - x \times \dfrac{T_C}{T_B}$

$$= 18 - 40.5 \times \dfrac{32}{20}$$

$$= -46.8 \text{ rpm (in the opposite direction of arm).}$$

Example 6.9: An epicyclic train of gears is arranged. How many revolutions does the arm, to which the pinions B and C are attached, makes:

i. When A makes one revolution clockwise and D makes half a revolution anticlockwise, and

ii. When A makes one revolution clockwise and D is stationary?

The no. of teeth on the gears A and D are 40 and 90 respectively.

Solution: Given $\qquad T_A = 40$

$$T_D = 90$$

Let d_A, d_B, d_C and d_D be the pitch circle diameters.

$$d_A + d_B + d_C = d_D$$

$$d_A + 2d_B = d_D$$

Thus

$$T_A + 2T_B = T_D$$

$$40 + 2T_B = 90$$

$\therefore \qquad T_B = 25$

and $\qquad T_C = 25$

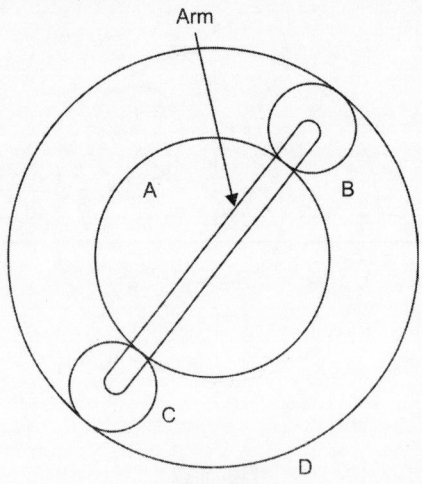

Fig. 6.21

Step no.	Conditions of motion	Revolutions of elements			
		Arm	Gear A	Compound gear B-C	Gear D
1.	Arm is fixed, gear A rotates through + 1 revolution	0	−1	$+\dfrac{T_A}{T_B}$	$+\dfrac{T_A}{T_D}$
2.	Arm is fixed, gear A rotates through + x revolution	0	−x	$+x\times\dfrac{+T_A}{T_B}$	$+x\times\dfrac{+T_A}{T_D}$
3.	Add − y revolutions to all elements	−y	−y	− y	− y
4.	Total motion	−y	−x−y	$x\times\dfrac{+T_A}{T_B}-y$	$x\times\dfrac{+T_A}{T_D}-y$

i. Speed of arm when A makes one revolution and D makes half revolution

Gear A makes one revolution, so

$$= -x - y = 1$$
$$= x + y = 1 \qquad \text{...(1)}$$

Also, the gear D makes half revolution, therefore

$$x \times \frac{T_A}{T_D} - y = \frac{1}{2}$$

$$\Rightarrow x \times \frac{40}{90} - y = \frac{1}{2}$$

or
$$x - 2.25y = 1.125 \qquad \text{...(2)}$$

From Eqs (1) and (2)

$$x = 1.04 - y = 0.04$$

Speed of arm
$$= -y = -(-0.04)$$
$$= 0.04 \text{ (anticlockwise)}$$

ii. Speed of arm when A makes one revolution clockwise and D is stationary

Gear A makes one revolution, so

$$-x - y = -1$$
$$x + y = 1 \qquad \text{...(3)}$$

Also the gear D is stationary, therefore

$$x \times \frac{T_A}{T_D} - y = 0$$

$$x \times \frac{40}{90} - y = 0$$

$$x - 2.25\, y = 0 \qquad \text{... (4)}$$

From Eqs (3) and (4)

$$x = 0.692 \text{ and } y = 0.308$$

∴
Speed of arm $= -y = -0.308$

Example 6.10: In an epicyclic gear train, the internal wheels A and B, and compound wheels C and D rotate independently about axis O. The wheels E and F rotate on pins fixed to the arm G. E gears with A and C and F with B and D. All the wheels have the same module and the no. of teeth are $T_C = 28$, $T_D = 26$, $T_E = T_F = 18$.

   i. Sketch the arrangement.

   ii. Find the no. of teeth on A and B.

   iii. If the arm G makes 100 rpm clockwise and A is fixed, find the speed of B.

   iv. If the arm G makes 100 rpm clockwise and wheel A makes 10 rpm counter-clockwise. Find the speed of wheel B.

Solution: i.

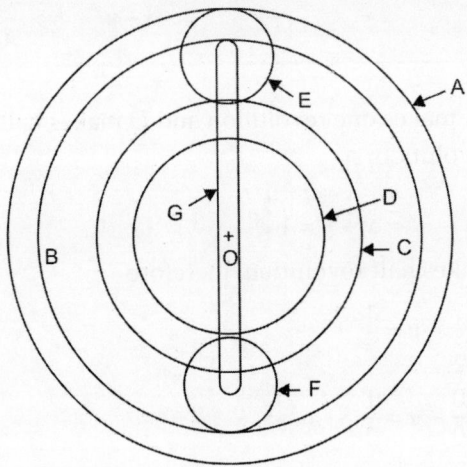

Fig. 6.22

ii.
$$T_C = 28$$
$$T_D = 26$$
$$T_E = T_F = 18$$

Let d_A, d_B, d_C, d_D, d_E and d_F be the pitch circle diameter.

$$d_A = d_C + 2d_E$$
$$d_B = d_D + 2d_F$$

Thus

$$T_A = T_C + 2T_E$$
$$= 28 + 2 \times 18$$
$$= 64$$

and
$$T_B = T_D + 2T_F$$
$$= 26 + 2 \times 18$$
$$= 62$$

Step no.	Conditions of motions	Revolutions of elements					
		Arm G	Wheel A	Wheel E	Compound wheel C-D	Wheel F	Wheel B
1.	Arm is fixed, wheel A rotates through +1 revolution	0	+ 1	$+\dfrac{T_A}{T_E}$	$-\dfrac{T_A}{T_E}$	$\dfrac{T_A}{T_C}\times\dfrac{T_D}{T_F}$	$\dfrac{T_A}{T_C}\times\dfrac{T_D}{T_B}$
2.	Arm is fixed, wheel A rotates through +x revolution	0	$+x$	$x\times\dfrac{T_A}{T_E}$	$-x\times\dfrac{T_A}{T_C}$	$x\times\dfrac{T_A}{T_C}\times\dfrac{T_D}{T_F}$	$x\times\dfrac{T_A}{T_C}\times\dfrac{T_D}{T_B}$
3.	Add +y revolution to all elements	$+y$	$+y$	$+y$	$+y$	$+y$	$+y$
4.	Total motion	$+y$	$x+y$	$y+x\dfrac{T_A}{T_E}$	$y-x\dfrac{T_A}{T_C}$	$y+x\times\dfrac{T_A}{T_C}\times\dfrac{T_D}{T_F}$	$y+x\times\dfrac{T_A}{T_C}\times\dfrac{T_D}{T_B}$

iii. Since the arm G makes 100 rpm, therefore

$$y = -100$$

Also, the wheel A is fixed, therefore

$$x + y = 0$$
$$x = -y = 100$$

∴   Speed of wheel $B = y + x \times \dfrac{T_A}{T_C} \times \dfrac{T_D}{T_B}$

$$= -100 + 100 \times \frac{64}{28} \times \frac{26}{62}$$

$$= -4.2 \text{ rpm} = 4.2 \text{ rpm clockwise.}$$

iv. Since the arm G makes 100 rpm clockwise,

$$y = -100$$

Also, the wheel A makes 10 rpm anticlockwise, therefore

$$x + y = 10$$

or $x = 110$

∴   Speed of wheel $B = y + x \times \dfrac{T_A}{T_C} \times \dfrac{T_D}{T_B}$

$$= -100 + 110 \times \frac{64}{28} \times \frac{26}{62}$$

$$= +5.4 \text{ rpm}$$

$$= -5.4 \text{ rpm (anticlockwise)}$$

Example 6.11: In an epicycle gear of the sun and planet type the pitch circle diameter of the internally toothed ring is to be 224 mm and the module 4 mm. When the ring D is stationary, the spider A which carries planet wheels C of equal size is to make one revolution in the same sense as the sun wheel B for every five revolutions of the driving spindle carrying the sun wheel B. Determine suitable no. of teeth for all wheels.

Solution: Given

$$d_D = 224 \text{ mm}$$

$$m = 4 \text{ mm}$$

$$N_A = N_B/5$$

Let T_B, T_C and T_D be the no. of teeth on the sun wheel B, planet wheels C and internally toothed ring D.

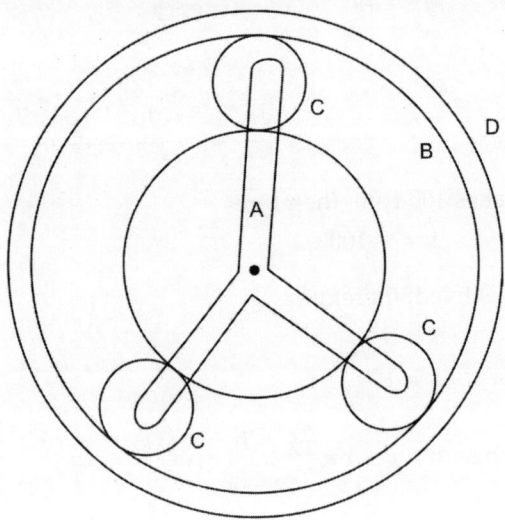

Fig. 6.23

Step no.	Conditions of motion	Revolutions of elements			
		Spider A	Sun wheel B	Planet wheel C	Internal gear D
1.	Spider A is fixed, wheel B rotates through $+1$ revolution	0	$+1$	$-\dfrac{T_B}{T_C}$	$-\dfrac{T_B}{T_D}$
2.	Spider A is fixed, sun sheel B rotates through $+x$ revolution	0	$+x$	$x \times -\dfrac{T_B}{T_C}$	$x \times -\dfrac{T_B}{T_D}$
3.	Add $+y$ revolution to all elements	$+y$	$+y$	$+y$	$+y$
4.	Total motion	$+y$	$x+y$	$y - x \times \dfrac{T_B}{T_C}$	$y - x \times \dfrac{T_B}{T_D}$

We know that when the sun wheel B makes $+ 5$ revolution, the spider A makes $+ 1$ revolution. Therefore

$$y = + 1 \text{ and } x + y = + 5$$

$\therefore$

$$x = 4$$

Since the internally toothed ring D is stationary, therefore

$$y - x \times \frac{T_B}{T_D} = 0$$

$$1 - 4 \times \frac{T_B}{T_D} = 0$$

$$\frac{T_B}{T_D} = \frac{1}{4}$$

$$T_D = 4T_B \qquad \qquad ...(1)$$

We know that

$$T_D = d_D/m = \frac{224}{4} = 56$$

$$T_B = T_D/4 = \frac{56}{4} = 14$$

Let d_B, d_C and d_D be the pitch circle diameter of sun wheel B, planet wheels C and internally toothed ring D respectively

$$d_B + 2d_C = d_D$$

Since $d \alpha T$

$$T_B + 2T_C = T_D$$
$$14 + 2T_C = 56$$
$$T_C = 21.$$

EXERCISE

6.1 Two wheels with standard involute teeth of 5 mm module are to gear together with a velocity ratio of 4.5, the pressure angle being 15°, find:
   (a) The minimum number of teeth in the pinion if interference is to be avoided.
   (b) The pressure between the teeth when such a pinion is transmitting a torque of 60 Nm.

6.2 Determine the number of pairs in contact at a given instant, if two equal involute gears of 18 teeth of pressure angle 19°–208° have addendum of 0.8 module.

6.3 In a spiral gear drive, the spiral of the teeth on the driving wheel has been fixed at 80°. The normal pitch of teeth is 0.3 cm and the driving wheel A runs at twice the speed of the driven wheel B. The shafts are at right angles and the shortest distance between their axes is approximately 20 cm. Determine the dimensions of suitable gears for this drive, giving for each wheel the following if the pitch circle diameters are equal.
   (i) The number of teeth
   (ii) The spiral angle of teeth

Also find the exact centre distance between the axes. If the friction angle is 6°, what is the efficiency of the wheels?

6.4 A 24 teeth pinion of 4.25 mm module pitch drives a rack. The addendum of both the rack and pinion is 6.25 mm. Determine the minimum pressure angle to avoid interference. Hence, determine the length of the arc of contact and the number of teeth in contact at a time.

6.5 A pair of spur wheels with 14 and 21 teeth are involute profile and pressure angle $16°$. Find the maximum addendum on the pinion and gear wheel to avoid interference in $m = 6$ mm. Also find the maximum velocity of sliding of pitch point if the pinion turns at 300 rpm.

6.6 In a spiral gear drive connecting two shafts, the approximate centre distance is 400 mm and speed ratio is 3. The angle between the two shafts is $50°$ and normal pitch is 18 mm. The spiral angles for the driving and driven wheels are required. Find:
   (i) Number of teeth on each wheel
   (ii) Exact centre distance
   (iii) Efficiency of drive, if friction angle is $6°$.

6.7 In an epicycle train, an annular wheel A having 54 teeth meshes with a planet wheel B which gears with a sun wheel C, the wheels A and C being coaxial. The wheel B is carried on a pin fixed on one end of arm P which rotates about the axis of the wheels A and C. If the wheel A makes 20 rpm in a clockwise sense and the arm P rotates at 100 rpm in the anticlockwise direction and the wheel C has 24 teeth, determine rpm and sense of rotation of the wheel C.

6.8 An arm A carries 4 gear wheels B, C, D and E. Gear wheel B meshes with gear C and gear wheel D meshes with gear wheel E. Gear wheels C and D form a compound gear. The number of teeth on gear wheel B = 20, that of gear wheel C = 15, that of gear wheel D = 35 and gear wheel E 20 teeth. If the speed of arm is 100 rpm clockwise and gear wheel E is fixed, calculate the speed of gear wheel B. Draw the sketch of the gear train.

6.9 Two parallel shafts are to be connected by spur gearing. The approximate distance between the shafts is 600 mm. If one shaft runs at 120 rpm and the other at 360 rpm, find the number of teeth in each wheel, if the module is 8 mm. Also determine the exact distance of the shafts.

6.10 In an epicycle gear train of the sun and planet type, the annular gear A has 48 teeth cut and meshes internally. Three planet wheels of equal size mesh with the annular gear A and sun wheel B. When the gear A is stationary, the spider C which carries the planet wheels, is to make one revolution for every five revolutions of the spindle carrying the sun wheel B. Determine the number of teeth on sun and planet wheels.

6.11 In an epicyclic gear train, an arm carries two wheels A and B having 36 and 45 teeth respectively. If the arm rotates at 150 rpm in the anticlockwise direction about the centre of the wheel A which is fixed, determine the speed of wheel B. If the wheel A, instead of being fixed, makes 300 rpm in the clockwise direction, what will be the speed of B?

6.12 An internal wheel B with 80 teeth is keyed to a shaft SA fixed internal wheel C with 82 teeth is concentric with B. A compound wheel D-E gears with the two internal wheels; D has 28 teeth and gears with C while E gears with B. The compound wheels revolved freely on a pin which projects from a disc keyed to a shaft A coaxial with S. If the wheels have the same pitch and shaft A makes 800 rpm, what is the speed of the shaft S. Sketch the arrangement.

OBJECTIVE TYPE QUESTIONS

6.1 The speed of gears used to connect two parallel coplanar shafts are
    (a) spur gearing
    (b) helical gearing
    (c) bevel gearing
    (d) spiral gearing

6.2 Two intersecting coplanar shafts are connected by
    (a) spur gearing
    (b) helical gearing
    (c) bevel gearing
    (d) spiral gearing

6.3 If the peripheral speed of gears exceeds 15 m/sec, they are said to be
    (a) low velocity gears
    (b) high speed gears
    (c) high velocity gears
    (d) none of the above

6.4 The module is the reciprocal of
    (a) circular pitch
    (b) diametral pitch
    (c) pitch circle diameter
    (d) none of the above

6.5 The value of contact ratio for gears is
    (a) less than unity
    (b) more than unity
    (c) zero
    (d) none of the above

6.6 Interference occurs in
    (a) involute profile
    (b) cycloidal profile
    (c) both (a) and (b)
    (d) none of the above

6.7 The product of diametral pitch and module is equal to
    (a) zero
    (b) infinity
    (c) unity
    (d) two

6.8 The value of pressure angle for spur gear is kept small
    (a) to reduce the axial thrust on bearing on which gears are mounted
    (b) to increase the force and the power transmission in gears
    (c) both (a) and (b)
    (d) none of the above

6.9 Which of the following profile of gears satisfy the law of gearing?
    (a) involute profiles of mating teeth
    (b) cycloidal profiles of mating teeth
    (c) conjugate profiles of mating teeth
    (d) all of the above

6.10 The interference can be avoided in involute gears by
  (a) changing the pressure angle, the centre distance can be varied
  (b) using modified involute
  (c) increasing the addendum on the smaller wheel and reduce it on the larger wheel
  (d) All of the above methods

6.11 In helical gears, the normal circular pitch is given by
  (a) $P_n = P_c \cos ae$
  (b) $P_n = P_c \sin ae$
  (c) $P_n = \dfrac{P_c}{\cos ae}$
  (d) None of the above

6.12 Pressure angle is constant for a gear pair having
  (a) involute tooth profile
  (b) cycloidal tooth profile
  (c) circular path profile
  (d) none of the above

6.13 From the strength point of view,
  (a) involute teeth are better
  (b) rack is better
  (c) cycloidal teeth are better
  (d) none of the above

6.14 Bevel gear pair allows transmission of motion when the angle between the axes of the shafts is
  (a) only 90°
  (b) 90°
  (c) 45°
  (d) none of the above

6.15 In a reverted gear train,
  (a) one gear is fixed
  (b) none of the gears is fixed
  (c) the axes of the first and the last gear are coaxial
  (d) the speed of the last gear is equal to the speed of the first gear

6.16 Hour and minute hands are connected in a clock mechanism by means of
  (a) simple gear train
  (b) epicyclic gear train
  (c) reverted gear train
  (d) none of the above

6.17 Simple gear train and epicyclic train are different in a way that
  (a) in epicyclic gear train the axes of the gears also rotate and in simple gear train the axes on the gears are fixed
  (b) epicyclic gear train has more than four gears and simple gear train has only two gears
  (c) epicyclic gear train is used for power transmission and simple gear train for motion transmission
  (d) all of the above

6.18 In a gear train of n wheels , the speed ratio is defined as
  (a) $\dfrac{N_1}{N_n}$
  (b) $\dfrac{N_n}{N_1}$

(c) $\dfrac{T_1}{T_n}$

(d) $\dfrac{T_n}{T_1}$

6.19 Train value of gear train is
   (a) always less than unity
   (b) always greater than unity
   (c) equal to reciprocal of speed ratio
   (d) equal to speed ratio of gear train

ANSWERS

6.1 (a)	6.2 (c)	6.3 (c)	6.4 (b)	6.5 (b)	6.6 (a)
6.7 (c)	6.8 (c)	6.9 (d)	6.10 (d)	6.11 (a)	6.12 (a)
6.13 (c)	6.14 (d)	6.15 (c)	6.16 (c)	6.17 (a)	6.18 (a)
6.19. (c)					

7

Force Analysis, Turning Moment and Flywheel

7.1 INTRODUCTION

In any machinery, forces are transmitted from one member to the other. While designing these members, it is necessary to know the magnitude and directions of forces transmitted, in order to select proper material and size. If a member is not of required size, it may fail during operations. Oversized members will add on the cost, weight and size.

Forces acting on members can be analysed in two ways.

  i. **Static force analysis:** Inertia forces are produced due to masses of accelerating parts. Inertia force is an imaginany force, which when acts on a rigid body brings it in equilibrium position.

$$\text{Inertia force} = -\text{accelerating force} = -ma$$

  If the magnitude of inertia forces are small compared to the externally applied loads, these can be neglected while analysing the mechanism. Such an analysis is known as static force analysis.

  ii. **Dynamic force analysis:** When inertia forces due to accelerating masses are significant and are considered, the analysis is known as dynamic force analysis. Machines in which operating speeds are high, dynamic forces are dominant. Therefore, it becomes necessary to carry out dynamic analysis. Moreover dynamic forces may lead to vibrations, noise, wear and even failure of machine parts.

A flywheel is a reservoir of energy, which stores energy when it is available in excess and gives away when required. It stores energy by virtue of its mass and speed. The function of flywheel is continues. Flywheel is used in automobiles, punching machines, shearing machines, etc.

7.2 STATIC EQUILIBRIUM

A body is said to be in static equilibrium if it remains in its state of rest or motion. The change in state occurs only when some external force is applied. For a rigid body to be in static equilibrium, the vector sum of all forces acting on the body is zero, i.e.

$$\Sigma F = 0$$

For coplanar force system, $\Sigma F_x = 0$, $\Sigma F_y = 0$

The vector sum of moments of all forces about any arbitrary point is zero,

$$\Sigma M = 0$$

7.3 EQUILIBRIUM OF TWO-FORCE SYSTEM

A body subjected to two forces will be in equilibrium if the forces are of same magnitude, opposite in directions and act along the same line.

$$F_1 = F_2$$

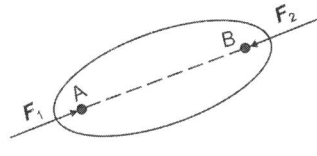

Fig. 7.1

7.4 EQUILIBRIUM OF THREE-FORCE SYSTEM

A body subjected to three non-parallel forces will be in equilibrium, if the resultant of the forces is zero and the lines of action of the forces intersect at a point.

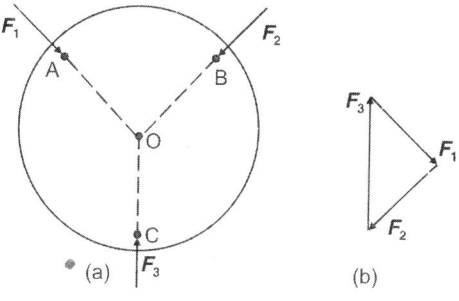

Fig. 7.2

A body is acted upon by three forces F_1, F_2 and F_3, lines of action of which intersect at a point O. Resultant of these forces is zero, therefore body is in equilibrium.

7.5 EQUILIBRIUM OF FOUR-FORCE SYSTEM

When four or more forces act on a body then the forces known completely are combined into a single force. This reduces the number of forces acting on a body to two or three.

A body subjected to four forces is shown in Fig. 7.3a. O_1 is the point of intersection of lines of action of forces F_1 and F_2 and O_2 is the point of intersection of lines of action of forces F_3 and F_4. Join O_1 and O_2. Resultant of forces F_1 and F_2 and the resultant of forces F_3 and F_4 is parallel to $O_1 O_2$. To make force polygon (Fig 7.3b), draw a line OA parallel to the line of action of force F_1 of given magnitude and in direction of F_1 (F_1 known fully). From 'a' draw a line parallel to the direction of force F_2 (direction known, magnitude unknown) and from 'O' draw a line parallel to the resultant $O_1 O_2$ to intersect the line of action of F_2 at b. Join ob and find $F_2 = ab$. From b draw a line parallel to the line of action of force F_3 and from O draw a line parallel to the line of action of force F_4. These lines intersect at point c. From closed polygon F_3 and F_4 can be determined by measurement.

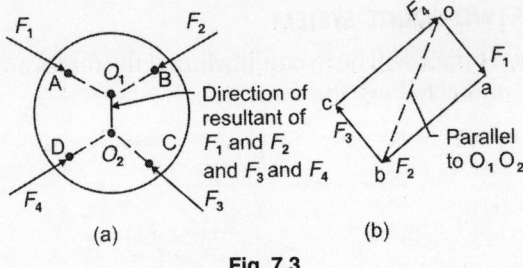

Fig. 7.3

7.6 SYSTEM OF TWO-FORCE AND A TORQUE

A body subjected to two forces and a torque will be in equilibrium if the forces are equal in magnitude, parallel in direction, opposite in sense and form a couple, moment of which is equal and opposite to the applied torque.

A body subjected to two equal forces F_1 and F_2 and a torque T is shown in Fig. 7.4.

$$F_1 = F_2, \; \vec{F_1} = -\vec{F_2}$$

For equilibrium, $\qquad T = F_1 \times h = F_2 \times h$

Moment of couple formed by F_1 and F_2 is anticlockwise.

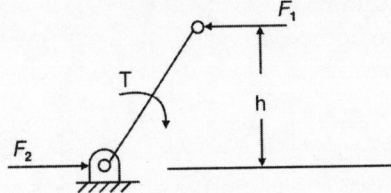

Fig. 7.4

Example 7.1. Figure 7.5 shows a quaternary link $ABCD$ under the action of forces F_1 F_2, F_3, and F_4 acting at A, B, C and D respectively. The link is in static equilibrium. Determine the magnitude of the forces F_2 and F_3 and the direction of F_3.

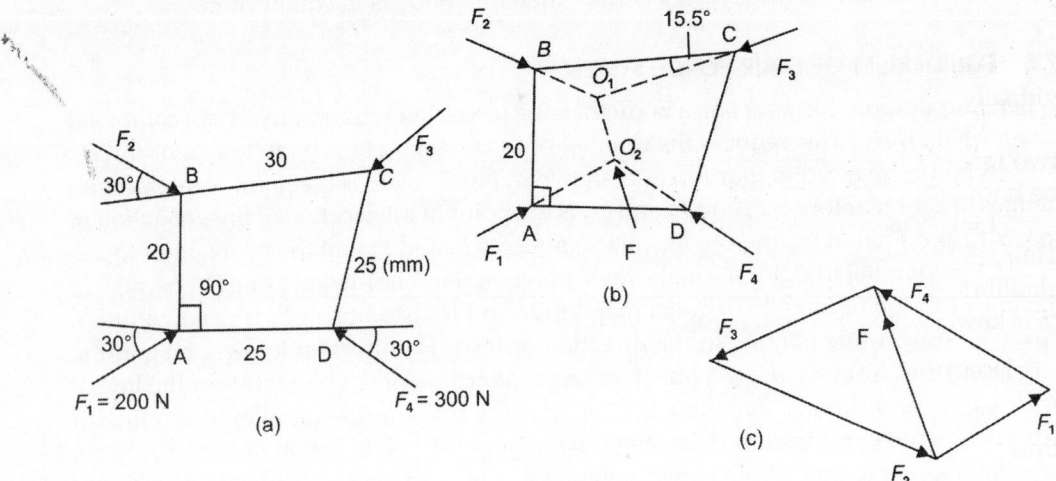

Fig. 7.5

Solution: The forces F_1 and F_4 can be combined into a single force F by obtaining their resultant (Fig. 7.5b and c). The force F acts through O_2, the point where lines of action of F_1 and F_4 meet.

Now, the four-force member *ABCD* is reduced to a three-force member under the action of forces F (completely known), F_2 (only the direction known) and F_3 (completely unknown).

Let F and F_2 meet at O_1. Then CO_1 is the line of action of force F_3. By completing the force triangle, obtain the magnitude of F_2 and F_3.

Magnitude of $F_2 = 380\ N$

Magnitude of $F_3 = 284\ N$

Line of action of force F_3 makes an angle of 15.5° with *CB*.

7.7 FORCE CONVENTION

The force applied by link *i* on link *j* is represented by F_{ij}.

7.8 FREE BODY DIAGRAM

Free body diagram (FBD) is a diagram showing all the forces and reactions acting on the body isolated from its surrounding to determine the nature and magnitude of unknown forces. Figure 7.6 shows free body diagrams of various members of a four-bar mechanism and slider crank mechanism.

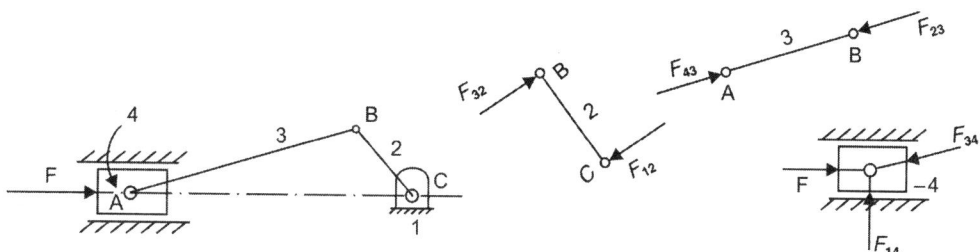

Fig. 7.6: Free body diagram of slider–crank mechanism

For equilibrium of a mechanism, each of its members must be in equilibrium individually.

Assume that the force F of member 4 is known completely. To know the other two forces acting on this member completely, the direction of one more force must be known.

Link 3 is a two-force member and for its equilibrium F_{23} and F_{43} must act along *BC*. Thus, F_{34} being equal and opposite to F_{43} also acts along *BC*. For member 4 to be in equilibrium, F_{14} passes through the intersection of F and F_{34}. By drawing a force triangle (F is knwon completely), magnitudes of F_{14} and F_{34} can be known (Fig. 7.7).

Now $$F_{34} = F_{43} = F_{23} = F_{32}$$

Member 2 will be in equilibrium if F_{12} is equal, parallel and opposite to F_{32}, thus

$$T = F_{12} \times h = F_{32} \times h$$

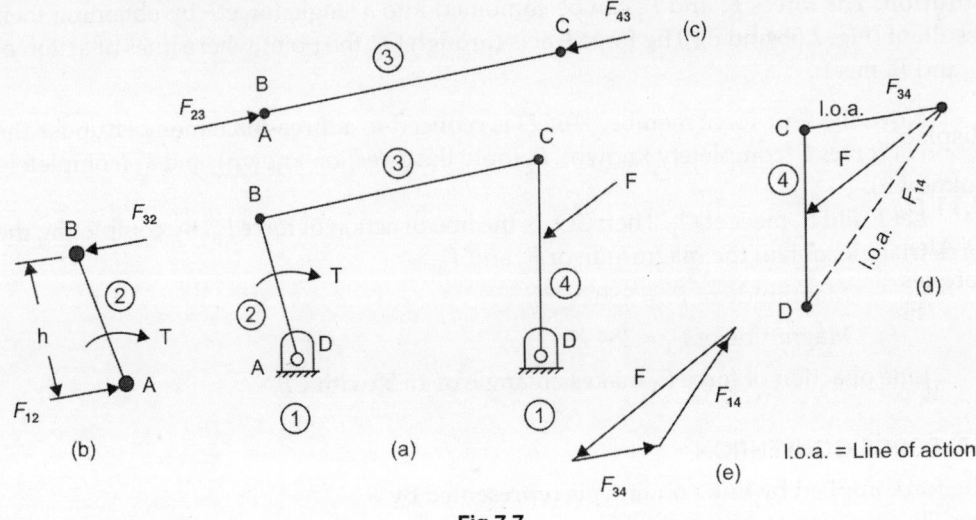

Fig 7.7

7.9 SUPERPOSITION

A linear system is one in which output force is directly proportional to the input force. In linear system, dry or coulomb friction is negleted. If a member of loads acts on a linear system, the net effect is equal to the sum of the effects of the individual loads taken one at a time.

7.10 PRINCIPLE OF VIRTUAL WORK

It states that the work done during a virtual displacement from the equilibrium is equal to zero. Virtual displacement is an imaginary infinitesimal displacement given to the system. Virtual displacement can be angular displacement or linear displacement. Principle of virtual work is applied to the system as a whole. In a slider–crank mechanism shown in Fig. 7.8, a virtual angular displacement $\delta\theta$ is given to the crank. Corresponding linear displacement of slider is δx.

Work done is considered positive if a force acts in the direction of the displacement and negative if it acts in the opposite direction.

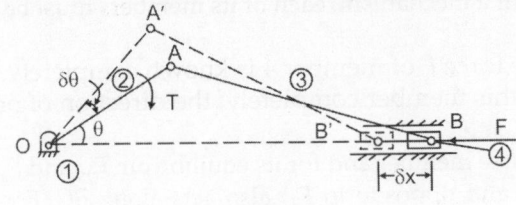

Fig. 7.8

Virtual work, $\qquad W = T\delta\theta + F\delta x = 0$

Taking limit $\qquad \delta\tau \to 0$

$$T\frac{d\theta}{dt} + F\frac{dx}{dt} = 0$$

$$T\omega + Fv = 0 \qquad\qquad \left[\frac{d\theta}{dt} = \omega \,(\text{angular velocity})\right]$$

$$T = -\frac{Fv}{w} \qquad \left[\frac{dx}{dt} = v(\text{linear velocity})\right]$$

Negative sign indicates that T must be applied in opposite direction to the angular displacement.

7.11 D'ALEMBERT'S PRINCIPLE

D'Alembert's principle states that the inertia forces and couples, the external forces and torques acting on a body together result in static equilibrium.

From Newton's second law, force acting on a body,

$$F = ma$$

It can be written as

$$F - ma = 0$$

or $\qquad\qquad F + F_i = 0 \qquad\qquad\qquad ...(7.1)$

Here $F_i = -m \cdot a = $ Inertia force

Inertia is a property of matter by virtue of which a body resists any change in its state of rest or motion.

Similarly, torque acting on a body, $T = I\alpha$

or $\qquad\qquad T - I\alpha = 0$

or $\qquad\qquad T + T_i = 0 \qquad\qquad\qquad ...(7.2)$

Here $T_i = -I\alpha = $ Inertia torque.

Inertia torque resists any change in the angular velocity. Therefore, according to D'Alembert's principle, the vector sum of forces and torques must be zero for equilibrium,

i.e. $\qquad\qquad \Sigma F + F_i = 0$

and $\qquad\qquad \Sigma T + T_i = 0$

Equations (7.1) and (7.2) are similar to the equations of static equilibrium. Therefore, inertia forces and couples can be treated as static loads and the problem of dynamic analysis is reduced to that of static analysis.

7.12 EQUIVALENT OFFSET INERTIA FORCE

In plane motions involving accelerations, the inertia force acts on a body through its centre of mass. However, if the body is acted upon by forces such that their resultant does not pass through the centre of mass, a couple also acts on the body. In graphical solutions, inertia force and inertia couple is replaced by equivalent offset inertia force which can account for both. This is done by displacing the line of action of the inertia force from the centre of mass by a distance h. The perpendicular displacement h of the force from the centre of mass is such that the torque so produced is equal to the inertia couple acting on the body and opposite in direction,

i.e. $\qquad\qquad T_i = C_i$

or $\qquad\qquad F_i \times h = C_i$

or $\qquad\qquad h = \dfrac{C_i}{F_i} = \dfrac{-I\alpha}{-ma} = \dfrac{mk^2\alpha}{ma} = \dfrac{k^2\alpha}{a}$

7.13 DYNAMIC ANALYSIS OF FOUR-LINK MECHANISM

For dynamic analysis of four-link mechanisms, the following procedure is adopted:

1. Draw the velocity and acceleration diagrams of the mechanism from the configuration diagram by usual methods.
2. Determine the linear acceleration of centre of masses of various links, and also the angular accelerations of the links.
3. Calculate the inertia forces and inertia couples from the relations $F_i = -ma$ and $C_i = -I\alpha$.
4. Replace F_i with equivalent offset inertia force to take into account F_i as well as C_i.
5. Assume equivalent offset inertia forces on the links as static forces and analyse the mechanism.

7.14 ANALYTICAL APPROACH TO DYNAMIC ANALYSIS OF SLIDER–CRANK MECHANISM

7.14.1 Velocity and Acceleration of Piston

Consider a slider–crank mechanism of a reciprocating steam engine. AB is connecting rod of length l and OA is crank of length r. Crank rotates in clockwise direction with a uniform angular velocity. When the crank has turned through an angle θ from the inner dead centre in time t sec., piston is displaced by a distance x.

Let ϕ = angle made by connecting rod AB with line of stroke

n = ratio of length of connecting rod to the radius of crank

$$= \frac{l}{r}$$

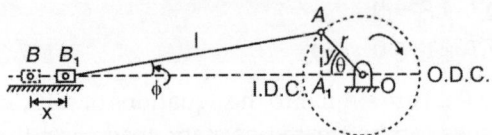

Fig 7.9: Motion of a crank and connecting rod

From the gemometry of Fig. 7.9

$$x = B_1B = BO - B_1O$$
$$= BO - (B_1A_1 + A_1O) \qquad (B_1O = B_1A_1 + A_1O)$$
$$= (l + r) - (l\cos\phi + r\cos\theta)$$
$$= (nr + r) - (nr\cos\phi + r\cos\theta) \qquad \left(\frac{l}{r} = n\right)$$
$$x = r[(n + 1)] - (n\cos\phi + \cos\theta) \qquad \dots(7.3)$$

Obtaining value of $\cos\phi$ in terms of angle θ

$$\cos\phi = \sqrt{1 - \sin^2\phi} = \sqrt{1 - \frac{y^2}{l^2}} \qquad \left[from\ \Delta B_1AA_1,\ \sin\phi = \frac{y}{l}\right]$$

$$= \sqrt{1 - \frac{(r\sin\theta)^2}{(nr)^2}} \qquad [from\ \Delta AA_1O,\ y = r\sin\theta\ and\ l = nr]$$

$$= \sqrt{1 - \frac{\sin^2\theta}{n^2}} = \frac{1}{n}\sqrt{n^2 - \sin^2\theta} \qquad \ldots(7.4)$$

Substituting value of $\cos\phi$ in Eq. (7.3),

$$x = r\left[(n+1) - \left(n \times \frac{1}{n}\sqrt{n^2 - \sin^2\theta} + \cos\theta\right)\right]$$

$$= r\left[(n+1) - \left(\sqrt{n^2 - \sin^2\theta} + \cos\theta\right)\right]$$

$$x = r\left[(1 - \cos\theta) + (n - \sqrt{n^2 - \sin^2\theta})\right] \qquad \ldots(7.5)$$

If the connecting rod is very large as compared to crank, n^2 will be very large and maximum possible value of $\sin^2\theta$ is 1.

Then $\sqrt{n^2 - \sin^2\theta} \to \sqrt{n^2} = n$

$$\therefore \qquad x = r[(1 - \cos\theta) + (n - n)]$$

$$x = r(1 - \cos\theta) \qquad \ldots(7.6)$$

Equation (7.6) is an expression of a simple harmonic motion. Hence, piston executes simple harmonic motion, if connecting rod is very large.

Velocity of Piston

The velocity of piston is obtained by differentiating displacement x [Eq. (7.5)] with respect to time t.

$$v = \frac{dx}{dt} = \frac{dx}{d\theta}\cdot\frac{d\theta}{dt} = \frac{d}{d\theta}\left[r\left\{(1 - \cos\theta) + \left(n - \sqrt{n^2 - \sin^2\theta}\right)\right\}\right]\cdot\omega$$

$$\left(\frac{d\theta}{dt} = \omega = \text{angular velocity}\right)$$

$$= r[0 + \sin\theta] + 0 - \frac{1}{2}\left[n^2 - \sin^2\theta\right]^{-\frac{1}{2}}\cdot[0 - 2\sin\theta\cos\theta]\omega$$

$$v = r\omega\left[\sin\theta + \frac{\sin^2\theta}{2\sqrt{n^2 - \sin^2\theta}}\right] \qquad \ldots(7.7)$$

If n^2 is large in comparsion to $\sin^2\theta$, then:

$$\sqrt{n^2 - \sin^2\theta} \to \sqrt{n^2} = n$$

$$v = r\omega\left[\sin\theta + \frac{\sin 2\theta}{2n}\right] \qquad \ldots(7.8)$$

If n is very large, then $\dfrac{\sin 2\theta}{2n}$ can be neglected

$$n = r\omega\sin\theta \qquad \ldots(7.9)$$

Acceleration of piston

Acceleration of piston (a) is obtained by differentiating velocity in Eq. (7.8) with respect to time t.

$$a = \frac{dv}{dt} = \frac{dv}{d\theta} \cdot \frac{d\theta}{dt}$$

$$= \frac{d}{d\theta}\left[r\omega\left(\sin\theta + \frac{\sin 2\theta}{2n}\right)\right]\omega$$

$$= r\omega\left[\cos\theta + \frac{2\cos 2\theta}{n}\right]\omega$$

$$a = r\omega^2\left[\cos\theta + \frac{\cos 2\theta}{n}\right] \qquad\qquad\qquad ...(7.10)$$

If n is very large, $\dfrac{\cos 2\theta}{n} \rightarrow 0$

$$a = r\,\omega^2\cos\theta \qquad\qquad\qquad ...(7.11)$$

Equation (7.11) is the equation for SHM.

When $\theta = 0°$, i.e. When crank is at I.D.C.,

$$a = r\omega^2\left(1 + \frac{1}{n}\right) \qquad\qquad\qquad [\textit{from} \text{ Eq. (7.10)}]$$

When $\theta = 180°$, i.e. when crank is at O.D.C.,

$$a = r\omega^2\left(-1 + \frac{1}{n}\right)$$

At 180°, when crank moves in reverse direction, i.e. towards I.D.C., sign is changed in above expression.

$$a = -r\omega^2\left(-1 + \frac{1}{n}\right)$$

$$a = r\omega^2\left(1 - \frac{1}{n}\right)$$

Expression of acceleration given above in Eq. (7.11) is not exact but approximate as it has been obtained by differentiating the approximate expression for velocity. To differentiate the exact expression of velocity is very cumbersome.

7.14.2 Angular Velocity and Angular Acceleration of Connecting Rod

Angular velocity of connecting rod is $\omega_c = \dfrac{d\phi}{dt}$

From Fig. 7.9 $\qquad\qquad y = l\sin\phi = r\sin\theta$

or $\qquad\qquad\qquad \sin\phi = \dfrac{r}{l}\times\sin\theta = \dfrac{\sin\theta}{n} \qquad \left(\dfrac{l}{r} = n\right)$

Differentiating above equation with respect to time 't'

$$\cos\phi\,\frac{d\phi}{dt} = \frac{l}{n}\cdot\cos\theta\,\frac{d\theta}{dt}$$

or

$$\cos\phi\cdot\omega_c = \frac{\cos\theta}{n}.\omega$$

or

$$\omega_c = \frac{\cos\theta}{n\cdot\cos\phi}.\omega$$

Substituting value of $\cos\theta$ from Eq. (7.4)

$$\omega_c = \frac{\cos\theta}{n\dfrac{1}{n}\sqrt{n^2 - \sin^2\theta}}\,\omega$$

or

$$\omega_c = \frac{\omega\cos\theta}{\sqrt{n^2 - \sin^2\theta}} \qquad\qquad ...(7.12)$$

Angular acceleration of the connecting rod, α_c

$$\alpha_c = \frac{d\omega_c}{dt} = \frac{d\omega_c}{d\theta}\cdot\frac{d\theta}{dt}$$

$$= \frac{d}{d\theta}\left[\frac{\omega\cos\theta}{\sqrt{n^2 - \sin^2\theta}}\right].\omega$$

$$= \omega^2\frac{d}{d\theta}\left[\cos\theta(n^2 - \sin^2\theta)^{-1/2}\right]$$

$$= \omega^2\left[\cos\theta.\left(-\frac{1}{2}\right)(n^2 - \sin^2\theta)^{-3/2}(0 - 2\sin\theta\cos\theta) + \right.$$

$$\left. (n^2 - \sin^2\theta)^{-1/2}(-\sin\theta)\right]$$

$$= \omega^2\sin\theta\left[\cos^2\theta(n^2 - \sin^2\theta)^{-3/2} - (n^2 - \sin^2\theta)^{-1/2}\right]$$

$$= \omega^2\sin\theta.(n^2 - \sin^2\theta)^{-3/2}[\cos^2\theta - (n^2 - \sin^2\theta)]$$

$$= \omega^2\sin\theta\frac{[\cos^2\theta - n^2 + \sin^2\theta]}{(n^2 - \sin^2\theta)^{3/2}}$$

$$= \omega^2\sin\theta\left[\frac{1 - n^2}{(n^2 - \sin^2\theta)^{3/2}}\right] \qquad (\sin^2\theta + \cos^2\theta = 1)$$

or

$$\alpha_c = -\omega^2\sin\theta\left[\frac{(n^2 - 1)}{(n^2 - \sin^2\theta)^{3/2}}\right] \qquad\qquad ...(7.13)$$

The negative sign shows that the sense of angular acceleration of connecting rod is such that it tends to reduce the angle ϕ. In present case, the angular acceleration of the connecting rod is clockwise.

If value of n^2 is very large, $\sin^2\theta$ may be neglectecd as $(\sin^2\theta)_{max} = 1$. Equations (7.12) and (7.13) reduces to

$$\omega_c = \frac{\omega \cos\theta}{n}$$

$$\alpha_c = \frac{-\omega^2 \sin\theta(n^2 - 1)}{(n^2)^{3/2}} = \frac{-\omega^2 \sin\theta(n^2 - 1)}{n^3}$$

$$\alpha_c = \frac{-\omega^2 \sin\theta.n^2}{n^3} = \frac{-\omega^2 \sin\theta}{n} \quad (n^2 - 1 \approx n^2 \text{ as } n^2 \text{ is very large})$$

7.15 ENGINE FORCE ANALYSIS

The various forces acting on parts of an engine, neglecting the weight and inertia effect of the connecting rod, are described below.

7.15.1 Piston Effort

The net force acting on the piston along the line of storke is known as piston effort (Fig. 7.10). Piston effort is also known as effective driving force.

Consider, A_1 = area of cover end

A_2 = area of the piston rod end

p_1 = pressure on the cover end

p_2 = pressure on the rod end

m = mass of reciprocating parts

a^1 = cross-sectional area or piston rod

Force on piston due to gas pressure, $F_P = p_1 A_1 - p_2 A_2$

or $\qquad F_P = p_1 A_1 - p_2(A_1 - a^1) \quad (A_2 = A_1 - a^1)$ ...(7.14)

F_P can also be determined by,

$$F_P = p \times \frac{\pi}{4} D^2$$

Where, D is the diameter of piston and p is net pressure of gas, i.e. the difference of pressure between cover end and piston end.

Intertia force, $\qquad F_i = -ma$

where, $\qquad a$ = acceleration of reciprocating parts (piston)

$$= r\omega^2 \left[\cos + \frac{\cos 2\theta}{n} \right]$$

$\therefore \qquad F_i = -mr\omega^2 \left[\cos\theta + \frac{\cos 2\theta}{n} \right]$ ...(7.15)

Net force on the piston (piston effort)

= Force due to gas pressure + Inertia force

$F = F_P + F_i$ ...(7.16)

Inertia force (F_i) is taken −ve during first half of the storke as it resists accelerating reciprocating masses. During the later half of the stroke, reciprocating masses decelerate and inertia force opposes this or acts in the direction of applied gas pressure. Therefore, in later half F_i is taken +ve.

If frictional resistance (F_R) is also taken into account, piston effort

$$F = F_P + F_i - F_R \qquad \qquad ...(7.17)$$

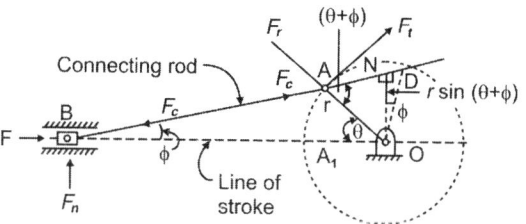

Fig. 7.10

In a vertical engine, when piston is moving downwards, weight of reciprocating parts $(+W)$ helps piston effort. And when the piston is moving upwards, weight $(-W)$ opposes the piston effort.

∴ Piston effort, $F = F_P + F_i - F_R \pm W \qquad \qquad ...(7.18)$

7.15.2 Force Acting along the Connecting Rod (F_C)

F_C = force acting along the connecting rod. Equating horizontal components of forces,

$$F_c \cos \phi = F$$

or
$$F_c = \frac{F}{\cos \phi} \qquad \qquad ...(7.19)$$

7.15.3 Thrust on the Sides of Cylinder Walls

It is the normal reaction on the cylinder walls.

$$F_n = F_c \sin \phi$$

$$= \frac{F}{\cos \phi} \cdot \sin \phi \qquad \qquad (\textit{from Eq. (7.19)})$$

$$F_n = F \tan \phi \qquad \qquad ...(7.20)$$

7.15.4 Crank Effort

The net force applied at the crank pin perpendicular to the crank is knwon as crank effort. Crank effort (F_t) provides the required turning moment to the crankshaft.

$$F_t = \text{component of } F_c \text{ perpendicular to crank}$$

$$= F_c \sin (\theta + \phi)$$

$$= \frac{F}{\cos \phi} \sin(\theta + \phi) \qquad \qquad (\textit{from Eq. 7.19})$$

$$F_t = \frac{F \sin(\theta + \phi)}{\cos \phi} \qquad \qquad ...(7.21)$$

7.15.5 Thrust on Crankshaft Bearings

The component of F_c along the crank is knwon as thrust on crankshaft bearings (F_b).

$$F_b = F_c \cos(\theta + \phi)$$

$$F_b = \frac{F}{\cos\theta} \cdot \cos(\theta + \phi) \qquad \qquad ...(7.22)$$

7.16 TURNING MOMENT OR TORQUE ON THE CRANK SHAFT

Turning moment or torque (T) on the crank shaft is equal to the product of crank effort (F_t) and crank radius (r).

$$T = F_t \times r$$

$$= \frac{F\sin(\theta + \phi)}{\cos\phi} \times r$$

$$= \frac{Fr}{\cos\phi}(\sin\theta\cos\phi + \cos\theta\sin\phi)$$

$$T = Fr\left[\sin\theta + \cos\theta.\frac{\sin\phi}{\cos\phi}\right]$$

Substituting value of $\sin\phi = \dfrac{\sin\theta}{n}$ and $\cos\phi = \dfrac{1}{n}\sqrt{n^2 - \sin^2\theta}$

$$T = Fr\left[\sin\theta + \cos\theta.\frac{\sin\theta}{n}.\frac{1}{\frac{1}{n}\sqrt{n^2 - \sin^2\theta}}\right]$$

$$= Fr\left[\sin\theta + \frac{2\sin\theta\cos\theta}{2\sqrt{n^2 - \sin^2\theta}}\right]$$

$$T = Fr\left[\sin\theta + \frac{\sin 2\theta}{2\sqrt{n^2 - \sin^2\theta}}\right] \qquad \qquad ...(7.23)$$

Also, from ΔAON and ΔNOD

$$ON = r\sin(\theta + \phi) = OD\cos\phi$$

$$T = F_t \times r$$

$$= \frac{F}{\cos\phi} \times \sin(\theta + \phi) \times r \qquad \qquad \text{[from Eq. (7.21)]}$$

$$= \frac{F}{\cos\phi} \times OD\cos\phi$$

or $$T = OD \times F \qquad \qquad ...(7.24)$$

Example 7.2: A single cylinder engine has a bore of 36 cm, a stroke of 36 cm and a 72 cm long connecting rod. The mass of reciprocating parts is 120 kg. If the speed of the engine is 360 rpm, determine the approximate turning moment of the crank shaft at the instant when the crank and connecting rod are perpendicular to each other. Net pressure on piston at the instant is 70 N/cm².

Solution: Speed of the engine, $N = 360$ rpm

$$\omega = \frac{2\pi N}{60} = \frac{2\pi \times 360}{60} = 37.7 \text{ rad/s}$$

$\theta = 90°$ (crank and connecting rod are perpendicular)
Bore, $D = 36$ cm = 0.36 m
Stroke, $L = 36$ cm = 0.36 m

Length of crank (crank radius), $r = \dfrac{L}{2} = 18$ cm = 0.18 m

Length of connecting rod, $l = 72$ cm = 0.72 m
Mass of reciprocating parts, $m = 120$ kg
Net pressure on piston, $P = 70$ N/cm²

$$n = \frac{l}{r} = \frac{72}{18} = 4$$

Piston area, $A = \dfrac{\pi}{4} D^2 = \dfrac{\pi}{4} \times (36)^2 = 1017.8$ cm²

Net Force on piston due to gas pressure, $F_p = P \cdot A = 70 \times 1017.8$
$= 71251.3$ N

Inertia force of reciprocating parts, $F_i = mr\omega^2 \left(\cos\theta + \dfrac{\cos 2\theta}{n} \right)$

or
$$F_i = -120 \times 0.18 \times (37.7)^2 \times \left[\cos 90 + \frac{\cos 2 \times 90}{4} \right]$$
$$F_i = +7675 \text{ N}$$

Net force, $F = F_p + F_i = 71251.3 + 7675 = 78926.3$ N

$$\sin\varphi = \frac{\sin\theta}{n} = \frac{\sin 90°}{4} = \frac{1}{4} = 0.25$$

∴ $\varphi = \sin^{-1}(0.25) = 14.47°$

Turning moment on the crank shaft, $T = \dfrac{F\sin(\theta + \varphi)}{\cos\varphi} \times r$

or
$$T = 78926.3 \times \frac{\sin(90 + 14.47)}{\cos 14.47} \times 0.18$$

$$T = 14206.7 \text{ N·m}$$

Example 7.3: A vertical petrol engine 150 mm diameter and 200 mm stroke has connecting rod 350 mm long. The mass of piston is 1.6 kg and engine speed is 1800 rpm. On the expansion stroke, at crank angle 30° from TDC, the gas pressure is 750 kN/m². Determine the net thrust on the engine.

Solution: $D = 150$ mm = 0.150 m.

Stroke, $L = 200$ mm $= 0.20$ m, $r = \dfrac{0.2}{2} = 0.1$

Length of connecting rod, $l = 350$ mm $= 0.35$ m

Mass of piston, $m = 1.6$ kg, $N = 1800$ rpm, $\dfrac{l}{r} = \dfrac{0.35}{0.1} = 3.5$

Gas pressure, $P = 750$ kN/m^2 $= 750 \times 10^3$ N/m^2

$$w = \dfrac{2\pi N}{60} = \dfrac{2\pi \times 1800}{60} = 188.4 \text{ rad/s}$$

Force due to gas pressure, $F_p = P \times \dfrac{\pi}{4} D^2 = 750 \times 10^3 \times \dfrac{\pi}{4} \times (0.15^2)$

$$F_p = 13246.875 \text{ N}$$

Inertia force, $\qquad F_i = -ma = -mrw^2 \left[\cos\theta + \dfrac{\cos 2\theta}{n} \right]$

or $\qquad F_i = -1.6 \times 0.1 \times (188.4^2) \times \left[\cos 30° + \dfrac{\cos 60°}{3.5} \right]$

$$= -5729.4 \text{ N}$$

Weight, $\qquad W = mg = 1.6 \times 9.8 = 15.68$ N

Net force, $\qquad F = F_p + F_i + W$

$$= 13246.875 - 5729.4 + 15.68$$

$$F = 7533.155 \text{ N}$$

Example 7.4: Determine the required input torque on the crank of a slider–crank mechanism for the static equilibrium when the applied piston load is 1500 N. The length of the crank and connecting rod are 40 mm and 100 mm respectively and the crank has turned through 45° from the inner-dead center.

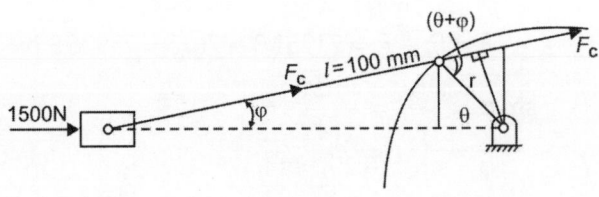

Fig. 7.11

Solution: $\qquad r = 40$ mm, $\theta = 45°$, $n = \dfrac{l}{r} = \dfrac{100}{40} = 2.5$

$$\sin\phi = \dfrac{\sin\theta}{n} = \dfrac{\sin 45°}{2.5} = 0.2828$$

or $\qquad\qquad \phi = 16.429°$

For static equilibrium, $F_c \cos\phi = 1500$

or $\qquad\qquad F_c = \dfrac{1500}{\cos\phi} = \dfrac{1500}{\cos 16.429°} = 1563.8$ N

Input torque,
$$T = F_c \times r \sin(\theta + \phi)$$
$$= 1563.8 \times 0.04 \times \sin(45° + 16.429°)$$
$$= 54.9 \approx 55 \text{ N·m}$$

7.17 DYNAMICALLY EQUIVALENT SYSTEM

Previously, mass of connecting rod was ignored. But in actual practice mass of the connecting rod is also significant, hence, inertia due to the same should also be taken into account. As neither the mass of the connecting rod is uniformly distributed nor the motion is linear, hence, it is difficult to find the inertia of the connecting rod. Inertia of connecting rod is determined by replacing it with a dynamically equivalent system of two point masses. The system of two point masses has the same motion as the connecting rod when subjected to the same force. The two mass system will be dynamically equivalent to the rigid body (connecting rod) if

    i. the sum of two masses is equal to the mass of the rigid body ($m_1 + m_2 = m$)

    ii. the combined centre of mass of two masses coincides with that of the rigid body ($m_1 L_1 = m_2 L_2$)

    iii. the total moment of inertia of two masses about perpendicular axis through their combined centre of mass is equal to that of the rigid body ($m_1 L_1^2 + m_2 L_2^2 = I$).

Figure 7.12 shows a rigid body of mass m with centre of mass at G.

Let m_1 = mass at B

    m_2 = mass at C

    L = distance between two masses, m_1 and m_2

    L_1 = distance of mass m_1 from G

    L_2 = distance of mass m_2 from G

    k = radius of gyration of rigid body about an axis through G.

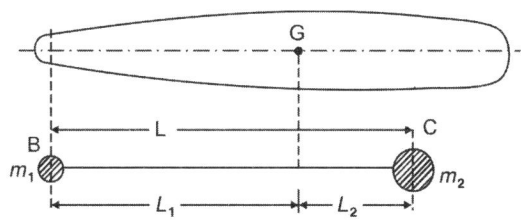

Fig. 7.12

For system of two masses to be dynamically equivalent, conditions are:

    i. $m_1 + m_2 = m$ ...(7.25)

    ii. $m_1 \times L_1 = m_2 \times L_2$ or $m_2 = \dfrac{m_1 L_1}{L_2}$ ...(7.25a)

    iii. $m_1(L_1)^2 + m_2(L_2)^2 = I = mk^2$ ...(7.25b)

From Eq. (7.25), substituting value of m_2 in Eq. (7.25a), we have

$$m_1 + \frac{m_1 L_1}{L_2} = m$$

or $$m_1\left(1+\frac{L_1}{L_2}\right) = m \text{ or } m_1\left(\frac{L_1+L_2}{L_2}\right) = m$$

$$m_1 = \frac{m \cdot L_2}{L_1+L_2} \qquad \qquad ...(7.26)$$

and $$m_2 = \frac{m_1 L_1}{L_2} = \frac{m \cdot L_2}{(L_1+L_2)} \times \frac{L_1}{L_2}$$

$$m_2 = \frac{m \cdot L_1}{L_1+L_2} \qquad \qquad ...(7.27)$$

Now substituting value of m_1 and m_2 in Eq. (7.25b), we have

$$\left[\frac{m \cdot L_2}{(L_1+L_2)}\right](L_1)^2 + \left[\frac{m \cdot L_1}{(L_1+L_2)}\right](L_2)^2 = mk^2$$

or $$mL_1L_2\left[\frac{(L_1+L_2)}{(L_1+L_2)}\right] = mk^2$$

or $$k^2 = L_1/L_2 \qquad \qquad ...(7.28)$$

Therefore, either L_1 or L_2 is choosen arbitrarily and other is determined from Eq. (7.28). If radius of gyration k is unknown, then to find the value of L_2, the body is suspended vertically so that it is free to swing about the axis passing through B. This body will have same period of oscillation as the simple pendulum. Length of simple pendulum, $L = L_1 + L_2$

$\therefore$ $\quad\quad\quad\quad\quad L_2 = (L - L_1)$ is determined.

7.18 CORRECTION COUPLE

It is convenient to consider two point masses at the two ends of the rigid body (in connecting rod, one mass at piston pin end and other mass at crank pin end) (Fig. 7.13). But, then condition iii of equivalence is not satisfied. Hence, the two-mass system is not dynamically equivalent to the rigid body. To make this system dynamically equivalent, the third condition should be satisfied.

Let two masses m_1 and m_3 are placed at B and A at a distance L_1 and L_3 from the centre of mass of the body.

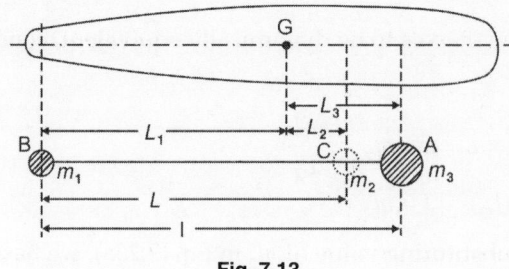

Fig. 7.13

Then, $\quad\quad\quad m_1 + m_3 = m \qquad \qquad ...(7.29)$
$\quad\quad\quad\quad\quad m_1 L_1 = m_3 L_3 \qquad \qquad ...(7.29a)$

From Eqs (7.29) and (7.29a)

$$m_1 = \frac{m \cdot L_3}{(L_1 + L_3)}$$

$$m_3 = \frac{m \cdot L_1}{(L_1 + L_3)}$$

New mass moment of intertia $I_1 = m_1 L_1^2 + m_3 L_3^2$.

Substituting values of m_1 and m_3 and simplifying

$$I_1 = m \cdot L_1 \cdot L_3 \qquad \qquad ...(7.29b)$$

Since $L_3 > L_2$, therefore $I_1 > I$

That means inertia torque $(T_1 - I_1 \alpha_c)$ is also increased. The difference of two torques T_1 and T $(= I\alpha_c)$ is known as correction couple. This correction couple must be applied in the opposite direction to that of the applied inertia torque to make the system dynamically equivalent (Eq. 7.14).

Correction couple, $\quad \Delta T = I_1 \alpha_c - I \alpha_c = (I_1 - I)\, \alpha_c$

$$= (mL_1 L_3 - mL_1 L_2)\, \alpha_c$$

$$\Delta T = mL_1 (L_3 - L_2)\, \alpha_c$$

$$\Delta T = mL_1 \alpha_c [(L_3 + L_1) - (L_2 + L_1)]$$

(adding L_1 to both L_3 and L_2)

or $\qquad\qquad \Delta T = mL_1\alpha_c [l - L] \qquad \begin{matrix} (L_1 + L_2 = L) \\ (L_1 + L_3 = l) \end{matrix} \qquad ...(7.30)$

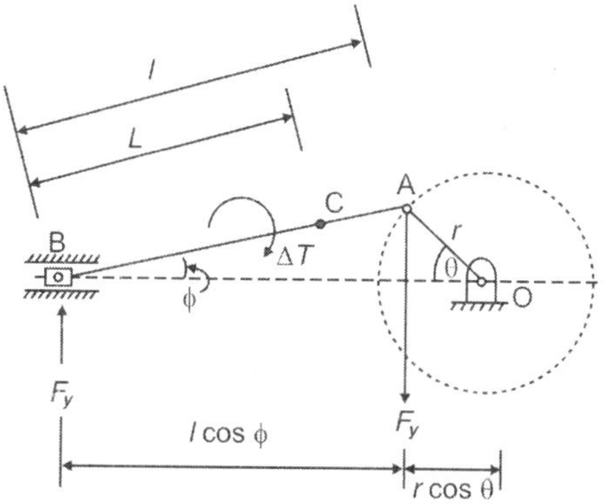

Fig. 7.14

Since direction of inertia torque is always opposite to the direction of the angular acceleration, hence, the direction of the correction couple will be same as that of angular acceleration (direction of decreasing angle ϕ).

The correction couple is produced by two equal, parallel and opposite forces F_y acting at the gudgeon pin end and crank pin end perpendicular to the line of stroke. Force (F_y) at gudgeon pin end is provided by the reaction of guides.

$$\Delta T = F_y \times l \cos \phi$$

Turning moment at crank shaft due to force F_y at A (correction torque)

$$T_c = F_y \times r \cos \theta$$

or
$$T_c = \frac{\Delta T}{l\cos\phi} \times r\cos\theta = \frac{\Delta T}{l/r} \cdot \frac{\cos\theta}{\cos\phi}$$

$$= \frac{\Delta T}{n} \times \frac{\cos\theta}{\dfrac{1}{n}\sqrt{n^2 - \sin^2\theta}}$$

or
$$T_c = \Delta T \times \frac{\cos\theta}{\sqrt{n^2 - \sin^2\theta}} \qquad ...(7.31)$$

Torque exerted at the crankshaft due to weight of the mass at A,

$$T_A = (m_3 g) \times r\cos\theta \qquad ...(7.32)$$

In vertical engines, a torque is exerted on the crankshaft due to weight of mass at B,

$$T_B = (m_1 g) \times r\left[\sin\theta + \frac{\sin 2\theta}{2\sqrt{n^2 - \sin^2\theta}}\right] \qquad ...(7.33)$$

From Eq. (7.33) net torque on crankshaft will be the algebraic sum of torque due to pressure (T), correction torque (T_c), T_A and T_B.

Example 7.5: The following data relates to the connecting rod of a reciprocating engine:

Mass = 50 kg

Distance between bearing centres = 900 mm

Diameter of big end bearing = 100 mm

Diameter of small end bearing = 80 mm

Time of oscillation when the connecting rod is suspended from

big end = 1.7 s

small end = 1.85 s

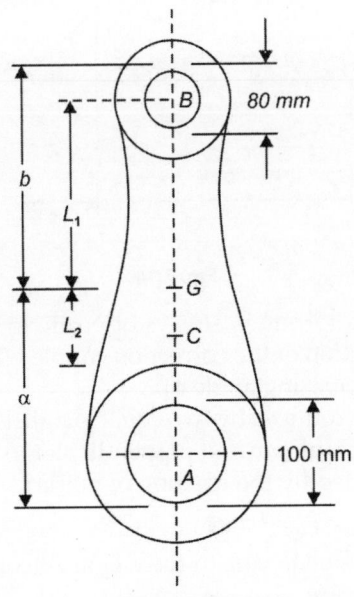

Fig. 7.15

Determine:

i.   radius of gyration k of the rod about an axis through centre of mass perpendicular to the plane of oscillation
ii. moment of inertia of the rod about the same axis, and
iii. dynamically equivalent system of the connecting rod comprising two masses, one at the small end bearing centre.

Solution:

Let L_a = length of equivalent simple pendulum when suspended from the top of big end bearing

L_b = length of equivalent simple pendulum when suspended from the top of small end bearing

a = distance of centre of mass G from top of big-end bearing

b = distance of centre of mass G from top of small-end bearing

$$t_a = 2\pi\sqrt{\frac{L_a}{g}} \quad \text{and} \quad t_b = 2\pi\sqrt{\frac{L_b}{g}}$$

or

$$1.7 = 2\pi\sqrt{\frac{L_a}{g}} \quad \text{and} \quad 1.85 = 2\pi\sqrt{\frac{L_b}{9.81}}$$

or

$$L_a = 0.7181 \text{ m and } L_b = 0.8505 \text{ m}$$

or

$$a + \frac{k^2}{a} = 0.7181 \text{ and } b + \frac{k^2}{b} = 0.8505 \text{ m}$$

or

$$k^2 = 0.7181a - a^2 = 0.8505b - b^2$$

But

$$a + b = 900 + \frac{100}{2} + \frac{80}{2} = 990 \text{ mm} = 0.99 \text{ m or } a = 0.99 - b$$

∴ (i) From the given values, we have

$$0.7181(0.99 - b) - (0.99 - b)^2 = 0.8505b - b^2$$

or $0.7109 - 0.718b - (0.9801 + b^2 - 1.98b) = 8505b - b^2$

or $$0.4115b = 0.2692$$

or $$b = 0.654 \text{ m and } a = 0.99 - 0.654 = 0.336 \text{ m}$$

$$k^2 = 0.8505 \times 0.654 - (0.654)^2 = 0.1286$$

or $$k = 0.358 \text{ m}$$

(ii) Moment of inertia $I = mk^2 = 50 \times (0.358)^2 = 6.4 \text{ kg·m}^2$

(iii) Distance of centre of mass of connecting rod from the centre of small end bearing.

$$L_1 = 654 - (80/2) = 614 \text{ mm}$$

Let the second mass be placed at D. Take $GD = L_2$ and m_2 = mass at C

Then,

$$L_2 = \frac{K^2}{L_1} = \frac{0.1285}{0.614} = 0.209 \text{ m}$$

$$m_2 = \frac{m \times L_1}{L_1 + L_2} = \frac{50 \times 0.614}{0.614 + 0.209} = 37.3 \text{ kg}$$

$$m_1 = 50 - 37.3 = 12.7 \text{ kg}$$

Example 7.6: A connecting rod 220 mm long between centres, has a mass of 2 kg and moment of inertia of 2×10^4 kg mm² about its centre of gravity. Centre of gravity is located at a distance of 150 mm from the small end centre. Determine the dynamically equivalent two mass system when one mass is located at the small end centre.

If the connecting rod is replaced by two masses located at the two centres, find the correction couple that must be applied for complete dynamical equivalence of the system, when the angular acceleration of the connecting rod is 20000 rad/s² clockwise.

Solution: Distance between centres, $l = 220$ mm

Mass, $m = 2$ kg; moment of inertia, $I = 2 \times 10^4$ kg mm²

Distance of CG from the small end centre $L_1 = 150$ mm

∴ Distance of CG of rod from big end centre = 220 – 150 = 70 mm

Let m_1 = first mass which is placed at small end centre

m_2 = second mass which is to be placed in such a way that the system becomes dynamically equivalent

L_2 = distance of mass m_2 from the centre of gravity of the rod.

Figure 7.16 shows the given connecting rod and the equivalent dynamical system of two masses m_1 and m_2.

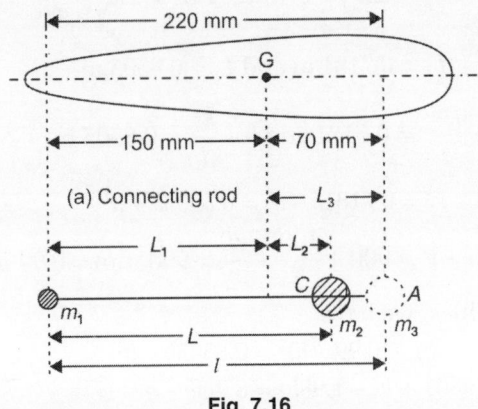

Fig. 7.16

For equivalent dynamical system,
$$k^2 = L_1 \times L_2 \qquad \qquad \text{...(i)}$$
Let us find the value of k from mass moment of inertia of the connecting rod.
$$I = mk^2$$
$$2 \times 10^4 = 2 \times k^2 \qquad \qquad [\because I = 2 \times 10^4 \text{ kg·mm}^2 \text{ and } m = 2 \text{ kg}]$$
$$k = 10^2 = 100 \text{ mm}$$
Substituting the value of k in Eq. (i), we get
$$100^2 = L_1 \times L_2$$
$$= 150 \times L_2 \qquad \qquad [\because L_1 = 150 \text{ mm}]$$

∴
$$= L_2 = \frac{100^2}{150} = 66.67 \text{ mm}$$

The value of m_1 is given as
$$m_1 = \frac{m \times L_2}{(L_1 + L_2)} = \frac{2 \times 66.67}{(150 + 66.67)} = 0.615 \text{ kg}$$

The value of m_2 is obtained from

$$m_1 + m_2 = m$$

or

$$m_2 = m - m_1 = 2 - 0.615 = 1.385 \text{ kg}$$

Correction couple (ΔT)

Angular acceleration of connecting rod = 20000 rad/s² clockwise

∴

$$a = 20000 \text{ rad/s}^2$$

When the two masses are located at the two centres of the connecting rod (i.e. mass m_1 at the small end centre and mass m_3 at the big end centre) then for complete dynamical equivalence of the system, the correction couple (ΔT) that must be applied is given as

$$\Delta T = m \times L_1 \times (l - L) \times \alpha_c$$

where L_1 = 150 mm

l = distance between two centres = 220 mm

L = distance between two masses which are dynamically equivalent to the given rigid body

$$= L_1 + L_2 = 150 + 66.67 = 216.67 \text{ mm}$$

$$\Delta T = 2 \times 150 \times (220 - 216.67) \times 20000 = 19980000 \text{ kg·mm}^2/\text{s}^2$$

$$= \frac{19980000}{1000 \times 1000} \text{ kg·m}^2/\text{s}^2 \qquad \left[\because \text{kg} \times \frac{\text{m}}{\text{s}^2} = \text{N} \right]$$

$$= 19.98 \text{ (kg·m/s}^2) \times \text{m} = 19.98 \text{ Nm.}$$

This correction couple has the same sense as angular acceleration, hence acts clockwise.

7.19 TURNING MOMENT DIAGRAMS

Turning moment diagram is a diagram that shows the variation of turning moment or torque acting on crank shaft for various positions of crank. Hence, it is a plot between T and θ. Turning moment is taken along y axis and crank angle θ is taken along x axis. The inertia effect of connecting rod is generally neglected while drawing these diagrams, but can be considered if required.

Varying torque (T) during one revolution of crankshaft of a steam engine or IC engine is given by

$$T - F_t \times r$$

or

$$T = F \times r \left[\sin \theta + \frac{\sin 2\theta}{2\sqrt{n^2 - \sin^2 \theta}} \right] \qquad \text{...(7.33a)}$$

Here, F = net piston effort and F_t = crank effort.

Crank effort diagram ($F_t V_s \theta$) is identical to a turning moment diagram.

The turning moment diagram for different types of engines are described below.

7.19.1 Single Cylinder Double-acting Steam Engine

The variation of turning moment with crank angle θ for single cylinder double-acting steam engine is 8shown in Fig. 7.17.

Turning moment is zero when θ = 0° and 180° and it is maximum when θ is little less than 90°.

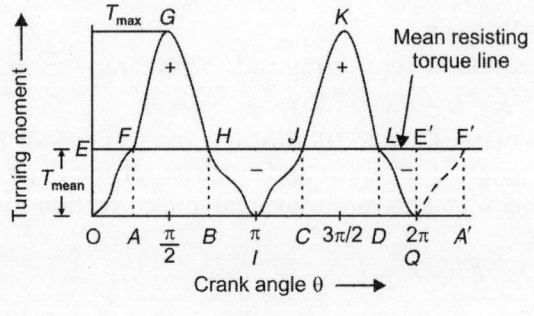

Fig. 7.17

Work done = $T \cdot \theta$

Work done per revolution

$$= T \times 2\pi \qquad\qquad (\theta = 2\pi)$$

= area of turning moment diagram for one revolution of crank

This relation implies that the area of the turning moment diagram is proportional to the work done per revolution. In the actual practice, the engine is assumed to work against the uniform mean resisting torque.

Mean resisting torque, $T_{mean} = \dfrac{\text{Area } OGIKQ}{2\pi} = OE$

Area of rectangle $OEE'Q = OE \times OQ$

$$= T_{mean} \times 2\pi \qquad\qquad (DE = T_{mean}, OQ = 2\pi)$$

$$= \text{Area } OGIKQ$$

$$= \text{Work done per revolution}$$

Hence, area of rectangle $OEE'Q$ represents the work done per revolution against mean resisting torque.

Let T = torque at any instant on the crankshaft.

If $T > T_{mean}$, the engine accelerates and the excess work is stored in the flywheel.

If $T < T_{mean}$, engine retards and flywheel gives up some of its energy to make up the required work.

7.19.2 Single Cylinder Four-Stroke Internal Combustion Engine

In four-stroke internal combustion engine, the turning moment diagram repeats itself after every two revolutions. The four strokes of IC engines are:

(i) Suction stroke for which angle θ varies from 0° to 180° (i.e. 0 to π).

(ii) Compression stroke for which θ varies from 180° to 360° (i.e. π to 2π)

(iii) Expansion stroke for which θ varies from 350° to 540° (i.e. 2π to 3π)

(iv) Exhaust stroke for which θ varies from 540° to 720° (i.e. 3π to 4π)

Figure 7.18 shows the turning moment diagram for a four-stroke internal combustion engine.

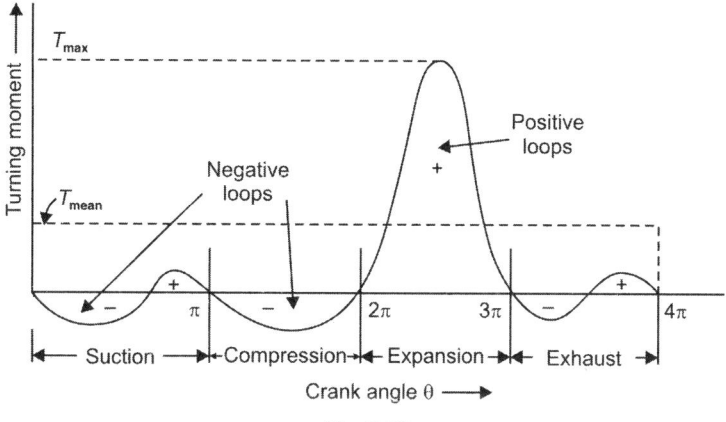

Fig. 7.18

During suction stroke, the pressure inside the cylinder is less than the atmospheric pressure and hence, the turning moment on the crank is negative for most of the suction stroke. Hence, a negative loop is formed as shown in Fig. 7.18. During compression stroke, the work is done by piston on the gases and hence, a large negative loop is obtained. During expansion stroke, the work is done by the gases on the piston and due to this, the turning moment on the crank is positive. Hence, during this stroke a large positive loop is obtained. During exhaust stroke, the work is done by the piston on gases and hence, turning moment on crank is negative for most of the exhaust stroke.

7.19.3 Multicylinder Engines

For multicylinder engines, the total turning moment at any instant is obtained by adding the turning moments developed by each cylinder at that instant.

The variation in turning moment is less with more no. of cylinders taken in engine. The speed of the engine will be maximum at crank positions *B*, *D* and *F* and minimum corresponding to *C*, *E* and *G*.

7.20 FLUCTUATION OF ENERGY

In Fig. 7.19, the areas a_1, a_3 and a_5 represent quantities of energies added to the flywheel. Similarly, areas a_2, a_4 and a_6 represent quantities of energies taken from the flywheel.

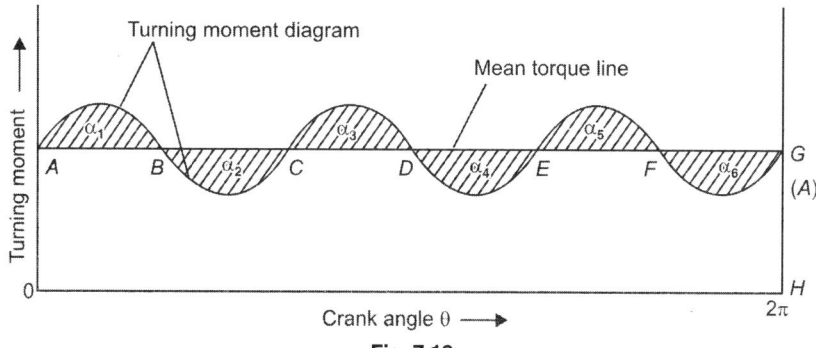

Fig. 7.19

Let E = Energy in the flywheel corresponding to point A.

The energies of the flywheel corresponding to points B, C, D, E, F and G are obtained from Fig. 7.19 as follows:

Energy at B = Energy at A + Area $a_1 = E + a_1$

Energy at C = Energy at B – Area $a_2 = E + a_1 - a_2$

Energy at D = Energy at $C + a_3 = E + a_1 - a_2 + a_3$

Energy at E = Energy at $D - a_4 = E + a_1 - a_2 + a_3 - a_4$

Energy at F = Energy at $E + a_5 = E + a_1 - a_2 + a_3 - a_4 + a_5$

Energy at G = Energy at $F - a_6 = E + a_1 - a_2 + a_3 - a_4 + a_5 - a_6$

The cycle repeats after G. Hence energy at G is equal to energy at A i.e., $E + a_1 - a_2 + a_3 - a_4 + a_5 - a_6 = E$.

The greatest of these energies is the maximum kinetic energy of the flywheel. And the least of these energies is the least kinetic energy of the flywheel. The difference between the greatest and least energies is known as maximum fluctuation of energy.

$\therefore$ Maximum fluctuation of energy = Greatest energy – least energy.

7.20.1 Coefficient of Fluctiation of Energy (K_e)

The ratio of maximum fluctuation of energy to the work done per cycle is known as coefficient of fluctuation of energy. It is represented by K_e. Mathematically, coefficient of fluctuation of energy is given by

$$K_e = \frac{\text{Maximum fluctuation of energy}}{\text{Work done per cycle}}$$

7.20.2 Fluctuation of Speed

When the crank moves from A to B, the flywheel starts absorbing energy. When the crank is at B, the maximum energy (this energy corresponds to area a_1) has been absorbed in the flywheel. Hence, the speed of the flywheel becomes maximum.

When the crank moves from B to C, the flywheel starts giving out the energy. When the crank is at C, the maximum energy (corresponding to area a_2) has been given out by the flywheel. Hence, the speed of the flywheel becomes minimum. Similarly, the speed of flywheel is maximum when crank is at D.

The greatest speed is the greater of the two maximum speeds and the least speed is the lesser of the two minimum speeds.

The *fluctuation of speed* is the difference between the greatest and the least speeds of the crank-shaft for one revolution.

7.20.3 Coefficient of Fluctuation of Speed (K_s)

The ratio of maximum fluctuation of speed to the mean speed is known as coefficient of fluctuation of speed. It is represented by K_s. Mathematically, the coefficient of fluctuation of speed is given by,

$$K_s = \frac{\text{Maximum fluctuation of speed}}{\text{Mean speed}}$$

ω_1, ω_2 and ω = Corresponding maximum angular speed, minimum and mean angular speeds during the cycle. Then coefficient of fluctuation of speed,

$$K_s = \frac{\text{Maximum speed} - \text{minimum speed}}{\text{Mean speed}} \quad \text{or} \left(\frac{\omega_1 - \omega_2}{\omega} \right)$$

$$= \frac{N_1 - N_2}{N} = \frac{N_1 - N_2}{\left(\dfrac{N_1 + N_2}{2} \right)} = \frac{2(N_1 - N_2)}{(N_1 + N_2)}$$

$$= \frac{2(\omega_1 - \omega_2)}{\omega_1 + \omega_2} \quad \left(\because \omega_1 = \frac{2\pi N_1}{60} \text{ and } \omega_2 = \frac{2\pi N_2}{60} \right)$$

7.21 FLYWHEEL

A flywheel is used to control the speed variations caused by the fluctuation of energy during each cycle of operation. It acts as a reservoir of energy which stores energy during the period when the supply of energy is more than the requirement and releases the energy during the period when the supply of energy is less than the requirement. When the flywheel absorbs the energy, the speed of flywheel increases whereas when the flywheel releases the energy, the speed of flywheel decreases.

Let N_1 = Maximum speed of flywheel in rpm during the cycle

$\quad\;\; N_2$ = Minimum speed during the cycle.

$\quad \omega_1$ and ω_2 = Corresponding maximum and minimum angular speeds

$\quad\quad\;\; I$ = Moment of inertia of flywheel

$\quad\;\; m$ = Mass of flywheel

$\quad\;\; k$ = Radius of gyration of the flywheel

$\quad \Delta E$ = Maximum fluctuation of energy

$\quad\; K_e$ = Coefficient of fluctuation of energy

$\quad\; K_s$ = Coefficient of fluctuation of speed

$$N = \text{Mean speed of flywheel during the cycle} = \frac{N_1 + N_2}{2}$$

$$\omega = \text{Mean angular speed of flywheel during the cycle} = \frac{\omega_1 + \omega_2}{2}$$

$\quad\; E$ = Kinetic energy of flywheel at mean speed.

We know that the kinetic energy of the flywheel corresponding to mean angular velocity is given by

$$E = \frac{1}{2} \times I \times \omega^2$$

$$= \frac{1}{2} \times mk^2 \times \omega^2 \quad\quad\quad \left(\because I = mk^2 \right)$$

The speed of flywheel changes from ω_1 to ω_2, hence, maximum fluctuation of energy is given by,

$$\Delta E = \text{Maximum KE} - \text{minimum KE}$$

$$= \frac{1}{2} \times I \times \omega_1^2 - \frac{1}{2} \times I \times \omega_2^2$$

$$= \frac{1}{2} I(\omega_1^2 - \omega_2^2) = \frac{1}{2} I(\omega_1 + \omega_2)(\omega_1 - \omega_2)$$

$$= I \times \omega \times (\omega_1 - \omega_2) \qquad\qquad \left(\frac{\omega_1 + \omega_2}{2} = \omega\right)$$

$$= I \times \omega^2 \left(\frac{\omega_1 - \omega_2}{\omega}\right) \quad ...\text{[multiplying and dividing by } \omega]$$

But $\dfrac{\text{Max. speed} - \text{min. speed}}{\text{Mean speed}} = \dfrac{\omega_1 - \omega_2}{\omega} = K_s$

∴ The above equation becomes as,

$$\Delta E = I \times \omega^2 \times K_s \qquad\qquad\qquad ...(7.34)$$

$$= \frac{1}{2} I \times \omega^2 \times 2 \times K_s \quad ...\text{[multiplying and dividing by 2]}$$

$$= E \times 2 \times K_s \qquad\qquad ...\left(\because \frac{1}{2}I\omega^2 = E\right)$$

$$= 2E \times K_s \qquad\qquad\qquad\qquad ...(7.35)$$

Equation (7.34) can be written as

$$\Delta E = I \times \omega^2 \times K_s = mk^2 \times \omega^2 \times K_s \ (\because I = mk^2) \qquad ...(7.36)$$

If the thickness of the rim of the flywheel is very small as compared to the diameter of flywheel, then radius of gyration may be taken equal to mean radius of the flywheel, i.e. $k = R$. Hence, substituting the value of $k = R$ in Eq. (7.36), we get

$$\Delta E = m \times R^2 \times \omega^2 \times K_s$$

$$= m \times v^2 \times K_s \qquad (\because v = \omega \times R) \qquad\qquad ...(7.37)$$

Example 7.7: The turning moment diagram for a petrol engine is drawn to a vertical scale of 1 mm to 6 N·m and a horizontal scale of 1 mm to 1°. The turning moment repeats itself after every half revolution of engine. The areas above and below the mean torque line are: 305, 710, 50, 350, 980 and 275 mm² respectively. The rotating parts amounts to a mass of 40 kg at a radius of gyration of 140 mm. Calculate the coefficient of fluctuation of speed if the speed of engine is 1500 rpm.

Solution: Let on x axis, 1 mm = 1° and on y axis 1 mm = 6 N·m

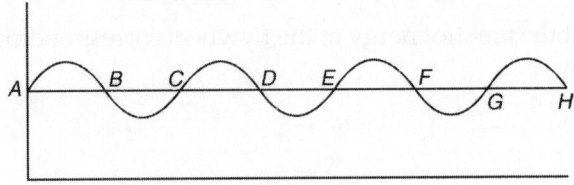

Fig. 7.20

Let energy at $A = E$, then

$E_A = E$

$E_B = E + 305$ (maximum KE)

$E_C = E + 305 - 710 = E - 405$

$$E_D = E + 405 + 50 = E - 355$$
$$E_E = E - 355 - 350 = E - 705 \text{ (minimum KE)}$$
$$E_F = E - 705 + 980 = E + 275$$
$$E_G = E + 275 - 275 = E$$
$$e = (\text{max. KE} - \text{min. KE}) = E + 305 - E + 705 = 1010 \text{ mm}^2$$

$$= 1010 \times 6 \times 1 \times \frac{\pi}{180} = 105.766 \text{ J}$$

$$m = 40 \text{ kg}, \, k = 140 \text{ mm}, \, N = 1500, \, \omega = \frac{2\pi N}{60} = \frac{2\pi \times 1500}{60} = 157.07$$

Coefficient of fluctuation of speed $K_s = \dfrac{e}{I\omega^2} = \dfrac{105.766}{mk^2\omega^2}$

or $$K_s = \frac{105.766}{40 \times (0.14)^2 \times (157.07)^2} = 0.00546$$

or $$= 0.546\% \approx 0.55\%$$

Example 7.8: The maximum and minimum speeds of a flywheel are 121 rpm and 119 rpm respectively. The mass of the flywheel is 2600 kg and radius of gyration is 1.8 m. Find: (i) mean speed of flywheel, (ii) maximum fluctuation of energy, and (iii) coefficient of fluctuation of speed.

Solution: $N_{max} = 121$ rpm, $N_{min} = 119$ rpm

$$m = 2600 \text{ kg}, \, k = 1.8 \text{ m}$$

$$\omega_{max} = \frac{2\pi N_{max}}{60} = \frac{2\pi \times 121}{60} = 12.671 \text{ rad/s}$$

$$\omega_{min} = \frac{2\pi N_{min}}{60} = \frac{2\pi \times 119}{60} = 12.4616 \text{ rad/s}$$

$$I = mk^2 = 2600 \times 1.8^2 = 8424 \text{ kg·m}^2$$

(i) $$N_{mean} = \frac{N_{max} + N_{min}}{2} = \frac{121 + 119}{2} = 120 \text{ rpm}$$

(ii) $$e = \frac{1}{2}I(\omega_{max^2} - \omega_{min^2}) = \frac{1}{2} \times 8424 \times (12.671)^2 - (12.4616)^2$$

$$= 22166.7 \text{ J}$$

(iii) Coefficient of fluctuation of speed $K_s = \dfrac{e}{2E} = \dfrac{22166.7}{2 \times \frac{1}{2}I\omega^2}$

or $$K_s = \frac{22166.7}{8424 \times \left(\dfrac{2\pi \times 120}{60}\right)^2} = 0.0166$$

Example 7.9: Find the maximum and minimum speeds of a flywheel of mass 3250 kg and radius of gyration 1.8 m when the fluctuation of energy is 112 kNm. The mean speed of engine is 240 rpm.

Solution: Given $m = 3250$ kg, $k = 1.8$ m, $e = 112$ kN·m $= 112 \times 1000$ N·m

$$N = 240 \text{ rpm}, \quad \omega = \frac{2\pi N}{60} = \frac{2\pi \times 240}{60} = 8\pi = 25.1 \text{ rad/s}$$

$$I = mk^2 = 3250 \times (1.8)^2 = 10530$$

$$e = \frac{1}{2}I\omega_1^2 - \frac{1}{2}I\omega_2^2 = \frac{1}{2}I(\omega_1^2 - \omega_2^2)$$

or
$$(\omega_1^2 - \omega_2^2) = \frac{2e}{I} = \frac{2 \times 112 \times 1000}{10530} = 21.27$$

or
$$(\omega_1 + \omega_2)(\omega_1 - \omega_2) = 21.27$$

$$\omega = \frac{\omega_1 + \omega_2}{2} = 25.1 \text{ or } (\omega_1 + \omega_2) = 2 \times 25.1 = 50.2$$

Substituting above, we have
$$50.2 \times (\omega_1 - \omega_2) = 21.27 \text{ or } \omega_1 - \omega_2 = 0.423$$
$$\omega_1 + \omega_2 = 50.2$$
$$\omega_1 - \omega_2 = 0.423$$
$$2\omega_1 = 50.623$$
$$\omega_1 = 25.31$$
$$\omega_2 = 50.2 - 25.31 = 24.88$$

$$N_1 = \omega_1 \times \frac{60}{2\pi} = 25.31 \times \frac{60}{2\pi} = 242 \text{ rpm}$$

$$N_2 = \omega_2 \times \frac{60}{2\pi} = 24.88 \times \frac{60}{2\pi} = 238 \text{ rpm}$$

SOLVED EXAMPLES

Example 7.10: A four-link mechanism with the following dimensions is acted upon by a force $80 \angle 150°$ on link *DC* (Fig. 7.21a).

$AD = 50$ mm, $AB = 40$ mm, $BC = 100$ mm, $DC = 75$ mm, $DE = 35$ mm

Determine the input torque T on the link *AB* for the static equilibrium of the mechanism for the given configuration.

Solution: As the mechanism is in static equilibrium, each of its members must also be in equilibrium individually.

Member 4 is acted upon by three forces F, F_{34} and F_{14}

Member 3 is acted upon by two forces F_{23} and F_{43}

Member 2 is acted upon by two forces F_{32} and F_{12} and a torque T.

Initially, the direction and the sense of some of the forces are not known.

Now, adopt the following procedure:

- Force F on member 4 is known completely. To know the other two forces acting on this member completely, the direction of one more force must be known. To know that, link 3 will have to be considered first which is a two-force member.

- As link 3 is a two-force member (Fig. 7.21b), for its equilibrium, F_{23} and F_{43} must act along *BC* (the sense of direction of forces F_{23} and F_{43} is not known). Thus, the line of action of F_{34} is also along *BC*.

- As force F_{34} acts through point C on link 4, draw a line parallel to BC through C by taking a free body of link 4 to represent the same. Now, as link 4 is three force member, the third force F_{14} passes through the intersection of F and F_{34} (Fig. 2.21c). By drawing a force triangle (F is completely known), magnitudes of F_{14} and F_{34} are known (Fig. 7.21d).

From force triangle, $F_{34} = 47.8$ N.

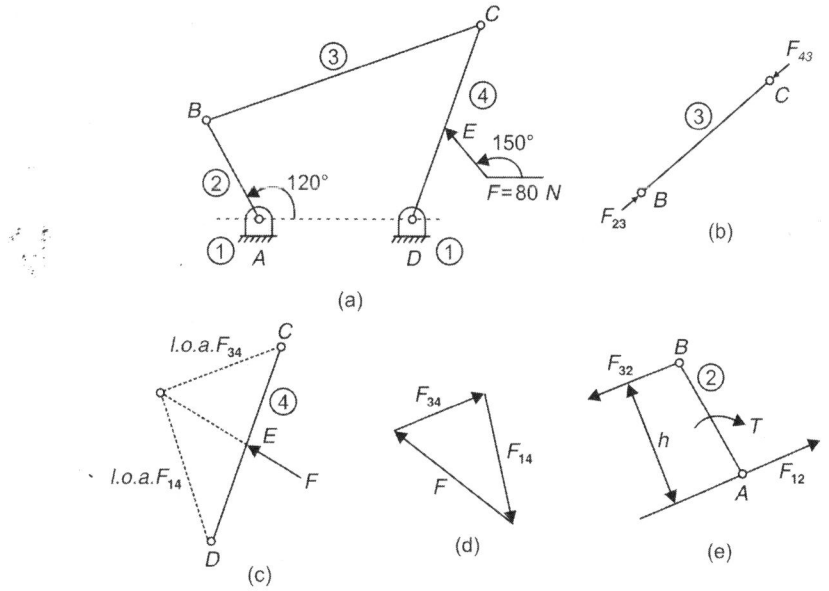

Fig. 7.21

Now, $$F_{34} = -F_{43} = F_{23} = -F_{32}$$

Member 2 will be in equilibrium (Fig. 7.21e) if F_{12} is equal, parallel and opposite to F_{32} and

$$T = -F_{32} \times h = 47.8 \times 39.3 = -18.78 \text{ N·mm}$$

The input torque has to be equal and opposite to this couple, i.e. $T = 18.78$ N·mm (clockwise).

Example 7.11: In a four-link mechanism shown in Fig. 7.22a, torque T_3 and T_4 have magnitude of 30 N·m and 20 N·m respectively. The link lengths are $AD = 800$ mm, $AB = 300$ mm, $BC = 700$ mm and $CD = 400$ mm. For the static equilibrium of the mechanism, determine the required input torque T_2.

Solution: The solution to this problem can be obtained by superposition of the solutions of subproblems a and b.

(i) *Subproblem a* (Fig. 7.22a): Neglecting torque T_3.

Torque T_4 on link 4 is balanced by a couple having two equal, parallel and opposite forces at C and D. As link 3 is a two-force member, F_{43} and therefore, F_{34} and F_{14} will be parallel to BC.

$$F_{34} = F_{14} \frac{T_4}{h_{4a}} = \frac{20}{0.383} = 52.2 \text{ N}$$

and

$$F_{34} = F_{43} = F_{23} = F_{32} = F_{12} = 52.2 \text{ N}$$
$$T_{2a} = F_{32} = h_{2a} = 52.2 \times 0.274$$
$$= 14.3 \text{ N-m anticlockwise}$$

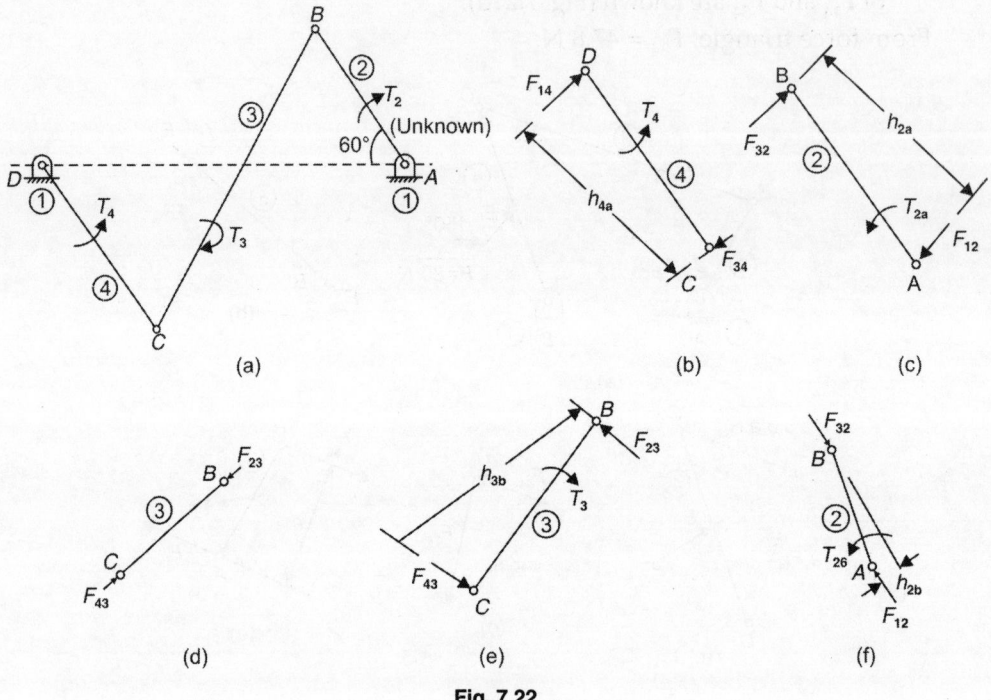

(a) (b) (c)

(d) (e) (f)

Fig. 7.22

(ii) *Subproblem b* (Fig. 7.22b): Neglecting torque T_4. F_{43} is along CD

$$F_{43} = F_{23} = \frac{T_3}{h_{3b}} = \frac{30}{0.67} = 44.8 \text{ N}$$

$$F_{23} = F_{32} = F_{12} = 44.8 \text{ N}$$
$$F_{2b} = F_{32} \times h_{2b} = 44.8 \times 0.042$$
$$= 1.88 \text{ N·m anticlockwise}$$
$$T_2 = T_{2a} + T_{2b} = 14.3 + 1.88$$
$$= 16.18 \text{ N (anticlockwise).}$$

Example 7.12: The lengths of crank and connecting rod of a vertical reciprocating engine are 300 mm and 1.5 m respectively. The crank is rotating at 200 rpm clockwise as shown in Fig. 7.23. Find analytically, (i) acceleration of piston, (ii) velocity of piston, and (iii) angular acceleration of the connecting rod when the crank has turned through 40° from the top dead centre and piston is moving downwards.

Solution: $r = 300 \text{ mm} = 0.3 \text{ m}; l = 1.5 \text{ m}, N = 200 \text{ rpm}$

$$\omega = \frac{2\pi N}{60} = \frac{2\pi \times 200}{60} = 20.944 \text{ rad/s}, \theta = 40°$$

Now, $$n = \frac{l}{r} = \frac{1.5}{0.3} = 5$$

(i) *Acceleration of piston*: Acceleration of piston

$$a = r\omega^2 \left(\cos\theta + \frac{\cos 2\theta}{n} \right)$$

$$= 0.3 \times \left(\frac{2\pi \times 200}{60} \right)^2 \left(\cos 40° + \frac{\cos(2 \times 40)}{5} \right)$$

$$= 0.3 \times 20.944^2 \times \left(0.766 + \frac{0.1736}{5} \right)$$

$$= 0.3 \times 438.65 \times 0.8 = 105.28 \text{ m/s}^2.$$

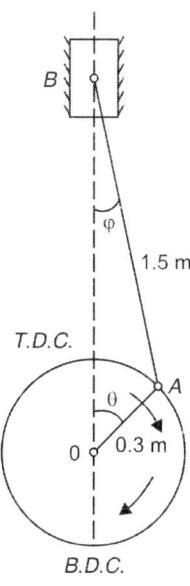

Fig. 7.23

(ii) *Velocity of piston*: The velocity of piston

$$v = r\omega \left(\sin\theta + \frac{\sin 2\theta}{2\sqrt{n^2 - \sin^2\theta}} \right)$$

$$= 0.3 \times \left(\frac{2\pi \times 200}{60} \right) \left(\sin 40° + \frac{\sin(2 \times 40)}{2 \times \sqrt{5^2 - \sin^2 40}} \right)$$

$$= 0.3 \times 20.944 \times \left(0.6428 + \frac{0.9848}{2\sqrt{25 - 0.413}} \right)$$

$$= 0.3 \times 20.944 \times 0.742 = 4.66 \text{ m/s}.$$

(iii) *Angular acceleration of the connecting rod*: Angular acceleration of the connecting rod

$$a_c = -\omega^2 \sin\theta \left[\frac{n^2 - 1}{(n^2 - \sin^2\theta)^{3/2}} \right]$$

$$= -20.944^2 \times \sin 40° \left[\frac{5^2 - 1}{(5^2 - \sin^2 40)^{3/2}} \right]$$

$$= -438.65 \times 0.6428 \left[\frac{24}{(25 - 0.4132)^{3/2}} \right]$$

$$= -438.65 \times 0.6428 \times \frac{24}{121.91} = -55.5 \text{ rad/s}^2.$$

(–ve sign shows that the sense of the acceleration of the connecting rod is such that it tends to reduce the angle ϕ. Hence, connecting rod is rotating clockwise).

Example 7.13: Attributes are same as Example 7.12, hence find (i) the crank angle at which the velocity of piston is maximum and (ii) value of maximum velocity of piston.

Solution: $r = 0.3$ m; $l = 1.5$ m; $\omega = 20.94$ rad/s; $n = 5$

(i) *Crank angle* (θ) *at which velocity of piston is maximum*: The velocity of piston

$$v = rw \left(\sin\theta + \frac{\sin 2\theta}{2\sqrt{n^2 - \sin^2\theta}} \right)$$

The value of $\sin^2\theta$ is small in comparison to n^2 and hence, it may be neglected. Then, the above expression becomes

$$v = rw \left(\sin\theta + \frac{\sin 2\theta}{2n} \right) \qquad \text{...(i)}$$

For maximum velocity of the piston, $\dfrac{dv}{d\theta}$ should be zero.

$$\frac{dv}{d\theta} = r\omega \left(\cos\theta + \frac{2\cos 2\theta}{2n} \right)$$

$\therefore$ For maximum velocity, $\dfrac{dv}{d\theta} = 0$

or $\qquad r\omega \left(\cos\theta + \dfrac{2\cos 2\theta}{2n} \right) = 0$

or $\qquad \cos\theta + \dfrac{\cos 2\theta}{n} = 0$

or $\qquad n\cos\theta + \cos 2\theta = 0$

or $\qquad n\cos\theta + (2\cos^2\theta - 1) = 0 \qquad\qquad (\because \cos 2\theta = 2\cos^2\theta - 1)$

or $\qquad 2\cos^2\theta + n\cos\theta - 1 = 0$

The above equation is a quadratic equation in $\cos\theta$.

$$\therefore \qquad \cos\theta = \frac{-n \pm \sqrt{n^2 + 4 \times 2 \times 1}}{2 \times 2}$$

$$= \frac{-5 \pm \sqrt{5^2 + 8}}{4} \qquad\qquad (\because n = 5)$$

$$= \frac{-5 \pm 5.744}{4} = \frac{0.744}{4} \qquad\qquad \text{(taking +ve value)}$$

$$= 0.186$$

$$\therefore \qquad \theta = \cos^{-1} 0.186 = 79.28°.$$

(ii) *Maximum velocity of piston*: Substituting the value of $\theta = 79.28°$ in equation (i), the value of max. velocity

$\therefore$

$$v_{max} = rw\left(\sin 79.28° + \frac{\sin(2 \times 79.28°)}{2 \times n}\right)$$

$$= 0.3 \times 20.94\left(0.9825 + \frac{0.3655}{2 \times 5}\right)$$

$$(\because n = 5, r = 0.3, \omega = 20.94)$$

$$= 6.4 \text{ m/s}.$$

Example 7.14: The lengths of crank and connecting rod of a horizontal reciprocating engine are 100 mm and 500 mm respectively. The crank is rotating at 400 rpm. When the crank has turned 30° from the inner dead centre, find analytically (i) acceleration of the piston, (ii) velocity of the piston, (iii) angular velocity of the connecting rod, and (iv) angular acceleration of the connecting rod.

Solution: $r = 100$ mm $= 0.1$ m; $l = 500$ mm $= 0.5$ m; $N = 400$ rpm

or $\omega = \dfrac{2\pi N}{60} = \dfrac{2 \times \pi \times 400}{60} = 41.88$ rad/s, $\theta = 30°$. The value of $n = \dfrac{l}{r} = \dfrac{0.5}{0.1} = 5$.

(i) *Acceleration of piston*: Acceleration of piston (a) is given as

$$a = rw^2\left(\cos\theta + \frac{\cos 2\theta}{n}\right)$$

$$= 0.1 \times 41.88^2\left(\cos 30° + \frac{\cos 60°}{5}\right)$$

$$= 0.1 \times 1753.93\left(0.866 + \frac{0.5}{5}\right)$$

$$= 0.1 \times 1753.93 \times 0.966 = 169.43 \text{ m/s}^2.$$

(ii) *Velocity of piston*: Velocity of piston

$$v = rw\left(\sin\theta + \frac{\sin 2\theta}{2\sqrt{n^2 - \sin^2\theta}}\right)$$

$$= 0.1 \times 41.88 \times \left(\sin 30° + \frac{\sin 60°}{2\sqrt{5^2 - \sin^2 30°}}\right)$$

$$= 0.1 \times 41.88 \times \left(0.5 + \frac{0.866}{2\sqrt{25 - 0.25}}\right)$$

$$= 0.1 \times 41.88 (0.5 + 0.087) = 2.458 \text{ m/s}.$$

(iii) *Angular velocity of connecting rod*: Angular acceleration of connecting rod

$$\omega_c = \omega\frac{\cos\theta}{\sqrt{n^2 - \sin^2\theta}}$$

$$= 41.88 \frac{\cos 30°}{\sqrt{5^2 - \sin^2 \theta}} = 41.88 \times \frac{0.866}{\sqrt{25 - 0.25}}$$

$$= 7.29 \text{ rad/s}.$$

(iv) *Angular acceleration of connecting rod*: Angular acceleration of connecting rod

$$\alpha_c = -\omega^2 \sin\theta \left[\frac{n^2 - 1}{(n^2 - \sin^2 \theta)^{3/2}}\right]$$

$$= -41.88^2 \times \sin 30° \left[\frac{5^2 - 1}{(5^2 - \sin^2 30°)^{3/2}}\right]$$

$$= -1753.93 \times 0.5 \left[\frac{25 - 1}{(25 - 0.25)^{3/2}}\right]$$

$$= -1753.93 \times 0.5 \left[\frac{24}{123.13}\right] = -170.93 \text{ rad/s}^2.$$

Example 7.15: Attributes are same as Example 7.14, hence find the position of the crank from the inner deed centre for zero acceleration of the piston.

Solution: $r = 0.1$ m, $l = 0.5$ m, $\omega = 41.88$ rad/s, $n = 5$.

Acceleration of the piston is given as

$$a = r\omega^2 \left(\cos\theta + \frac{\cos 2\theta}{n}\right)$$

For zero acceleration, the value of 'a' should be zero in the above equation. Hence, we get

$$0 = r\omega^2 \left(\cos\theta + \frac{\cos 2\theta}{n}\right)$$

$$\cos\theta + \frac{\cos 2\theta}{n} = 0$$

$$n \cos\theta + \cos 2\theta = 0$$

$$n \cos\theta + (2\cos^2\theta - 1) = 0 \qquad\qquad (\because \cos 2\theta = 2\cos^2\theta - 1)$$

or $\qquad 2\cos^2\theta + n \cos\theta - 1 = 0$

or $\qquad 2\cos^2\theta + 5\cos\theta - 1 = 0 \qquad\qquad (\because n = 5)$

The solution of the above equation is

$$\cos\theta = \frac{-5 + \sqrt{5^2 + 4 \times 2 \times 1}}{2 \times 2} = \frac{-5 \pm \sqrt{25 + 8}}{4}$$

$$= \frac{-5 \pm 5.744}{4}$$

$$= \frac{0.744}{4} = 0.186 \qquad\qquad \text{[taking +ve value]}$$

$$\therefore \qquad\qquad \theta = \cos^{-1} 0.186 = 79.28°.$$

Example 7.16: The lengths of crank and connecting rod of a horizontal reciprocating engine are 200 mm and 1.0 m respectively. The crank is rotating at 400 rpm. When the crank has turned 30° from the inner dead centre, the difference of pressure between the cover end and piston end is 0.4 N/mm². If the mass of the reciprocating parts is 100 kg and cylinder bore is 0.4 m, then calculate: (i) Inertia force, (ii) Force on piston, (iii) Piston effort, (iv) Thrust on the sides of cylinder walls, (v) Thrust in the connecting rod, (vi) Crank effort, and (vii) Turning moment on the crank shaft. Neglect the effect of piston rod diameter and frictional resistance.

Solution:
$$r = 200 \text{ mm} = 0.2 \text{ m}; \ l = 1.0 \text{ m}; \ N = 400 \text{ rpm}$$

or
$$\omega = \frac{2\pi N}{60} = \frac{2\pi \times 400}{60} = 41.88 \text{ rad/s};$$

$$\theta = 30°, \ (p_1 - p_2) = 0.4 \text{ N/mm}^2; \ m = 100 \text{ kg}; \ D = 0.4 \text{ m}$$

(i) *Inertia force*: Inertia force
$$F_i = -m \times a \qquad \text{(−ve sign is due to the fact that inertia force opposes the accelerating force)}$$

where a = acceleration of the piston which is given as

$$= rw^2 \left(\cos\theta + \frac{\cos 2\theta}{n} \right) \text{ where } \theta = 30° \text{ and } n = \frac{l}{r} = \frac{1.0}{0.20} = 5$$

$$= 0.2 \times 41.88^2 \left(\cos 30° + \frac{\cos(2 \times 30)}{5} \right) = 0.2 \times 1753.93 \left(0.866 + \frac{0.5}{5} \right)$$

$$= 0.2 \times 1753.93 \times 0.966 = 338.86 \text{ m/s}^2$$

As the value of a is +ve, hence, piston is accelerating

∴
$$F_i = -m \times a$$
$$= -100 \times 338.86 = -33886 \text{ N.}$$

(ii) *Force on piston*: The force on piston by steam or gas is given by,

$$F_P = P \times \frac{\pi}{4}D^2$$

where p = difference of pressure = $p_1 - p_2 = 0.4 \text{ N/mm}^2$
and D = dia. of piston or bore of cylinder = 0.4 m = 400 mm

∴
$$F_P = 0.4 \times \frac{\pi}{4}(400)^2 \ N = 50265 \text{ N.}$$

(iii) *Piston effort*: For a horizontal reciprocating engine, the piston effort neglecting frictional resistance is given as
$$F^* = F_P + F_i$$
$$= 50265 - 33886$$
$$= 16379 \text{ N.}$$

(iv) *Thrust on the sides of cylinder walls or normal reaction*: The thrust on the sides of cylinder walls (or normal reaction)
$$F_N = F \tan\phi,$$
where F = Piston effort = 16379 N
ϕ = Angle made by connecting rod with line of stroke.

The value of ϕ in terms of θ is given as

$$\sin\theta = \frac{r}{l}\sin\theta$$

$$= \frac{0.2}{1.0} \times \sin 30° \qquad (\because r = 0.2, l = 1.0 \text{ and } \theta = 30°)$$

$$= \frac{0.2}{1.0} \times 0.5 = 0.1$$

$\therefore \qquad \phi = \sin^{-1} 0.1 = 5.739°$

$\therefore \qquad F_N = F \tan\theta = 16379 \times \tan 5.739°$

$$= 16379 \times 0.1005 = 1646.1 \text{ N}.$$

(v) *Thrust in the connecting rod* (F_c): The thrust in the connecting rod is given by,

$$F_c = \frac{F}{\cos\phi}$$

$$= \frac{16379}{\cos 5.739°} \qquad (\because \phi = 5.739°)$$

$$= 16461.5 \text{ N}.$$

(vi) *Crank effort* (F_T) *or Tangential force on crank-pin*: The crank effort

$$F_T = F_c \sin(\theta + \phi) = 16461.5 \sin(30 + 5.739)$$

$$= 16461.5 \times \sin 35.739 = 9615 \text{ N}.$$

(vii) *Turning moment (on torque) on the crank shaft*: Turning moment on the crank-shaft is given by

$$T = F_T \times r = 9615 \times 0.2 = 1923 \text{ N·m}.$$

Example 7.17: A vertical petrol engine 150 mm diameter and 200 mm stroke has a connecting rod 350 mm long. The mass of the piston is 1.6 kg and engine speed is 1800 rpm. On the expansion stroke with crank angle 30° from the top dead centre, the gas pressure is 750 kN/m². Determine the net thrust on the engine.

Solution: Engine diameter $D = 150$ mm $= 0.15$ m; stroke $L = 200$ mm $= 0.2$ m or radius of crank.

$$r = \frac{L}{2} = \frac{0.2}{2} = 0.1\text{m}; \text{ connecting rod length } l = 350 \text{ mm} = 0.35 \text{ m}; \text{ mass of piston}$$

$m = 1.6$ kg; engine speed $N = 1800$ rpm

$$\text{angular velocity } \omega = \frac{2\pi N}{60} = \frac{2\pi \times 1800}{60}$$

$$= 188.49 \text{ rad/s, crank angle } \theta = 30°,$$

Gas pressure $p = 750$ kN/m² $= 75 \times 10^3$ N/m²

Engine is vertical.

The net thrust for vertical engine is given as

$$F = F_P + F_i \pm W \qquad \qquad ...(i)$$

Now F_p = Force exerted on the piston by the pressure of gas

$$= p \times \text{Area of piston}$$

$$= p \times \frac{\pi}{4} D^2 = (750 \times 10^3) \times \left(\frac{\pi}{4} \times 0.15^2\right) N$$

$$= 750000 \times 0.01767 = 13253.59 \text{ N}$$

F_i = Inertia force

$$= -(\text{mass of piston}) \times \text{acceleration of piston}$$

The acceleration of piston

$$a = \omega^2 r \left(\cos\theta + \frac{\cos 2\theta}{n}\right)$$

$$= (188.49)^2 \times 0.1 \left(\cos 30° + \frac{\cos 60°}{3.5}\right)$$

$$\left[\because r = \frac{L}{2} = 0.1 \text{ m and } n = \frac{l}{r} = \frac{0.35}{0.1} = 3.5\right]$$

$$= 3552.848 \left(0.866 + \frac{0.5}{3.5}\right) = 3584.31 \text{ m/s}^2$$

$\therefore \quad F_i = -m \times a = -1.6 \times 3584.31 = -5734.89 \text{ N}$

W = weight of piston or weight of reciprocating parts

$$= m \times g$$

$$= 1.6 \times 9.81 = 15.696 \text{ N}$$

As $\theta = 30°$, hence, piston position is during first half of stroke. During this position, piston is accelerating and inertia force opposes the accelerating force. Hence, inertia force acts opposite to the acceleration. The piston is moving downward when $\theta = 30°$. Hence, weight is acting downward, i.e. in the direction of acceleration

$\therefore \qquad\qquad\qquad F = F_p + F_i + W$

Substituting the values of F_p, F_i and W, we get

$$F = 13253.59 + (-5734.89) + 15.696$$

$$= 13253.59 - 5734.89 + 15.696$$

$$= 7534.396 \text{ N}$$

$\therefore$ Net thrust on piston is 7534.396 N.

Example 7.18: The crank and connecting rod of a vertical petrol engine, running at 1800 rpm are 60 mm and 270 mm respectively. The diameter of the piston is 100 mm and the mass of the reciprocating parts is 1.2 kg. During the expansion stroke when the crank has a turned 20° from the top dead centre, the gas pressure is 650 kN/m². Determine

    (i) net force on the piston
    (ii) net load on the gudgeon pin
    (iii) thrust on the cylinder walls
    (iv) spread at which the gudgeon pin load is reversed in direction.

Solution: $r = 0.06$ m, $l = 0.27$ m, $N = 1800$ rpm, $p = 650$ kN/m²,

$$m = 1.2 \text{ kg}, d = 0.1 \text{ m}, \theta = 20°$$

$$n = l/r = 0.27/0.06 = 4.5$$

$$\omega = \frac{2\pi \times 1800}{60} = 188.5 \text{ rad/s}$$

$$\sin \phi = \frac{\sin \theta}{n} = \frac{\sin 20°}{4.5} = 0.076 \text{ or } \phi = 4.36°$$

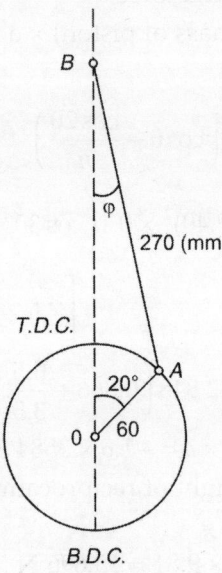

Fig. 7.24

Force due to gas pressure, $F_p = $ Area $\times$ pressure

$$= \frac{\pi}{4}(d)^2 \times p = \frac{\pi}{4}(0.1)^2 \times 650 \times 10^3 = 5105 \text{ N}$$

Inertia force, $F_i = mr\omega^2 \left(\cos \theta + \frac{\cos 2\theta}{n} \right)$

$$= 1.2 \times 0.06 \times (188.5)^2 \left(\cos 20° + \frac{\cos 40°}{4.5} \right) = 2840 \text{ N}$$

(i) Net (effective) force on the piston, $F = F_p - F_i + mg$

$$= 5105 - 2840 + 1.2 \times 9.81$$

$$= 2276.8 \text{ N}$$

(ii) Net load on the gudgeon pin = Force in the connecting rod

$$= \frac{F}{\cos \phi} = \frac{2276.8}{0.9971} = 2283.4 \text{ N}$$

(iii) Thrust on the cylinder walls

$$= F \tan \phi = 2276.8 \tan 4.36° = 173.5 \text{ N}$$

(iv) Speed at which the gudgeon pin load is reversed in direction,

$$F = F_p - mr\omega^2 \left(\cos \theta + \frac{\cos 2\theta}{n} \right) + mg$$

$$0 = 5105 - 1.2 \times 0.06\,\omega^2 \left(\cos 20° + \frac{\cos 40°}{4.5} \right) + 1.2 \times 9.81$$

$$0.079\ 91\ \omega^2 = 5116.8$$

$$\omega = 253.04, \frac{2\pi N}{60} = 253.04, N = 2416.3 \text{ rpm}$$

Example 7.19: In a vertical double-acting steam engine, the connecting rod is 4.5 times the crank. Weight of reciprocating parts is 120 kg and the stroke of the piston is 440 mm. The engine runs at 250 rpm. If the net load on the piston due to steam pressure is 25 kN when the crank has turned through an angle of 120° from the top dead centre, determine
  (i) thrust in the connecting rod
  (ii) pressure on slide bars
  (iii) tangential force on the crank pin
  (iv) thrust on the bearings
  (v) turning moment on the crankshaft.

Solution: $r = 0.44/2 = 0.22$ m, $N = 250$ rpm, $F = 25$ kN, $m = 120$ kg, $\theta = 120°$, $n = 4.5$

$$\omega = \frac{2\pi \times 250}{60} = 26.18 \text{ rad/s}$$

$$\sin \phi = \frac{\sin \theta}{n} = \frac{\sin 120°}{4.5} = 0.1925$$

or $$\phi = 11.1°$$

Accelerating force, $F_i = mr\omega^2 \left(\cos\theta + \frac{\cos 2\theta}{n} \right)$

$$= 120 \times 0.22 \times (26.18)^2 \left(\cos 120° + \frac{\cos(240°)}{4.5} \right)$$

$$= -11058 \text{ N}$$

Force on the piston; $F = F_p + mg - F_i$

$$= 25000 + 120 \times 9.81 - (-11058)$$

$$= 37235 \text{ N}$$

(i) Thrust in the connecting rod

$$F_c = \frac{F}{\cos\phi} = \frac{37235}{\cos 11.1} = 37945 \text{ N}$$

(ii) Pressure on slide bars

$$F_n = F \tan\phi = 37235 \tan 11.1° = 7305 \text{ N}$$

(iii) Tangential force on the crank pin

$$F_t = F_c \sin(\theta + \phi) = 37945 \times \sin(120° + 11.1°)$$

$$= 28594 \text{ N}$$

(iv) Thrust on the bearings

$$F_r = F_c \cos(\theta + \phi) = 37945 \times \cos(120° + 11.1°)$$

$$= -24944 \text{ N}$$

(v) Turning moment on the crankshaft

$$T = F_t \times r = 28594 \times 0.22 = 6290.7 \text{ N·m}$$

Example 7.20: The crank and the connecting rod of a vertical single cylinder gas engine running at 1800 rpm are 60 mm and 240 mm respectively. The diameter of the piston is 80 mm and the mass of the reciprocating parts is 1.2 kg. At a point during the power stroke when the piston has moved 20 mm from the top dead centre position, the pressure on the piston is 800 kN/m². Determine

   (i) net force on the piston
  (ii) thrust in the connecting rod
 (iii) thrust on the sides of cylinder walls
 (iv) engine speed at which the above values are zero.

Solution: $r = 0.06$ m, $l = 0.24$ m, $N = 1800$ rpm, $m = 1.2$ kg, $n = 0.24/0.06 = 4$, $d = 0.08$ m

$$\omega = \frac{2\pi \times 1800}{60} = 188.5 \text{ rad/s}$$

Draw the configuration for the given position to some scale (Fig. 7.25) and obtain angle θ which is found to be 43.5°.

$$\sin \varphi = \frac{\sin \theta}{n} = \frac{\sin 43.5°}{4} = 0.1721$$

or

$$\varphi = 9.91°$$

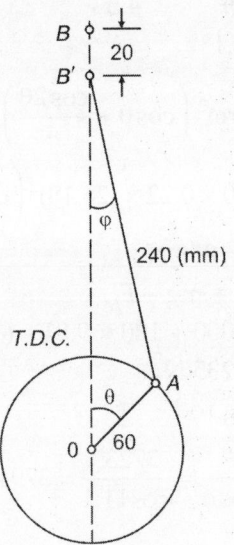

Fig. 7.25

Force due to gas pressure,

$$F_p = \text{Area} \times \text{pressure} = \frac{\pi}{4}(d)^2 \times p$$

$$= \frac{\pi}{4}(0.08)^2 \times 800 \times 10^3$$

$$= 4021 \text{ N}$$

Accelerating force, $F_b = mr\omega^2 \left(\cos\theta + \dfrac{\cos 2\theta}{n} \right)$

$$= 1.2 \times 0.06 \times (188.5)^2 \left(\cos 43.5° + \dfrac{(\cos 87°)}{4} \right)$$

$$= 1189 \text{ N}$$

(i) Force on the piston,

$$F = F_p + mg - F_b$$
$$= 4021 + 1.2 \times 9.81 - 1889 = 2144 \text{ N}$$

(ii) Thrust in the connecting rod,

$$F_c = \dfrac{F}{\cos\varphi} = \dfrac{2144}{\cos 9.91°} = 2176 \text{ N}$$

(iii) Thrust on the sides of cylinder walls,

$$F_n = F \tan\varphi = 2176 \tan 9.91° = 380 \text{ N}$$

(iv) The above values are zero at the speed when the force on the piston F is zero.

$$F = F_p - mr\omega^2 \left(\cos\theta + \dfrac{\cos 2\theta}{n} \right) + mg$$

$$0 = 4021 - 1.2 \times 0.06\,\omega^2 \left(\cos 43.5° + \dfrac{\cos 87°}{4} \right) + 1.2 \times 9.81$$

$$0.05317\omega^2 = 4032.8, \ \omega = 75849$$

$$\dfrac{2\pi N}{60} = 275.4, \ N = 2630 \text{ rpm.}$$

Example 7.21: The following data relates to a horizontal reciprocating engine:

Mass of reciprocating parts = 120 kg, crank length = 90 mm,
engine speed = 600 rpm

Connecting rod:

Mass = 90 kg

Length between centres = 450 mm

Distance of centre of mass from big end centre = 180 mm

Radius of gyration about an axis through centre of mass = 150 mm

Find the magnitude and the direction of the inertia torque on the crankshaft when the crank has turned 30° from the inner-dead centre.

Solution: It is required to find the inertia torque or turning moment on the crankshaft due to the inertia of the piston as well as of the connecting rod. This can be obtained as follows:

$$\omega = \dfrac{2\pi N}{60} = \dfrac{2\pi \times 600}{60} = 62.8 \text{ rad/s}$$

Divide the mass of the connecting rod into two parts (Fig. 7.26).

Mass at crank pin, $m_3 = 90 \times \left(\dfrac{450 - 180}{450} \right) = 54 \text{ kg}$

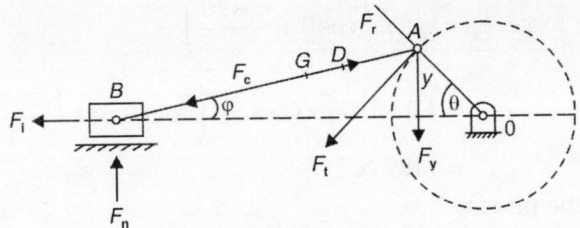

Fig. 7.26

Mass at gudgeon pin, $m_1 = 90 - 54 = 36$ kg

Total mass of reciprocating parts, $m = 120 + 36 = 156$ kg

Acceleration of the reciprocating parts, $a = r\omega^2 \left(\cos\theta + \dfrac{\cos 2\theta}{n} \right)$

As θ is less than $90°$, it is towards right and thus, the inertia force is towards left.

Inertia force, $\qquad F_i = ma = mr\omega^2 \left(\cos\theta + \dfrac{\cos 2\theta}{n} \right)$

$$= 156 \times 0.09 \times (62.8)^2 \left(\cos 30° + \frac{\cos 60°}{5} \right)$$

$$= 53490 \text{ N}$$

Inertia torque due to reciprocating parts,

$$T = Fr \left(\sin\theta + \frac{\sin 2\theta}{2\sqrt{n^2 - \sin^2\theta}} \right)$$

$$= 53490 \times 0.09 \left(\sin 30° + \frac{\sin 60°}{2\sqrt{(5)^2 - \sin^2 30°}} \right)$$

$$= 2826 \text{ N·m}$$

(counterclockwise as inertia force is towards left)

Correction couple due to assumed second mass of connecting rod at A,

$$\Delta T = m\alpha_c (l - L) L_1$$

where $L_1 = 450 - 180 = 270$ mm, $l = 450$ mm

and $\qquad L = L_1 + \dfrac{k^2}{L_1} = 270 + \dfrac{(150)^2}{270} = 353.3$ mm

$$\alpha_c = -\omega^2 \sin\theta \left[\frac{n^2 - 1}{(n^2 - \sin^2\theta)^{3/2}} \right]$$

$$= -(62.8)^2 \sin 30° \left[\frac{5^2 - 1}{(25 - \sin^2 30°)^{3/2}} \right]$$

$$= -384.7 \text{ rad/s}^2$$

$\therefore \qquad \Delta T = 90 \times (-384.7) \times 0.27 \times (0.45 - 0.3533)$

$$= -903.97 \text{ N·m}$$

The direction of the correction couple will be the same as that of angular acceleration, i.e. in the direction of decreasing angle ϕ. Thus, it is clockwise.

∴ Correction torque on the crankshaft,

$$T_c = \Delta T \frac{\cos\theta}{\sqrt{n^2 - \sin^2\theta}}$$

$$= -903.97 \times \frac{\cos 30°}{\sqrt{25 - \sin^2 30°}}$$

$$= -157.4 \text{ N·m}$$

Correction torque is to be deducted from the inertia torque on the crankshaft or as the force F_y due to ΔT (which is clockwise) is towards left of the crankshaft, the correction torque is counterclockwise.

Torque due to weight of mass at

$$A, T_a = (m_3 g) \, r \cos\theta$$
$$= 54 \times 9.81 \times 0.09 \times \cos 30°$$
$$= 41.3 \text{ N·m (counterclockwise)}$$

∴ Total inertia torque on the crankshaft $= T_b - T_c + T_a$
$$= 2826 - (-157.4) + 41.3$$
$$= 3024.7 \text{ N·m (counterclockwise)}$$

Example 7.22: The turning moment diagram for a multicylinder engine has been drawn to a scale of 1 mm = 325 Nm vertically and 1 mm = 3° horizontally. The areas above and below the mean torque line are -26, $+378$, -256, $+306$, -302, $+244$, -380, $+261$ and -225 mm² respectively. The engine is running at a mean speed of 600 rpm. The total fluctuation of speed is not to exceed $\pm 1.8\%$ of the mean speed. If the radius of flywheel is 0.7 m, find the mass of the flywheel.

Solution: Turning moment scale: 1 mm = 325 Nm

Crank angle scale, 1 mm $= 3° = 3 \times \dfrac{\pi}{180} = 0.05236$ rad

∴ 1 mm² on turning moment diagram
$$= 325 \times 0.05236 \text{ Nm} = 17.017 \text{ Nm}$$

Mean speed, $N = 600$ rpm

∴ Mean angular speed, $\omega = \dfrac{2\pi N}{60} = \dfrac{2\pi \times 600}{60} = 62.84$ rad/s

Radius, $R = 0.7$ m

As the total fluctuation of speed is not to exceed $\pm 1.8\%$ of the mean speed,

hence, $(\omega_1 - \omega_2) = (1.8 \times 2)\%$ of ω, i.e. $\dfrac{\omega_1 - \omega_2}{\omega} = \dfrac{3.6}{100} = 0.036.$

∴ Coefficient of fluctuation of speed $K_s = 0.036.$

Let 1 mm² on turning moment diagram $= x$ Nm

∴ $\qquad\qquad\qquad x = 17.017$ Nm $\qquad$ (∵ 1 mm² = 17.017 Nm) $\qquad\qquad$ ...(i)

Then the areas above and below the mean torque line in work units (i.e. Nm) will be $-26x$, $+378x$, $-256x$, $+306x$, $-302x$, $+244x$, $-380x$, $+261x$ and $-225x$ respectively.

Let E = total energy at A

Then energy at $\quad B = E - 26x$

Energy at $\quad\quad\quad C = E - 26x + 378x = E + 352x$

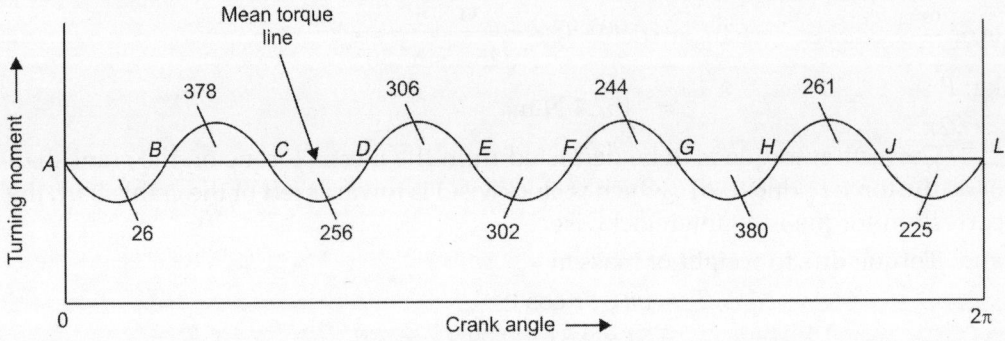

Fig. 7.27

Energy at

$\quad D = E + 352x - 256x = E + 96x$

$\quad E = E + 96x + 306x = E + 402x \quad\quad\quad\quad$ ...(max. energy)

$\quad F = E + 402x - 302x = E + 100x$

$\quad G = E + 100x + 244x = E + 344x$

$\quad H = E + 344x - 380x = E - 36x \quad\quad\quad\quad$ ...(min. energy)

$\quad J = E - 36x + 261x = E + 225x$

$\quad L = E + 225x - 225x = E = $ Energy at A

The cycle repeats after L and hence, energy at L is equal to energy at A.

The maximum energy is at E and minimum energy is at H.

$\therefore$ Max. energy = $E + 402x$

Min. energy = $E - 36x$

$\therefore$ Max. fluctuation of energy,

$$\Delta E = \text{Max. energy} - \text{Min. energy}$$
$$= (E + 402x) - (E - 36x) = 438x$$
$$= 438 \times 17.017 \text{ Nm} \quad\quad [\because \textit{from Eq. (i)}, x = 17.017 \text{ Nm}]$$
$$= 7453.446 \text{ Nm}$$

Let m = mass of flywheel.

Now using equation $\Delta E = mR^2 \times \omega^2 \times K_s$ [In this equation, $k = R$ which is true if thickness of rim of flywheel is very small as compared to diameter of wheel]

But $\Delta E = 7453.446$ Nm, $R = 0.7$ m, $\omega = 62.84$ rad/s and $K_s = 0.036$. Hence,

$$7453.446 = m \times 0.7^2 \times 62.84^2 \times 0.036$$

or

$$m = \frac{7453.446}{0.7^2 \times 62.84^2 \times 0.036} = 107.00.$$

Example 7.23: The radius of gyration of a flywheel is 1 meter and the fluctuation of speed is not to exceed 1% of the mean speed of the flywheel. If the mass of the flywheel is 3340 kg and the steam engine develops 150 kW at 135 rpm, then find:

(i) Maximum fluctuation of energy and

(ii) Coefficient of fluctuation of energy.

Solution: $k = 1$ m; fluctuation of speed = 1% of mean speed or $(\omega_2 - \omega_1)$ = 1% of ω or $\dfrac{\omega_2 - \omega_1}{\omega} = \dfrac{1}{100} = 0.01$ or coefficient of fluctuation of speed. $K_s = 0.01$; $m = 3340$ kg; $P = 150$ kW = 150×1000 W = 150000 W; $N = 135$ rpm or mean angular speed,

$$\omega = \frac{2\pi N}{60} = \frac{2\pi \times 135}{60} = 14.137 \text{ rad/s.}$$

Let ΔE = Max. fluctuation of energy

and $\qquad K_e$ = Coefficient of fluctuation of energy.

The maximum fluctuation of energy is given as,

$$\Delta E = mk^2 \times \omega^2 \times K_s$$
$$= 3340 \times 1^2 \times 14.137^2 \times 0.01$$
$$(\because m = 3340 \text{ kg}; k = 1 \text{ m}; \omega = 14.137 \text{ rad/s and } K_s = 0.01)$$
$$= 6675.13 \text{ Nm.}$$

The coefficient of fluctuation of energy (K_e) is given by

$$K_e = \frac{\text{Maximum fluctuation of energy}}{\text{Work done per cycle}} \qquad \text{...(i)}$$

Work done per cycle = $T_{mean} \times \theta$

$$= T_{mean} \times 2\pi \qquad (\because \theta = 2\pi \text{ radians for steam engine})$$

The value of T_{mean} is obtained from

$$P = T_{mean} \times \omega$$

$\therefore \qquad\qquad T_{mean} = \dfrac{P}{\omega} = \dfrac{150000}{14.137} \qquad (\because P = 150000 \ W \text{ and } \omega = 14.137 \text{ rad/s})$

$$= 10610.45 \text{ Nm}$$

$\therefore \quad$ Work done per cycle = $T_{mean} \times 2\pi = 10610.45 \times 2\pi$

$$= 66667.42 \text{ Nm/cycle}$$

Substituting the value in Eq. (i), we get

$$K_e = \frac{\Delta E}{66667.42} = \frac{6675.13}{66667.42} = 0.1001 \approx 0.1.$$

Example 7.24: The areas above and below the mean torque line for an IC engine are -25, $+200$, -100, $+150$, -300, $+150$ and -75 mm² taken in order. The scale for the turning moment diagram is 1 mm vertical scale = 10 Nm and 1 mm horizontal scale = 1.5°. The mass of the rotating parts are 45 kg with a radius of gyration of 150 mm. If the engine speed is 1500 rpm, find the coefficient of fluctuation of speed.

Solution: Turning moment scale: 1 mm = 10 Nm and crank angle scale 1 mm = 1.5°

$$= 1.5 \times \frac{\pi}{180} \text{ rad} = 0.02618 \text{ rad}; \ m = 45 \text{ kg}, \ k = 150 \text{ mm} = 0.15 \text{ m}; \ N = 1500 \text{ rpm or } \omega$$

$$= \frac{2\pi N}{60} = \frac{2\pi \times 1500}{60} = 50\pi \text{ rad/s}.$$ Now 1 mm² on turning moment diagram = 10 × 0.02618 = 0.2618 Nm.

Let 1 mm² on turning moment diagram = x Nm

Then, $x = 0.2618$ Nm

The areas above and below the mean torque line in works units (i.e. in Nm) will be $-25x, +200x, -100x, +150x, -300x, +150x$ and $-75x$ respectively.

Let E = Total energy at A. Then

Energy at

$$B = E - 25x$$
$$C = E - 25x + 200x = E + 175x$$
$$D = E + 175x - 100x = E + 75x$$
$$E = E + 75x + 150x = E + 225x \qquad \text{...(max. energy)}$$
$$F = E + 225x - 300x = E - 75x \qquad \text{...(min. energy)}$$
$$G = E - 75x + 150x = E + 75x$$
$$H = E + 75x - 75x = E = \text{Energy at } A$$

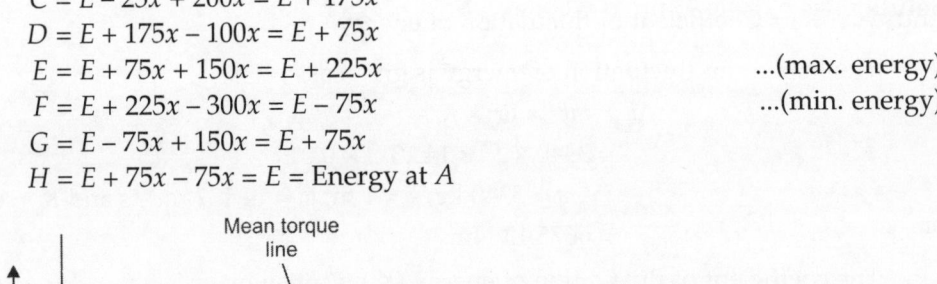

Fig. 7.28

The cycle repeats after H and hence, energy at H is equal to energy at A.

The maximum energy is at E whereas the minimum energy is at F.

∴ Max. energy = $E + 225x$ and min. energy = $E - 75x$

∴ Max. fluctuation of energy,

$$\Delta E = \text{Max. energy} - \text{Min. energy}$$
$$= (E + 225x) - (E - 75x)$$
$$= 300x = 300 \times 0.2618 \text{ Nm}$$

$$[\because \textit{from} \text{ Eq. (i)}, x = 0.2618 \text{ Nm}]$$

$$= 78.54 \text{ Nm}$$

Let K_s = Coefficient of fluctuation of speed

$$\Delta E = mk^2 \times \omega^2 \times K_s$$

$$K_s = \frac{\Delta E}{mk^2 \times \omega^2} = \frac{78.54}{45 \times 0.15^2 \times (50\pi)^2} = 0.0031$$

$$K = \frac{e}{I\omega^2} = \frac{e}{mk^2\omega^2} = \frac{23038}{55 \times 2.1^2 \times \left(\dfrac{2\pi \times 1600}{60}\right)^2}$$

$$K = 0.0034 \text{ or } 0.34\%$$

Example 7.25: The turning moment diagram for a multicylinder engine has been drawn to a vertical scale of 1 mm = 650 N·m and a horizontal scale of 1 mm = 4.5°. The areas above and below the mean torque line are $-28, +380, -260, +310, -300, +242, -380, +265$ and -229 mm² respectively.

The fluctuation of speed is limited to $\pm$ 1.8% of the mean speed which is 400 rpm. Density of the rim material is 7000 kg/m³ and width of the rim is 4.5 times its thickness. The centrifugal stress (hoop stress) in the rim material is limited to 6 N/mm². Neglecting the effect of the boss and arms, determine the diameter and cross-section of the flywheel rim.

Solution: $\rho = 7000$ kg/m³, $\sigma = 6 \times 10^6$ N/m², $N = 400$ rpm,

$$K = 0.018 + 0.018 = 0.036, b = 4.5t$$

From the consideration of strength of materials,

$$\sigma = \rho v^2 \text{ or } 6 \times 10^6 = 7000 \times v^2$$
$$v = 29.28 \text{ m/s}$$

or

$$\frac{\pi d n}{60} = \frac{\pi \times d \times 400}{60} = 29.28 \text{ or } d = 1.398 \text{ m}$$

Let flywheel KE at

$a = E$
$b = E - 28$
$c = E - 28 + 380 = E + 352$
$d = E + 352 - 260 = E + 92$
$e = E + 92 + 310 = E + 402$
$f = E + 402 - 300 = E + 102$
$g = E + 102 + 242 = E + 344$
$h = E + 344 - 380 = E - 36$
$j = E - 36 + 265 = E + 229$
$k = E + 229 - 229 = E$

Maximum energy = $E + 402$ (at e),
Minimum energy = $E - 36$ (at h)

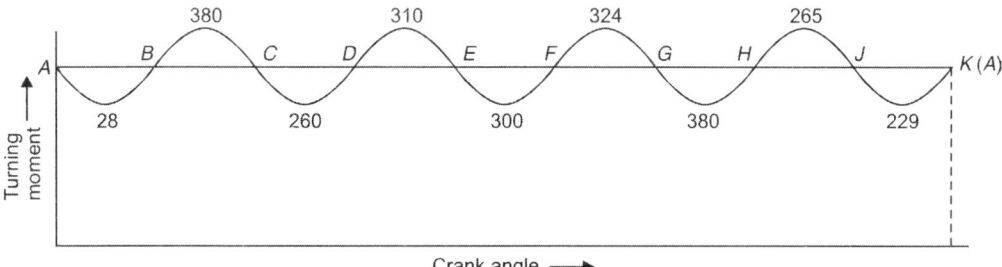

Crank angle ⟶

Fig. 7.29

Maximum fluctuation of energy,

$$\Delta E = (E + 402) - (E - 36) \times \text{Hor. scale} \times \text{Vert. scale}$$

$$= 438 \times \left(4.5 \times \frac{\pi}{180}\right) \times 650 = 22360 \text{ N·m}$$

$$K = \frac{\Delta E}{I\omega^2} = \frac{\Delta E}{mk^2\omega^2}$$

$$0.036 = \frac{22360}{m\left(\dfrac{1.398}{2}\right)^2\left(\dfrac{2\pi \times 400}{60}\right)^2} \quad m = 724.5 \text{ kg}$$

or density × volume = 724.5 or $\rho \times (\pi d) \times t \times 4.5t = 724.5$

or $7000 \times \pi \times 1.398 \times t \times 4.5t = 724.5$

or $t = 0.0512$ m or 51.2 mm, $b = 4.5 \times 51.2 = 230.3$ mm.

EXERCISE

7.1 What are conditions for a body to be in equilibrium under the action of two, three forces and two forces and a torque?

7.2 What is free body diagram? How is it helpful in finding the various forces acting on members of a mechanism?

7.3 What is the principle of virtual work? Explain with examples.

7.4 Explain the superposition theorem for a system of forces.

7.5 Define inertia force and inertia torque.

7.6 State and explain D' Alembert's principle.

7.7 Explain equivalent offset inertia force.

7.8 Derive the relation of angular acceleration for the connecting rod.

7.9 What is piston effort and crank effort?

7.10 Describe the graphical and analytical methods of finding the inertia torque on the crank shaft of horizontal reciprocating engine.

7.11 What is the function of the flywheel? How does it differ from that of a governor?

7.12 What is the function of flywheel? Explain the terms fluctuation of energy and fluctuation of speed as applied to flywheels.

7.13 Derive the relation for acceleration of piston for slider–crank mechanism.

7.14 Distinguish clearly between the terms 'crank pin effort' and 'crank effort'. Derive expression for these. State how the size of a flywheel is calculated.

7.15 Describe analytical method of dynamic force analysis for any planar mechanism.

7.16 Differentiate between the functions of a flywheel and a governor. Derive an expression for moment of inertia of a flywheel in terms of maximum fluctuation of kinetic energy, mean engine speed and maximum fluctuation of speed.

7.17 Derive a relation for the turning moment at the crankshaft in terms of piston effort and the angle turned by the crank.

7.18 What is meant by a dynamically equivalent system?

7.19 What do you mean by turning moment diagram? Why are they drawn?

7.20 Define the terms 'coefficient of fluctuation of energy' and 'coefficient of fluctuation of speed'.

7.21 What do you mean by mean resisting torque? What are the conditions for a flywheel of a steam engine to accelerate and retard?

7.22 Derive a relation for the coefficient of fluctuation of speed in terms of maximum fluctuation of energy and the kinetic energy of the flywheel at mean speed.

7.23 How the inertia of the connecting rod of a reciprocating engine is taken into account?

7.24 Derive an expression for the correction couple and correction torque to be applied to a crank shaft if the connecting rod of a reciprocating engine is replaced by two point masses at the piston pin and the crank pin respectively.

7.25 In a four-link mechanism ABCD, the link, AB revolves with an angular velocity of 10 rad/s and angular acceleration of 25 rad/s^2 at the instant when it makes an angle of 45° with AD, the fixed link. The length of the links are:

$AB = CD = 800$ mm, $BC = 1000$ mm, and $AD = 1500$ mm.

The mass of the links is 4 kg/m length. Determine the torque required to overcome the inertia forces, neglecting the gravitational effects. Assume all links to be of uniform cross-sections. (*Ans.* 82.2 N·m)

7.26 The effective steam pressure on the piston of a vertical steam engine is 200 kN/m^2 when the crank is 40° from the inner-dead centre on the down stroke. The crank length is 300 mm and the connecting rod length 1200 mm. The diameter of the cylinder is 800 mm. What will be the torque on the crankshaft if the engine speed is 300 rpm and the mass of the reciprocating parts 250 kg? (*Ans.* 9916 N·m)

7.27 The dimensions of a four link mechanism are: AB = 400 mm, BC = 600 mm, $CD = 500$ mm, $AD = 900$ mm, and $\angle DAB = 60°$. AD is the fixed link. E is a point on link BC such that $BE = 400$ mm and $CE = 300$ mm (BEC clockwise). A force of 150 $\angle 45°$ N acts on DC at a distance of 250 mm from D. Another force of magnitude 100 $\angle 180°$ N acts at point E. Find the required input torque on link AB for static equilibrium of the mechanism. (*Ans.* 0.35 N·m clockwise)

7.28 The effective steam pressure on the piston of a vertical stream engine is 20 N when the crank is 40° from the inner dead centre on the down stroke. The crank length is 300 mm and the connecting rod length 1200 mm. The diameter of the cylinder is 800 mm. What will be the torque on the crankshaft if the engine speed is 300 rpm and the mass of reciprocating parts 250 kg?

7.29 The length of the connecting rod of a gas engine is 500 mm and its centre of gravity lies at 165 mm from the crank pin centre. The rod has a mass of 80 kg and a radius of gyration of 182 mm about an axis through the centre of mass. The stroke of piston is 225 mm and the crank speed is 300 rpm. Determine the inertia force on the crankshaft when the crank has turned (a) 30° and (b) 135° from the inner-dead centre.

7.30 The connecting rod of an IC engine is 450 mm long and has a mass of 2 kg. The centre of mass of the rod is 300 mm from the small end and its radius of gyration about an axis through this centre is 175 mm. The mass of the piston and the gudgeon pin is 2.5 kg and the stroke is 300 mm. The cylinder diameter is 115 mm. Determine the magnitude and the direction of the torque applied on the crankshaft when the crank is at 40° and the piston is moving away from the inner dead centre under an effective gas pressure of 2 N/mm^2. The engine speed is 1000 rpm. (*Ans.* 1014 N·m)

7.31 The connecting rod of a vertical high-speed engine is 600 mm long between centres and has a mass of 3 kg. Its centre of mass lies at 200 mm from the big-end bearing. When suspended as a pendulum from the gudgeon pin axis, it makes 45 complete oscillations in 30 second. The piston stroke is 250 mm. The mass of the reciprocating parts is 1.2 kg. Determine the inertia torque on the crankshaft when the crank makes an angle of 140° with top-dead centre. The engine speed is 1500 rpm. (*Ans.* 361.7 N·m)

7.32 The length of crank and connecting rod of a vertical reciprocating engine are 150 mm and 750 mm respectively. The crank is rotating at 400 rpm clockwise. Find analytically: (i) acceleration of the piston, (ii) velocity of the piston, and (iii) angular acceleration of the connecting rod when the crank has turned through 40° from the TDC and piston is moving downwards. (*Ans.* 210.54 m/s^2; 4.66 m/s; –222 rad/s^2)

7.33 The length of crank and connecting rod of a reciprocating engine are 100 mm and 400 mm respectively. The crank is rotating at a uniform speed of 240 rpm. Using Klein's construction find: (i) the acceleration of the piston, (ii) the acceleration of the middle point of the connecting road, and (iii) angular acceleration of the connecting rod when the crank has turned through 45° from the inner dead centre.
(*Ans.* 45.92 m/s^2; 51.92 m/s^2; 110.51 rad/s^2)

7.34 The length of crank and connecting rod of horizontal steam engine are 300 mm and 1.2 m respectively. When the crank has moved 30° from the inner dead centre, the acceleration of piston is 35 m/s^2. The average frictional resistance to the motion of piston is equivalent to a force of 550 N and net effective steam pressure on piston is 500 kN/m^2. The diameter of piston is 0.3 m and mass of reciprocating parts is 160 kg. Determine:
  (i) Reaction on the cross-head guides;
  (ii) Thrust on the crank-shaft bearings; and
  (iii) Torque on the crank shaft.   (*Ans.* 3677.56 N, 23441.8 N, 5334.52 Nm)

7.35 The length of the connecting rod of a vertical double acting steam engine is 1.5 m. The diameter of the cylinder is 400 mm and stroke of the engine is 600 mm. The crank is rotating at 200 rpm in the clockwise direction. The crank has turned through 40° from the top dead centre and piston is moving downwards. The steam pressure above the piston is 0.6 N/mm^2 and below the piston is 0.05 N/mm^2. The mass of the reciprocating parts is 200 kg. The diameter of piston rod is given as 50 mm. Find the thrust on guide bars and crankshaft bearing and also turning moment on crankshaft.

7.36 The turning moment diagram for a petrol engine is drawn to a vertical scale of 1 mm = 500 N·m and a horizontal scale of 1 mm = 3°. The turning moment diagram repeats itself after every half revolution of the crankshaft. The area above and below the mean torque line are 260, – 580, 80, – 380, 870 and – 250 mm^2 respectively. The rotating parts have a mass of 55 kg and radius of gyration of 2.1 m. If the engine speed is 1600 rpm, determine the coefficient of fluctuation of speed.   (*Ans.* 34%)

7.37 The maximum and minimum speed of a flywheel are 242 rpm and 238 rpm respectively. The mass of flywheel is 2600 kg and radius of gyration is 1.8 m. Find (i) mean speed of flywheel, (ii) maximum fluctuation of energy, and (iii) coefficient of fluctuation of speed.
(*Ans.* 240 rpm, 88911 N·m, 0.0167)

OBJECTIVE TYPE QUESTIONS

7.1 A pair of action and reaction forces acting on a body are known as:
  (a) applied forces
  (b) constraint forces
  (c) accelerating forces
  (d) inertia forces

7.2 In static equilibrium the vector sum of all the forces acting on the body and all the moment about .... point in zero.
  (a) a fixed
  (b) a particular
  (c) any arbitrary
  (d) a permanent

7.3 The point of intersection of lines of action of three or more forces is known as the ......
point.
  (a) equilibrium point
  (b) central
  (c) zero
  (d) concurrency

7.4 A part isolated from the mechanism .... be in equilibrium.
  (a) may
  (b) may or may not
  (c) must

7.5 The displacement of the piston in a reciprocating steam engine is given by:
  (a) $\omega \cdot r \left(\sin\theta + \dfrac{\cos 2\theta}{n} \right)$

  (b) $r(1 - \cos\theta) + l(1 - \cos\phi)$

  (c) $\omega^2 \cdot r \left(\sin\theta + \dfrac{\cos 2\theta}{n} \right)$

  (d) none of the above

7.6 The velocity of the piston in reciprocating steam engine is given by:
  (a) $\omega \cdot r \left(\sin\theta + \dfrac{\sin 2\theta}{2n} \right)$

  (b) $\omega^2 \cdot r \left(\cos\theta + \dfrac{\cos 2\theta}{n} \right)$

  (c) $\omega^2 \cdot r \left(\sin\theta + \dfrac{\cos 2\theta}{2n} \right)$

  (d) none of the above

7.7 The acceleration of the piston in a reciprocating steam engine is given by:
  (a) $\omega \cdot r \left(\sin\theta + \dfrac{\sin 2\theta}{n} \right)$

  (b) $\omega^2 \cdot r \left(\cos\theta + \dfrac{\cos 2\theta}{n} \right)$

  (c) $\omega^2 \cdot r \left(\sin\theta + \dfrac{\sin 2\theta}{2n} \right)$

  (d) none of the above

7.8 In a reciprocating horizontal engine the inertia forces due to reciprocating mass help
the piston effort at:
  (a) $\theta = 45°$
  (b) $\theta = 30°$
  (c) $\theta = 120°$
  (d) $\theta = 180°$

7.9 Crank effort is the net force applied at the crank pin .... to the crank that gives the
required turning moment on the crankshaft.
  (a) parallel
  (b) perpendicular
  (c) at 45°
  (d) at 135°

7.10 In a dynamically equivalent system, uniformly distributed mass is divided into .... point masses.
  (a) two
  (b) three
  (c) four
  (d) five

7.11 Any distributed mass can be replaced by two point masses to have the same dynamical properties if:
  (a) the sum of two masses is equal to the total mass
  (b) the combined centre of mass coincides with that of rod
  (c) the moment of inertia of two point masses about perpendicular axis through their combined centre of mass is equal to that of rod
  (d) all of the above

7.12 When the crank is as the inner dead centre, in a horizontal reciprocating steam engine, then the velocity of the piston will be:
  (a) zero
  (b) minimum
  (c) maximum
  (d) none

7.13 In an engine, the work done by inertia forces in a cycle is:
  (a) positive
  (b) zero
  (c) negative
  (d) none

7.14 State whether true or false.
  (a) The inertia force is equal in magnitude and opposite in direction to accelerating force
  (b) The magnitude of intetia force is given by the expression
  $$F_i = m_r \omega^2 r \left(\cos\theta + \frac{\cos 2\theta}{n} \right)$$ where symbols have their usual meanings
  (c) The resultant force acting on a body together with the reversed effective force, are in equilibrium
  (d) D'Alembert's principle is used to reduce a dynamic problem into an equivalent static problem
  (e) all are correct

7.15 A flywheel is:
  (a) an essential element of every prime mover
  (b) used in storing up energy and gives up whenever required during a cycle
  (c) a device for coordination between the prime mover and the external resistance
  (d) used for all the above purposes

7.16 A flywheel absorbs energy during those periods of crank rotation when:
  (a) the turning moment is greater than the resisting moment
  (b) the turning moment is equal to the resisting moment
  (c) the turning moment is less than the resisting moment
  (d) absorbs energy during all periods of crank rotation

7.17 Absorption of energy is accompanied:
  (a) by decrease of speed
  (b) at all speeds
  (c) an increase of speed
  (d) no relation with speed

7.18 Which is the correct statement?
  (a) the flywheel influences the mean speed of prime mover
  (b) the flywheel influences the variation of load demand on prime mover
  (c) the flywheel influences the cyclic variation of turning moment
  (d) the flywheel influences the mean torque developed by the prime mover

7.19 In a flywheel safe stress is 7×10^6 N/m^2 and density of flywheel material 700 kg/m^3, maximum peripheral velocity will be:
  (a) 100 m/sec
  (b) 10 m/sec
  (c) 10^4 m/sec
  (d) none of the above

7.20 The maximum fluctuation of energy of flywheel:
  (a) is directly proportional to coefficient of fluctuation of speed
  (b) is directly proportional to square of angular velocity of flywheel
  (c) is directly proportional to moment of inertia of flywheel
  (d) all of the above

7.21 The ratio of maximum fluctuation of energy to the work done per cycle is called:
  (a) coefficient of fluctuation of energy
  (b) coefficient of fluctuation of speed
  (c) all of the above

7.22 The mass of the flywheel is concentrated in the rims because then it will:
  (a) store much energy
  (b) store less energy
  (c) store zero energy
  (d) make the flywheel stronger

7.23 The maximum fluctuation of energy is the:
  (a) ratio of maximum and minimum energies
  (b) sum of maximum and minimum energies
  (c) difference of maximum and minimum energies
  (d) difference of maximum and minimum energies from mean energy

7.24 The maximum fluctuation of energy of a flywheel is equal to
  (a) $I\omega(\omega_1 - \omega_2)$
  (b) $I\omega^2 k$
  (c) 2 KE
  (d) all
  (e) none

7.25 The essential condition of placing the two masses, so that the system becomes dynamically equivalent is
  (a) $l_1 l_2 = k_G^2$
  (b) $l_1 l_2 = k_G$
  (c) $l_1 = k_G$
  (d) $l_2 = k_G$
  where l_1 and l_2 are distance of two masses from the centre of gravity of the body, and k_G is radius of gyration of the body.

7.26 The ratio of the maximum fluctuation of speed to the mean speed is called:
  (a) fluctuation of speed
  (b) maximum fluctuation of speed
  (c) coefficient of fluctuation of speed
  (d) none

7.27 In a turning moment diagram, the variations of energy above and below the mean resisting torque line is called:
  (a) fluctuation of energy
  (b) maximum fluctuation of energy
  (c) coefficient of fluctuation of energy
  (d) none

ANSWERS

7.1 (b)	7.2 (c)	7.3 (d)	7.4 (c)	7.5 (b)	7.6 (a)
7.7 (b)	7.8 (c)	7.9 (b)	7.10 (a)	7.11 (c)	7.12 (a)
7.13 (a)	7.14 (e)	7.15 (b)	7.16 (a)	7.17 (c)	7.18 (c)
7.19 (a)	7.20 (d)	7.21 (a)	7.22 (a)	7.23 (c)	7.24 (d)
7.25 (a)	7.26 (c)	7.27 (a)			

8

Balancing of Machines

8.1 INTRODUCTION

There are some moving parts (rotary or reciprocating) in all the machines. The inertia forces acting on moving parts produce imbalance. If the centre of mass of revolving part lies on its axis of rotation, forces are balanced. And if the centre of mass is at a distance (eccentric) from the axis of rotation, an unbalanced force is produced. For example, in high speed machines like rotary compressors, steam turbine rotors, centrifugal pumps, etc., if forces are not properly balanced, dynamic (time varying) unbalanced forces act on frames. These unbalanced forces:

    i. Impart vibratory motion to the machines and produce unpleasant noise.

    ii. Cause human discomfort and reduce their performance.

    iii. Increase the load on the bearings.

    iv. Increase the stress in machine members that affects machine performance.

Balancing is defined as the process of designing or modifying a machine in a way that unbalanced forces are reduced to minimum acceptable level. To avoid the detrimental effects of unbalanced forces, all the rotating and reciprocating parts of high speed machines should be as completely balanced as possible.

8.2 STATIC BALANCING

A system of rotating masses is said to be statically balanced if the combined centre of mass of the system lies on its axis of rotation or if the resultant of all centrifugal forces is zero, i.e. $\Sigma F_c = 0$.

8.3 DYNAMIC BALANCING

A system of rotating masses is said to be dynamically balanced if resultant centrifugal force and resultant couple acting on the system is equal to zero, i.e.

$$\Sigma F_c = 0$$
$$\Sigma_c = 0$$

If a system is dynamically balanced, the condition of static balance is automatically satisfied.

8.4 BALANCING OF ROTATING MASSES

An unbalanced mass rotating in a circular path can be balanced by attaching another mass to the opposite side of the shaft at such a distance that the certifugal force of both the masses are equal and opposite. This process of attaching second mass is known as balancing of rotating masses.

8.4.1 Balancing of a Single Rotating Mass

It can be balanced in two ways.

i. Balance mass rotating in the same plane

Consider a disturbing mass m_1, attached to the shaft rotating with anglular velocity ω (rad/s) (Fig. 8.1).

Let r_1 = distance of centre of gravity of mass m_1 from the axis of rotation

m_2 = balancing mass in the same plane of rotation

r_2 = distance of centre of gravity of mass m_2 from the axis of rotation

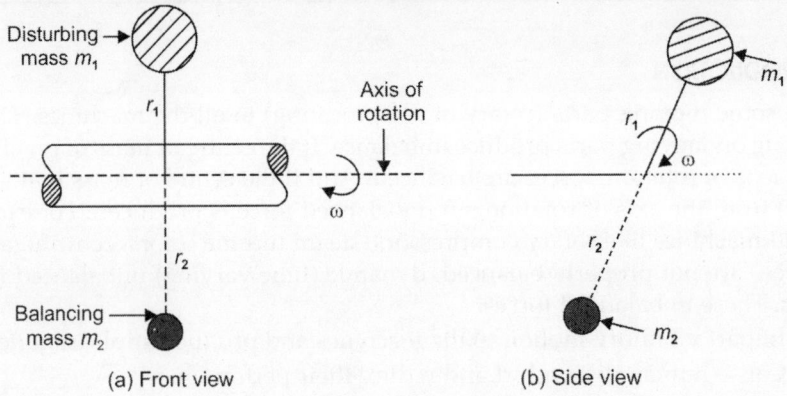

(a) Front view　　　　(b) Side view

Fig. 8.1: Balancing of a single rotating mass

Centrifugal force (disturbing force) acting radially outwards on mass m_2, $(F_{c_1}) = m_1 r_1 \omega^2$.

This force produces bending moment on the shaft.

In order to counter act the disturbing force, a balancing mass m_2 is attached to the shaft in the same plane such that the centrifugal forces of the two masses are equal and opposite.

Balancing force, $\qquad F_{c_2} = m_2 r_2 \omega^2$

For balancing, $\qquad F_{c_1} = F_{c_2}$

or $\qquad\qquad m_1 r_1 \omega^2 = m_2 r_2 \omega^2$

or $\qquad\qquad m_1 r_1 = m_2 r_2$

The value of r_2 is kept larger to reduce the balancing mass m_2.

ii. Balance mass rotating in different planes

If the disturbing mass and balancing mass lie in different planes, disturbing mass cannot be balanced by a single mass. As the lines of action of centrifugal forces acting on two masses are different, they form a disturbing couple. Since a couple can be balanced by a couple, hence a second balancing mass is required to be attached to shaft to produce a couple of opposite sense. So three masses are arranged in a way that the resultant force and couple on the shaft is zero.

This is possible if the lines of action of centrifugal forces due to three masses are parallel and the algebraic sum of their moments about any point in the same plane is zero.

Let m = mass of disturbing body in plane A

   m_1 = mass of balancing weight in plane B

   m_2 = mass of 2nd balancing weight in plane C

   l_1 = distance between planes A and B

   l_2 = distance between planes A and C

$l(l_1 + l_2)$ = distance between planes band C, r, r_1 and r_2 are distance's of centre of masses of m, m_1 and m_2 respectively from the axis of rotation.

Each mass experiences a centrifugal force acting radially outward from the axis of rotation.

$$F_c = mr\omega^2$$
$$F_{c_1} = m_1 r_1 \omega^2$$
$$F_{c_2} = m_2 r_2 \omega^2$$

For balancing of the system shown in Fig. 8.2,

$$F_c = F_{c_1} + F_{c_2}$$

or $\qquad m r \omega^2 = m_1 r_1 \omega^2 + m_2 r_2 \omega^2$

or $\qquad mr = m_1 r_1 + m_2 r_2$ $\qquad\qquad$ ...(8.1)

and sum of moments about any point (P) on the shaft should be zero.

$$F_{c_1} \times (l_1 + l_2) = F_c \times l_2$$

or $\qquad m_1 r_1 \omega^2 \times l = m r \omega^2 \times l_2$ $\qquad (l_1 + l_2 = 1)$

$$m_1 r_1 = mr \frac{l_2}{l} \qquad\qquad ...(8.2)$$

Taking moment about Q,

$$F_{c_2} \times (l_1 + l_2) = F_c \times l_1$$

or $\qquad m_2 r_2 \omega^2 \times l = m r \omega^2 l_1$

or $\qquad m_2 r_2 = mr \frac{l_1}{l} \qquad\qquad ...(8.3)$

Similar procedure can be adopted to find the balancing masses in case when the planes of balancing masses lie on the same side of the plane of the disturbing mass.

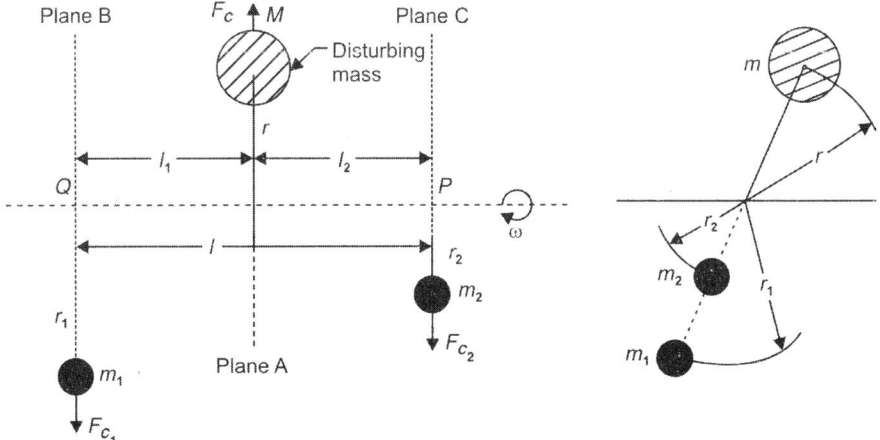

Fig. 8.2: Balancing of a single rotating mass by two rotating masses in different planes

8.4.2 Balancing of Several Masses Rotating in the Same Plane

Consider a number of point masses m_1, m_2, m_3 and m_4 which are rigidly attached to a shaft at radii r_1, r_2, r_3 and r_4 and lie in the same transverse plane. The shaft is revolving at uniform angular speed ω rad/s about an axis passing through O.

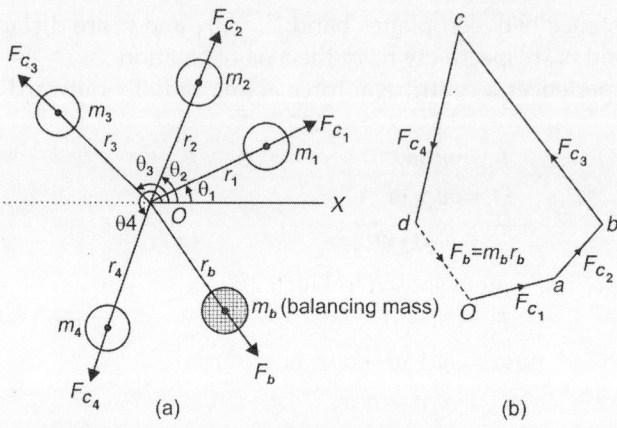

Fig. 8.3

The angles made by these masses with the horizontal line OX are θ_1, θ_2, θ_3 and θ_4 respectively. While revolving, each mass experiences a centrifugal force acting radially outward from the axis of rotation.

The vector sum of all the centrifugal forces,

$$F = m_1 r_1 \omega^2 + m_2 r_2 \omega^2 + m_3 r_3 \omega^2 + m_4 r_4 \omega^2$$

The shaft will be statically balanced if the sum of forces is zero. If sum of forces is not zero; then a, balancing mass m_b is attached to the shaft at radius r_b to balance the shaft. Therefore,

$$m_1 r_1 \omega^2 + m_2 r_2 \omega^2 + m_3 r_3 \omega^2 + m_4 r_4 \omega^2 + m_b r_b \omega^2 = 0$$

or
$$m_1 r_1 + m_2 r_2 + m_3 r_3 + m_4 r_4 + m_b r_b = 0 \qquad \qquad ...(8.4)$$

Either of m_b or r_b can be selected and other can be determined. Equation (8.4) can be written as

$$\Sigma mr + m_b r_b = 0 \qquad \qquad ...(8.5)$$

To solve Eq. (8.5) mathematically, each force is resolved horizontally and vertically.

$$m_1 r_1 \sin\theta_1 + m_2 r_2 \sin\theta_2 + m_3 r_3 \sin\theta_3 + m_4 r_4 \sin\theta_4 + m_b r_b \sin\theta_b = 0$$

and
$$m_1 r_1 \cos\theta_1 + m_2 r_2 \cos\theta_2 + m_3 r_3 \cos\theta_3 + m_4 r_4 \cos\theta_4 + m_b r_b \cos\theta_b = 0$$

or
$$\Sigma mr \sin\theta + m_b r_b \sin\theta_b = 0$$

and
$$\Sigma mr \cos\theta + m_b r_b \cos\theta_b = 0$$

$$m_b r_b \sin\theta_b = -\Sigma mr \sin\theta \qquad \qquad ...(8.6)$$

$$m_b r_b \cos\theta_b = -\Sigma mr \cos\theta \qquad \qquad ...(8.7)$$

Squaring and adding Eqs. (8.6) and (8.7), we get

$$m_b^2 r_b^2 \sin^2\theta_b + m_b^2 r_b^2 \cos^2\theta_b = (\Sigma mr \sin\theta)^2 + (\Sigma mr \cos\theta)^2$$

or
$$m_b r_b = \sqrt{(\Sigma mr \sin\theta)^2 + (\Sigma mr \cos\theta)^2} \qquad \qquad ...(8.8)$$

Dividing Eq. (8.6) by Eq. (8.7), we get

$$\tan \theta_b = \frac{-\sum mr \sin \theta}{-\sum mr \cos \theta} \qquad \text{...(8.9)}$$

Here, θ_b is the angle made by balancing force with the horizontal. Sign of numerator and denominator determines quadrant of angle. In graphical method (Fig 8.3b) *Od* represents the magnitude and direction of balancing force.

Example 8.1: Three masses are attached to a shaft with the following properties:

$$
\begin{array}{llll}
m_1 = 3 \text{ kg} & r_1 = 30 \text{ mm} & \theta_1 = 30° & \text{plane A} \\
m_2 = 4 \text{ kg} & r_2 = 20 \text{ mm·} & \theta_2 = 120° & \text{plane A} \\
m_3 = 2 \text{ kg} & r_3 = 25 \text{ mm} & \theta_3 = 270° & \text{plane A}
\end{array}
$$

Find the amount of counter mass at a radial distance of 35 mm for the static balance.

Solution: Let balancing mass, m_b be attached at angular position θ_b.

For static balance, conditions are

$$m_1 r_1 \sin \theta_1 + m_2 r_2 \sin \theta_2 + m_3 r_3 \sin \theta_3 + m_b r_b \sin \theta_b = 0$$

or $3 \times 30 \sin 30° + 4 \times 20 \sin 120° + 2 \times 25 \sin 270° + m_b \times 35 \sin \theta_b = 0$

or $$45 + 69.28 - 50 + 35 \, m_b \sin \theta_b = 0$$

$$m_b \sin \theta_b = -1.837 \qquad \text{...(i)}$$

and $m_1 r_1 \cos \theta_1 + m_2 r_2 \cos \theta_2 + m_3 r_3 \cos \theta_3 + m_b r_b \cos \theta_b = 0$

or $3 \times 30 \cos 30° + 4 \times 20 \cos 120° + 2 \times 25 \cos 270° + m_b \times 35 \cos \theta_b = 0$

or $$77.94 - 40 + 0 + 35 \, m_b \cos \theta_b = 0$$

$$m_b \cos \theta_b = -1.084 \qquad \text{...(ii)}$$

Squaring and adding Eqs. (i) and (ii)

$$m_b^2 \sin^2 \theta_b + m_b^2 \cos^2 \theta_b = (-1.837)^2 + (-1.084)^2$$

or $$m_b{}^2 = 4.55 \text{ or } m_b = 2.13 \text{ kg}$$

Dividing Eqs. (i) by (ii)

$$\frac{m_b \sin \theta_b}{m_b \cos \theta_b} = \frac{-1.837}{-1.084}$$

or $$\tan \theta_b = 1.695$$

or $$\theta_b = 59.4°$$

But as values of $\sin \theta_b$ and $\cos \theta_b$ are negative, θ_b lies in the third quadrant. Therefore,
$$\theta_b = 180° + 59.4° = 239.4°$$

8.4.3 Balancing of Several Masses Rotating in Different Planes

Consider masses m_1, m_2 and m_3 attached to the shaft at radii r_1, r_2 and r_3 respectively. The shaft is revolving with a uniform angular velocity ω rad/s. The masses are revolving in planes A_1, A_2 and A_3 respectively. A reference plane B_1 is taken at O such that the distances of planes A_1, A_2, and A_3 are l_1, l_2, and l_3 respectively.

This method is based on the fact that a force acting on a rigid body at a point is equivalent to an equal and parallel force acting at another fixed point together with a couple, magnitude of which is equal to the product of the force and the distance between the two points. Therefore, transfer of each unbalanced force to the reference plane will introduce like number of forces and couples. Couple vectors are drawn parallel to the respective force vectors.

For complete balancing of masses, the resultant force in the reference plane and the resultant couple about the reference plane must be zero.

$\therefore$ $\qquad$ $m_1 r_1 \omega^2 + m_2 r_2 \omega^2 + m_3 r_3 \omega^2 = 0$ $\qquad$ ...(8.10)

and $\qquad$ $m_1 r_1 l_1 \omega^2 + m_2 r_2 l_2 \omega^2 + m_3 r_3 l_3 \omega^2 = 0$ $\qquad$ ...(8.11)

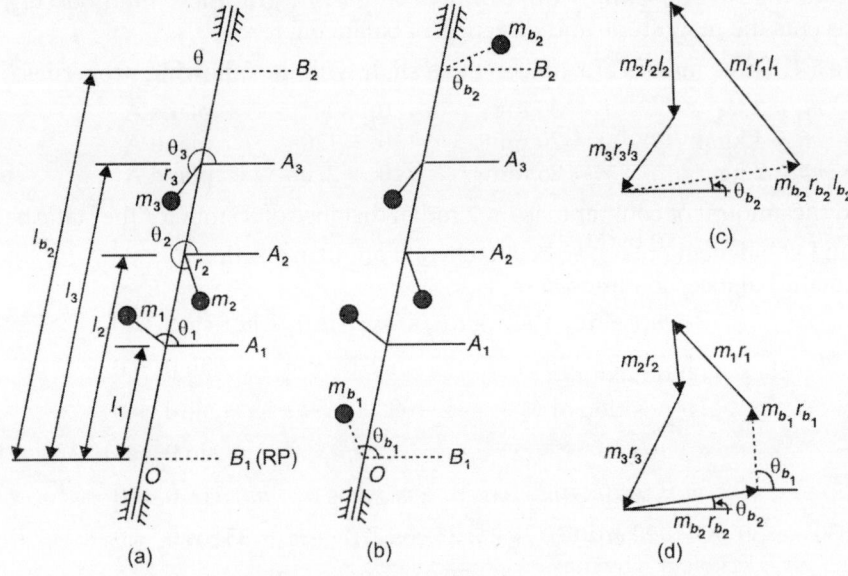

Fig. 8.4

An unbalanced force can be satisfied by attaching a mass to the shaft in reference plane but unbalanced couple can be satisfied only by attaching two masses in different planes. Therefore, to satisfy the unbalanced force and couple, two balancing masses m_{b_1} and m_{b_2} are attached at radii r_{b_1} and r_{b_2} respectively in planes B_1 and B_2.

$\therefore$ $\qquad$ $m_1 r_1 \omega^2 + m_2 r_2 \omega^2 + m_3 r_3 \omega^2 + m_{b_1} r_{b_1} \omega^2 + m_{b_2} r_{b_2} \omega^2 = 0$

or $\qquad$ $m_1 r_1 + m_2 r_2 + m_3 r_3 + m_{b_1} r_{b_1} + m_{b_2} r_{b_2} = 0$

or $\qquad$ $\Sigma mr + m_{b_1} r_{b_1} + m_{b_2} r_{b_2} = 0$ $\qquad$ ...(8.12)

Taking moments about 'O'

$\therefore$ $\qquad$ $m_1 r_1 l_1 \omega^2 + m_2 r_2 l_2 \omega^2 + m_3 r_3 l_3 \omega^2 + m_{b_2} r_{b_2} l_{b_2} \omega^2 = 0$

or $\qquad$ $\Sigma mrl + m_{b_2} r_{b_2} l_{b_2} = 0$ $\qquad$ ...(8.12)

To solve Eqs. (8.12) and (8.13), each force and couple is resolved horizontally and vertically.

$\therefore$ $\qquad$ $\Sigma mr \, l \sin \theta + m_{b_2} r_{b_2} l_{b_2} \sin \theta_{b_2} = 0$

and $\qquad$ $\Sigma mr \, l \cos \theta + m_{b_2} r_{b_2} l_{b_2} \cos \theta_{b_2} = 0$

or $\qquad$ $m_{b_2} r_{b_2} l_{b_2} \sin \theta_{b_2} = - \Sigma mrl \sin \theta$ $\qquad$ ...(a)

and $\qquad$ $m_{b_2} r_{b_2} l_{b_2} \cos \theta_{b_2} = - \Sigma mrl \cos \theta$ $\qquad$ ...(b)

Squaring and adding Eqs. (8.12a) and (8.12b) and taking square root, we have

$$m_{b_2} r_{b_2} l_{b_2} = \sqrt{(\Sigma mrl \sin \theta)^2 + (\Sigma mrl \cos \theta)^2}$$ $\qquad$ ...(8.14)

Dividing Eq. (a) by Eq. (b),

$$\tan\theta_{b_2} = \frac{-\Sigma mrl\sin\theta}{-\Sigma mrl\cos\theta} \qquad\qquad ...(8.15)$$

Now Eq. (8.12) is solved

$$\Sigma mr\sin\theta + m_{b_1}r_{b_1}\sin\theta_{b_1} + m_{b_2}r_{b_2}\sin\theta_{b_2} = 0$$

or
$$m_{b_1}r_{b_1}\sin\theta_{b_1} = -(\Sigma mr\sin\theta + m_{b_2}r_{b_2}\sin\theta_{b_2}) \qquad ...(8.15a)$$

and $\quad \Sigma mr\,os\,\theta + m_{b_1}r_{b_1}\cos\theta_{b_1} + m_{b_2}r_{b_2}\cos\theta_{b_2} = 0$

or
$$m_{b_1}r_{b_1}\cos\theta_{b_1} = -(\Sigma mr\cos\theta + m_{b_2}r_{b_2}\cos\theta_{b_2}) \qquad ...(8.15b)$$

Squaring and adding Eqs (8.15a) and (8.15b) and taking square root

$$m_{b_1}r_{b_1} = \sqrt{(\Sigma mr\sin\theta + m_{b_2}r_{b_2}\sin\theta_{b_2})^2 + (\Sigma mr\cos\theta + m_{b_2}r_{b_2}\cos\theta_{b_2})^2} \qquad ...(8.16)$$

Dividing Eq. (8.15a) by Eq. (8.15b), we have

$$\tan\theta_{b_1} = \frac{-(\Sigma mr\sin\theta + m_{b_2}r_{b_2}\sin\theta_{b_2})}{-(\Sigma mr\cos\theta + m_{b_2}r_{b_2}\cos\theta_{b_2})} \qquad\qquad ...(8.17)$$

In graphical method, couple polygon is made first. The closing vector $m_{b_2}r_{b_2}l_{b_2}$ in this, gives the angular position of balancing mass m_{b_2}. Now the force polygon is made in which closing vector $m_{b_1}r_{b_1}$ gives the magnitude and direction of the balancing mass m_{b_1}.

Example 8.2: A shaft carries four rotating masses A, B, C and D in this order along its axis. The mass A may be assumed to be concentrated at a radius of 12 cm, B at 15 cm, C at 14 cm and D at 18 cm. The masses of A, C and D are 15 kg, 10 kg and 8 kg respectively. The planes of revolution of A and B are 15 cm apart and of B and C are 18 cm apart. The angle between the radii of A and C is 90°. If the shaft is in complete dynamic balance, determine: (i) the angles between the radii of A, B and D, (ii) the distance between the planes of revolution of C and D and (iii) the mass B.

Solution:
Given,

$r_A = 12$ cm	$m_A = 15$ kg
$r_B = 15$ cm	$m_B = ?$
$r_C = 14$ cm	$m_C = 10$ kg
$r_D = 18$ cm	$m_D = 8$ kg

Let x = Distance between the planes C and D. Assume plane C as the reference plane.

Plane	mass (m) kg	Radius (r) cm	Centrifugal force (ω^2) (mr) kg·cm	Distance from plane C (l) cm	Couple (mr·l) kg.m^2
1	2	3	4	5	6
A	15	12	180	− 33	−5940
B	m_B	15	$15m_B$	− 18	$-270\,m_B$ $= -270 \times 7.2$ $= 1944$
C (RP)	10	14	140	0	0
D	8	18	144	$+x$	$144x$

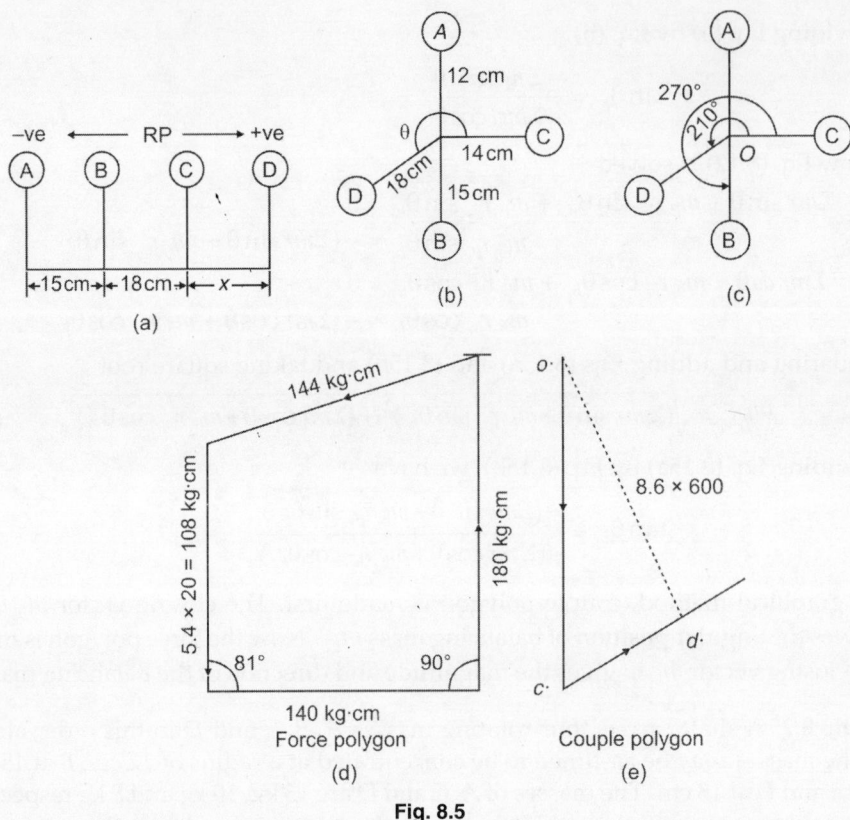

Fig. 8.5

Here, the angular displacement position is not clear and one data is missing. Assuming that the angle between radii of C and B is 270°.

Draw force polygon.

Scale 20 kg·cm = 1 cm

Force polygon

$$15m_B = 108 \text{ kg}$$
$$m_B = 7.2 \text{ kg}$$

Angle between C and $D = 20°$

Drawing couple polygon,

$$8.6 \times 600 = 144x$$
$$x = 35.8 \text{ cm}$$

Example 8.3: A shaft carries four rotating masses A, B, C and D which are completely balanced. The masses B, C and D are 50 kg, 80 kg and 70 kg respectively. The masses C and D make angles of 90° and 195° respectively with mass B in the same sense. The masses A, B, C and D are concentrated at radius 75 mm, 100 mm, 50 mm and 90 mm respectively. The plane of rotation of masses B and C are 250 mm apart. Determine:

    (i) the mass A and its angular position
    (ii) the position of planes of A and D.

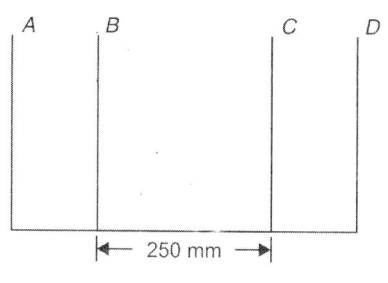

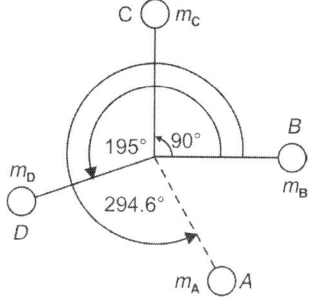

(a) Position of planes   (b) Angular position of masses

Fig. 8.6

Solution: Masses, $m_B = 50$ kg; $m_C = 80$ kg; $m_D = 70$ kg.
Radius of masses,

$$r_A = 75 \text{ mm}; r_B = 100 \text{ mm}; r_C = 50 \text{ mm}; r_D = 90 \text{ mm}.$$

Angles between masses,

$$\angle \text{BOC} = 90° \text{ and } \angle \text{BOD} = 195° \text{ or } \theta_C = 90° \text{ and } \theta_D = 195°$$

Distance between planes B and $C = 250$ mm.
Find: (i) m_A and θ_A; (ii) Positions of planes A and D.

Analytical method: Figure 8.6a shows the position of planes A, B, C and D whereas Fig. 8.6b shows the angular position of masses, B, C and D in which the angular position of mass m_B is assumed in horizontal direction.

Hence, $\theta_B = 0°$.

(i) The mass A and its angular position.

Let m_A = Mass A

θ_A = Angular position of mass A with respect to mass B.

The four masses are completely balanced, hence resultant force should be zero or $\Sigma m \times r \times \cos \theta = 0$ and $\Sigma m \times r \times \sin \theta = 0$.

Let us first find the product of known mass and corresponding radius.

$\therefore$
$$m_B \times r_B = 50 \times 100 = 5000$$
$$m_C \times r_C = 80 \times 50 = 4000$$
$$m_D \times r_D = 70 \times 90 = 6300$$

Also we know that $\theta_B = 0, \theta_C = 90°$ and $\theta_D = 195°$.

Now for $\Sigma m \times r \times \cos \theta = 0$, we have

$$m_A \times r_A \times \cos \theta_A + m_B \times r_B \times \cos \theta_B + m_C \times r_C \times \cos \theta_C + m_D \times r_D \times \cos \theta_D = 0$$

or $m_A \times 75 \times \cos \theta_A + 5000 \times \cos 0° + 4000 \times \cos 90° + 6300 \times \cos 195° = 0$

$75m_A \times \cos \theta_A + 5000 + 0 + (-6085.3) = 0$

$75m_A \times \cos \theta_A = 6085.3 - 5000 = 1085.3$ ...(i)

For $\Sigma m \times r \times \sin \theta = 0$, we have

$$m_A \times r_A \times \sin \theta_A + m_B \times r_B \times \sin \theta_B + m_C \times r_C \times \sin \theta_C + m_D \times r_D \times \sin \theta_D = 0$$

or $m_A \times 75 \times \sin \theta_A + 5000 \times \sin 0° + 4000 \times \sin 90° + 6300 \times \sin 195° = 0$

or $75m_A \times \sin \theta_A + 0 + 4000 + (-1630.5) = 0$

$75m_A \times \sin \theta_A = -4000 + 1630.5 = -2369.5$ ...(ii)

Squaring and adding Eqs (i) and (ii), we get

$$75^2 \times m_A^2 \times (\cos^2 \theta_A + \sin^2 \theta_A) = (1085.3)^2 + (-2369.5)^2$$

or

$$5625 \times m_A^2 = 1177876 + 5614530 = 6792406$$

$$m_A = \sqrt{\frac{6792406}{5625}} = 34.75 \text{ kg}$$

Dividing Eqs (ii) by (i), we get

$$\frac{75 \times m_A \times \sin \theta_A}{75 \times m_A \times \cos \theta_A} = \frac{-2369.5}{1085.3}$$

or

$$\frac{\sin \theta_A}{\cos \theta_A} = \tan \theta_A = \frac{-2369.5}{1085.3}$$

In the above equation, the numerator (i.e., $\sin \theta_A$) is −ve whereas the denominator (i.e., $\cos \theta_A$) is +ve. Hence θ_A lies in fourth quadrant as sine of an angle is −ve and cosine of the angle is +ve in fourth quadrant.

$$\therefore \qquad \theta_A = \tan^{-1} = \frac{-2369.5}{1085.3} = \tan^{-1}(-2.183)$$

$$= -65.4° = 360 - 65.4 = 294.6°.$$

(ii) Position of planes A and D.

Take plane A as the reference plane.

Then $l_A = 0$

l_B = distance of plane B from plane A

l_C = distance of plane C from plane $A = l_B + 250$

l_D = distance of plane D from plane A

As the four masses are completely balanced, hence the resultant couple about the reference plane should be zero

or $\quad \Sigma m \times r \times l \cos \theta = 0$ and $\Sigma m \times r \times l \times \sin \theta = 0$

For $\Sigma m \times r \times l \cos \theta = 0$ about plane A, we have

$$m_B \times r_B \times l_B \times \cos \theta_B + m_C \times r_C \times l_C \times \cos \theta_C + m_D \times r_D \times l_D \times \cos \theta_D = 0$$

or $\quad 5000 \times l_B \times \cos 0° + 4000 \times (l_B + 250) \cos 90° + 6300 \times l_D \times \cos 195° = 0$

$\quad (\because m_B \times r_B = 5000; m_C \times r_C = 4000; m_D \times r_D = 6300; l_C = l_B + 250)$

$$5000\, l_B + 0 + 6300 \times l_D \times (-0.966) = 0$$

$$5000\, l_B - 6085.3\, l_D = 0$$

or $\quad l_B = \dfrac{6085.3\, l_D}{5000} = 1.217\, l_D$ \qquad\qquad ...(iii)

$\Sigma m \times r \times l \sin \theta = 0$ about plane A, we have

$$m_B \times r_B \times l_B \times \sin \theta_B + m_C \times r_C \times l_C \times \sin \theta_C + m_D \times r_D \times l_D \times \sin \theta_D = 0$$

or $\quad 5000 \times l_B \times \sin 0° + 4000 \times (l_B + 250) \sin 90° + 6300 \times l_D \times \sin 195° = 0$

$\qquad\qquad\qquad\qquad\qquad\qquad\qquad\qquad (\because l_C = l_B + 250)$

or $\quad 0 + 4000\, (l_B \times 250) + 6300 \times l_D \times (-0.2588) = 0$

or $4000\, l_B + 1000000 - 1630.5\, l_D = 0$

or $4000\, (1.217\, l_D) + 1000000 - 1630.5\, l_D = 0$

$[\because l_B = 1.217\, l_D \text{ from Eq. (iii)}]$

or $4868\, l_D + 1000000 - 1630.5\, l_D = 0$

or $3237.5\, l_D = -1000000$

$$l_D = \frac{-1000000}{3237.5} = -308.8 \text{ mm}$$

From Eq. (iii),

But $l_C = l_B + 250 = -375.8 + 250 = -125.8 \text{ mm}$

(iii) The correct position of planes, *A, B, C* and *D* are shown in Fig. 8.7 whereas the angular position of mass m_A is shown in Fig. 8.7b.

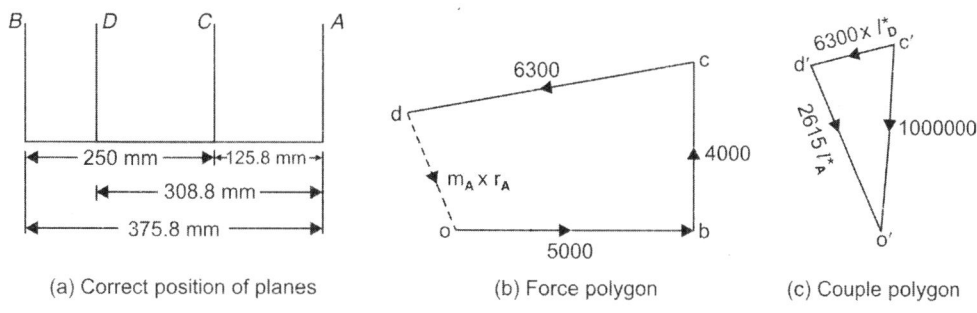

(a) Correct position of planes   (b) Force polygon   (c) Couple polygon

Fig. 8.7

Graphical method: We know that

$m_B \times r_B = 5000; m_C \times r_C = 4000;$ and $m_D \times r_D = 6300.$

(i) The mass m_A and its angular position

The four masses are completely balanced, hence, the force polygon should be a closed figure. The force polgon is drawn as shown in Fig. 8.7b by the method given below:

    a. Take any point *o*. From *o*, draw vector *ob* = 5000 and parallel to *OB* of Fig. 8.6b to some suitable scale.

    b. From point *b*, draw vector *bc* = 4000 and parallel to *OC*.

    c. From point *c*, draw vector *cd* = 6300 and parallel to *OD*. Join point *d* to *o*.

The closing side *do* represents the product of mass and radius of balancing mass (i.e., $m_A \times r_A$). Measure *do*. On measurement, *do* = 2615.

$$m_A \times r_A = 2615$$

$\therefore$ $$m_A = \frac{2615}{r_A} = \frac{2615}{75} = 34.86 \text{ kg.}$$

To find the value of θ_A (i.e. angle made by mass m_A with mass m_B), draw the line *OA* in Fig. 8.6b parallel to line *do* from Fig. 8.7b. On measurement angle $\theta_A = 294.6°$.

$$\theta_A = 294.6°$$

(ii) The position of planes of *A* oand *D*

Take plane B of Fig. 8.7c as the reference plane.

Then $l_D^* =$ distance of plane D from plane B

$\quad\quad l_C^* =$ distance of plane C from plane $B = 250$ mm (given)

$\quad\quad l_A^* =$ distance of plane A from plane B

We know that,

$$m_C \times r_C \times l_C^* = 4000 \times 250$$
$$= 1000000; \, m_D \times r_D \times l_D^* = 6300 \times l_D^*$$

and $\quad\quad\quad m_C \times r_C \times l_A^* = 2615 \times l_A^*$

($\therefore m_C \times r_C = 4000$, $l^* c = 250$, $m_D r_D = 6300$ and $m_A \times r_A$ from force polygon $= 2615$).

The product $m_B \times r_B \times l_B^* = O$ about reference plane B.

The four masses are completely balanced, hence, the couple polygon should be a closed figure. About reference plane B, there are only three couples. Hence, the couple polygon will be triangle. Also the couples are acting in the direction of forces. The couple polygon is drawn as shown in Fig. 8.7c by the method as given below:

(iii) Take any point o'. From o', draw vector $o'c' = m_C \times r_C \times l_C^* = 1000000$ and parallel to OC. From point c', draw a vector $c'd'$ parallel to OD. And from point o', draw a vector $o'd'$, parallel to OA. The vector $c'd'$ and $o'd'$ intersects at d'. Measure $c'd'$ and $o'd'$.

On measurement, $c'd' = 425000$ and $o'd' = 983000$.

But $\quad\quad\quad\quad c'd' = 6300 \times l_D^*$ and $o'd' = 2615 \times l_A^*$

$\therefore \quad\quad\quad\quad c'd' = 6300 \times l_D^* = 425000$

or $\quad\quad\quad\quad l_D^* = \dfrac{425000}{6300} = 67.46$ mm. Ans.

and $\quad\quad\quad\quad o'd' = 2615 \times l_A^* = 983000$

or $\quad\quad\quad\quad l_A^* = \dfrac{983000}{2615} = 375.9$ mm. Ans.

The value of l_A^* and l_D^* by analytical method are:

$\quad\quad l_A^* = 375.8$ mm and $l_D^* = 375.8 - 308.8 = 67$ mm.

The values given by analytical method and graphical method are closely equal.

Example 8.4: A rotating shaft carries four unbalanced masses 18 kg, 14 kg, 16 kg and 12 kg at radii 5 cm, 6 cm, 7 cm and 6 cm respectively. The 2nd, 3rd and 4th masses revolve in planes 8 cm, 16 cm and 28 cm respectively measured from the plane of the first mass and are angularly located at 60°, 135° and 270° respectively measured anticlockwise from the first mass looking from this mass end of the shaft. The shaft is dynamically balanced by two masses, both located at 5 cm radii and revolving in planes midway between those of 1st and 2nd masses and midway between those of 3rd and 4th masses. Determine graphically or otherwise, the magnitudes of the masses and their respective angular positions.

Solution:

$$m_1 = 18 \text{ kg}, r_1 = 5 \text{ cm} = 0.05 \text{ m}, \theta_1 = 0°$$
$$m_2 = 14 \text{ kg}, r_2 = 6 \text{ cm} = 0.06 \text{ m}, \theta_2 = 60°$$
$$m_3 = 16 \text{ kg}, r_3 = 7 \text{ cm} = 0.07 \text{ m}, \theta_3 = 135°$$
$$m_4 = 12 \text{ kg}, r_4 = 6 \text{ cm} = 0.06 \text{ m}, \theta_4 = 270°$$

Let the two balancing masses are m_A and m_B.

The position of planes and angular position of masses are shown in Figs. 8.8a and b. m_A is placed between plane 1 and 2 m_B is placed between planes 3 and 4. It can be seen that B and A are at a distance of 22 cm and 4 cm from plane 1 respectively.

We assume that plane A is the reference plane. The distance to the right of this plane are taken as +ve while to the left as –ve.

Table 8.1

Plane	mass (m) kg	Radius (r) m	Centrifugal force (ω^2) (m·r) kg·m	Distance from RP	Couple (mr·l) kg·m²
1	18	0.05	0.90	–0.04	–0.036
A(RP)	m_A	0.05	$0.05\, m_A$	0	0
2	14	0.06	0.84	0.04	0.0336
3	16	0.07	1.12	0.12	0.1344
B	m_B	0.05	$0.05\, m_B$	0.18	$0.009\, m_B$
4	12	0.06	0.72	0.24	0.1728

Couple polygon:

1. First of all, couple polygon is drawn from the data given in Table 8.1 (Fig. 8.8c).
2. Draw $o'a'$ parallel to OA and $oa' = -0.036$. It will be in the reverse direction of OA.
3. Draw $a'b'$ parallel to OB starting from point a' and $a'b' = 0.0336$
4. From b' draw $b'c' = 0.1344$, parallel to OC.
5. From c' draw $c'd' = 1728$, parallel to OD.
6. Join d' to o'. Vector $o'd'$ is equal to $0.009\, m_B$.

$$o'd' = 0.009\, m_B$$

Thus, we find $m_B = \dfrac{o'd'}{0.009} = \dfrac{0.12}{0.09} = 13.33 \text{ kg}$ (by measurement $o'd' = 0.12$).

The angular position of mass m_B is from mass $m_1 \cdot \theta_B = 25°$.

Force Polygon

1. Now draw the force polygon from the data in Table 8.1 (Fig. 8.8d).
2. Draw $oa = 0.90$, parallel to OA.
3. From a draw $ab = 0.84$, parallel to OB.
4. From b draw $bc = 1.12$, parallel to OC.
5. From c draw $cd = 0.72$, parallel to OD.
6. From d draw $de = 0.05\, m_B = 0.05 \times 13.33$ parallel to $o'd'$ in couple polygon.

7. Join e with o. Then $oe = 6.3 \text{ cm} \Leftrightarrow \dfrac{6.3}{4} = 0.05\, m_A$

$$(\therefore oe = \text{balanced force} = 0.05\, m_A)$$

So mass $m_A = \dfrac{1.575}{0.05} = 31.5 \text{ kg}$ and $\theta_A = 220°$ from plane 1.

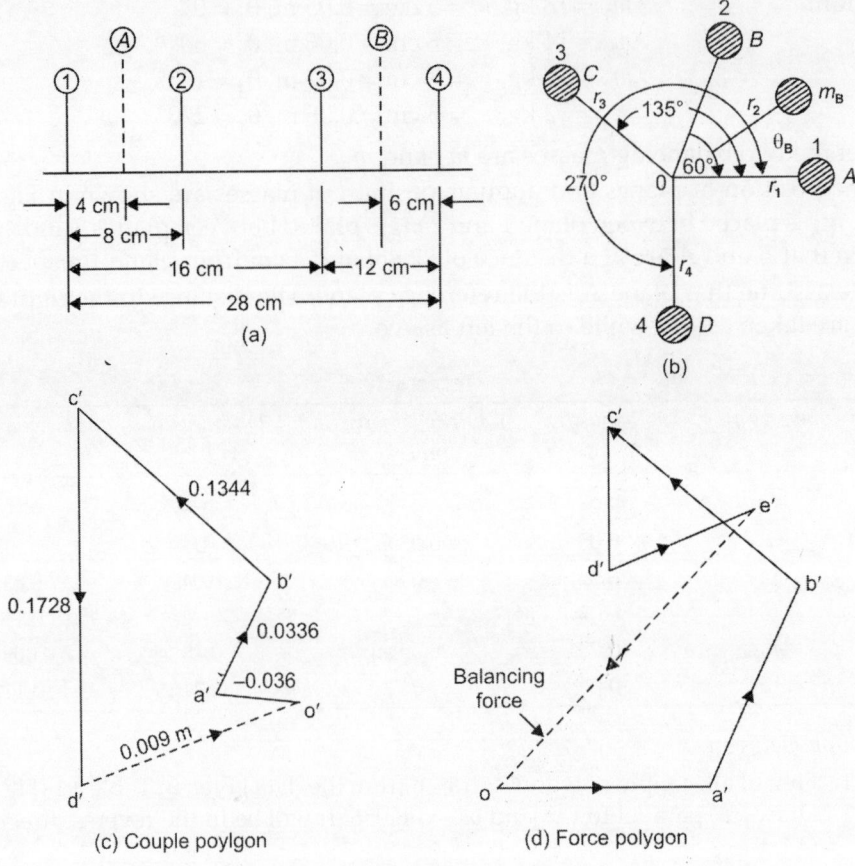

(a)

(b)

(c) Couple poylgon

(d) Force polygon

Fig. 8.8

8.5 BALANCING OF RECIPROCATING MASS

Acceleration of the slider (reciprocating mass) of a slider–crank mechanism is given as

$$a = r\omega^2 \left[\cos\theta + \frac{\cos 2\theta}{n} \right]$$

Force required to accelerate the reciprocating mass

$$F = m \times a = mr\omega^2 \left[\cos\theta + \frac{\cos 2\theta}{n} \right]$$

Inertia force is equal and opposite to accelerating force, i.e.

$$F_i = -F = -mr\omega^2 \left[\cos\theta + \frac{\cos 2\theta}{n} \right]$$

or

$$F_i = -\left[mr\omega^2 \cos\theta + mr\omega^2 \frac{\cos 2\theta}{n} \right] \tag{8.18}$$

$$F_i = -[F_P + F_S]$$

Here, $F_P = mr\omega^2 \cos\theta$ = primary unbalanced force

$$F_S = mr\omega^2 \frac{\cos 2\theta}{n} = \text{secondary unbalanced force}$$

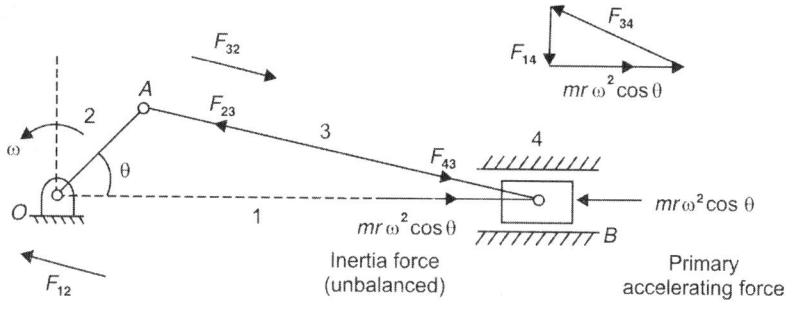

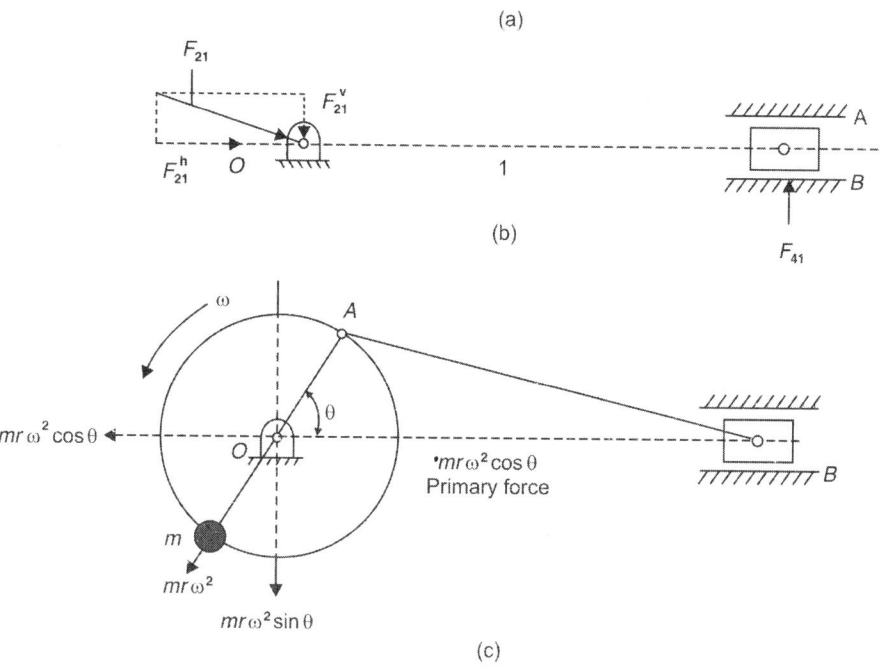

Fig. 8.9

Maximum value of primary unbalanced force, $F_{P(max)} = mr\,\omega^2$

Maximum value of secondary unbalanced force, $F_{s(max)} = \dfrac{mr\,\omega^2}{n}$

Since the inertia force acts along the line of stroke, hence primary and secondary unbalanced force will also act along the line of stroke.

As the value of n is much greater than unity, the secondary unbalanced force is small compared with the primary unbalanced force, it is neglected for slow speed engines.

The unbalanced force due to reciprocating mass (F_P and F_S) is constant in direction (acting along the line of stroke) but varies in magnitude (due to variation in θ), whereas the unbalanced force due to rotating mass is constant in magnitude ($mr\omega^2$) but varies in direction.

8.6 PARTIAL BALANCING OF UNBALANCED PRIMARY FORCE

Due to inertia force (F_i), a force F_{21} is exerted by the crank shaft on the main bearings. F_{21} has two components F_{21}^h and F_{21}^v. Force $F_{21}^h = mr\omega^2 \cos\theta$ is primary unbalanced force that acts from O to B. Vertical forces F_{21}^v and F_{41}^v form a couple known as unbalanced shaking couple. The magnitude and direction of unbalanced primary force and couple keep on changing with crank angle θ. The unbalanced force produces linear vibrations in horizontal direction and unbalanced couple produces oscillatory vibrations. The approach of balancing unbalaned force is that the primary unbalanced foree ($mr\omega^2 \cos\theta$) is considered as the component of centrifugal force produced by a rotating mass m placed at radius r. To counter this, a balancing mass m is attached at radius r, directly opposite to crank.

The horizontal component of the centrifugal force due to the balancing mass is $mr\omega^2 \cos\theta$ along the line of stroke from B to O. This force balances the unbalanced primary force. But the component of centrifugal force perpendicular to the line of stroke ($mr\omega^2 \sin\theta$) remains unbalanced. The unbalanced force is zero at $\theta = 0°$ and $180°$, which are the ends of the stroke and maximum at the middle when $\theta = 90°$ and $270°$, maximum value of unbalanced force is still same, i.e. $mr\omega^2$. Effect of this vertical unbalanced force is that engine starts jumping up and down.

As a compromise to minimise the effect of the unbalanced force, a fraction (let c fraction) of the reciprocating mass is balanced.

Primary force balanced by the mass = $cmr\omega^2 \cos\theta$

∴ Unbalanced force along the line of stroke = $mr\omega^2\cos\theta - cmr\omega^2 \cos\theta$

or $= (1-c) mr\omega^2 \cos\theta$

Unbalanced force perpendicular to the line of stroke = $cmr\omega^2\sin\theta$ ...(8.19)

∴ Resultant unbalanced forced at any instant

$$= \sqrt{[(1-c)mr\omega^2 \cos\theta]^2 + [cmr\omega^2 \sin\theta]^2}$$

$$= mr\omega^2 \sqrt{[(1-c)\cos^2\theta + c^2 \sin^2\theta}$$ ...(8.20)

The value of c is taken from $\dfrac{1}{2}$ to $\dfrac{3}{4}$. The resultant unbalanced force is minimum when $c = \dfrac{1}{2}$.

Minimum unbalanced force = $mr\omega^2 \sqrt{\left(1-\dfrac{1}{2}\right)^2 \cos^2\theta + \left(\dfrac{1}{2}\right)^2 \sin^2\theta} = \dfrac{mr\omega^2}{2}$

Example 8.5: A single cylinder reciprocating engine has the following data:
Speed of engine = 120 rpm; stroke = 320 mm; mass of reciprocating parts = 45 kg; mass of revolving parts = 35 kg at crank radius. If 60% of the reciprocating parts and all the revolving parts are to be balanced, then find:

(i) The balance mass required at a radius of 300 mm
(ii) The unbalanced force when the crank has rotated 60° from top dead centre.

Solution:

$$N = 120 \text{ rpm or } \omega = \frac{2\pi N}{60} = \frac{2\pi \times 120}{60} = 4\pi \text{ rad/s}$$

stroke = 320 mm = 0.32 m or crank radius $r = \dfrac{0.32}{2} = 0.16$ m;

mass of reciprocating parts, $m_R = 45$ kg;

mass of revolving parts, $M = 35$ kg;

radius of revolving parts, $r = 160$ mm = 0.16 m;

fraction of reciprocating parts to be balanced $= 60\% = \dfrac{60}{100} = 0.6$ or $c = 0.6$;

all revolving parts are to be balanced.

(i) Balance mass required at a radius of 300 mm

Let m_b = balance mass required

r^* = radius of rotation of balance mass = 300 mm = 0.3 m

We have

$$m_b \times r^* = (M + c \times m_R) \times r$$
$$m_b \times 0.3 = (35 + 0.6 \times 45) \times 0.16 = (35 + 27) \times 0.16$$

or
$$m_b = \frac{(35+27) \times 0.16}{0.3} = 33.06 \text{ kg}$$

(ii) Unbalanced force when crank has rotated 60° from top dead centre

Here $\theta = 60°$ (given)

The unbalanced force at any instant is given by

$$= m_R \times \omega^2 \times r \sqrt{(1-c)^2 \cos^2\theta + c^2 \sin^2\theta}$$

$$= 4.5 \times (4\pi)^2 \times 0.16 \sqrt{(1-0.6)^2 \times \cos^2 60° + 0.6^2 \sin^2 60°}$$

$$= 1136.98 \sqrt{0.16 \times 0.25 + 0.6 \times 0.75}$$

$$= 1136.98 \sqrt{0.04 + 0.45}$$

$$= 1136.98 \times 0.7 = 795.9$$

8.7 PARTIAL BALANCING OF LOCOMOTIVES

Locomotives are of two types—coupled and uncoupled. If two or more pairs of wheels are coupled together to enhance the adhesion between the wheels and the track, it is known as coupled locomotive, otherwise it is called as uncoupled locomotive. In uncoupled locomotives, there are four planes, two of cylinders and two of the driving wheels, whereas in coupled locomotives there are six planes for consideration, two of cylinders, two of coupling rods and two of the wheels. Most of the locomotives have two cylinders of same dimensions with cranks placed at right angle to each other. Depending upon the position of cylinders, locomotives are of two types:

(i) **Outside cylinder locomotive:** When the two cylinders are placed outside the driving wheels, one on each side, it is called outside cylinder locomotive.

(ii) **Inside cylinder locomotive:** When the two cylinders are placed in between the planes of the driving wheels, it is called inside cylinder locomotive.

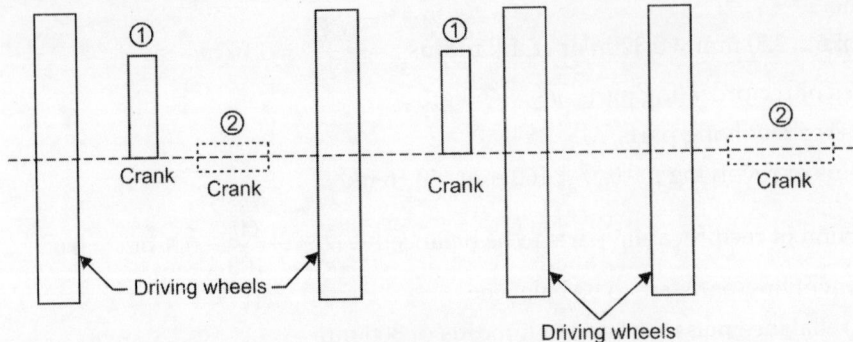

Fig. 8.10

8.8 EFFECTS OF PARTIAL BALANCING IN LOCOMOTIVES

The reciprocating parts of an engine are partially balanced to minimise the effect of unbalanced force. This means there is always an unbalanced primary force along the line of stroke and unbalanced force perpendicular to the line of stroke. The unbalanced primary force along the line of stroke produces:

(i) Variation of tractive force along the line of stroke

(ii) Swaying couple

The primary unbalanced force perpendicular to the line of stroke produces hammer blow.

8.8.1 Variation of Tractive Force

Variation in tractive force (effort) of an engine is caused by the unbalanced portion of the primary force acting along the line of stroke of a locomotive engine. Variation of tractive force is the resultant unbalanced force along the line of stroke.

Let the crank of first cylinder be inclined at an angle θ with the line of stroke.

Crank of second cylinder is at right angle to the crank of first cylinder.

Hence, angle of inclination of crank of second cylinder with line of stroke = $(90° + \theta)$

If 'c' fraction of reciprocating mass is balanced, then unbalanced primary force along the line of stroke for cylinder 1, $F_1 = (1 - c) \, mr\omega^2 \cos\theta$.

Unbalanced primary force along the line of stroke for cylinder 2,

$$F_2 = (1 - c) \, mr\omega^2 \cos(90° + \theta)$$
$$= -(1 - c) \, mr\omega^2 \sin\theta$$

Variation in tractive force = resultant unbalanced force along the line of stroke

$$= F_1 + F_2$$
$$= (1 - c)mr\omega^2\cos\theta - (1 - c) \, mr\omega^2\sin\theta$$
$$= (1 - c) \, mr\omega^2(\cos\theta - \sin\theta)$$

This is maximum when $(\cos\theta - \sin\theta)$ is maximum,

i.e. when $\dfrac{d}{d\theta}(\cos\theta - \sin\theta) = 0$

or $-\sin\theta - \cos\theta = 0$

or $\sin\theta = -\cos\theta$

$\tan\theta = -1$

for which $\theta = 135°$ or $315°$

When $\theta = 135°$,

the maximum variation in tractive force $= (1 - c)\, mr\omega^2\, (\cos 135° - \sin 135°)$

$$= (1 - c)\, mr\omega^2 \left[\frac{-1}{\sqrt{2}} - \frac{1}{\sqrt{2}} \right]$$

$$= -\sqrt{2}\, (1 - c)\, mr\omega^2$$

When $\theta = 315°$,

the maximum varition in tractive force $= (1 - c)\, mr\omega^2\, (\cos 315° - \sin 315°)$

$$= (1 - c)\, mr\omega^2 \left[\frac{1}{\sqrt{2}} + \frac{1}{\sqrt{2}} \right] \qquad (\sin 315° = -\frac{1}{\sqrt{2}})$$

$$= \sqrt{2}\, (1 - c)\, mr\omega^2$$

$\therefore$ Maximum variation in tractive force $= \pm \sqrt{2}\, (1 - c)\, mr\omega^2$ ...(8.21)

8.8.2 Swaying Couple

The unbalanced primary forces along the line of stroke act at a distance (say l) between the lines of stroke of two cylinders. These forces constitute a couple which is known as *swaying couple*. Swaying couple tends to make the leading wheels sway from side to side.

Swaying couple = Moments of unbalanced forces about the engine centre line.

$$= [(1- c)\, mr\omega^2 \cos \theta] \times \frac{l}{2} - [(1- c)\, mr\omega^2 \cos (90 + \theta) \times \frac{l}{2}]$$

$$= (1- c)\, mr\omega^2 (\cos \theta + \sin \theta) \times \frac{l}{2}$$

The swaying couple will be maximum when $(\cos\theta + \sin\theta)$ is maximum. and $\cos \theta + \sin \theta$ will be maximum when

$$\frac{d}{d\theta} (\cos \theta + \sin \theta) = 0$$

or $\qquad -\sin \theta + \cos \theta = 0$

or $\qquad \tan \theta = 1$

or $\qquad \theta = 45° \text{ or } 225°$

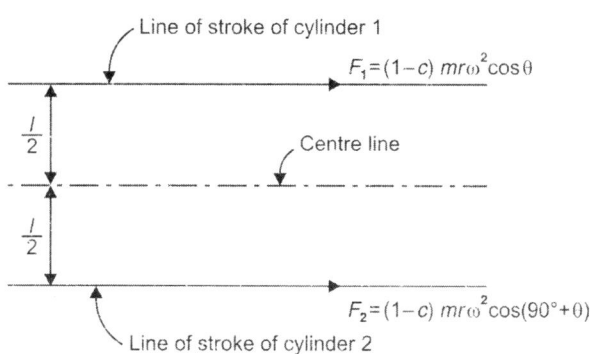

Fig. 8.11

When $\theta = 45°$, maximum swaying couple $= (1 - c) \, mr\omega^2 \, (\cos 45° + \sin 45°) \times \dfrac{l}{2}$

$$= \left[\frac{1}{\sqrt{2}} + \frac{1}{\sqrt{2}} \right] \times \frac{l}{2}$$

$$= \frac{1}{\sqrt{2}} \, (1 - c) \, mr\omega^2 \, l$$

When $\theta = 225°$, maximum swaying couple $= (1 - c) \, mr\omega^2 \, [\cos 225° + \sin 225°) \times \dfrac{l}{2}$

$$= (1 - c) \, mr\omega^2 \left[\frac{-1}{\sqrt{2}} - \frac{1}{\sqrt{2}} \right] \times \frac{l}{2}$$

$$= \frac{-1}{\sqrt{2}} \, (1 - c) \, mr\omega^2 \, l$$

$\therefore$ Maximum swaying couple $= \pm \dfrac{1}{\sqrt{2}} \, (1 - c) \, mr\omega^2 \, l$ ...(8.22)

8.8.3 Hammer Blow

The maximum unbalanced force perpendicular to the line of stroke, caused by the mass provided to balance the reciprocating masses is known as hammer blow. Unbalanced force perpendicular to line of stroke $= m_b r\omega^2 \sin\theta$ (m_b = balancing mass)

This force will be maximum when $\sin\theta$ is maximum at $\theta = 90°$ and $270°$ and its value is $+1$ and -1 respectively.

$\therefore$ Hammer blow $= \pm m_b r\omega^2$ (8.23)

Hammer blow varies as square of speed (ω^2). At high speed, hammer blow can exceed the static load on the wheels and can cause lifting off of the wheels from the rails when the direction of hammer blow is vertically upward.

Due to hammer blow, there will be variation in pressure between the wheel and the rail. This variation for one revolution of wheel is shown in Fig. 8.12.

Static load on each wheel, $W = mg$ (acting downward)

$\therefore$ Net pressure between the wheel and rail $= W \pm m_b r\omega^2$ (8.24)

If $(W - m_b r\omega^2)$ is negative, the wheel will be lifted from the rails. The limiting condition when the wheel does not lift from the rails is

$$W - m_b r\omega^2 = 0$$

or $W = m_b r\omega^2$

or $\omega = \sqrt{\dfrac{W}{m_b r}}$ (8.25)

This gives permissible value of angular speed (ω).

Example 8.6: The following data apply to two cylinder locomotive with cranks at right angles:

     Reciprocating mass per cylinder = 300 kg

     Crank radius = 0.3 m

     Driving wheel diameter = 1.8 m

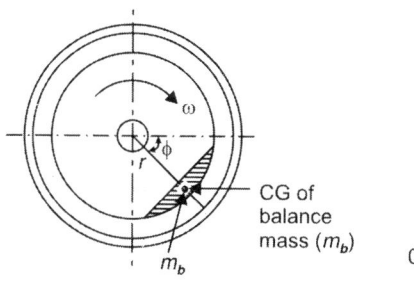
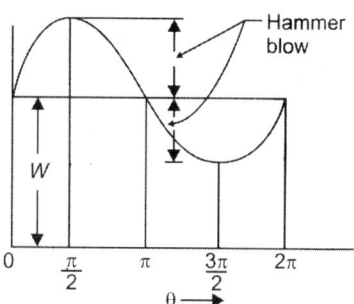

Fig. 8.12

Distance between cylinder centres = 0.65 m

Distance between driving wheel centres = 1.6 m

Determine:

   (a) fraction of reciprocating masses to be balanced if the hammer blow is not to exceed 45 kN at 100 km/hr.

   (b) variation in tractive effort.

   (c) maximum swaying couple.

Solution: Cranks of two cylinder are at 90°.

   $m_s = 300$ kg; $r = 0.3$ m

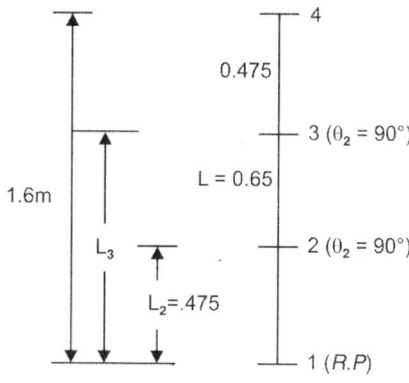

Fig. 8.13

Driving wheel diameter, $D = 1.8$ m

Distance between cylinder centres, $l = 0.65$ m

(a) $c = ?$

Hammer blow = 45 kN = 45×10^3 N

$$v = 100 \text{ km/hr} = 100 \times \frac{1000}{3600} \text{ m/s} = \frac{1000}{36}$$

$$\omega = \frac{v}{D/2} = \frac{1000}{36 \times 0.9} \approx 31 \text{ rad/s}$$

Hammer blow = $m_b r_b \omega^2$

or $45 \times 10^3 = m_b r_b \times (31)^2 \Rightarrow m_b r_b = 46.82$ ...(i)

Now, taking moment about RP (1)

$$m_2 r_2 l_2 \cos\theta_2 + m_3 r_3 l_3 \cos\theta_3 + m_4 r_4 l_4 \cos\theta_4 = 0$$

$$300 \times c \times 0.3 \times 0.475 \cos 0 + 300 \times c \times 0.3 \times 1.125 \cos 90°$$
$$+ m_b r_b \times 1.6 \times \cos\theta_b = 0$$

or
$$m_b r_b \cos\theta_b = -26.7c \qquad \text{...(ii)}$$

Similarly,
$$300 \times c \times 0.3 \times 1.125 \sin 90° + m_b r_b \times 1.6 \sin\theta_b = 0$$

or
$$m_b r_b \sin\theta_b = -63.28c \qquad \text{...(iii)}$$

Squaring and adding Eqs (ii) and (iii) and taking root, we have

$$m_b r_b = \sqrt{(26.7c)^2 + (63.28c)^2} = 68.68c \qquad \text{...(iv)}$$

From Eqs (i) and (iv), $68.68c = 46.82$ or $c = 0.681$

(b) Variation in tractive effort $= \pm\sqrt{2}\,(1-c)\,m_s r\omega^2$

$$= \pm\sqrt{2}\,(1-0.681) \times 300 \times 0.3 \times (31)^2$$

$$= \pm 38931.4 \text{ N}$$

(c) Maximum swaying couple $= \pm\dfrac{1}{\sqrt{2}}\,(1-c)\,m_s r\omega^2 l$

$$= \pm\dfrac{1}{\sqrt{2}}\,(1-0.681) \times 300 \times 0.3 \times (31)^2 \times (0.65)$$

$$= \pm 12681 \text{ N} \cdot \text{m}$$

Example 8.7: The following data refer to a two cylinder uncoupled locomotive:

Rotating mass per cylinder	= 280 kg
Reciprocating mass per cylinder	= 300 kg
Distance between wheels	= 1400 mm
Distance between cylinder centres	= 600 mm
Diameter of treads of driving wheels	= 1800 mm
Crank radius	= 300 mm
Radius of centre of balance mass	= 620 mm
Locomotive speed	= 50 km/hr
Angle between cylinder cranks	= 90°
Dead load on each wheel	= 3.5 tonne

Determine:

  (i) balancing mass required in the planes of driving wheels if whole of the revolving and two-third of the reciprocating mass are to be balanced.

 (ii) swaying couple.

(iii) varition in the tractive force.

(iv) maximum and minimum pressure on the rails.

 (v) maximum speed of locomotive without lifting the wheels from the rails.

Solution: Total mass to be balanced $= m_p + cm = 280 + \dfrac{2}{3} \times 300 = 480$ kg.

(i) Taking 1 as the reference plane and angle $\theta_2 = 0°$ (Fig. 8.14). Writing the couple equations, we have

$$m_2 r_2 l_2 \cos\theta_2 + m_3 r_3 l_3 \cos\theta_3 + m_4 r_4 l_4 \cos\theta_4 = 0$$

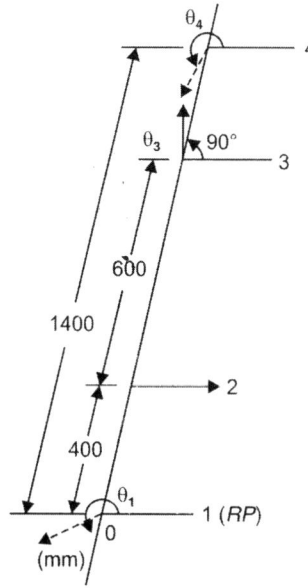

Fig. 8.14

or $\quad\quad\quad\quad 480 \times 300 \times 400 \cos 0° + 480 \times 300 \times 1000 \cos 90°$
$$+ m_4 \times 620 \times 1400 \cos \theta_4 = 0$$

or $\quad\quad\quad\quad\quad\quad\quad\quad\quad\quad\quad\quad m_4 \cos \theta_4 = -66.36 \quad\quad ...(i)$

and $\quad\quad 480 \times 300 \times 400 \sin 0° + 480 \times 300 \times 1000 \sin 90°$
$$+ m_4 \times 620 \times 1400 \sin \theta_4 = 0$$

or $\quad\quad\quad\quad\quad\quad\quad\quad\quad\quad\quad\quad m_4 \sin \theta_4 = -165.9 \quad\quad ...(ii)$

Squaring and adding Eqs (i) and (ii), $m_4 = 178.7$ kg

Dividing Eqs (ii) by (i), $\tan \theta_4 = \dfrac{-165.9}{-66.36} = 2.5$

$$\theta_4 = 248.2°$$

Taking 4 as the reference plane and writing the couple equation,

$$m_2 r_2 l_2 \cos \theta_2 + m_3 r_3 l_3 \cos \theta_3 + m_1 r_1 l_1 \cos \theta_1 = 0$$

or $\quad\quad\quad 480 \times 300 \times 1000 \cos 0° + 480 \times 300 \times 400 \cos 90°$
$$+ m_1 \times 620 \times 1400 \cos \theta_1 = 0$$

or $\quad\quad\quad\quad\quad\quad\quad\quad\quad\quad\quad\quad m_1 \cos \theta_1 = -165.9 \quad\quad ...(iii)$

Similarly, $\quad\quad\quad\quad\quad\quad\quad\quad\quad\quad\quad m_1 \sin \theta_1 = -66.36 \quad\quad ...(iv)$

From Eqs (iii) and (iv), $m_1 = 178.7$ kg $= m_4$

$$\tan \theta_1 = \frac{-66.36}{-165.9} = 0.4 \text{ or } \theta_1 = 201.8°$$

This shows that the magnitude of m_1 could have directly been written equal to m_4.

(ii) $\quad\quad\quad\quad\quad\quad \omega = \dfrac{50 \times 1000 \times 1000}{60 \times 60} \times \dfrac{1}{\dfrac{1800}{2}} = 15.43 \text{ rad/s}$

Swaying couple $= \pm \dfrac{1}{\sqrt{2}}(1-c)mr\omega^2 l$

$= \pm \dfrac{1}{\sqrt{2}}\left(1-\dfrac{2}{3}\right) \times 300 \times 0.3 \times (15.43)^2 \times 0.6$

$= 3030.3$ N·m

(iii) Variation in tractive force $= \pm \sqrt{2}(1-c)mr\omega^2$

$= \pm \sqrt{2}\left(1-\dfrac{2}{3}\right) \times 300 \times 0.3 \times (15.43)^2$

$= 10100$ N

(iv) Balance mass for reciprocating parts only $= 178.7 \times \dfrac{\dfrac{2}{3} \times 300}{480} = 74.46$ kg

Hammer blow $= m_b r\omega^2$

$= 74.46 \times 0.62 \times (15.43)^2$

$= 10991$ N

Dead load $= 3.5 \times 1000 \times 9.81 = 34335$ N

Maximum pressure on rails $= 34335 + 10991 = 45326$ N

Maximum pressure on rails $= 64335 - 10991 = 23344$ N

(v) Maximum speed of the locomotive without lifting the wheels from the rails will be when the dead load becomes equal to the hammerblow.

i.e., $\qquad 74.46 \times 0.62 \times \omega^2 = 34335$

or $\qquad\qquad \omega = 27.27$ rad/s

Velocity of wheels $= \omega r = \left(27.27 \times \dfrac{1.80}{2}\right)$ m/s

$= \left(27.27 \times \dfrac{18}{2} \times \dfrac{60 \times 60}{1000}\right)$ km/hr $= 88.36$ km/hr.

Example 8.8: The following data refer to a four-coupled wheel locomotive with two inside cylinders:

Pitch of cylinders	= 600 mm
Reciprocating mass/cylinder	= 315 kg
Revolving mass/cylinder	= 260 kg
Distance between driving wheels	= 1.6 m
Distance between coupling rods	= 2 m
Diameter of driving wheels	= 19 m
Revolving parts for each coupling rod crank	= 130 kg
Engine crank radius	= 300 kg
Coupling rod crank radius	= 240 mm
Distance of centre of balance mass in planes of driving wheels from axle centre	= 750 mm
Angle between engine cranks	= 90°
Angle between coupling rod crank with adjacent engine crank	= 180°

The balanced mass required for the reciprocating parts is equally divided between each pair of coupled wheels. Determine:

(i) the magnitude and position of the balance mass required to balance two-third of reciprocating and whole of the revolving parts.

(ii) the hammer blow and the maximum variation of tractive force when the locomotive speed is 80 km/hr.

Solution:

(i) *Leading wheels*: Balance mass on each leading wheel

$$= mp + \frac{1}{2} cm = 260 + \frac{1}{2}\left(\frac{2}{3} \times 315\right) = 365 \text{ kg}$$

Taking plane 2 as the reference plane and $\angle\theta_3 = 0°$ (refer Fig. 8.15)

$$m_1 = m_6 = 130 \text{ kg}; m_3 = m_4 = 365 \text{ kg}$$
$$r_1 = r_6 = 0.24 \text{ m}; r_2 = r_5 = 0.75 \text{ m}; r_3 = r_4 = 0.3 \text{ m}$$
$$l_1 = -0.2 \text{ m}; l_3 = 0.5 \text{ m}; l_4 = 1.1 \text{ m}; l_5 = 1.6 \text{ m}; l_6 = 1.8 \text{ m}$$
$$m_1 r_1 l_1 = 130 \times 0.24 \times (-0.2) = -6.24$$
$$m_3 r_3 l_3 = 365 \times 0.3 \times 0.5 = 54.75$$
$$m_4 r_4 l_4 = 365 \times 0.3 \times 1.1 = 120.45$$
$$m_5 r_5 l_5 = m_5 \times 0.75 \times 1.6 = 1.2 \, m_5$$
$$m_6 r_6 l_6 = 130 \times 0.24 \times 1.8 = 56.16$$

$$1.2m_5 = \begin{bmatrix} (-6.24\cos180° + 54.75\cos0° + 120.45\cos90° \\ + 56.16\cos270°)^2 + (-6.24\sin180° + 54.75\sin0° \\ + 120.45\sin90° + 56.16\sin270°)^2 \end{bmatrix}^{1/2}$$

$$= [(60.99)^2 + (64.29)^2]^{1/2} = 88.62$$

$$m_5 = 73.85 \text{ kg}$$

$$\tan\theta_5 = \frac{-64.29}{-60.99} = 1.054 \text{ or } \theta_5 = 226.5°$$

From symmetry of the system,

$$m_2 = m_5 = 73.85 \text{ kg}$$

and

$$\tan\theta_2 = \frac{-60.99}{-64.29} = 0.949$$

or

$$\theta_2 = 223.5°$$

Trailing wheels: The arrangement remains the same except that only half of the required reciprocating masses have to be balanced at the cranks.

i.e.,

$$m_3 = m_4 = \frac{1}{2}\left(\frac{2}{3} \times 315\right) = 105 \text{ kg}$$

Then,

$$m_3 r_3 l_3 = 105 \times 0.3 \times 0.5 = 15.75$$

and

$$m_4 r_4 l_4 = 105 \times 0.3 \times 1.1 = 34.65$$

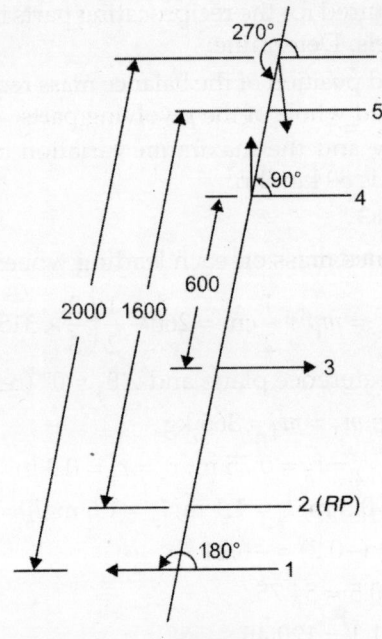

Fig. 8.15

$$1.2m_5 = \left[\begin{array}{l} (-6.24\cos180° + 54.75\cos0° + 34.65\cos90° \\ + 56.16\cos270°)^2 + (-6.24\sin180° + 15.75\sin0° \\ \qquad + 34.65\sin90° + 56.16\sin270°)^2 \end{array} \right]^{1/2}$$

$$= [(21.99)^2 + (-21.51)^2]^{1/2} = 30.76$$

$$m_5 = 25.63 \text{ kg}$$

$$\tan\theta_5 = \frac{-(-21.51)}{-21.99} = \frac{+21.51}{-21.99} = -0.978$$

or $\qquad\qquad \theta_5 = 135.4°$

By symmetry $\qquad m_2 = m_5 = 25.63$ kg

and $\qquad\qquad \tan\theta_2 = \dfrac{-21.99}{+21.51} = -1.022$ or $\theta_2 = 314.4°$

(ii) Hammer blow = $m_b r\omega^2$, where m_b is the balance mass for reciprocating parts only and neglecting m_1 and m_6 in the above calculations.

Thus $\qquad m_1 r_1 l_1 = m_6 r_6 l_6 = 0$

$$1.2m_5 = \left[\begin{array}{l} (15.75\cos0° + 34.65\cos90°)^2 \\ + (15.75\sin0° + 34.65\sin90°)^2 \end{array} \right]^{1/2}$$

$$= [(15.75)^2 + (34.65)^2]^{1/2} = 38.06$$

$$m_5 = 31.72 \text{ kg} = m_b$$

$$\omega = \frac{80 \times 1000}{60 \times 60} \times \frac{1}{1.9/2} = 23.39 \text{ rad/s}$$

Hammer blow = $31.72 \times 0.75 \times (23.39)^2 = 13015$ N

Maximum variation of tractive force

$$= \pm \sqrt{2} (1-c) mr\omega^2$$

$$= \pm \sqrt{2} \left(1 - \frac{2}{3}\right) \times 315 \times 0.3 \times (23.39)^2 = \pm 24372 \text{ N}$$

8.9 SECONDARY BALANCING

The secondary force due to accelerating reciprocating mass is given as

$$F_S = mr\omega^2 \frac{\cos 2\theta}{n}$$

$$F_S = m \frac{r}{4n} (2\omega)^2 \cos (2\omega t) \qquad\qquad (\theta = \omega t) \quad ...(8.26)$$

This is identical with primary force acting along the line of stroke for mass m attached to an imaginary crank of length $r/4n$, revolving at twice the speed of actual crank (2ω).

Actual and imaginary cranks are compared below in Fig. 8.16.

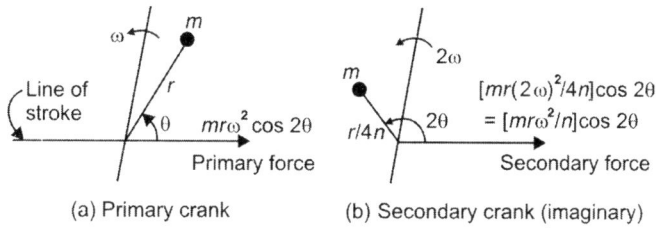

(a) Primary crank   (b) Secondary crank (imaginary)

Fig. 8.16

Imaginary crank coincides with the actual at inner dead centre (IDC). At other positions, it makes an angle with the line of stroke equal to twice that of the actual crank (i.e. 2θ).

For secondary balance, sum of secondary forces and secondary couples must be equal to zero, i.e. secondary force and secondary couple polygon must close.

Complete Balancing of Reciprocating Parts

For complete balancing of reciprocating parts of an engine, the following conditions must be satisfied:

  i. Primary forces must balance, i.e. primary force polygon is closed.
  ii. Primary couples must balance, i.e. primary couple polygon is closed.
  iii. Secondary forces must balance, i.e. secondary force polygon is closed.
  iv. Secondary couples must balance, i.e. secondary couple polygon is closed.

Generally, it is not possible to satisfy all the conditions for a multicylinder engine. Some unbalanced force or couple or both always remain in reciprocating engines.

8.10 BALANCING OF IN-LINE ENGINES

Consider a shaft consisting of three equal cranks unsymmetrically spaced. The equivalents of three unequal reciprocating masses are transferred at crankpins. Then

Primary force $= \Sigma mr\omega^2 \cos \theta$
Primary couple $= \Sigma mr\omega^2 \, l \cos \theta$

Secondary force $= \Sigma mr \dfrac{(2\omega)^2}{4n} \cos 2\theta = \Sigma mr\omega^2 \dfrac{\cos 2\theta}{n}$

Secondary couple $= \Sigma \left[mr\omega^2 \dfrac{\cos 2\theta}{n} \times l \right]$

To solve these equations graphically, first primary force polygon ($\Sigma mr \cos \theta$) is drawn (ω^2 is common to all). The axial component of the resultant force ($m_r r_r \cos \theta$) multiplied by ω^2 gives the primary unbalanced force on the system at that moment. This unbalanced force is zero when $\theta = 90°$ and maximum when $\theta = 0°$.

If the force polygon is closed, resultant force will be zero ($\Sigma F_{Ph} = 0, \Sigma F_{Ph} = 0$) and the system will be in primary balance.

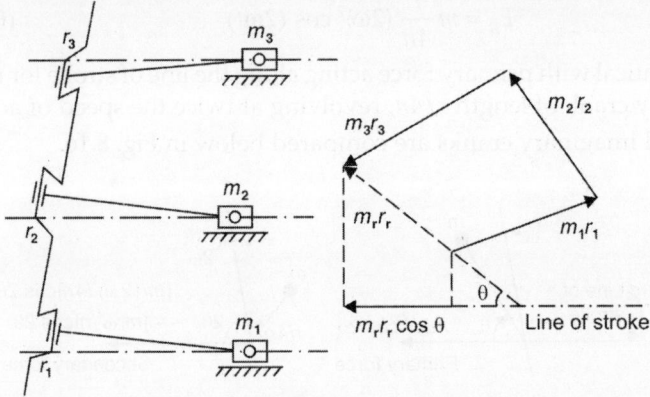

Fig. 8.17

To find the secondary unbalance force, first find the positions of the imaginary secondary cranks. Then transfer the reciprocating masses at the crank pins. Draw the *mr* polygon. Measure the resultant axial component and multiply the same by $(2\omega)^2/4n$ or ω^2/n) to get the secondary force.

In the same way, primary and secondary couple (*mrl*) polygons can be drawn for primary and secondary couples.

8.10.1 In-line Two Cylinder Engine

Consider a two cylinder engine shown in Fig. 8.18, cranks of which are 180° apart and has equal reciprocating masses. Take a plane through the centre line as the reference plane.

Primary force $= mr\omega^2 [\cos \theta + \cos (180° + \theta) = 0$

Primary couple $= mr\omega^2 \left[\dfrac{l}{2}\cos\theta + \left(-\dfrac{l}{2}\right)\cos(180° + \theta) \right]$

$= mr\omega^2 \, l\cos\theta$

Maximum values are $mr\omega^2 \, l$ at $\theta = 0°$ and 180°.

Secondary force $= \dfrac{mr\omega^2}{n} [\cos 2\theta + \cos (360° + 2\theta)]$

$= 2\dfrac{mr\omega^2}{n} \cos 2\theta$

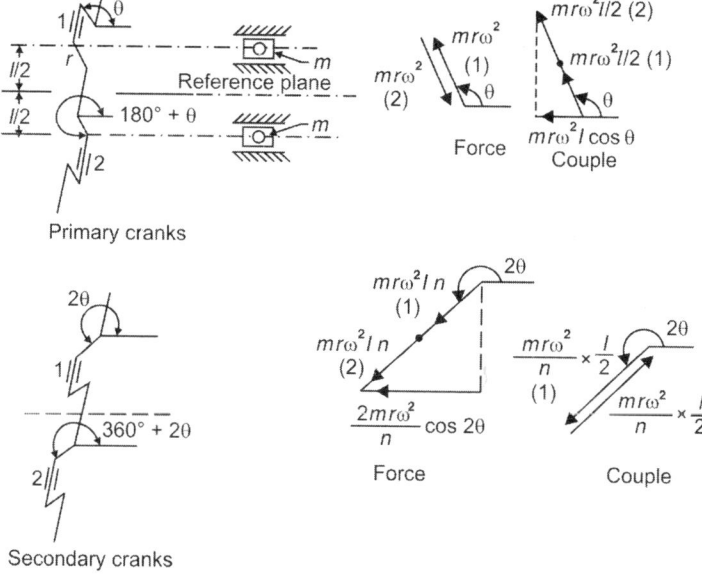

Fig. 8.18

Maximum values are $\dfrac{2mr\omega^2}{n}$ when $2\theta = 0°$, $180°$, $360°$ and $540°$

or

$$\theta = 0°, 90°, 180°, \text{ and } 270°$$

Secondary couple $= \dfrac{mr\omega^2}{n}\left[\dfrac{l}{2}\cos 2\theta + \left(-\dfrac{l}{2}\right)\cos(360° + 2\theta)\right] = 0$

Maximum values of forces and couples depend upon the positions of crankshafts. Graphical solution is also shown in Fig. 8.18.

8.10.2 Four-cylinder Four-stroke in-line engine

This engine has two outer and two inner cranks in line as shown in Fig. 8.19. The inner cranks are at 180° to the outer cranks. The angular positions of cranks are

θ for cylinder 1

$(180° + \theta)$ for cylinder 2

$(180° + \theta)$ for cylinder 3

θ for cylinder 4.

For convenience, the plane passing through the middle bearing is chosen as reference plane.

Primary force $= mr\omega^2 [\cos\theta + \cos(180° + \theta) + \cos(180° + \theta) + \cos\theta] = 0$

Primary couple $= mrw^2\left[\dfrac{3l}{2}\cos\theta + \dfrac{l}{2}\cos(180° + \theta) + \left(-\dfrac{l}{2}\right)\cos(180° + \theta)\right.$

$$\left. + \left(\dfrac{-3l}{2}\right)\cos\theta\right] = 0$$

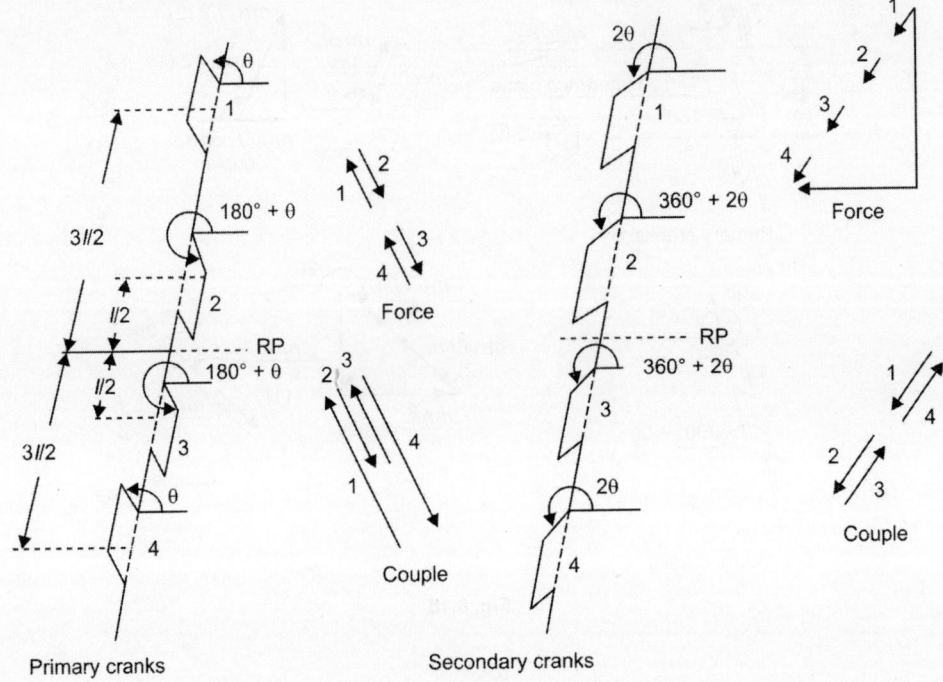

Fig. 8.19

$$\text{Secondary force} = \frac{mr\omega^2}{n}\left[\cos 2\theta + \cos(360° + 2\theta) + \cos(360° + 2\theta) + \cos 2\theta\right]$$

$$= \frac{4mr\omega^2}{n}\cos 2\theta$$

$$\text{Maximum value} = \frac{4mr\omega^2}{n} \text{ at } 2\theta$$

$$= 0°, 180°, 360° \text{ and } 540° \text{ or } \theta = 0°, 90°, 180° \text{ and } 270°$$

$$\text{Secondary couple} = mrw^2\left[\frac{3l}{2}\cos 2\theta + \frac{l}{2}\cos(360° + 2\theta) + \left(-\frac{l}{2}\right)\cos(360° + 2\theta)\right.$$

$$\left. + \left(-\frac{3l}{2}\right)\cos 2\theta\right] = 0$$

Graphical solution is shown in Fig. Thus, this engine is not balanced in secondary forces.

8.11 BALANCING OF V-ENGINES

Radial engines are also known as V-engines, in these engines the cylinders are arranged along radial lines. The centre lines of the cylinders form V. These cylinders have a common crank. Since crank revolves in the same plane, balancing of primary or secondary couples is not required in radial engines (Fig. 8.20).

Consider two symmetrically placed cylinders. A common crank OA is operated by two connecting rods AB and AC. The lines of stroke OB and OC are inclined to X-axis at an angle α. The crank OA is moved by an angle θ from the X-axis.

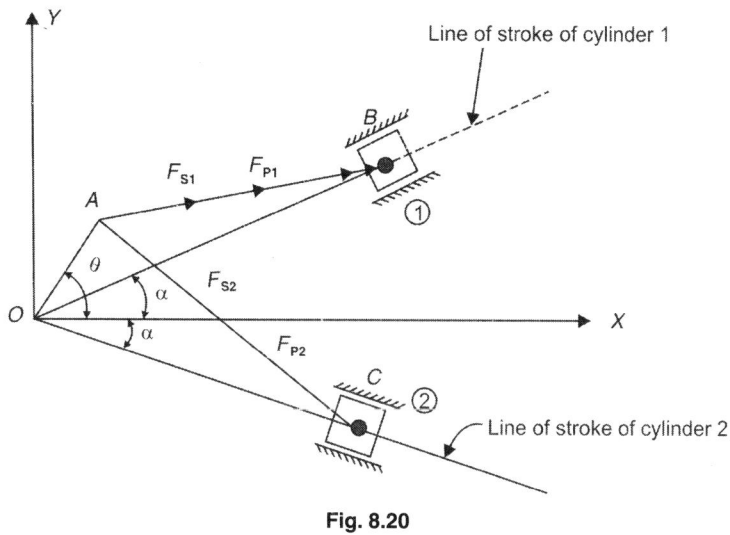

Fig. 8.20

Let m = mass of reciprocating parts per cylinder

l = length of each connecting rod

r = radius of crank

$$n = \frac{l}{r}$$

ω = angular velocity of crank.

Primary Force

Primary force of cylinder 1 along the line of stroke $OB = m\omega^2 r \cos(\theta - \alpha)$

Primary force of cylinder 1 along X-axis

$$F_{H1} = m\omega^2 r \cos(\theta - \alpha) \cos\alpha \qquad \qquad ...(i)$$

Primary force of cylinder 2 along the line of stroke $OC = m\omega^2 r \cos(\theta + \alpha)$

Primary force of cylinder 2 along X-axis

$$F_{H2} = m\omega^2 r \cos(\theta + \alpha) \cos\alpha \qquad \qquad ...(ii)$$

Total primary force along X-axis, $F_H = F_{H1} + F_{H2}$

$$
\begin{aligned}
F_H &= mr\omega^2 \cos\alpha \left[\cos(\theta - \alpha) + \cos(\theta + \alpha)\right] \\
&= mr\omega^2 \cos\alpha \left[(\cos\theta\cos\alpha + \sin\theta\sin\alpha) \right. \\
&\qquad\qquad\qquad\qquad \left. + (\cos\theta\cos\alpha - \sin\theta\sin\alpha)\right] \\
&= 2mr\omega^2 \cos\alpha\cos\theta\cos\alpha \\
&= 2mr\omega^2 \cos^2\alpha\cos\theta
\end{aligned}
$$

Similarly, total primary force along Y-axis

$$
\begin{aligned}
F_V &= mr\omega^2 \left[\cos(\theta - \alpha)\sin\alpha - \cos(\theta + \alpha)\sin\alpha\right] \\
&= mr\omega^2 \sin\alpha \begin{bmatrix} (\cos\theta\cos\alpha - \sin\theta\sin\alpha) \\ -(\cos\theta\cos\alpha - \sin\theta\sin\alpha) \end{bmatrix} \\
&= mr\omega^2 \sin\alpha \cdot 2\sin\theta\sin\alpha \\
&= 2mr\omega^2 \sin^2\alpha\sin\theta
\end{aligned}
$$

Resultant primary force, $F_P = \sqrt{F_H^2 + F_V^2}$

$$= \sqrt{(2mr\omega^2\cos^2\alpha\cos\theta)^2 + (2mr\omega^2\sin^2\alpha\cos\theta)^2}$$

$$= 2mr\omega^2\sqrt{(\cos^2\alpha\cos\theta)^2 + (\sin^2\alpha\sin\theta)^2}$$

It will be at an angle β with the X-axis, given by

$$\tan\beta = \frac{\sin^2\alpha\sin\theta}{\cos^2\alpha\cos\theta}$$

If $\qquad\qquad 2\alpha = 90°,$

resultant force $= 2mr\omega^2\sqrt{(\cos^2 45°\cos\theta)^2 + (\sin^2 45°\sin\theta)^2}$

$$= mr\omega^2$$

$$\tan\beta = \frac{\sin^2 45°\sin\theta}{\cos^2 45°\cos\theta} = \tan\theta,$$

i.e. $\beta = \theta$ or it acts along the crank and, therefore, can be completely balanced by a mass at a suitable radius diametrically opposite to the crank such that $m_r r_r = mr$.

For a given value of α, the resultant primary force is maximum when

$\qquad (\cos^2\alpha\cos\theta)^2 + (\sin^2\alpha\sin\theta)^2$ is maximum

or $\qquad (\cos^4\alpha\cos^2\theta) + (\sin^4\alpha\sin^2\theta)$ is maximum

or $\qquad \dfrac{d}{d\theta}(\cos^4\alpha\cos^2\theta + \sin^4\alpha\sin^2\theta) = 0$

or $\qquad -\cos^4\alpha \cdot 2\cos\theta\sin\theta + \sin^4\alpha\, 2\sin\theta\cos\theta = 0$

or $\qquad -\cos^4\alpha \cdot \sin 2\theta + \sin^4\alpha\sin 2\theta = 0$

or $\qquad \sin 2\theta\,(\sin^4\alpha - \cos^4\alpha) = 0$

As α is not zero, therefore, for a given value α, the resultant primary force is maximum when θ is zero degree.

Secondary Force

Secondary force of cylinder 1 along

$$OB = \frac{mr\omega^2}{n}\cos 2\,(\theta - \alpha)$$

$$x\text{-axis} = \frac{mr\omega^2}{n}\cos 2\,(\theta - \alpha)\cos\alpha$$

Secondary force of cylinder 2 along

$$OB_1 = \frac{mr\omega^2}{n}\cos 2(\theta + \alpha)$$

$$x\text{-axis} = \frac{mr\omega^2}{n}\cos 2(\theta + \alpha)\cos\alpha$$

Total secondary force along x-axis

$$= \frac{mr\omega^2}{n}\cos\alpha\,[\cos 2(\theta - \alpha) + \cos 2(\theta + \alpha)]$$

$$= \frac{mr\omega^2}{n}\cos\alpha\,[(\cos 2\theta\cos 2\alpha + \sin 2\theta\sin 2\alpha)$$

$$+ (\cos 2\theta\cos 2\alpha - \sin 2\theta\sin 2\alpha)]$$

$$= \frac{2mr\omega^2}{n} \cos\alpha \cos 2\theta \cos 2\alpha$$

Similary, secondary force along Y-axis

$$= \frac{2\,mr\omega^2}{n} \sin\alpha \sin 2\theta \sin 2\alpha$$

Resultant secondary force

$$= \frac{2\,mr\omega^2}{n} \sqrt{(\cos\alpha \cos 2\theta \cos 2\alpha)^2 + (\sin\alpha \sin 2\theta \sin 2\alpha)^2}$$

and,
$$\tan\beta' = \frac{\sin\alpha \sin 2\theta \sin 2\alpha}{\cos\alpha \cos 2\theta \cos 2\alpha} = \tan\alpha \tan 2\theta \tan 2\alpha$$

If $2\alpha = 90°$ or $\alpha = 45°$,

$$\text{secondary force} = \frac{2\,mr\omega^2}{n} \sqrt{\left(\frac{\sin 2\theta}{\sqrt{2}}\right)^2}$$

$$= \sqrt{2}\,\frac{mr\omega^2}{n}\sin 2\theta$$

$$\tan\beta' = \infty \Rightarrow \beta' = 90°$$

This means that the force acts along Y-axis and is a harmonic force and special methods are needed to balance it.

8.12 DIRECT AND REVERSE CRANK METHOD OF BALANCING

This method is useful for balancing of radial or V-engines. Connecting rods are attached to a common crank. In this case, plane of rotation of the cranks is the same, so there is no unbalanced primary or secondary couple. Only the primary and secondary forces are balanced.

Figure 8.21 shows a reciprocating engine in which crank OA rotates uniformly at ω (rad/s) in clockwise direction. Let OA makes an angle θ with OB at any instant. The reverse crank OA' is the image of direct crank OA. It can be stated that when OA rotates in clockwise direction, OA' rotates in anticlockwise direction. OA and OA' are called as direct and reverse cranks respectively. Let mass of the reciprocating parts at B is m.

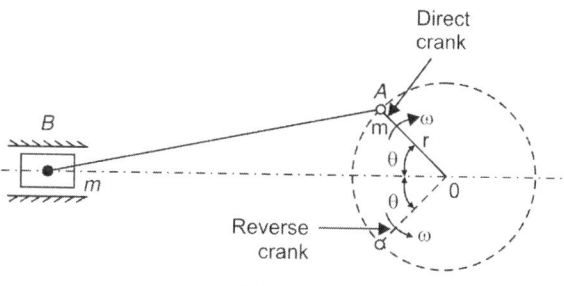

Fig. 8.21

The primary force is equal to $m\omega^2 r \cos\theta$. If we place a mass m at crank pin A, it produces a centrifugal force of magnitude $m\omega^2 r$. The horizontal component of this force is $m\omega^2 r \cos\theta$. Thus, both the forces are equal.

Now, let mass m be divided equally and placed at A and A' as skhown in Fig. 8.22.

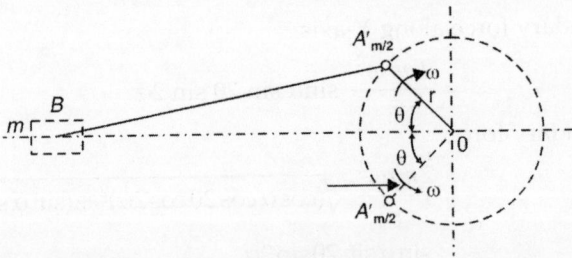

Fig. 8.22

The horizontal components of centrifugal forces due to masses $m/2$ placed at A and A' will be $\dfrac{m}{2}\omega^2 r\cos\theta$ each. Their combined effect (force) will be

$$\frac{m}{2}\omega^2 r\cos\theta + \frac{m}{2}\omega^2 r\cos\theta = m\omega^2 r\cos\theta$$

$$F_p = \text{primary force}$$

Primary Force

In this case, we have put each revolving mass one-half of the reciprocating mass to determine the primary force. The reciprocating mass m is replaced by two revolving masses $m/2$ each placed at crank pins A and A'.

Secondary Force

Secondary force F_s is given by:

$$F_s = \frac{m\omega^2 r}{n}\cos 2\theta$$

or

$$= m(2\omega)^2 \frac{r}{4n}\cos 2\theta$$

As in case of primary force, in this case also mass $m/2$ is placed at crank pins A and A'. The length of the cranks OA and OA' is $\dfrac{r}{4n}$ and cranks make an angle 2θ with OB as shown in Fig. 8.23.

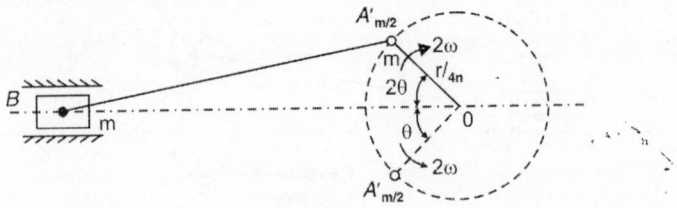

Fig. 8.23

Example 8.9: Each crank and the connecting rod of a four-crank in-line engine are 200 mm and 800 mm respectively. The outer cranks are set at 120° to each other and each has a reciprocating mass of 200 kg. The spacing between adjacent planes of cranks are 400 mm, 600 mm and 500 mm. If the engine is in complete primary balance, determine the reciprocating mass of the inner cranks and their relative angular positions. Also find the secondary unbalanced force if the engine speed is 210 rpm.

Solution: $$\omega = \frac{2\pi \times 210}{60} = 22 \text{ rad/s}, n = 800/200 = 4$$

Figure 8.24 represents the relative position of the cylinders and the cranks.

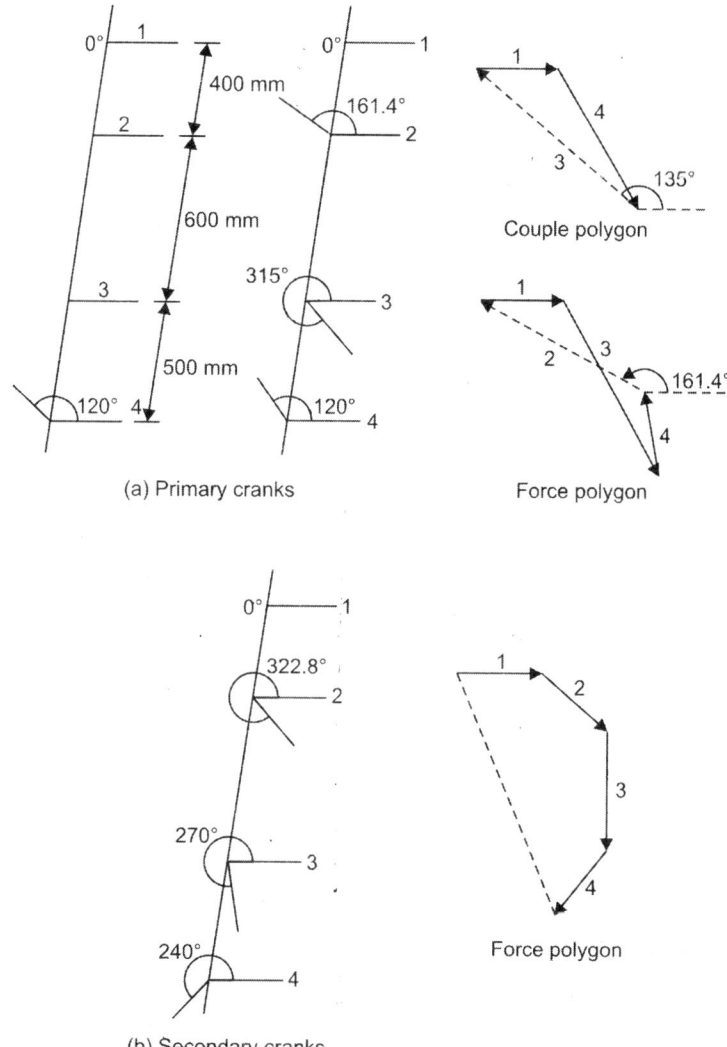

(a) Primary cranks

Couple polygon

Force polygon

(b) Secondary cranks

Force polygon

Fig. 8.24

Taking 2 as the reference plane,
primary couples about *RF*,

$$m_1 r_1 l_1 = 200 \times 0.2 \times 0.4 = 16$$
$$m_2 r_2 l_2 = 0$$
$$m_3 r_3 l_3 = m_2 \times 0.2 \times (-0.6) = -0.12m_3$$
$$m_4 r_4 l_4 = 200 \times 0.2 \times (-1.1) = -44$$

The couple polygon in drawn in Fig. 8.24.

$m_3 r_3 l_3$ of crank 3 from the diagram = 53.7 at 135°

∴ $m_3 r_3 l_3 = m_3 \times 0.12 = 53.7$ or $m_3 = 448$ kg

As its direction is negative, its direction is (135° + 180°) or 315°.

Primary force (mr) along each of outer cranks = 200 × 0.2 = 40

Primary force (mr) along crank 3 = 448 × 0.2 = 89.6

The force polygon is drawn in Fig. 8.24.

$m_2 r_2$ of crank 2 from the diagram = 87.6 at 161.4°

∴ $m_2 r_2 = m_2 \times 0.2 = 87.6$ or $m_2 = 438$ kg

Its angular position is 161.4°.

Figure 8.24 represents the relative position of the cylinders and the cranks.

From secondary unbalanced force polygon, $mr = 198$

Maximum unbalanced force $= 198 \times \dfrac{\omega^2}{n} = 198 \times \dfrac{22^2}{n} = 23.985$ N.

Example 8.10: The pistons of a 60° twin V-engine has strokes of 120 mm. The connecting rods driving a common crank has a length of 200 mm. The mass of the reciprocating parts per cylinder is 1 kg and the speed of the crank shaft is 2500 rpm. Determine the magnitude of the primary and secondary forces.

Solution: Stroke = $2r$ = 120 mm

$r = 60$ mm = 0.06 m, $l = 200$ mm = 0.20 m

$m = 1$ kg, $N = 2500$ rpm

$$\omega = \frac{2\pi N}{60} = \frac{2\pi \times 2500}{60} = 261.799 \text{ rad/s}$$

$$= 261.8 \text{ rad/s}$$

$$2\alpha = 60° \Rightarrow \alpha = 30°; n = \frac{l}{r} = \frac{0.20}{0.06} = 3.33$$

Primary force due to reciprocating parts are:

$F_{P1} = m\omega^2 r\cos(\theta + 30°)$ along the centre line of cylinder 1

$F_{P2} = m\omega^2 r\cos(\theta - 30°)$ along the centre line of cylinder 2

We can find the vertical component of these forces as

$$F_{VP} = F_{P1}\cos 30° + F_{P2}\cos 30°$$
$$= m\omega^2 r\cos(\theta + 30°)\cos 30° + m\omega^2 r\cos(\theta - 30°)\cos 30°$$
$$= m\omega^2 r\cos 30°[\cos(\theta + 30°) + \cos(\theta - 30°)]$$
$$= m\omega^2 r[\cos 30° \cos\theta\cdot\cos 30° - \cos 30°\sin\theta\sin 30°$$
$$+ \cos 30°\cdot\cos 30°\cos\theta + \cos 30°\sin 30°\sin\theta)$$
$$= m\omega^2 r[0.75\cos\theta - 0.433\sin\theta + 0.75\cos\theta + 0.433\sin\theta]$$

$$F_{VP} = 1.5\, m\omega^2 r\cos\theta \text{ N}$$

Similarly, the horizontal component of these forces is

$$F_{HP} = 0.5\, m\omega^2 r\sin\theta \text{ N}$$

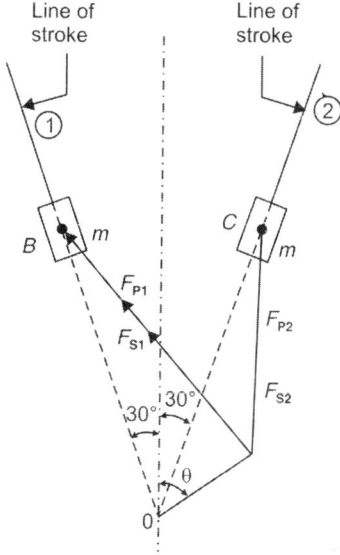

Fig. 8.25

Resultant primary force,

$$F_P = \sqrt{F_V^2 + F_H^2}$$

$$F_P = m\omega^2 r \left(2.25\cos^2\theta + 0.25\sin^2\theta\right)^{1/2}$$

Secondary forces, $\quad F_{S1} = \dfrac{m\omega^2 r}{n}\left[\cos 2(\theta + 30°)\right]$

$$F_{S2} = \dfrac{m\omega^2 r}{n}\left[\cos 2(\theta - 30°)\right]$$

The vertical component of the secondary forces is

$$F_{VS} = \cos 30° \, F_{S1} + F_{S2}\cos 30°$$

$$= \cos 30° \cdot \dfrac{m\omega^2 r}{n}\left[\cos(2\theta + 60°) + \cos(2\theta - 60°)\right]$$

$$= \dfrac{m\omega^2 r}{n} \times 0.866\left[\cos 2\theta \cos 60° - \sin 2\theta \sin 60°\right.$$

$$\left. + \cos 2\theta \cos 60 + \sin 2\theta \sin 60°\right]$$

$$= m\omega^2 \dfrac{r}{n} \cdot \cos 2\theta \times 0.866$$

$$F_{VS} = 0.866 \times \dfrac{m\omega^2 r}{n}\cos 2\theta$$

Similarly, horizontal component of secondary forces is

$$F_{HS} = 0.866\dfrac{m\omega^2 r}{n}\sin 2\theta$$

Resultant secondary force, $F_S = \sqrt{F_{VS}^2 + F_{HS}^2}$

$$= 0.866 \frac{m\omega^2 r}{n} \sqrt{(\cos 2\theta)^2 + (\sin 2\theta)^2}$$

$$= 0.866 \frac{m\omega^2 r}{n} = \frac{0.866 \times 1 \times (261.8)^2 \times 0.06}{3.33}$$

$$= 1069.5 \text{ N}.$$

Example 8.11: The four masses m_1, m_2, m_3 and m_4 having their radii of rotation as 200 mm, 150 mm, 250 mm and 300 mm are 200 kg, 300 kg, 240 kg and 260 kg in magnitude respectively. The angles between the successive masses are 45°, 75° and 135° respectively. Find the position and magnitude of the balance mass required, if its radius of rotation is 200 mm.

Solution:

$\theta_1 = 0, m_1 = 200$ kg, $r_1 = 200$ mm $= 0.2$ m

$\theta_2 = 45°, m_2 = 300$ kg, $r_2 = 150$ mm $= 0.15$ mm

$\theta_3 = 45° + 75° = 120°, m_3 = 240$ kg, $r_3 = 250$ mm $= 0.25$ m

$\theta_4 = 120° + 135° = 255°, m_4 = 260$ kg, $r_4 = 300$ mm $= 0.30$ m

$r_b = 200$ mm $= 0.20$ m

Let m_b = balancing mass

θ_b = the angle made by the balancing mass to m_1

We know that the centrifugal force is proportional to the product of mass and radius of rotation of each part, so

$$m_1 r_1 = 200 \times 0.2 = 40 \text{ kg·m}$$
$$m_2 r_2 = 300 \times 0.15 = 45 \text{ kg·m}$$
$$m_3 r_3 = 240 \times 0.25 = 60 \text{ kg·m}$$
$$m_4 r_4 = 260 \times 0.30 = 78 \text{ kg·m}$$

Balancing force $= m_b \times r_b = m_b \times 0.2 = 0.2 m_b$ kg·m

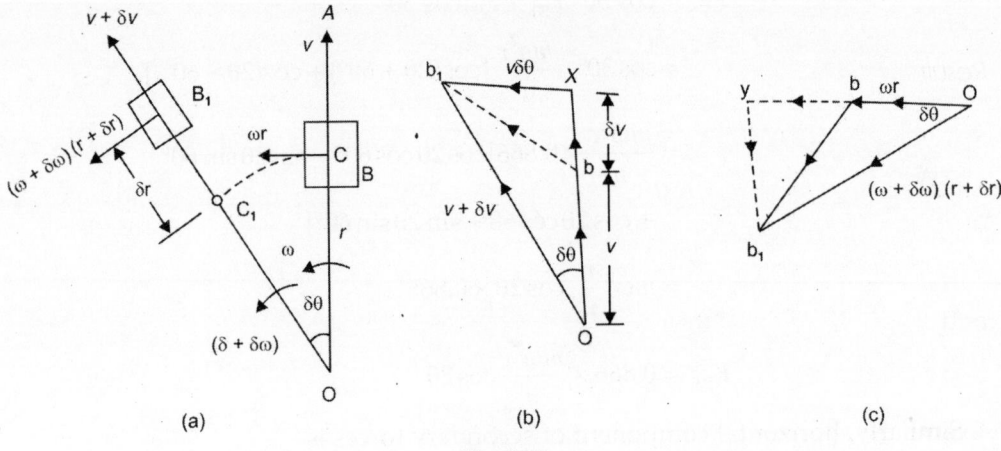

(a) (b) (c)

Fig. 8.26

The problem can be solved by graphical and analytical methods.

Graphical method

1. Main diagram of the configuration is drawn to scale showing the angular positions of all the masses. Refer to Fig. 8.26.
2. Calculate $m_1r_1, m_2r_2, m_3r_3, m_4r_4$ and so on. In our case,
   $m_1r_1 = 40$ kg·m, $m_2r_2 = 45$ kg·m, $m_3r_3 = 60$ kg·m, $m_4r_4 = 78$ kg·m.
3. Draw the vector diagram with the above calculated values as shown in Fig. 8.26c.
   i. Draw ab parallel to $OA = m_1r_1$.
   ii. From b draw bc parallel to $OB = m_2r_2$.
   iii. From C draw cd parallel to $OC = m_3r_3$.
   iv. From d draw de parallel to $OD = m_4r_4$.
   v. Join e to a.
   vi. ae represents resultant force.
4. We know that the balancing force is equal to the resultant force and opposite in direction. So $m_b r_b = ae$

$$m_b \times 0.2 = ae \text{ (vector)} = 22 \text{ kg·m}$$

$$(m_b) = \frac{22}{0.2} = 110 \text{ kg}$$

By measurement, we also find that
$$\theta = 201°.$$

Analytical method

Resolving the forces vertically, we get

$$\Sigma F_V = m_1r_1 \sin\theta_1 + m_2r_2 \sin\theta_2 + m_3r_3 \sin\theta_3 + m_4r_4 \sin\theta_4$$
$$= 40 \times \sin 0° + 45 \sin 45° + 60° \sin 120° + 78 \sin 255°$$
$$= 31.82 + 51.96 - 75.34 = 8.43 \text{ kg·m}$$

and resolving the forces horizontally, we get

$$\Sigma F_H = m_1r_1 \cos\theta_1 + m_2r_2 \cos\theta_2 + m_3r_3 \cos\theta_3 + m_4r_4 \cos\theta_4$$
$$= 40 \times \cos 0° + 45 \cos 45° + 60° \cos 120° + 78 \cos 255°$$
$$= 40 + 31.82 - 30 - 20.18 = 21.63 \text{ kg·m}$$

Resultant force

$$F = \sqrt{F_V^2 + F_H^2} = \sqrt{(8.43)^2 + (21.63)^2} = 23.2 \text{ kg·m}$$

We know that $m_b \times r_b = 23.2$

$$m_b = \frac{23.2}{0.20} = 116 \text{ kg}$$

and
$$\tan\theta = \frac{\Sigma F_V}{\Sigma F_H} = \frac{8.43}{21.63}$$

$$\theta = 201.5°$$

Example 8.12: The cranks of a two-cylinder, uncoupled inside cylinder locomotive, are at right angles and 325 mm long. The cylinders are 675 mm apart. The rotating mass per cylinder is 200 kg at the crank pin and the mass of the reciprocating parts per cylinder is 240 kg. The wheel centre lines are 1.5 m apart. The whole of the rotating and two-third of the reciprocating masses are to be balanced and the balance masses are to be placed in the planes of the rotation of the driving wheels at a radius of 800 mm. Find:

    i. the magnitude and direction of the balancing masses;

    ii. the magnitude of hammer blow;

    iii. variation in tractive force; and

    iv. maximum swaying couple at a crank speed of 240 rpm.

Solution: Inside cylinder locomotive means the two cylinders are placed symmetrically between the wheels. Let the two cylinders be A and B.

Crank radius, $r_A = r_B = 325$ mm $= 0.325$ m; distance between the cylinders, $a = 675$ mm $= 0.675$ m; rotating mass, $M = 200$ kg; reciprocating mass, $m_R = 240$ kg; distance between the centre lines of wheel $= 1.5$ m; radius of balance mass, $r = 800$ mm $= 0.8$ m; $N = 240$ rpm

$$\omega = \frac{2\pi N}{60} = \frac{2\pi \times 240}{60} = 8\pi \text{ rad/s}$$

Fraction of the reciprocating mass to be balanced $= \dfrac{2}{3}$ or $c = \dfrac{2}{3}$.

Now, the total equivalent revolving mass per cylinder at crank radius, which has to be balanced $= M + c \times m_R$

$$= 200 + \frac{2}{3} \times 240 = 200 + 160 = 360 \text{ kg}$$

∴ $m_A = m_B = 360$ kg

Let A and B be the planes of rotation of the crank and x and y be the planes of rotation of wheels and balance masses. Balance masses are placed at a radius of 0.8 m. Hence $r_x = r_y = 0.8$ m.

(i) Magnitude and direction of the balancing masses:

Let m_x = Magnitude of balance mass placed in plane x.

    m_y = Magnitude of balance mass placed in plane y.

    θ_x = Angular position of the balancing mass m_x from the crank A

    θ_y = Angular position of the balancing mass m_y from the crank A

To determine the magnitude and direction of the balancing masses, the couple polygon and force polygon are drawn as given below:

    a. Draw the space diagram to show the positions of the planes of the wheels and the cylinders as shown in Fig. 8.27a. The distance between cylinders A and B is 675 mm whereas the distance between wheels x and y is 1500 mm. The cylinders are placed symmetrically, hence, distance between planes x and A is equal to the distance between planes B and y. Hence,

$$xA = By = \frac{1500 - 675}{2} = \frac{825}{2} = 412.5 \text{ mm.}$$

    b. The cranks of the cylinder are at right angles. Assume the position of the crank of the cylinder A in the horizontal direction. Then the position of crank of the cylinder B will be at right angles as shown in Fig. 8.27b.

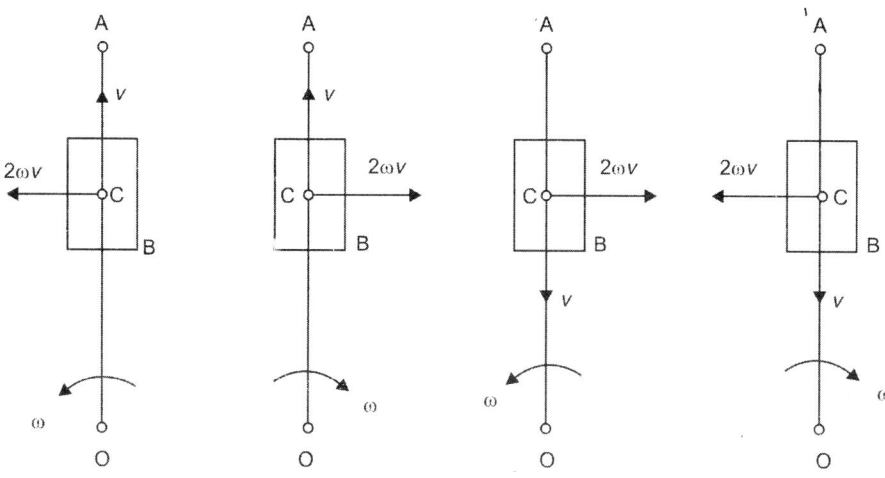

Fig. 8.27

c. Assuming the plane x as the reference plan, so as to eliminate the couple due to unknown mass in that plane, the data can be tabulated as shown in Table 8.2.

			Table 8.2		
Plane	Mass in (m) kg	Radius in (r) m	Cent. force (ω^2) (mr) kg·m	Distance from reference plane $x(l)$ m	Couple $(mr \cdot l)$ kg·m²
(1)	(2)	(3)	(4)	(5)	(6)
x(RP)	m_x	$r_x = 0.8$	$0.8m_x$	0	0
A	360	$r_A = 0.325$	117	0.4125	48.2625
B	360	$r_B = 0.325$	117	$412.5 + 675 = 1.0875$	127.28
y	m_y	$r_y = 0.8$	$0.8m_y$	1.5	$1.2 \times m_y$

d. From column (6) of the Table 8.2, the couple polygon can be drawn to some suitable scale as shown in Fig. 8.27c. Take any point o'. Draw $o'a' = 48.2625$ parallel to OA. From a' draw $a'b' = 127.28$ parallel to OB. The closing side $b'o'$ represents the balancing couple and it is proportional to $0.8m_y$. By measurement,

$$0.8m_y = \text{vector } b'o' = 136.12 \text{ kg·m}^2$$

or $\quad m_y = \dfrac{136.12}{1.2} = 113.43 \text{ kg}$

The vector $b'o'$ can also be obtained by calculation as

$$b'o' = \sqrt{(o'a')^2 + (a'b')^2}$$

$$= \sqrt{48.2625^2 + 127.28^2} = \sqrt{2329.27 + 16200} = 136.12 \text{ kg·m}^2$$

and $\quad \tan \alpha = \dfrac{127.28}{48.2625} = 2.637$

$\therefore \qquad \alpha = \tan^{-1} 2.637 = 69.23°$

e. The angular position of the balancing mass (m_y) is obtained by drawing the line OY parallel to vector $b'o'$ in Fig. 8.27b (or making angle $\alpha = 69.23°$ as shown in Fig. 8.27b).

f. The balancing mass at x (i.e. m_x) is obtained by drawing a force polygon from column 4 of Table 8.2 as shown in Fig. 8.27d. Take any point o. From o draw $oa = 117$ and parallel to OA. From a draw $ab = 117$ and parallel to OB. From b draw $by = 0.8 \times m_y = 0.8 \times 113.43 = 90.75$ and parallel to OY. Join Y to o. The closing side yo represents the balancing force and it is equal to $0.8m_x$. By measurement,

$$yo = 90.75$$
$$\therefore \quad 0.8 \times m_x = 90.75$$
$$\text{or} \quad m_x = \frac{90.75}{0.8} = 113.43 \text{ kg.}$$

g. The angular position of the balancing mass (m_x) is obtained by drawing the line OX in Fig. 8.27b parallel to vector yo. The angle β made by OX with x-axis is $20.77°$.

(ii) Magnitude of hammer blow: Hammer blow is the maximum unbalanced force perpendicular to the line of stroke and it is given as

$$\text{hammer blow} = \pm m_b \times r \times \omega^2$$

where m_b = part of balance mass which is required for balancing reciprocating parts only.

The unbalanced force perpendicular to the line of stroke is due to the part of the reciprocating mass which has been balanced.

Now total balancing mass for each cylinder (i.e. m_x or m_y) = 113.43 kg.

Part of this total balancing mass is used for balancing rotating parts and the remaining part is used for balancing reciprocating parts.

Total equivalent mass which has to be balanced per cylinder is given as

$$= (M + c \times m_R) = 200 + \frac{2}{3} \times 240 = 200 + 160 = 360 \text{ kg.}$$

Out of 360 kg mass, only $\frac{2}{3}$ of 240 = 160 kg mass is used for balancing reciprocating parts.

Hence, out of total balancing mass of 113.43 kg, the mass required for balancing reciprocating parts will be $\frac{160}{360} \times 113.43 = 50.41$ kg.

The vertical unbalanced force will be due to 50.41 kg.

$$\therefore \qquad m_b = 50.41 \text{ kg}$$
$$\therefore \qquad \text{Hammer blow} = \pm m_b \times r \times \omega^2 \ (\because r = 0.8 \text{ m and } \omega = 8\pi \text{ rad/s})$$
$$= \pm 50.41 \times 0.8 \times (8\pi)^2 = \pm 25473.37 \text{ N.}$$

(iii) Variation in tractive force: Maximum variation of tractive force is given as

$$= \pm \sqrt{2} \, (1 - c) \, m_R \times \omega^2 \times r$$
$$= \pm \sqrt{2} \left(1 - \frac{2}{3}\right) 240 \times (8\pi)^2 \times 0.325 \text{ N} \ (\because m_R = 240 \text{ kg}; r = 0.325 \text{ m})$$
$$= \pm 23225.6 \text{ N.}$$

(iv) Swaying couple:

$$\text{Max. swaying couple} = \pm \frac{a}{\sqrt{2}}(1-c)m_R \times \omega^2 \times r$$

$$= \pm \frac{0.675}{\sqrt{2}}\left(1 - \frac{2}{3}\right) \times 240 \times (8\pi)^2 \times 0.325 \quad (\because a = 0.675 \text{ m})$$

$$= \pm 7838.6 \text{ N·m.}$$

Analytical method

The magnitude and direction of the balancing masses (i.e. m_x, m_y, θ_x and θ_y) can also be determined by analytical method as given below:

(i) For complete balancing, the resultant force and resultant couple both should be zero, i.e.

$$F_R = 0 \text{ and } C_R = 0.$$

where F_R = Resultant force = $\Sigma m \times r \times \omega^2$ $\qquad (\because mr\omega^2 = \text{centrifugal force})$

C_R = Resultant couple = $\Sigma m \times r \times l \times \omega^2$

$$(\because mr\omega^2 \times l = \text{force} \times \text{distance} = \text{couple})$$

As the resultant force is zero, the components of the resultant force (and of resultant couple in x and y direction should also be zero.

$\therefore \quad (F_R)_x = 0$ or $\Sigma m \times r \times \omega^2 \times \cos\theta = 0$ or $\Sigma m \times r \times \cos\theta = 0$ ...(i)

$(F_R)_y = 0$ or $\Sigma m \times r \times \omega^2 \times \sin\theta = 0$ or $\Sigma m \times r \times \sin\theta = 0$ ...(ii)

$(C_R)_x = 0$ or $\Sigma m \times r \times l \times \omega^2 \times \cos\theta = 0$ or $\Sigma m \times r \times l \times \cos\theta = 0$ ...(iii)

and $(C_R)_y = 0$ or $\Sigma m \times r \times l \times \omega^2 \times \sin\theta = 0$ or $\Sigma m \times r \times l \times \sin\theta = 0$ ...(iv)

Assume the angular position of the crank of the cylinder A in the horizontal direction. Then the position of the crank of the cylinder B will be at right angles as shown in Fig. 8.27b.

The magnitude and direction of balance mass m_y (i.e. in the plane y):

Take the plane x of Fig. 8.27a as the reference plane so as to eliminate the couple due to the unknown mass in the plane. Now from Eq. (iii), we get

$$\Sigma m \times r \times l \times \cos\theta = 0$$

or

$$m_A \times r_A \times l_A \times \cos\theta_A + m_B \times r_B \times l_B \times \cos\theta_B$$
$$+ m_y \times r_y \times l_y \times \cos\theta_y = 0$$

But

$$m_A = m_B = 360 \text{ kg}; r_A = r_B = 0.325 \text{ m}; r_y = 0.8 \text{ m};$$
$$l_A = 0.4125 \text{ m}; l_B = 1.0875 \text{ m}; l_y = 1.5 \text{ m}$$
$$\theta_A = 0 \text{ and } \theta_B = 90°$$

$\therefore \quad 360 \times 0.325 \times 0.4125 \times \cos\theta + 360 \times 0.325 \times 1.0875 \times \cos 90°$
$$+ m_y \times 0.8 \times 1.5 \times \cos\theta_y = 0$$

or $\qquad\qquad 48.2625 \times 1 + 127.28 \times 0 + 1.2 m_y \times \cos\theta_y = 0$

or $\qquad\qquad 48.2625 + 0 + 1.2 m_y \cos\theta_y = 0$

or $\qquad\qquad 1.2 m_y \times \cos\theta_y = -48.2625$

or $\qquad\qquad m_y \times \cos\theta_y = \dfrac{-48.2625}{1.2} = -40.21875 \text{ ...(v)}$

Similarly from Eq. (iv), we have

$$\Sigma m \times r \times l \times \sin\theta = 0 \text{ or } m_A \times r_A \times l_A \times \sin\theta_A + m_B \times r_B \times l_B \times \sin\theta_B$$
$$+ m_y \times r_y \times l_y \times \sin\theta_y = 0$$

or
$$360 \times 0.325 \times 0.4125 \times \sin\theta + 360 \times 0.325 \times 1.0875$$
$$\times \sin 90° + m_y \times 0.8 \times 1.5 \times \sin\theta_y = 0$$

or $\quad 0 + 127.28 + 1.2 m_y \times \sin\theta_y = 0$

or $\qquad m_y \times \sin\theta_y = \dfrac{-127.28}{1.2} = -106.066$ ...(vi)

Dividing Eq. (vi) by Eq. (v), we get

$$\frac{m_y \times \sin\theta_y}{m_y \times \cos\theta_y} = \frac{-106.066}{-40.21875} \text{ or } \frac{\sin\theta_y}{\cos\theta_y} = \frac{-106.066}{-40.21875} \qquad \text{...(vii)}$$

In Eq. (vii), the sign of numerator and denominator will give the quadrant in which θ_y will lie. The numerator corresponds to sine and denominator corresponds to cosine. As sine is –ve and also cosine is –ve, hence θ_y will lie in third quadrant.

From Eq. (vii),

$$\tan\theta_y = 2.6372 \text{ or } \theta_y = \tan^{-1} 2.6372 = 69.23°$$

As θ_y will lie in the third quadrant, hence angular position of mass m_y will be at angle of $180° + 69.23° = 249.23°$ with respect to the angular position of crank A.

Squaring and adding Eqs (v) and (vi), we get

$$m_y^2 \cos^2\theta_y + m_y^2 \sin^2\theta_y = (-40.21875)^2 + (-106.066)^2$$
$$= 1617.5 + 11245 = 12862.5$$

or $\quad m_y^2 (\cos^2\theta_y + \sin^2\theta_y) = 12862.5$

or $\quad m_y^2 = 12862.5 \text{ or } m_y = \sqrt{12862.5} = 113.41 \text{ kg}$

Magnitude and direction of balance mass in the plane x:

Taking the plane y of Fig. 8.27a as the reference plane and writing the couple equations, i.e. Eqs (iii) and (iv) as

$$m_A \times r_A \times l_A^* \times \cos\theta_A + m_B \times r_B \times l_B^* \cos\theta_B + m_x \times r_x \times l_x^* \times \cos\theta_x = 0$$

where $\qquad l_B^* = 0.4125 \text{ m}; l_A^* = 0.4125 + 0.675 = 1.0875 \text{ m}; l_x^* = 1.5 \text{ m}$

∴ $\qquad 360 \times 0.325 \times 1.0875 \times \cos 0° + 360 \times 0.325 \times 0.4125 \times \cos 90°$

or $\qquad\qquad\qquad + m_x \times 0.8 \times 1.5 \times \cos\theta_x = 0$

or $\qquad\qquad\qquad 127.28 + 0 + 1.2 m_x \times \cos\theta_x = 0$

or $\qquad\qquad\qquad m_x \times \cos\theta_x = \dfrac{-127.28}{1.2} = -106.066$ (viii)

From couple Eq. (iv), we have:

$$m_A \times r_A \times l_A^* \times \sin\theta_A + m_B \times r_B \times l_B^* \sin\theta_B + m_x \times r_x \times l_x^* \times \sin\theta_x = 0$$

or $360 \times 0.325 \times 1.0875 \times \sin 0° + 360 \times 0.325 \times 0.4125 \times \sin 90°$

$$+ m_x \times 0.8 \times 1.5 \times \sin\theta_x = 0$$

or $\qquad\qquad\qquad 0 + 48.2625 + m_x \times 1.2 \times \sin\theta_x = 0$

or $\qquad\qquad\qquad m_x \times \sin\theta_x = \dfrac{-48.2625}{1.2} = -40.21875$ ...(ix)

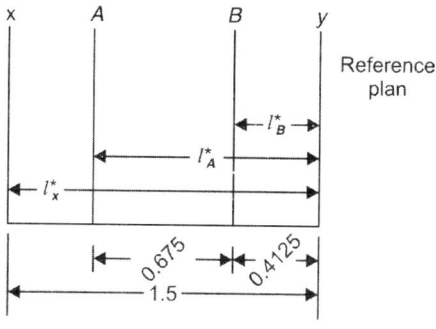

Fig. 8.28

Dividing Eq. (ix) by Eq. (viii), we get

$$\frac{m_x \times \sin\theta_x}{m_x \times \cos\theta_x} = \frac{-40.21875}{-106.066} \quad \text{or} \quad \frac{\sin\theta_x}{\cos\theta_x} = \frac{-40.21875}{-106.066}$$

As the numerator is –ve and also denominator is –ve in the above equation, hence θ_x lies in the third quadrant.

Also
$$\tan\theta_x = \frac{\sin\theta_x}{\cos\theta_x} = \frac{-40.21875}{-106.066} = 0.3791$$

or
$$\theta_x = \tan^{-1} 0.3791 = 20.76°$$

As θ_x lies in the third quadrant, hence, angular position of mass m_x will be at an angle of $180° + 20.76° = 200.76°$ with respect to angular position of crank A.

Squaring and adding Eqs (viii) and (ix), we get

$$m_x^2 [\cos^2\theta_x + \sin^2\theta_x] = (-106.066)^2 + (-40.21875)^2$$
$$= 11245 + 1617.5 = 12862.5$$

or
$$m_x = \sqrt{12862.5}$$
$$= 113.4 \text{ kg}$$

The correct angular position of balancing masses (m_x and m_y) are shown in Fig. 8.29.

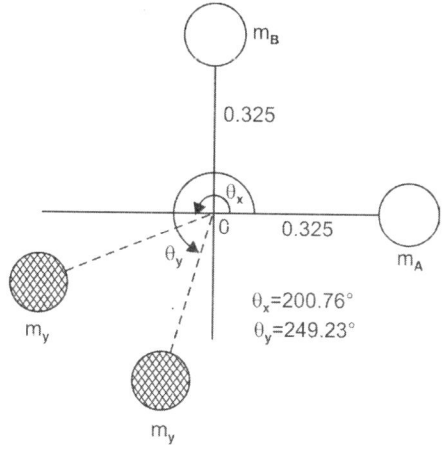

Fig. 8.29

Example 8.14: A single cylinder reciprocating engine has a reciprocating mass of 60 kg. The crank rotates at 60 rpm and the stroke is 320 mm. Mass of the revolving parts at 160 mm radius is 40 kg. If two-thirds of the reciprocating parts and the whole of the revolving parts are to be balanced, determine:

    i. The balance mass required at a radius of 350 mm.

    ii. The unbalanced force when the crank has turned 50° from the top-dead centre.

Solution: Mass of slider, $m_s = 60$ kg

$$N = 60 \text{ rpm}, \omega = \frac{2\pi N}{60} = \frac{2\pi \times 60}{60}$$

$$= 2\pi = 6.28 \text{ rad/s}$$

Stroke length $= 320$ mm, $r = \dfrac{320}{2} = 160$ mm $= 0.16$ m

Revolving mass, $m_r = 40$ kg at radius $r = 160$ mm

$$c = 2/3$$

Let balance mass $= m_b$ at radius, $r_b = 350$ mm

$$\theta = 50°$$

Total mass to be balanced, $m = m_r + Cm_s$

or
$$= 40 + \frac{2}{3} \times 60 = 80 \text{ kg}$$

For balancing,

$$m_b r_b = mr$$

or
$$m_b = \frac{mr}{r_b} = \frac{80 \times 160}{350} = 36.57 \text{ kg}$$

Unbalanced force,

$$F = mr\omega^2 \sqrt{(1-c)^2 \cos^2\theta + c^2 \sin^2\theta}$$

$$= 60 \times 0.60 \times (6.28)^2 \times \sqrt{\left(1 - \frac{2}{3}\right)^2 \cos^2 50° + \left(\frac{2}{3}\right)^2 \sin^2 50°}$$

$$F = 209.9 \text{ N}$$

Example 8.15: A four-cylinder oil engine is in complete primary balance. The arrangement of the reciprocating masses in different planes is shown in Fig. 8.30a. The stroke of each piston is $2r$ mm. Determine the reciprocating mass of the cylinder 2 and the relative crank positions.

Solution: Crank length $= 2r/2 = r$.

Take 2 as the reference plane and $\theta_3 = 0°$

$$m_1 r_1 l_1 = 380r \times (-1.3) = -494r$$
$$m_1 r_1 = 380r$$
$$m_3 r_3 l_3 = 590r \times 2.8 = 1652r$$
$$m_3 r_3 = 590r$$
$$m_4 r_4 l_4 = 480r \times (2.8 + 1.3) = 1968r$$

$$m_4 r_4 = 480r$$

$$-494r \cos\theta_1 + 1652r \cos 0° + 1968r \cos\theta_4 = 0 \qquad \text{...(1)}$$

and $\qquad -494r \sin\theta_1 + 1652r \sin 0° + 1968r \sin\theta_4 = 0$

or $\qquad\qquad\qquad\qquad 494 \sin\theta_1 = 1968 \sin\theta_4 \qquad \text{...(ii)}$

Squaring and adding Eqs (i) and (ii)

$$(494)^2 = (1652 + 1968 \cos\theta_4)^2 + (1968 \sin\theta_4)^2$$
$$= (1652)^2 + (1968)^2 \cos^2\theta_4 + 2 \times 1652 \times 1968 \cos\theta_4$$
$$+ (1968)^2 \sin^2\theta_4$$
$$= (1652)^2 + (1968)^2 + 2 \times 1652 \times 1968 \cos\theta_4$$

$$\cos\theta_4 = -0.978 \text{ or } \theta_4 = 167.9° \text{ or } 192.1°$$

Choosing one value, say $\theta_4 = 167.9°$

Dividing Eq. (ii) by Eq. (i), $\tan\theta_1 = \dfrac{1968 \sin 167.9°}{1652 + 1968 \cos 167.9°}$

$$= \frac{+412.53}{-272.28} = -1.515$$

$$\theta_1 = 123.4°$$

Writing the force equation (r is common),

$$380 \cos 123.4° + m_2 \cos\theta_2 + 590 \cos 0° + 480 \cos 167.9° = 0$$

or $\qquad\qquad\qquad\qquad\qquad m_2 \cos\theta_2 = 88.5 \qquad \text{...(iii)}$

and $\qquad 380 \sin 123.4° + m_2 \sin\theta_2 + 590 \sin 0° + 480 \sin 167.9° = 0$

or $\qquad\qquad\qquad\qquad\qquad m_2 \sin\theta_2 = -417.9 \qquad \text{...(iv)}$

Squaring and adding Eqs (iii) and (iv), $m_2 = 427.1$ kg

Dividing Eq. (iii) by Eq. (iv), $\tan\theta_2 = \dfrac{-417.9}{+88.5} = -4.72$

or $\qquad\qquad\qquad\qquad\qquad \theta_2 = 282°$

Figure 8.30b shows the relative crank positions.

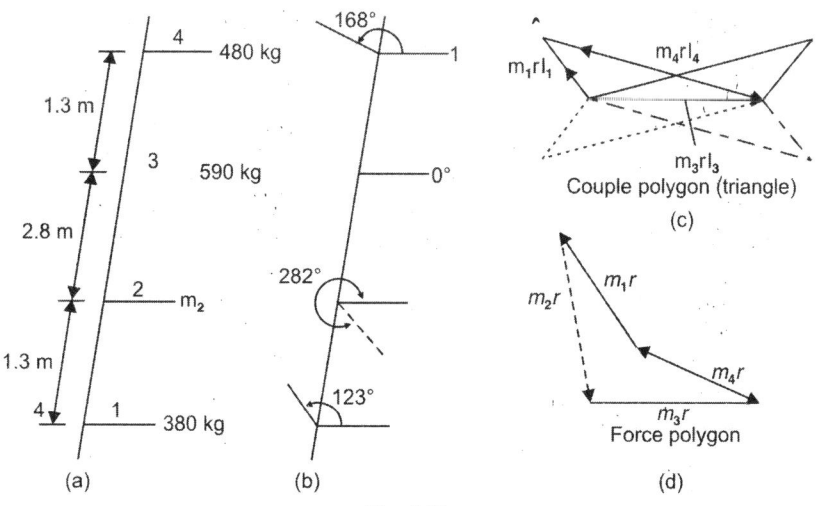

(a)   (b)   (c)   (d)

Fig. 8.30

For $\theta_4 = 192.1°$, a different set of values of m_2, θ_1 and θ_2 would have come.

To solve the problem graphically, draw the couple polygon (triangle) as shown in Fig. 8.30c from the three known values. This provides the relative direction of the masses m_1, m_3 and m_4. Now, complete the force polygon (Fig. 8.30d) and obtain the magnitude and direction of m_2. The results obtained are $\theta_4 = 168°$, $\theta_1 = 123°$, $\theta_2 = 282°$

Also $\qquad\qquad\qquad m_2 r = 427r$ or $m_2 = 427$ kg

$m_1 r_1 l_1$ is negative and, therefore, its direction is reversed in the diagram.

Example 8.16: The intermediate cranks of a four-cylinder symmetrical engine, which is in complete primary balance, are at 90° to each other and each has a reciprocating mass of 400 kg. The centre distance between intermediate cranks is 600 mm and between extreme cranks 1800 mm. Lengths of the connecting rods and the cranks are 900 mm and 200 mm respectively. Calculate the masses fixed to the extreme cranks with their relative angular positions. Also, find the magnitude of the secondary forces and couples about the centre line of system if the engine speed is 150 rpm.

Solution: $l = 0.9$ m, $m_2 = m_3 = 400$ kg, $r = 0.2$ m, $n = \dfrac{0.9}{0.2} = 4.5$.

The engine is in complete primary balance. Take 1 as the reference plane (Fig. 8.31).

$$m_2 r_2 l_2 = 400 \times 0.2 \times 0.6 = 48$$
$$m_3 r_3 l_3 = 400 \times 0.2 \times 1.2 = 96$$
$$m_4 r_4 l_4 = m_4 \times 0.2 \times 1.8 = 0.36 m_4$$

$$0.36 m_4 = \sqrt{(48\cos 0° + 96\cos 90°)^2 + (48\sin 0° + 96\sin 90°)^2}$$

$$= \sqrt{(48)^2 + (96)^2} = 107.33$$

$$m_4 = 298 \text{ kg}$$

$$\tan \theta_4 = \frac{-48}{-96} = 0.5; \theta_4 = 243.4°$$

By symmetry, $\qquad m_1 = 298$ kg

and $\qquad\qquad \tan \theta_1 = \dfrac{-48}{-96} = 0.5; \theta_1 = 206.6°$

The positions of the cranks for secondary forces and couples will be such that the angles are doubled (Fig. 8.31).

$$\omega = \frac{2\pi \times 150}{60} = 15.7 \text{ rad/s}$$

Secondary force

$$= \frac{r\omega^2}{n} \left[\begin{array}{l} 298(\cos 53.2° + \cos 126.8°) + 400(\cos 0° + \cos 180°)^2 \\ +298(\sin 53.2° + \sin 126.8°) + 400(\sin 0° + \sin 180°)^2 \end{array} \right]^{1/2}$$

$$= \frac{0.2 \times (15.7)^2}{4.5} (\sin 53.2° + \sin 126.8°) \times 298$$

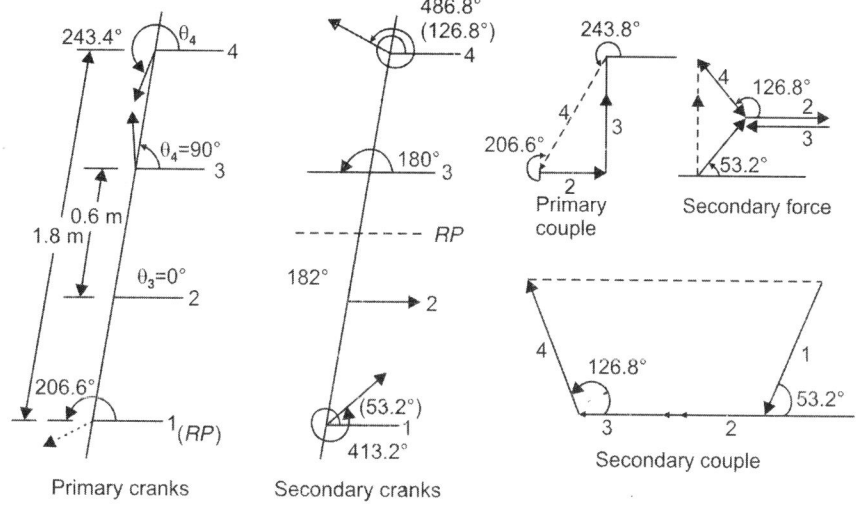

Fig. 8.31

$$= 5233.6 \text{ N}$$

Secondary couple about the centre line

$$= \frac{r\omega^2}{n} \begin{bmatrix} 298\,(-0.9\cos 53.2° + 0.9\cos 126.8°) \\ + 400(-0.3\cos 0° + 0.3\cos 180°)^2 \\ + 298\,(-0.9\sin 53.2° + 0.9\sin 126.8°) \\ + 400(-0.3\sin 0° + 0.3\sin 180°)^2 \end{bmatrix}^{1/2}$$

$$= \frac{0.2 \times (15.7)^2}{4.5}\,[298 \times (-0.9\cos 53.2° + 0.9\cos 126.8°) + 400\,(-0.6)]$$

$$= 6155 \text{ N·m}$$

Example 8.17: A V-twin engine has the cylinder axes at right angles and the connecting rods operate a common crank as shown in Fig. 8.32. The reciprocating mass per cylinder is 10 kg, the crank is 7 cm and the connecting rods 35 cm. Show that the engine may be balanced for primary effects by means of a revolving balance mass.

If the speed is 500 rpm, what is the maximum value of the resultant secondary force and in which direction does it act?

Solution: From Fig. 8.32, $2\alpha = 90°$; $\alpha = 45°$, $m = 10$ kg

$$r = 7 \text{ cm} = 0.07 \text{ m}, \, l = 35 \text{ cm} = 0.35 \text{ m}$$

$$n = l/r = \frac{35}{7} = 5, \, N = 500 \text{ rpm}$$

$$\omega = \frac{2\pi N}{60} = \frac{2\pi \times 500}{60}, \, \omega = 52.36 \text{ rad/s}$$

We know that the resultant primary force is given by

$$F_P = 2m\omega^2 r \sqrt{(\cos^2 \alpha \cos\theta)^2 + (\sin^2 \alpha \sin\theta)^2}$$

$$= 2m\omega^2 r \sqrt{(\cos^2 45° \cos\theta)^2 + (\sin^2 45° \sin\theta)^2}$$

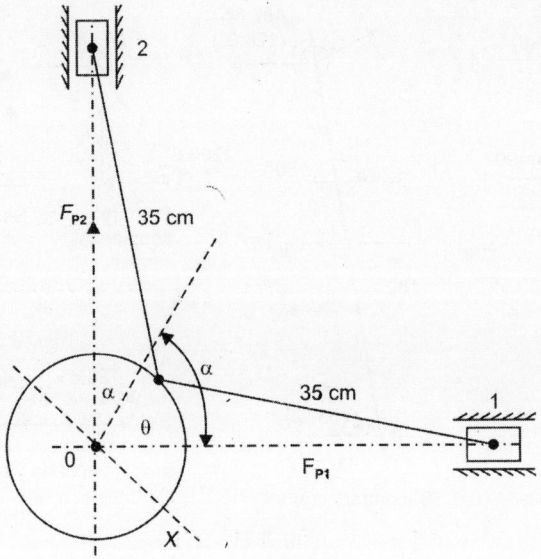

Fig. 8.32

$$F_P = 2m\omega^2 r \cdot \frac{1}{2}\sqrt{(\cos^2\theta + \sin^2\theta)}$$

$$= m\omega^2 r = \text{centrifugal force}$$

As F_P is equal to centrifugal force produced by a mass m at crank radius r when rotating at ω rad/s, primary forces can be balanced by revolving masses.

Maximum resultant secondary force

$$F_{HS} = \frac{m\omega^2 r}{n}\cos 2\theta \quad \text{(in horizontal direction)}$$

$$F_{VS} = \frac{m\omega^2 r}{n}\cos(180° - 2\theta)$$

$$F_{VS} = -\frac{m\omega^2 r}{n}\cos 2\theta \quad \text{(in vertical direction)}$$

The resultant of these forces,

$$F_s = \sqrt{F_{HS}^2 + F_{VS}^2} = \frac{m\omega^2 r}{n}\sqrt{2}\cos 2\theta$$

$$= 10 \times (52.36)^2 \times 0.07 \times \sqrt{2}/5 = 542.8 \text{ N}$$

It will be maximum when $\cos 2\theta = \pm 1$, i.e. $\theta = 0, 90°, 180°, 270°$. The horizontal and vertical components are equal in magnitude, so at the above angular positions, the force is at 45° to the crank.

Example 8.18: The cylinders of a twin V-engine are set at 60° angle with both pistons connected to a single crank through their respective connecting rods. Each connecting rod is 600 mm long and the crank radius is 120 mm. The total rotating mass is equivalent to 2 kg at the crank radius and the reciprocating mass is 1.2 kg per piston. A balance mass is also fitted opposite to the crank equivalent to 2.2 kg at a radius of

150 mm. Determine the maximum and minimum values of the primary and secondary forces due to inertia of the reciprocating and the rotating masses if the engine speed is 1050 rpm.

Solution: $m = 1.2$ kg, $M = 2$ kg, $l = 600$ mm, $m' = 2.2$ kg, $r' = 150$ mm, $N = 1050$ rpm.

$$\omega = \frac{2\pi N}{60} = \frac{2\pi \times 1050}{60} = 110 \text{ rad/s}$$

$$n = \frac{400}{80} = 5$$

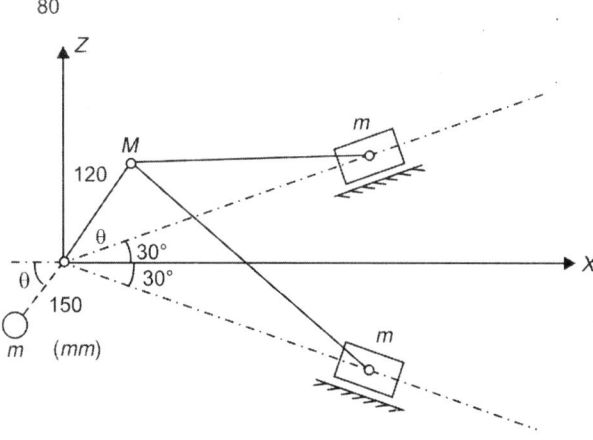

Fig. 8.33.

Primary force:
Total primary force along X-axis $= 2mr\omega^2 \cos^2\alpha \cos\theta$.
Centrifugal force due to rotating mass along X-axis $= Mr\omega^2 \cos\theta$
Centrifugal force due to balancing mass along X-axis
$$= -m'r'\omega^2 \cos\theta$$
Total unbalanced force along X-axis
$$= 2mr\omega^2 \cos^2\alpha \cos\theta + Mr\omega^2 \cos\theta - m'r'\omega^2 \cos\theta$$
$$= \omega^2\cos\theta\,(2mr\cos^2\alpha + Mr - m'r')$$
$$= 110^2 \times \cos\theta(2 \times 1.2 \times 0.12 \cos^2 30° + 2 \times 0.12 - 2.2 \times 0.15)$$
$$= 110^2 \times \cos\theta(0.216 + 0.24 - 0.33) = 1524.6 \cos\theta \text{ N}$$
Total primary force along Y-axis $= 2mr\omega^2 \sin^2\alpha \sin\theta$
Centrifugal force due to rotating mass along Y-axis $= Mr\omega^2 \sin\theta$
Centrifugal force due to balancing mass along Y-axis
$$= m'r'\omega^2 \sin\theta$$
Total unbalanced force along Y-axis
$$= 2mr\omega^2 \sin^2\alpha \sin\theta + Mr\omega^2 \sin\theta - m'r'\omega^2 \sin\theta$$
$$= \omega^2\sin\theta(2mr\sin^2\alpha + Mr - m'r')$$
$$= 110^2 \times \sin\theta\,(2 \times 1.2 \times 0.12 \sin^2 30° + 2 \times 0.12 - 2.2 \times 0.15)$$
$$= 110^2 \times \sin\theta(0.072 + 0.24 - 0.33) = -217.8 \sin\theta \text{ N}$$

Resultant primary force $= \sqrt{1524^2 \cos^2 \theta + (-217.8)^2 \sin^2 \theta}$

$$= \sqrt{2322576 \cos^2 \theta + 47437 \sin^2 \theta}$$

$$= \sqrt{2275139 \cos^2 \theta + 47437 \cos^2 \theta + 47437 \sin^2 \theta}$$

$$= \sqrt{2275139 \cos^2 \theta + 47437}$$

This is maximum when θ is $0°$ and minimum when $\theta = 90°$.

Maximum primary force $= \sqrt{2275139 + 47437} = 1524$ N

Minimum primary force $= \sqrt{47437} = 217.8$ N

Secondary force: The rotating masses do not affect the secondary forces as they are only due to second harmonics of the piston acceleration.

Resultant secondary force

$$= \frac{2mr\omega^2}{n} \sqrt{(\cos \alpha \cos 2\theta \cos 2\alpha)^2 + (\sin \alpha \sin 2\theta \sin 2\alpha)^2}$$

$$= \frac{2 \times 1.2 \times 0.12 \times 110^2}{5} \sqrt{\cos 30° \cos 2\theta \cos 60°)^2 + (\sin 30° \sin 2\theta \sin 60°)^2}$$

$$= 696.96 \sqrt{(0.433 \cos 2\theta)^2 + (0.433 \sin 2\theta)^2}$$

This is maximum when θ is $0°$ and minimum when θ is $90°$.

Maximum primary force $= 696.96 \times 0.433 = 301.8$ N

Minimum primary force $= 696.96 \times 0.433 = 301.8$ N

Thus, the secondary force has the same value for maximum and minimum.

EXERCISE

8.1 Why is balancing of the rotating parts of high speed engines necessary?

8.2 What do you understand by the static balancing and dynamic balancing? Write the necessary conditions to achieve them.

8.3 Explain how a single revolving mass is balanced by two masses revolving in different planes.

8.4 Discuss the method of balancing of different masses revolving in the same plane.

8.5 Explain how the different masses rotating in different planes are balanced.

8.6 Two masses in different planes are necessary to rectify the dynamic unbalance. Discuss.

8.7 Discuss the method of finding the counter masses in two planes to balance the dynamic unbalance of rotating masses.

8.8 What do you understand by primary and secondary unbalanced forces?

8.9 Discuss the partial balancing of unbalanced primary force in a reciprocating engine.

8.10 What are the effects of partial balancing in a reciprocating engine?

8.11 Derive the expression for variation in tractive force.

8.12 Derive the expression for swaying couple and hammer blow.

8.13 Discuss the balancing of four cylinder in line engine.

8.14 Prove that the resultant unbalanced force is minimum when half of the reciprocating masses are balanced by rotating masses.

8.15 Considering a two-cylinder uncoupled locomotive and discuss the following:
  (i) Variation of tractive force
  (ii) Swaying couple
  (iii) Hammer blow

8.16 Considering balancing of any engine you have studied. Derive the expressions for primary and secondary unbalanced forces.

8.17 What are in-line engines? State clearly how in-line four stroke and two stroke engines are balanced.

8.18 How are the different masses rotating in different planes balanced? Also, explain why only a part of the unbalanced force due to reciprocating masses is balanced by revolving masses.

8.19 Explain the terms, 'static balancing' and 'dynamic balancing'.

8.20 Derive the following expressions for an uncoupled two-cylinder locomotive engine:
  i. Variation in tractive force
  ii. Swaying couple

8.21 What do you mean by primary and secondary unbalance in reciprocating engines?

8.22 Three masses of 8 kg, 12 kg and 15 kg attached at radial distances of 80 mm, 100 mm and 60 mm respectively to a disc on a shaft are in complete balance. Determine the angular positions of the masses 12 kg and 15 kg relative to 8 kg mass.
(***Ans.*** 132.6° and 281° or 227.4° and 79°)

8.23 A circular disc mounted on a shaft carries three attached masses 4 kg, 3 kg and 2.5 kg at radial distances 75 mm, 85 mm and 50 mm and at the angular positions of 45°, 135° and 240° respectively. The angular positions are measured anticlockwise from the reference line along X-axis. Determine the amount of the balancing mass at a radial distance of 75 mm required for the static balance. (***Ans.*** 3.81 kg, 276.2°)

8.24 A rotor has the following properties:

Mass	Magnitude	Radius	Angle	Axial distance from 1st mass
1	9 kg	100 mm	0°	
2	7 kg	120 mm	60°	160 mm
3	8 kg	140 mm	135°	320 mm
4	6 kg	120 mm	270°	560 mm

If the shaft is balanced by two countermasses located at 100 mm radii and revolving in planes midway of planes 1 and 2, and midway of 3 and 4, determine the magnitude of the masses and their respective angular positions. (***Ans.*** 6.9 kg, 23°, 15.8 kg, 22.6°)

8.25 Four masses A, B, C and D are completely balanced. Masses C and D make angles of 90° and 210° respectively with B in the same sense. The planes containing B and C are 300 mm apart. Masses A, B, C and D can be assumed to be concentrated at radii of 360, 480, 240 and 300 mm respectively. The masses B, C and D are 15 kg, 25 kg and 20 kg respectively. Determine:
  i. the mass A and its angular position
  ii. the positions of planes A and D.
(***Ans.*** 10 kg, 236°; A is 985 mm towards right and D is 378 mm towards left of plane B)

8.26 Four masses are attached to a shaft at planes A, B, C and D at equal radii. The distance of planes B, C and D from A are 50 cm, 60 cm and 130 cm respectively. The masses at A, B and C are 60 kg, 55 kg and 80 kg respectively. If the system is in complete balance, determine the mass at D and the position of masses at B, C and D with respect to A.
(***Ans.*** 47 kg, 135°, 213°, 6°)

8.27 The cranks of a three-cylinder locomotive are set at 120°. The reciprocating masses are 450 kg for the inside cylinder and 390 kg for each outside cylinder. The pitch of the cylinder is 1.2 m and the stroke of each piston 500 mm. The planes of rotation of the balance masses are 960 mm from the inside cylinder. If 40% of the reciprocating masses are to be balanced, determine the magnitude and the position of the balancing masses required at a radial distance of 500 mm, and the hammer blow per wheel when the axle rotates at 350 rpm. (*Ans.* 86.25 kg each at 24° and 215°, 57.93 kN)

8.28 A four-cylinder vertical engine is in complete primary balance. The length of the cranks are 150 mm. The planes of rotation of the first, second and fourth cranks are 400 mm, 200 mm and 200 mm respectively from the third crank and their reciprocating masses are 50 kg, 60 kg and 50 kg respectively. Find: (i) the mass of the reciprocating parts for the third cylinder and (ii) the relative angular positions of the crank.
(*Ans.* 60 kg, 160°, 227°, 26° and 0°)

8.29 A shaft supported in bearing 1.6 m apart projects 400 mm beyond bearings at each end. It carries three pulleys one at each end at one at the centre of its length. The masses of the end pulleys are 40 kg and 22 kg and their centres of mass are at 12 mm and 18 mm respectively from the shaft axes. The mass of the centre pulley is 38 kg and its centre of mass is 15 mm from the shaft axis. The pulleys are arranged in a manner that they give static balance. Determine:
   i. the relative angular positions of the pulleys
   ii. the dynamic forces developed on the bearings when the shaft rotates at 210 rpm.

8.30 The crank of a four-cylinder marine oil engine are arranged at angular intervals of 90°. The engine speed is 70 rpm and the reciprocating mass per cylinder is 800 kg. The inner cranks are 1 m apart and are symmetrically arranged between the outer cranks which are 2.6 m apart. Each crank is 400 mm long. Determine the firing order of the cylinder for the best balance of reciprocating masses and also the magnitude of the unbalanced primary couple for that arrangement.

8.31 Three cranks of a three-cylinder locomotive are all on the same axle and are set at 120°. The pitch of the cylinders is 1 m and the stroke of each piston is 0.6 m.
The reciprocating masses are 200 kg for inside cylinder and 180 kg for each outside cylinder and the planes of rotation of the balance weights are 0.7 m from the inside crank. If $\frac{2}{3}$ of the reciprocating parts are to be balanced, find:
   (a) The magnitude and position of the balancing masses required at a radius of 0.5 m.
   (b) The hammer blow per wheel when the axle makes 6 rps.

8.32 Two outer cranks of a four-crank engine are set at 120° to each other with each reciprocating mass as 400 kg. The spacing between the planes of rotation of adjacent cranks are 450 mm, 750 mm and 600 mm. Determine the reciprocating mass and the relative angular position of each of the inner cranks if the engine is to be in complete primary balance. Also determine the maximum secondary unbalanced force is the length of the crank and the connecting rod are 300 mm and 1200 mm respectively and the speed is 240 rpm. (*Ans.* 878 kg, 314°, 853 kg, 162°, 90000 N)

8.33 Each crank of a four cylinder vertical engine is 225 mm. The reciprocating masses of the first, second and the third cranks are 100 kg, 120 kg and 100 kg and the planes of rotation are 600 mm, 300 mm and 300 mm from the plane of rotation of the third crank. Determine the mass of the reciprocating parts of the third cylinder and the relative angular positions of the cranks if the engine is in complete primary balance.
(*Ans.* 120 kg; $\theta_1 = 0°$, $\theta_2 = 157.7°$, $\theta_3 = 229.5°$, $\theta_4 = 27.2°$)

8.34 The cylinder axes of a V-engine are at right angle to each other. The weight of each piston is 2 kg and of each connecting rod 2.8 kg. The weight of the rotating parts like crank webs and the crank pin is 1.8 kg. The connecting rod is 400 mm long and its centre of mass is 100 mm from the crank pin centre. The stroke of the piston is 160 mm. Show that the engine can be balanced for the revolving and the primary force by a revolving countermass. Also, find the magnitude and the position if its centre of mass from the crank shaft centre is 100 mm. What is the value of the resultant secondary force if the speed is 840 rpm?

8.35 The cylinders of a V-engine are set at an angle of 40° with both cylinders connected to a common crank. The connecting rod is 300 mm long and the crank radius is 60 mm. The reciprocating mass is 1 kg per cylinder whereas the rotating mass at the crank pin is 1.5 kg. A balance mass equivalent to 1.8 kg is also fitted opposite to the crank at a radius of 800 mm. Determine the maximum and the minimum values of the primary and secondary forces due to inertia of the reciprocating and rotating masses if the engine rotates at 900 rpm. *(Ans.* 461.4 N, 354.9 N, 153.4 N, 46.9 N)

OBJECTIVE TYPE QUESTIONS

8.1 Static balancing involves balancing of
  (a) forces
  (b) couples
  (c) forces as well as couples
  (d) masses

8.2 For dynamic balancing of a shaft,
  (a) the net dynamic force is equal to zero
  (b) the net couple is equal to zero
  (c) both (a) and (b)
  (d) none

8.3 In case of rotating masses, the magnitude of the balancing mass is .... when the speed of the shaft is doubled.
  (a) doubled
  (b) halved
  (c) unaffected
  (d) quadrupled

8.4 The balancing of rotating and reciprocating parts of an engine is necessary when it runs at
  (a) slow speed
  (b) medium speed
  (c) high speed

8.5 If a rotating system is dynamically balanced, it is statically ....
  (a) balanced
  (b) unbalanced
  (c) partially balanced

8.6 For complete dynamic balance, at least .... mass/masses are necessary.
  (a) two
  (b) three
  (c) four
  (d) one

8.7 The magnitude of the secondary force is ,......... the primary force.
  (a) more than
  (b) less than
  (c) equal to
8.8 To balance the several masses revolving in different planes
  (a) the resultant force must be zero
  (b) the resultant couple must be zero
  (c) both the resultant force and couple must be zero
  (d) none of the above
8.9 In reciprocating engines, the primary unbalanced force
  (a) cannot be balanced
  (b) can be fully balanced
  (c) can be partially balanced
  (d) none of the above
8.10 The primary unbalanced force is maximum when the angle of crank with the line of stroke is ....
  (a) 45°
  (b) 90°
  (c) 135°
  (d) 180°
8.11 The partial balancing means
  (a) balancing partially the revolving masses
  (b) balancing partially the reciprocating masses
  (c) best balancing of engines
  (d) all of the above
  (e) none
8.12 The swaying couple is due to the
  (a) primary unbalanced force
  (b) secondary unbalanced force
  (c) two cylinders of locomotive
  (d) partial balancing
8.13 The swaying couple is maximum or minimum when the angle of inclination of the crank to the line of stroke (θ) is equal to
  (a) 45° and 135°
  (b) 90° and 135°
  (c) 135° and 225°
  (d) 45° and 225°
8.14 In a locomotive, the ratio of length of connecting rod to the crank radius is kept very large in order to
  (a) start the locomotive quickly
  (b) minimise the effects of primary forces
  (c) minimise the effects of primary forces
  (d) none of the above
8.15 By distributing balancing of reciprocating parts between coupled wheels, the hammer blow is
  (a) completely eliminated
  (b) reduced to half
  (c) increased considerably
  (d) constant

8.16 Two cylinders uncoupled locomotive have their cranks at
    (a) 270°
    (b) 90°
    (c) 180°
    (d) 20°

8.17 Hammer blow is maximum when
    (a) $\theta = 0°$
    (b) $\theta = 90°$ or $270°$
    (c) $\theta = 180°$ or $360°$
    (d) $\theta = 180°$ or $315°$

8.18 In certain aircraft engines, balancing masses for balancing secondary force produced by reciprocating parts are provided on
    (a) there is no need for balancing the secondary forces
    (b) secondary shaft running at twice the engine speed
    (c) primary shaft running at twice the engine speed
    (d) secondary shaft running at same speed as that of the engine

8.19 Field balancing of a single plane rotor without measurements of phase can be completed by carrying out
    (a) four test runs
    (b) three test runs
    (c) two test runs
    (d) one test run

8.20 In dynamic balancing, the following condition will hold
    (a) only the force polygon will be closed
    (b) only the couple polygon will be closed
    (c) both force and couple polygons will be closed
    (d) none of the above

8.21 The resultant unbalanced force is minimum in reciprocating engines
    (a) when one third of the reciprocating masses are balanced by rotating masses
    (b) when half the reciprocating masses are balanced by rotating masses
    (c) when three fourth of the reciprocating masses are balanced by rotating masses
    (d) none of the above

8.22 For complete balancing of the reciprocating parts, the condition arrived at is
    (a) primary force polygon must close
    (b) secondary force polygon must close
    (c) primary couple polygon must close
    (d) all of the above

8.23 The maximum magnitude of the unbalanced force along the perpendicular to the line of stroke is known as
    (a) hammer blow
    (b) tractive force
    (c) swaying couple
    (d) none of the above

8.24 Field balancing of a long rotor with measurement of phase can be completed by carrying out
    (a) four test runs
    (b) three test runs
    (c) two test runs
    (d) one test run

8.25 The entire reciprocating mass of a slider–crank system is never balanced because
  (a) reciprocating mass does not require balancing
  (b) it would only change the direction of unbalanced force
  (c) it is not possible to do so
  (d) it gives rise to unbalanced couple

8.26 If there are several unbalanced masses in a rotor in different planes, the minimum number of balancing masses required is
  (a) one
  (b) two
  (c) three
  (d) four

8.27 In a multicylinder in-line engine, number of cylinders are chosen as even, so that
  (a) primary forces are balanced
  (b) secondary forces are balanced
  (c) primary and secondary couples are balanced
  (d) none of the above

8.28 The frequency of secondary force as compared to primary force for ratio of connecting rod length to crank radius of 4 is
  (a) half
  (b) twice
  (c) four times
  (d) sixteen times

8.29 A balance mass of value 2/3 m is placed diametrically opposite to the crank at crank radius r. The unbalanced force along the line of stroke of a reciprocating engine is ......
  (a) $1/3\ mr\omega^2 \cos\theta$
  (b) $\sqrt{1/3}\, mr\omega^2\ \cos\theta$
  (c) $\sqrt{1/2}\, mr\omega^2\ \cos\theta$
  (d) none of the above

8.30 In partial balancing of locomotives, the maximum variation of tractive effort is
  (a) $2/3\ m\omega^2 r$
  (b) $\sqrt{2/3}\, m\omega^2 r$
  (c) $\sqrt{3/2}\, m\omega^2 r$
  (d) $1/2\ m\omega^2 r$

8.31 While considering balancing of coupled locomotives, one has to consider
  (a) two planes, one of the cylinders and the other of the driving wheels
  (b) four planes, two of the cylinders and the other of the driving wheels
  (c) six planes, two of the cylinders, two of the coupling rods and two of the driving wheels containing balancing weights
  (d) none of the above

8.32 In order to facilitate the starting of locomotive in any position, the cranks of a locomotive, with two cylinders, are placed at .... to each other
  (a) 45°
  (b) 90°
  (c) 120°
  (d) 180°

8.33 The tractive force is maximum or minimum when the angle of inclination of the crank to the line of stroke (θ) is equal to :
  (a) 90° and 225°
  (b) 135° and 180°
  (c) 180° and 225°
  (d) 135° and 315°

8.34 In a locomotive, the ratio of length of connecting rod to the crank radius is kept very large in order to
  (a) start the locomotive quickly
  (b) minimise the effects of primary forces

(c) minimise the effect of secondary forces

(d) none of the above

8.35 The effect of hammer blow in a locomotive can be reduced by

    (a) decreasing the speed

    (b) using two or three pairs of wheels coupled together

    (c) balancing whole of the reciprocating parts

    (d) both (a) and (b)

8.36 Multicylinder engines are desirable because

    (a) only balancing problems are reduced

    (b) only flywheel size is reduced

    (c) both (a) and (b)

    (d) none of these

8.37 When the primary direct crank of a reciprocating engine makes an angle θ with the line of stroke, then the secondary direct crank will make an angle of .... with the line of stroke.

    (a) $\theta/2$

    (b) θ

    (c) 2θ

    (d) 4θ

8.38 Secondary forces due to reciprocating mass on engine frame are

    (a) of same frequency as of primary forces

    (b) twice the frequency as of primary forces

    (c) four times the frequency as of primary forces

    (d) none of the above

8.39 The secondary unbalanced force produced by the reciprocating parts of a certain cylinder of a given engine with crank radius r and connecting rod length l can be considered as equal to primary unbalanced force produed by the same weight having

    (a) an equivalent crank radius $r^2/4l$ and rotating at twice the speed of the engine

    (b) $r^2/4l$ as equivalent crank radius and rotating at engine speed

    (c) equivalent crank length of $r^2/4l$ and rotating at engine speed

    (d) none of the above

8.40 Which of the following statements is correct?

    (a) in any engine, 100% of the reciprocating masses can be balanced dynamically

    (b) in the case of balancing multicylinder engine, the value of secondary force is higher than the value of the primary force.

    (c) in the case of balancing of multimass rotating system, dynamic balancing can be directly started without static balancing done to the system

    (d) none of the above.

ANSWERS

8.1 (a)	8.2 (c)	8.3 (c)	8.4 (c)	8.5 (a)	8.6 (a)
8.7 (b)	8.8 (c)	8.9 (c)	8.10 (d)	8.11 (b)	8.12 (a)
8.13 (d)	8.14 (c)	8.15 (b)	8.16 (d)	8.17 (b)	8.18 (b)
8.19 (a)	8.20 (c)	8.21 (b)	8.22 (e)	8.23 (a)	8.24 (b)
8.25 (d)	8.26 (b)	8.27 (c)	8.28 (b)	8.29 (a)	8.30 (b)
8.31 (c)	8.32 (b)	8.33 (d)	8.34 (c)	8.35 (d)	8.36 (c)
8.37 (c)	8.38 (b)	8.39 (a)	8.40 (c)		

9

Brakes and Dynamometers

9.1 INTRODUCTION

A brake is a device by which frictional resistance is applied to a moving machine member to stop or retard it by absorbing its kinetic energy. Heat produced is dissipated to surrounding air or water through the brake drums so that the excessive heat does not affect the brake lining. The capacity of brake depends upon the unit pressure between the braking surfaces, coefficient of friction between them, velocity of brake drum, heat dissipation capacity of brake, etc.

A dynamometer is a device used to measure the applied frictional resistance. The frictional resistance is obtained by applying the brake. Therefore, dynamometer is also defined as a brake used to determine the power developed by the machine, while maintaining its speed at a certain value.

The functional difference between a clutch and a brake is that a clutch connects two moving members of a machine, whereas a brake connects a moving member with a stationary member.

9.2 TYPES OF BRAKES

The brakes are classified as:
   (i) Hydraulic brakes, e.g. pumps or hydrodynamic brakes.
   (ii) Electric brakes, e.g. generators and eddy current brakes.
   (iii) Mechanical brakes

Hydraulic and electric brakes are used in laboratory dynamometers, highway trucks and electric locomotives.

Mechanical brakes are of the following types:
   (i) Block or shoe brake:
      (a) Single block or shoe brake
      (b) Double block or shoe brake
   (ii) Band brake
   (iii) Band and block brake
   (iv) Internal expanding shoe brake.

9.3 BLOCK OR SHOE BRAKE

A block or shoe brake consists of a block or shoe that is pressed against a rotating drum or wheel as shown in Fig. 9.1. The force on the drum is applied through a lever at one end. The other end of the lever is pivoted on a fixed fulcrum O.

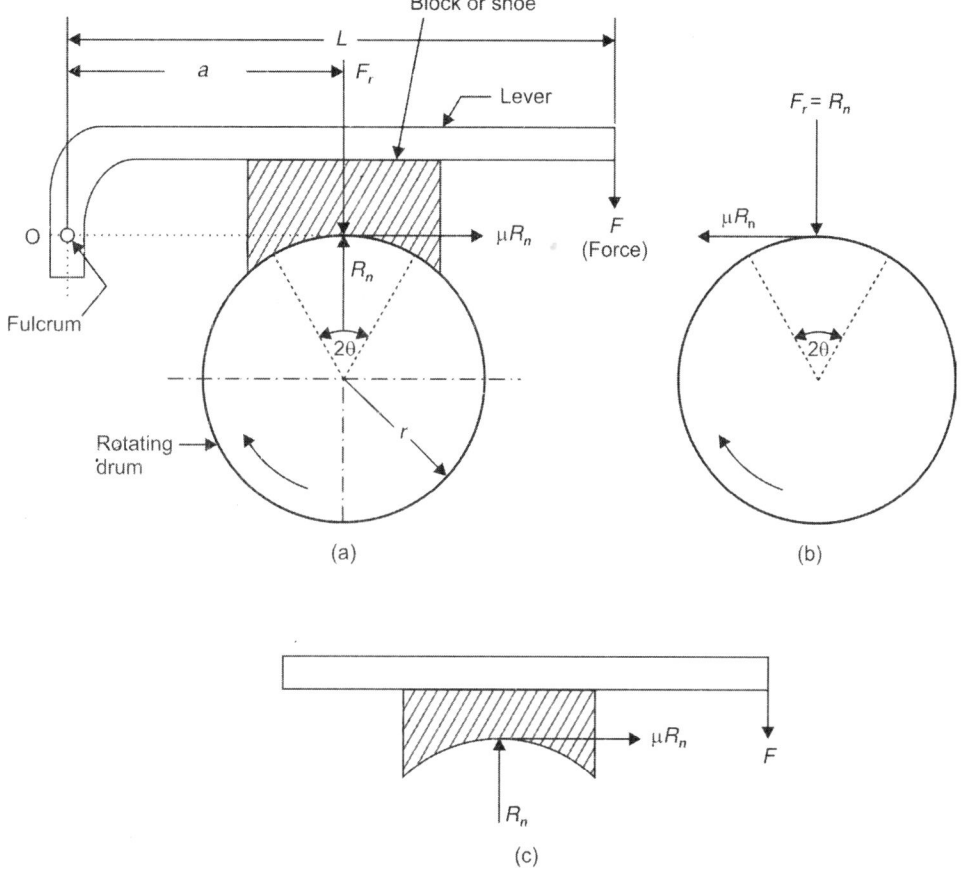

Fig. 9.1

As the force is applied on the lever, the block rigidly fixed to lever is pressed against the rotating drum. The material of the block or shoe is softer than that of drum or wheel so that it can be easily replaced on wearing. For light and slow moving vehicles, wood and rubber are used, whereas for heavy and fast moving vehicles, cast steel is used.

This brake may consists of single or double shoe or block. In case only one block is used, a side thrust will act on the bearing of the shaft. This drawback can be removed by using two blocks on the two sides of the drum (Refer to Fig. 9.6). This will double the braking torque also (Eq. 9.1).

Let r = radius of the drum

F = force applied at the lever end

R_n = normal reaction on the block

μ = coefficient of friction

f = frictional force = μR_n

T_B = braking torque

2θ = angle of contact of the block at the centre of the drum

F_r = radial force acting on the drum = normal reaction on the block (R_n)

Considering the clockwise rotation of the brake drum.

Braking torque on the drum = frictional force × radius

or
$$T_B = f \times r = \mu R_n \times r \qquad \qquad ...(9.1)$$

Value of R_n is obtained by considering the equilibrium of the block. The friction force on the drum will act in opposite direction (anticlockwise), whereas on block it is in the same direction.

Let line of action of frictional force (μR_n) passes through the fulcrum O of the lever.

Taking moment about the pivot O

$$R_n \times a = F \times L$$

or
$$R_n = \frac{F \times L}{a}$$

Substituting value of R_n in Eq. (9.1), we have

$$T_B = \mu \times \frac{F \times L}{a} \times r \qquad \qquad ...(9.2)$$

i. If the line of action of the frictional force (μR_n) is at a distance b below the fulcrum O and the drum rotates clockwise as shown in Fig. 9.2, the forces acting on the block are: (a) R_n, (b) μR_n and (c) F.

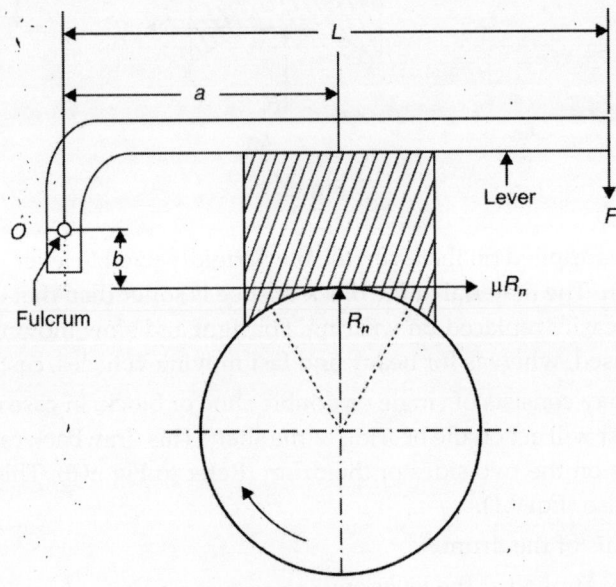

Fig. 9.2

Taking moments about the flucrum O.

$$R_n \times a + \mu R_n \times b = F \times L$$

or
$$R_n (a + \mu \times b) = F \times L$$

∴
$$R_n = \frac{F \times L}{(a + \mu b)}$$

Substituting this value of R_n in Eq. (9.1)

$$T_B = \mu \times \frac{F \times L}{(a + \mu b)} \times r$$

$$= \frac{\mu \times F \times L \times r}{(a + \mu b)} \qquad \qquad \text{...(9.3)}$$

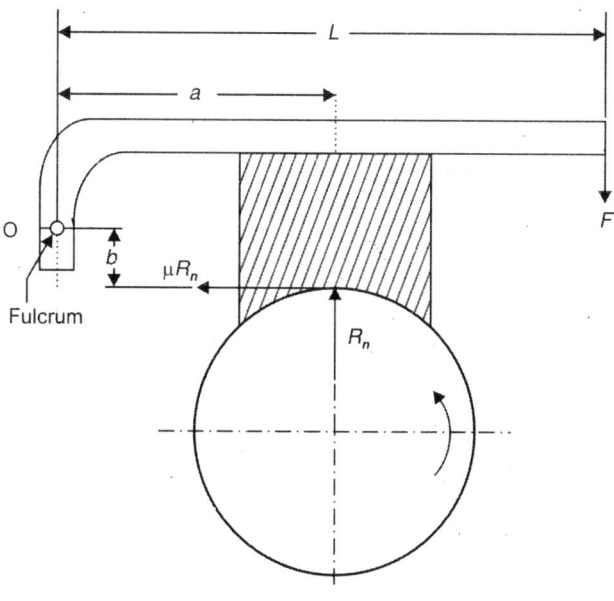

Fig. 9.3

If the drum is rotating in anticlockwise direction as shown in Fig. 9.3, then the frictional force $\mu \times R_n$ will also be acting in anticlockwise direction. Again taking moment about O,

$$R_n \times a = F \times L + \mu R_n \times b \qquad \qquad \text{...(9.4)}$$

or $\qquad R_n \times a - \mu R_n \times b = F \times L$

or $\qquad R_n (a - \mu b) = F \times L$

$$R_n = \frac{F \times L}{(a - \mu b)} \qquad \qquad \text{...(9.4a)}$$

Substituting the above value in Eq. (9.1), braking torque

$$T_B = \mu \times \frac{F \times L}{(a - \mu b)} \times r = \frac{\mu \times F \times L \times r}{(a - \mu b)} \qquad \qquad \text{...(9.5)}$$

Self-locking brake: From Eq. (9.4), the force (F) required to apply the brake

$$F = \frac{R_n (a - \mu b)}{L}$$

If $a \le \mu b$, then F will be negative or zero. This means that no external force is required to apply the brake and hence, the brake is *self-locking*. Hence, the condition for the brake to be self-locking is

$$a \le \mu b$$

Self-energised brake: In Eq. (9.4a), the moment of frictional force f about O is in the same direction as that of the applied force F. Hence, frictional force helps in applying the brake. Such a brake is known as self-energised brake. In practice, the brake should be self-energising and not self-locking.

ii. The line of action of the frictional force (μR_n) is at a distance b above the fulcrum O and the drum rotates clockwise as shown in Fig. 9.4.

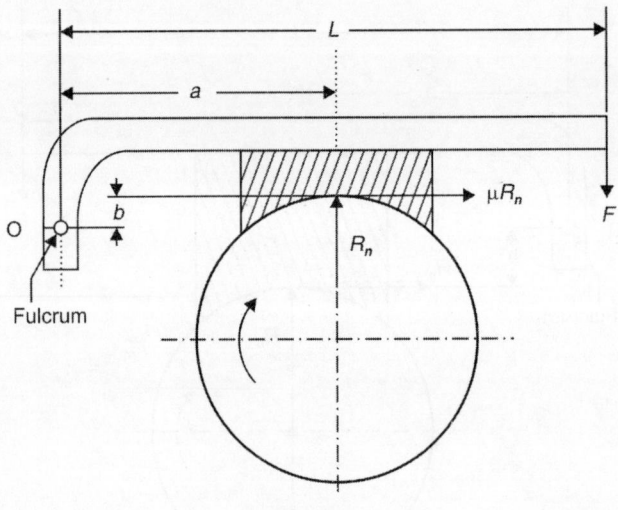

Fig. 9.4

Taking the moments about the fulcrum,

$$R_n \times a = F \times L + \mu R_n \times b$$

or

$$R_n \times a - \mu R_n \times b = F \times L + \mu R_n \times b$$

or

$$R_n (a - \mu \times b) = F \times L$$

$$\therefore \qquad R_n = \frac{F \times L}{(a - \mu b)}$$

Substituting this value of R_n in Eq. (9.1),

$$T_B = \mu \times \frac{F \times L}{(a - \mu b)} \times r$$

$$= \mu \times \frac{F \times L \times r}{(a - \mu b)} \qquad \qquad ...(9.6)$$

In this case also, the brake may be self-locking or self-energised. If $F = 0$, the brake is self-locking and if $F > 0$, the brake is self-energised.

If the drum is rotating in anticlockwise direction as shown in Fig. 9.5, then frictional force (μR_n) will also be acting in anticlockwise direction.

Taking the moments of all forces about the fulcrum,

$$R_n \times a + \mu R_n \times b = F \times L$$

or

$$R_n (a + \mu b) = F \times L$$

or

$$R_n = \frac{F \times L}{(a - \mu b)}$$

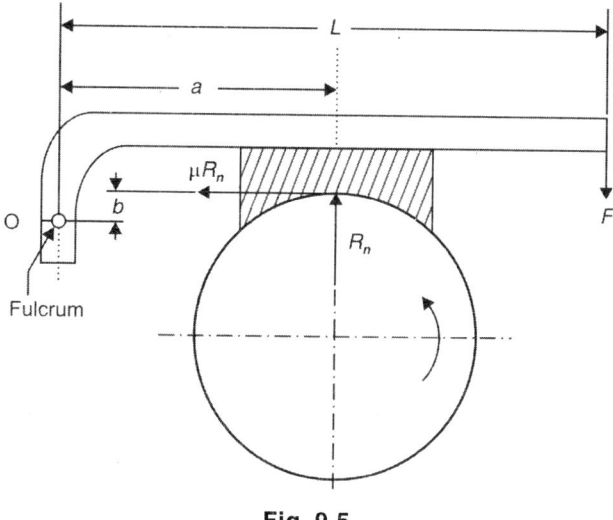

Fig. 9.5

Substituting this value in Eq. (9.1),

$$T_B = \mu \times \frac{F \times L}{(a + \mu b)} \times r$$

$$= \frac{\mu \times F \times L \times r}{(a + \mu b)} \qquad ...(9.7)$$

For all the above expressions, the normal reaction (R_n) and force of friction (μR_n) are assumed to be acting at the mid-point of the block. This is true only if the angle made by contact surface of the block at the centre of the rotating drum is less than or equal to 40°, i.e., $2\theta \le 40°$. But if the angle of contact is more than 40°, the normal pressure is less at the ends than at the centre. In that case, μ has to be replaced by an equivalent coefficient of friction μ', given by

$$\mu' = \frac{4\mu \sin\theta}{2\theta + \sin 2\theta} = \mu \left[\frac{4 \sin\theta}{2\theta + \sin 2\theta} \right] \qquad ...(9.8)$$

Double block or shoe brake: It consists of two blocks or shoes rigidly fixed to two levers which are pressed against the rotating drum at its opposite ends. This eliminates or reduces the unbalanced force on the shaft. This type of brake is used in electric cranes wherein the force F is produced by an electromagnet or solenoid.

Brake torque,

$$T_B = \mu R_{n_1} \times r \times \mu R_{n_2} \times r$$

$$T_B = (\mu R_{n_1} + \mu R_{n_2}) \times r \qquad (9.9)$$

The values of R_{n1} and R_{n2} can be obtained by taking moments of forces acting on blocks about fulcrum O_1 and O_2 respectively.

If ω is the width of brake shoe, then

projected bearing area of one shoe, $A = (2r \sin\theta) \times \omega$

bearing pressure on lining material, $p = \dfrac{R_n}{A}$

where R_n is maximum normal force on any shoe.

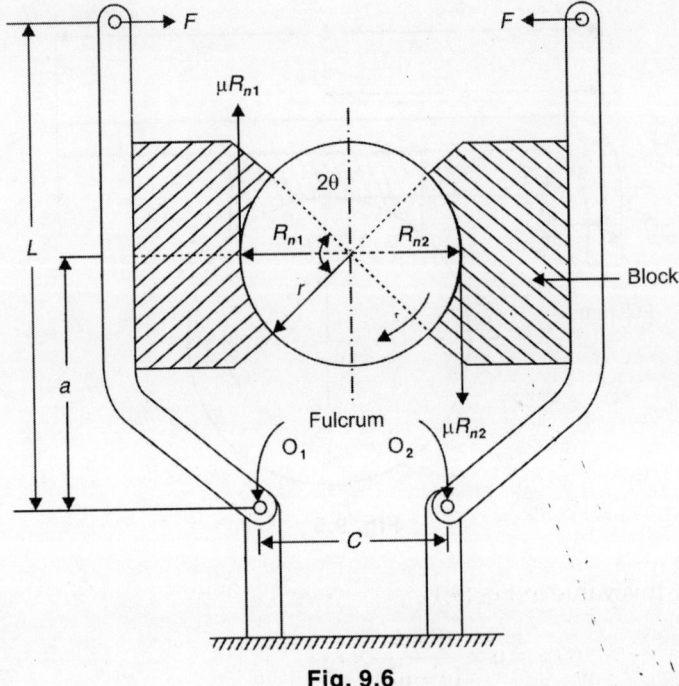

Fig. 9.6

Example 9.1: The brake drum of a single block brake is rotating at 500 r.p.m. in the clockwise direction. The diameter of the drum is 400 mm and the single block brake is as shown in Fig. 9.2. The force required at the end of the lever to apply the brake is 200 N. If angle of contact is 30° and $L = 1$ m, $a = 300$ mm and $b = 25$ mm, then determine the braking torque. The coefficient of friction is equal to 0.25.

Solution:

Speed $N = 500$ rpm

Dia. of drum $D = 400$ mm $= 0.4$ m

∴ Radius of drum $r = \dfrac{400}{2} = 200$ mm $= 0.2$ m

Force at the end of lever $F = 200$ N

Angle of contact $2\theta = 30°$

Length of lever from fulcrum $L = 1$ m

Distance of centre of the block from fulcrum,

$$a = 300 \text{ mm} = 0.3 \text{ m}$$

Perpendicular distance between line of action of frictional force and fulcrum,

$$b = 25 \text{ mm} = 0.025 \text{ m}$$

Rotation of drum = clockwise

Coefficient of friction, $\mu = 0.35$

Taking the moments of all forces about fulcrum O,

$$R_n \times a + \mu R_n \times b = F \times L$$

or $R_n \times 0.3 + 0.35 \times R_n \times 0.025 = 200 \times 1$

$$R_n (0.3 + 0.35 \times 0.025) = 200$$

$$R_n = \frac{200}{0.3 + 0.35 \times 0.025}$$

$$= \frac{200}{0.30875} = 647.77 \, \text{N}$$

Braking torque (T_B) is given as

$$T_B = \mu R_n \times r = 0.35 \times 647.77 \times 0.2$$
$$= 45.34 \, \text{N·m}$$

Example 9.2: A single block brake is shown in Fig. 9.7. The diameter of the drum is 250 mm and the angle of contact is 60°. If the operating force of 400 N is applied at the end of a lever and the coefficient of friction between the drum and the lining is 0.30, determine the torque that may be transmitted by the block brake.

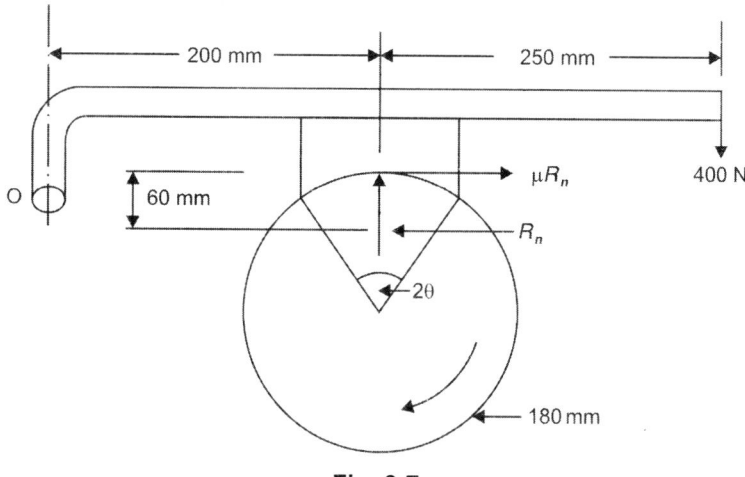

Fig. 9.7

Solution: Diameter of drum d = 180 mm

∴ Radius r = 90 mm = 0.09 m

$$2\theta = 60° = \frac{\pi}{180} \times 60 = \frac{\pi}{3}$$

$\theta = 30°$, $\mu - 0.30$, $F = 400 \, \text{N}$

The value of angle of contact is more than 40°, so equivalent coefficient of friction μ' is given by

$$\mu' = \frac{4\mu \sin \theta}{2\theta + \sin 2\theta}$$

$$= \frac{4 \times 0.30 \times \sin 30°}{\frac{\pi}{3} + \sin 60°} = 0.313$$

Taking moments about the fulcrum O,

$$400\,(250 + 200) + \mu' R_n \times 60 = R_n \times 200$$
$$400 \times 450 + 0.313 \times R_n \times 60 = 200 R_n$$
$$R_n = 993.3 \, \text{N}$$

Braking force $= \mu'R_n = 0.313 \times 993.3 = 310.9$ N

Torque transmitted by the block brake,

$$T_B = \mu'R_n \cdot r$$
$$= 310.9 \times 0.09 = 27.98 \text{ N·m}$$

Example 9.3: A double block brake can absorb a torque of 1200 N·m. The diameter of brake drum is 300 mm and the angle of contact for each shoe is 100°. If the coefficient of friction between the brake drum and lining is 0.4, find: (i) spring force necessary to set the brake, (ii) the width of the brake shoes, if the bearing pressure on the lining material is not to exceed 0.3 N/mm².

Solution:

$T_B = 1200$ N·m $= 1200 \times 10^3$ N·mm, $c = 40$ mm

$d = 300$ mm so $r = 150$ mm, $2\theta = 100° = 1.75$ rad

$\mu = 0.4$, $p = 0.3$ N/mm², $a = 200$ mm, $L = 450$ mm

(i) Since the angle of contact is greater than 40°, hence, equivalent coefficient of friction

$$\mu' = \frac{4\mu \sin\theta}{2\theta + \sin 2\theta} = \frac{4 \times 0.4 \times \sin 50°}{1.75 + \sin 100°} = 0.45$$

Taking moment about fulcrum O_1,

$$F \times 450 + \mu'R_{n_1} \times (150 - 20) = R_{n_1} \times 200$$

or $\qquad 200R_{n_1} - \mu'R_{n1} \times 130 = 450F$

or $\qquad \mu'R_{n_1}\left[\dfrac{200}{\mu'} - 130\right] = 450F$

or $\qquad \mu'R_{n_1}\left[\dfrac{200}{0.45} - 130\right] = 450F$ or $\mu'R_{n_1}[314.44] = 450F$

or $\qquad \mu'R_{n_1} = \dfrac{450\,F}{314.44} = 1.43F$

Now taking moment about O_2,

$$F \times 450 = R_{n_2} \times 200 + \mu'R_{n_2} \times (150 - 20)$$

or $\qquad \mu'RL_{n_2}\left[\dfrac{200}{\mu'} + 130\right] = 450F$

or $\qquad \mu'RL_{n_2}\left[\dfrac{200}{0.45} + 130\right] = 450F$ or $\mu'RL_{n_2}[574.44] = 450F$

or $\qquad \mu'R_{n_2} = \dfrac{450\,F}{574.44} = 0.783F$

Braking torque $\qquad T_B = (\mu'R_{n_1} + \mu'R_{n_2}) \times r$

(ii)

or $\qquad 1200 \times 10^3 = (1.43F + 0.783F) \times 150$

or $\qquad F = 3615$ N

Projecting bearing area of one shoe, $A = (2r \sin\theta) \times \omega$

or
$$A = (2 \times 150 \times \sin 50°) \times \omega = 229.8\omega \text{ mm}^2$$

Normal force on left hand side of shoe,

$$R_{n_1} = \frac{f_1}{\mu'} = \frac{\mu' R_{n_1}}{\mu'} = \frac{1.43F}{0.45}$$

$$= 3.18F = 3.18 \times 3615 = 11495.7 \text{ N}$$

Normal force on left hand side of shoe,

$$R_{n_2} = \frac{f_2}{\mu'} = \frac{\mu' R_{n_2}}{\mu'} = \frac{0.783F}{0.45}$$

$$= 1.74F = 1.74 \times 3615 = 6290.1 \text{ N}$$

Bearing pressure on lining material,

$$p = \frac{R_n}{A} \ (R_n \text{ is maximum of } R_{n_1} \text{ and } R_{n_2})$$

$$0.3 = \frac{3.18 \times 3615}{229.8\,\omega} = \frac{11495.7}{229.8\,\omega}$$

or
$$\omega = \frac{11495.7}{229.8 \times 0.3}$$

or
$$\omega = 166.75 \text{ mm.}$$

9.4 BAND BRAKE

A band brake consists of rope, leather belt or flexible steel band lined with friction material and controlled by a lever to apply the brake. When the force is applied at the free end of a lever to apply the brake, band is pressed against the external surface of a cylindrical drum. Band brake can be of two types:

1. Simple band brake,
2. Differential band brake.

9.4.1 Simple Band Brake

A band brake shown in Fig. 9.8 is known as simple band brake in which one end of the band is attached with fulcrum of the lever and the other end is attached to the lever at a distance 'a' from the fulcrum.

The force F is applied at the free end of the lever which turns about the fulcrum O. This tightens the band on the drum and hence, the brakes are applied. The braking force is provided by the friction between the band and the drum. The force F at the end of the lever for clockwise rotation and anticlockwise rotation of drum is obtained as explained below.

Let θ = angle of lap of the band on the drum
T_1 = tension in the tight side of the band
T_2 = tension in the slack side of the band
r = radius of the drum
μ = coefficient of friction between band and the drum
t = thickness of band

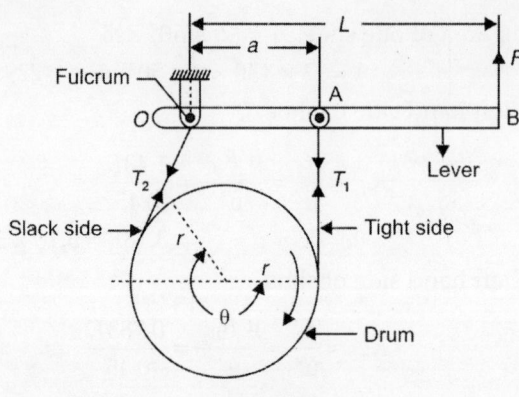

Fig. 9.8

r_e = effective radius of the drum = $\left(r + \dfrac{t}{2} \right)$, and

F = force at the end of the lever

Limiting ratio of tensions is given by

$$\frac{T_1}{T_2} = e^{\mu\theta} \qquad\qquad ...(9.10)$$

or $\qquad\qquad 2.3 \log \left(\dfrac{T_1}{T_2} \right) = \mu\theta$

Braking torque on the drum is given by

$$T_B = (T_1 - T_2) \times r \qquad \text{(if thickness of belt is neglected)}$$
$$= (T_1 - T_2) \times r_e \qquad \text{(if thickness of belt is considered)}$$

(i) **Value of F for clockwise rotation of drum:** For clockwise rotation of drum as shown in Fig. 9.8, the end of the band connected to the fulcrum O will be slack side with tension T_2 and the end of the band attached to A will be tight side with tension T_1.

Taking moments about the fulcrum O,

$$F \times L = T_1 \times a \qquad (\because T_2 \text{ passes through } O) \qquad\qquad ...(9.11)$$

where L = distance OB and a = perpendicular distance from O to the line of action of T_1.

(ii) **Value of F for anticlockwise rotation of drum:** For anticlockwise rotation of the drum as shown in Fig. 9.9, the end of the band connected to the fulcrum O will be tight side with tension T_1 and the end of the band attached to A will be slack wide with tension T_2. Taking the moments about the fulcrum O,

$$F \times L = T_2 \times a \qquad (\because T_1 \text{ passes through } O) \qquad\qquad ...(9.12)$$

9.4.2 Differential Band Brake

In differential band brake, ends of band are connected at A and B on different sides of fulcrum 'O'.

The effectiveness of the force F applied at the end of lever depends upon the following:

(i) The direction of rotation of drum

(ii) The direction of the applied force F

(iii) The ratio of lengths a and b, i.e. $a > b$ or $a < b$

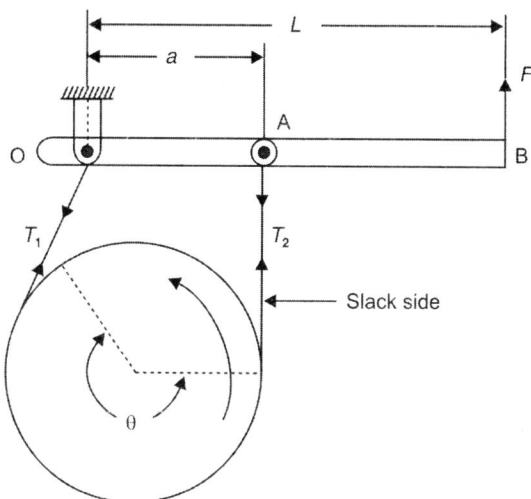

Fig. 9.9: Anticlockwise rotation

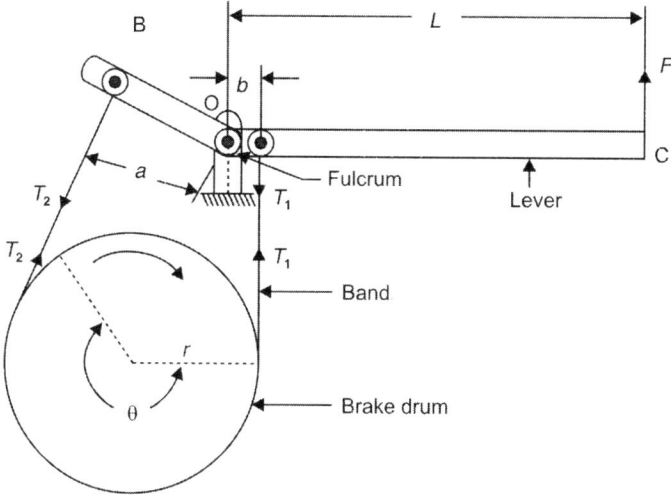

Fig. 9.10: Clockwise rotation

To apply the brake, the band is tightened on the drum if:

(a) Force F is applied in the downward direction when $a > b$

(b) Force F is applied in the upward direction when $a < b$.

When the drum rotates in the clockwise direction, the end of band attached at A will be tight with tension T_1 and the end of the band attached at B will be slack with tension T_2 as shown in Fig. 9.10.

Taking moments of all forces about pivot O

$$F \times L + T_1 \times b = T_2 \times a$$

or $\qquad\qquad F \times L = T_2 \times a - T_1 \times b$

or $\qquad\qquad\qquad F = \dfrac{T_2 \times a - T_1 \times b}{L}$ $\qquad\qquad$ ...(9.13)

As T_1 is always greater than T_2, force F will be positive if $T_2 \times a > T_1 \times b$

or $$\frac{T_1}{T_2} < \frac{a}{b} \qquad \qquad ...(9.14)$$

Force F will be zero or negative if $\dfrac{T_2}{T_1} \leq \dfrac{b}{a}$.

This is the condition of self-locking.

Figure 9.11 shows a differential band brake in which brake drum is rotating anti-clockwise. The end of the band connected to B will be tight with tension T_1, whereas the end of the band connecting to A will be slack with tension T_2.

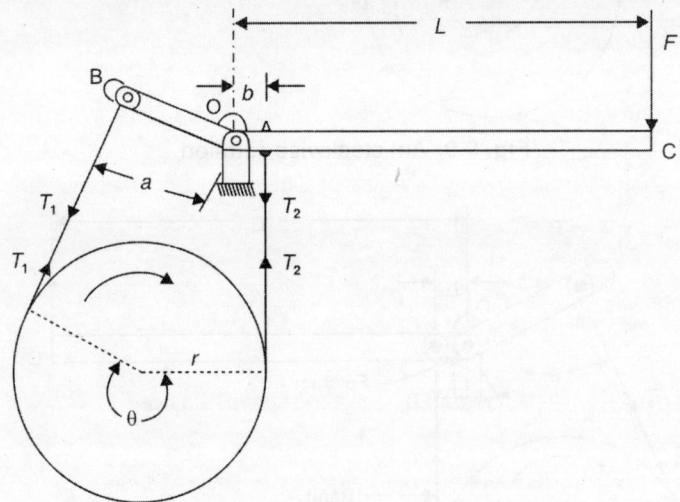

Fig. 9.11: Anticlockwise rotation (differential band brake)

Taking the moments about the fulcrum O,

$$F \times L + T_2 \times b = T_1 \times a$$

or $$F \times L = T_1 \times a - T_2 \times b$$

or $$F = \frac{T_1 \times a - T_2 \times b}{L}$$

For self-locking of the brake, the force F should be zero or negative. But force F will be zero or negative if

$$T_1 \times a \leq T_2 \times b$$

or $$\frac{T_1}{T_2} \leq \frac{b}{a} \qquad \qquad ...(9.15)$$

(i) If either a or b is made zero, a simple band brake is obtained. Simple band brake can neither have self-energising properties nor it can be self-locked.

(ii) The brake is more effective when maximum braking force is applied with the least effort F.

(iii) Self-locking brakes are used in hoists and conveyers where motion is permissible in one direction only.

Example 9.4: A simple band brake shown in Fig. 9.12 is applied to a shaft carrying a flywheel of mass 250 kg and of radius of gyration 300 mm. The shaft speed is 200 rpm. The drum diameter is 200 mm and the coefficient of friction 0.25. The dimensions a and l are 100 mm and 280 mm respectively and the angle $\beta = 135°$. Determine:

   (i) The brake torque when a force of 120 N is applied at the lever end

   (ii) The number of turns of the flywheel before it comes to rest

   (iii) The time taken by the flywheel to come to rest.

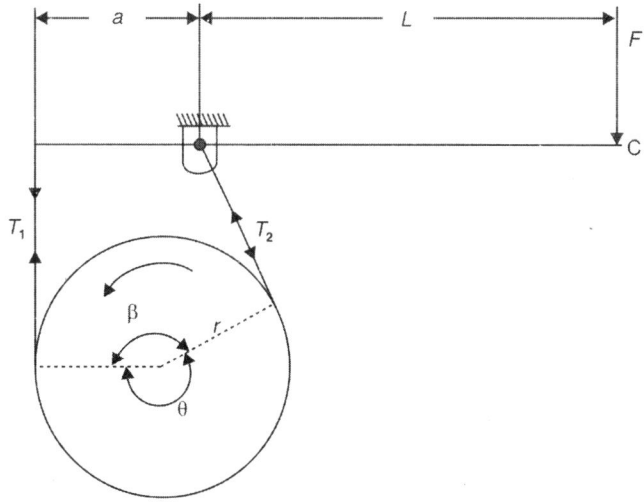

Fig. 9.12

Solution:
$$m = 250 \text{ kg, } \mu = 0.25, k = 300 \text{ mm} = 0.3 \text{ m}$$
$$r = 100 \text{ mm} = 0.1 \text{ m, } a = 100 \text{ mm} = 0.1 \text{ m}$$
$$\beta = 135°, l = 280 \text{ mm} = 0.28 \text{ m}$$

(i) Angle of contact $\theta = 360° - \beta = 360° - 135° = 225°$

Tension ratio $\dfrac{T_1}{T_2} = e^{\mu\theta} = e^{0.25 \times 225 \times \frac{\pi}{180}} = 2.67$ ...(i)

Taking moments about the fulcrum O,
$$F \cdot l = T_1 \cdot a$$
$$120 \times 0.28 = T_1 \times 0.10$$
$$T_1 = 336 \text{ N}$$

From Eq. (i),
$$T_1 = 2.67 T_2$$
$$336 = 2.67 \times T_2$$
$$T_2 = 125.84 \text{ N}$$

Braking torque T_B is given as
$$T_B = (T_1 - T_2)\, r$$
$$= (336 - 125.84) \times 0.10 = 21.01 \text{ N·m}$$

(ii) Kinetic energy of flywheel,

$$KE = \frac{1}{2}I\omega^2 = \frac{1}{2}mK^2\omega^2$$

$$= \frac{1}{2} \times 250 \times (0.30)^2 \left(\frac{2\pi \times 200}{60}\right)^2 = 4934.80 \text{ N·m}$$

Let the kinetic energy of the flywheel be used in n revolutions by the braking torque. So,

$$KE \text{ of flywheel} = T_B\omega$$

$$4934.8 = 21.01 \times \omega = 21.01 \times 2\pi n \ (\because \omega = 2\pi n)$$

where n = number of revolutions = 37.38 revolutions

(iii) For uniform retardation, average speed $= \dfrac{200 + 0}{2} = 100$ rpm

Time taken $= \dfrac{n}{N} = \dfrac{37.38}{100} = 0.3738$ minutes

Example 9.5: A simple band brake is operated by a lever of length 500 mm long. The brake drum has a diameter of 500 mm and the brake band embraces 5/8 of the circumference. One end of the band is attached to the fulcrum of the lever while the other is attached to a pin on the lever 100 mm from the fulcrum. If the effort applied to the end of the lever is 2000 N and the coefficient of friction is 0.25, find the maximum braking torque on the drum.

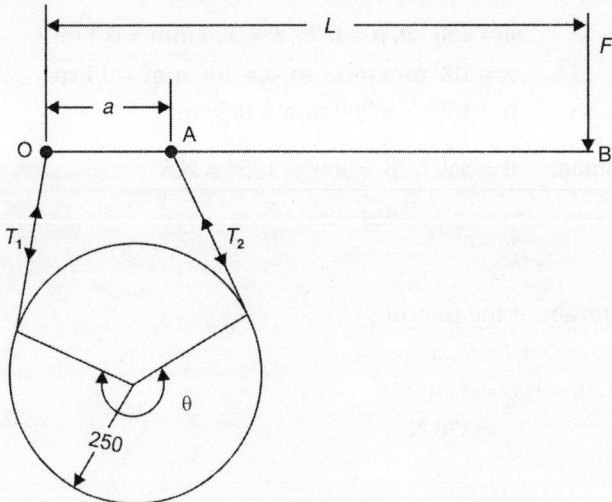

Fig. 9.13

Solution: Length of lever, $L = 500$ mm $= 0.50$ m, $a = 0.10$ m

Diameter of the drum, $d = 500$ mm

Radius, $r = 250$ mm $= 0.25$ m

Coefficient of friction, $\mu = 0.25$

Angle of contact $\theta = \dfrac{5}{8} \times 2\pi = 3.927$ rad

External applied force $F = 2000$ N

$$\frac{T_1}{T_2} = e^{\mu\theta} = e^{0.25 \times 3.927} = 2.669 \tag{i}$$

Taking moment about the fulcrum O,

$$F \times L = T_2 \times a$$
$$2000 \times 0.50 = T_2 \times 0.10$$
$$T_2 = 10000 \text{ N}$$

From Eq. (i)

$$T_1 = T_2 \times 2.669 = 10000 \times 2.669$$
$$= 26690 \text{ N}$$

Braking torque, $T_B = (T_1 - T_2)\, r$
$$= (26690 - 10000) \times 0.25$$
$$= 4172.5 \text{ N·m}$$

Example 9.6: A differential band brake has a drum of diameter 800 mm. The two ends of the band are fixed to the pins on the opposite sides of the fulcrum of the lever at distances of 40 mm and 200 mm from the fulcrum. The angle of contact is 270° and the coefficient of friction 0.2. Determine the brake torque when a force of 600 N is applied to the lever at a distance of 800 mm from the fulcrum.

Solution:

$F = 600$ N·$l = 800$ mm, $r = 400$ mm, $\theta = 270°$, $\mu = 0.2$
Let, $a = 40$ mm and $b = 200$ mm
As $a < b$, F must be upwards to apply the brake.
For counterclockwise rotation of the drum,

$$\frac{T_1}{T_2} = e^{\mu\theta} = e^{0.2 \times 270 \times \frac{\pi}{180}} = 2.57$$

Taking moments about the fulcrum O,
$$Fl + T_1 a - T_2 b = 0$$
$$600 \times 800 + 2.57\, T_2 \times 40 - T_2 \times 200 = 0$$

or
$$600 \times 800 = T_2\, (200 - 2.57 \times 40)$$
$$T_2 = 4938 \text{ N}, T_1 = 4938 \times 2.57 = 12691 \text{ N}$$

Braking torque, $T_B = (T_1 - T_2)\, r$
$$= (12691 - 4938) \times 0.4 = 3101 \text{ N·m}$$

For clockwise rotation of the drum,

$$\frac{T_2}{T_1} = 2.57$$

Taking moments about 0,
$$600 \times 800 + T_1 \times 40 - 2.57\, T_1 \times 200 = 0$$

or
$$T_1 = 1012.7 \text{ N and } T_2 = 1012.7 \times 2.57 = 2602.5 \text{ N}$$
$$T_B = (2602.5 - 1012.7) \times 0.4 = 636 \text{ N·m}$$

Thus, the brake is more effective when the drum rotates counterclockwise.

9.5 BAND AND BLOCK BRAKE

A band and block brake consists of a number of blocks or other material lined inside a flexible steel band (Fig. 9.14). The friction between the blocks and the drum provides the braking action. When the brake is applied, blocks are pressed against the external surface of the drum. This brake is more effective and economical as wooden blocks have higher coefficient of friction and can be easily replaced on being worn out.

Let there be n blocks, each subtending an angle 2θ at the centre of the drum. If the drum rotates in clockwise direction, the end of band attached to A will be tight with tension T_n and other side will be slack with tension T_0. The frictional force on the blocks acts in clockwise direction.

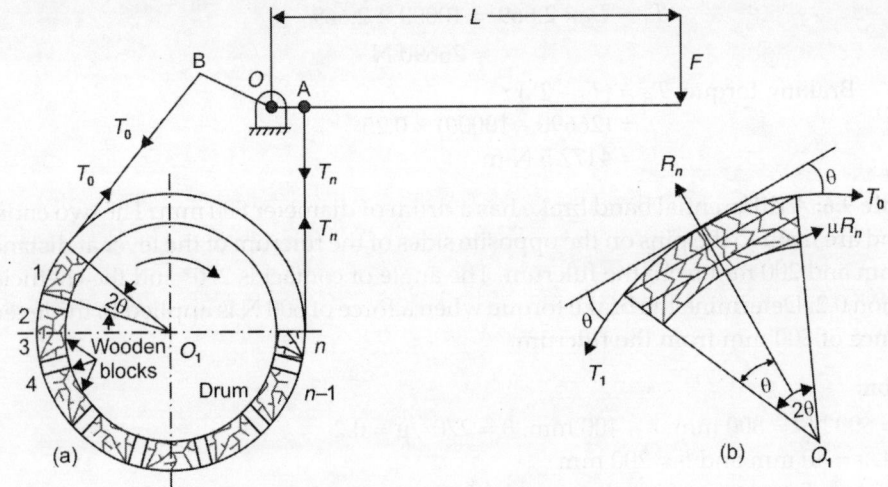

Fig. 9.14: Band and block brake

Let T_0 = tension on the slack side

   T_1 = tension on the tight side after one block

   T_2 = tension on the tight side after two blocks

   T_n = tension on the tight side after n blocks

   μ = coefficient of friction

   R_n = normal reaction on the block

For equilibrium of one block of the brake.

Resolving the forces tangentially,

$$T_1 \cos\theta - T_0 \cos\theta = \mu R_N \quad (\because T_1 \text{ is more than } T_0)$$

or $$(T_1 - T_0)\cos\theta = \mu R_N \qquad \qquad ...(9.15a)$$

Resolving the forces radially,

$$T_1 \sin\theta + T_0 \sin\theta = R_N$$

or $$(T_1 + T_0)\sin\theta = R_N \qquad \qquad ...(9.15b)$$

Dividing Eq. (i) by Eq. (ii),

$$\frac{(T_1 - T_0)\cos\theta}{(T_1 + T_0)\sin\theta} = \frac{\mu R_N}{R_N}$$

or $\dfrac{(T_1 - T_0)}{(T_1 + T_0)} \times \dfrac{1}{\tan\theta} = \mu$

or $\dfrac{(T_1 - T_0)}{(T_1 + T_0)} = \dfrac{\mu\tan\theta}{1}$

or $\dfrac{(T_1 - T_0) + (T_1 + T_0)}{(T_1 - T_0) - (T_1 + T_0)} = \dfrac{\mu\tan\theta + 1}{\mu\tan\theta - 1}$

$$-\dfrac{2T_1}{2T_0} = -\dfrac{1 + \mu\tan\theta}{1 - \mu\tan\theta}$$

$$\dfrac{T_1}{T_0} = \dfrac{1 + \mu\tan\theta}{1 - \mu\tan\theta}$$

Similarly, $\dfrac{T_2}{T_1} = \dfrac{1 + \mu\tan\theta}{1 - \mu\tan\theta}$ and so on.

For the nth block, the ratio of tensions will be

$$\dfrac{T_n}{T_n - 1} = \dfrac{1 + \mu\tan\theta}{1 - \mu\tan\theta}$$

Hence, the ratio of tensions in the tight and slack sides of the complete band and block brake

$$\dfrac{T_n}{T_0} = \dfrac{T_n}{T_n - 1} \times \dots \times \dfrac{T_3}{T_2} \times \dfrac{T_2}{T_1} \times \dfrac{T_1}{T_0}$$

$$= \left(\dfrac{1 + \mu\tan\theta}{1 - \mu\tan\theta}\right) \times \dots \times \left(\dfrac{1 + \mu\tan\theta}{1 - \mu\tan\theta}\right) \times \left(\dfrac{1 + \mu\tan\theta}{1 - \mu\tan\theta}\right) \times \left(\dfrac{1 + \mu\tan\theta}{1 - \mu\tan\theta}\right)$$

$$= \left(\dfrac{1 + \mu\tan\theta}{1 - \mu\tan\theta}\right)^n \qquad \qquad \dots(9.16)$$

Now the braking torque on the drum will be

$$T_B = (T_n - T_0) \times r$$

where r = effective radius of band.

Example 9.7: A band and block brake, having 14 blocks each of which subtends an angle of 15° at the centre, is applied to a drum with effective diameter 1 m. The drum and flywheel mounted on the same shaft has a mass of 2000 kg and a combined radius of gyration of 500 mm. The two ends of the band are attached to pins on opposite sides of the brake lever at distances of 30 mm and 120 mm from the fulcrum. If a force of 200 N is applied at a distance of 750 mm from the fulcrum, find:

(a) Maximum braking torque
(b) Angular retardation of the drum
(c) Time taken by the system to come to rest from the speed of 360 rpm.

Solution: $n = 14$, $2\theta = 15° \Rightarrow \theta = 7\dfrac{1}{2}°$, $m = 2000$ kg

$d = 1$ m, $r = 0.5$ m, $k = 500$ mm $= 0.5$ m, $\mu = 0.25$

$L = 750$ mm $= 0.75$ m, $N = 360$ rpm, $F = 200$ N

(a) The braking torque T_B will be maximum when the following conditions are satisfied:

    i. $b > a$, i.e. $OA > OB$

    ii. Brake drum rotates anticlockwise

    iii. The applied force acts upwards

Here T_2 and T_1 are the tensions in the band on slack side and tight side respectively. The system is shown in Fig. 9.15

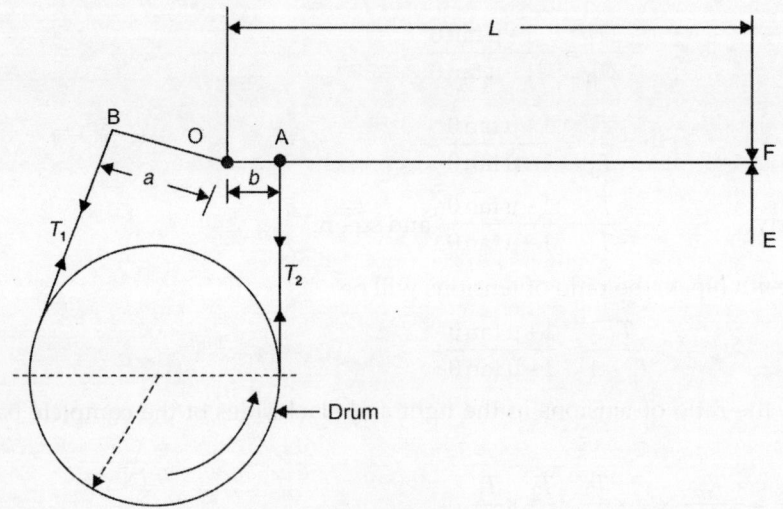

Fig. 9.15

$$\frac{T_1}{T_2} = \left(\frac{1+\mu\tan\theta}{1-\mu\tan\theta}\right)^n$$

$$= \left(\frac{1+0.25\tan 7.5°}{1-0.25\tan 7.5°}\right)^{14} = 2.5135 \qquad \text{...(i)}$$

Taking moments about O,

$$FL + T_1\, a = T_2 b$$
$$200 \times 0.75 + T_1 \times 0.03 = T_2 \times 0.12$$
$$150 + 0.03\, T_1 = 0.12 T_2$$
$$15000 + 3 T_1 = 12 T_2$$
$$5000 + T_1 = 4 T_2 \qquad \text{...(ii)}$$

Solving Eqs (i) and (ii)

$$T_2 = 3363.60 \text{ N}$$
$$T_1 = 8454.40$$

Braking torque T_B is given by

$$T_B = (T_1 - T_2)r = (8454.4 - 3363.6) \times 0.50 = 2545.4 \text{ N–m}$$
$$T_B = I\alpha \text{ and } I = mK^2$$

where, I = moment of inertia

    α = angular retardation

(b) $$I = 2000 (0.50)^2 = 500 \text{ kg·m}^2$$
$$\therefore \quad 2545.4 = 500 \, \alpha$$
$$\alpha = 5.0908 \text{ rad/s}^2$$

(c) We know that, $\omega_2 = \omega_1 + \alpha t = \omega_1 - \alpha t$ (α is negative)
$$\omega_2 = 0$$
$$\omega_1 = \alpha t \quad \text{or} \quad t = \frac{\omega_1}{\alpha} = \frac{2\pi N}{60 \times \alpha}$$

or
$$t = \frac{2\pi \times 360}{60 \times 5.0908} = 7.41 \text{ s.}$$

Example 9.8: In a band and block brake having 10 blocks, each block subtends an angle of 15° at the centre of wheel. Determine the maximum force required at the end of the lever for the brake to absorb 250 kW of power at 280 rpm. The effective diameter of the drum is 840 mm. Take $\mu = 0.35$, $a = 200$ mm, $b = 40$ mm and $L = 300$ mm.

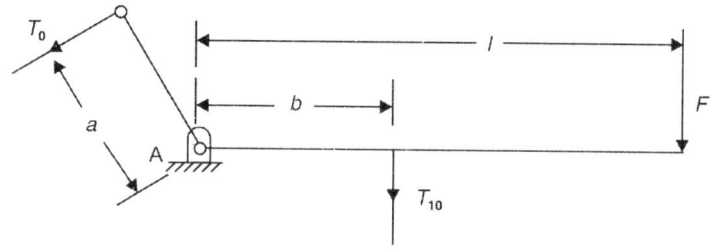

Fig. 9.16

Solution: $n = 10$, $2\theta = 15°$ or $\theta = \dfrac{15°}{2} = 7.5°$

$\mu = 0.35$, $P = 250$ kW $= 250 \times 10^3$ W

$N = 280$ rpm, $\omega = \dfrac{2\pi N}{60} = \dfrac{2\pi \times 280}{60}$

$d = 0.84$ m, $a = 200$ mm, $b = 40$ mm, $L = 300$ mm

We know that

$$\frac{T_{10}}{T_0} = \left[\frac{1 + \mu \tan\theta}{1 - \mu \tan\theta}\right]^n = \left[\frac{1 + 0.35 \tan 7.5°}{1 - 0.35 \tan 7.5°}\right]^{10}$$

$$= \left[\frac{1.046078}{.95392}\right]^{10} = (1.0966)^{10} = 2.5148$$

or
$$T_{10} = 2.52 T_0$$

Since $a > b$, F will be downward for clockwise rotation.

$$\text{Power, } P = T_B \times \omega = \left(T_{10} - T_0\right) \times \frac{d}{2} \times \left(\frac{2\pi \times 280}{60}\right)$$

or
$$250 \times 1000 = \left(2.52 \, T_0 - T_0\right) \times \frac{0.84}{2} \times \left(\frac{2\pi \times 280}{60}\right)$$

or
$$T_0 = 13355.5 \text{ N}$$
$$T_{10} = 33655.88 \text{ N}$$

Taking moment about A, $\Sigma M_A = 0$

$F \times l + T_{10} \times b - T_0 \times a = 0$

or $\quad F = \dfrac{T_0 \times a - T_{10} \times b}{l} = \dfrac{13355.5 \times 0.2 - 33655.88 \times 0.4}{0.3}$

$F = 4416$ N.

9.6 INTERNAL EXPANDING SHOE BRAKE

Band brakes have been replaced by internal expanding shoe brake as they are enclosed in a drum and not exposed to dust and water. Internal expanding shoe brake provides large braking power as they have at least one self-energising shoe per wheel producing tremendous friction. Internal expanding shoe brakes shown in Fig. 9.17 are commonly used in motor cars and light trucks.

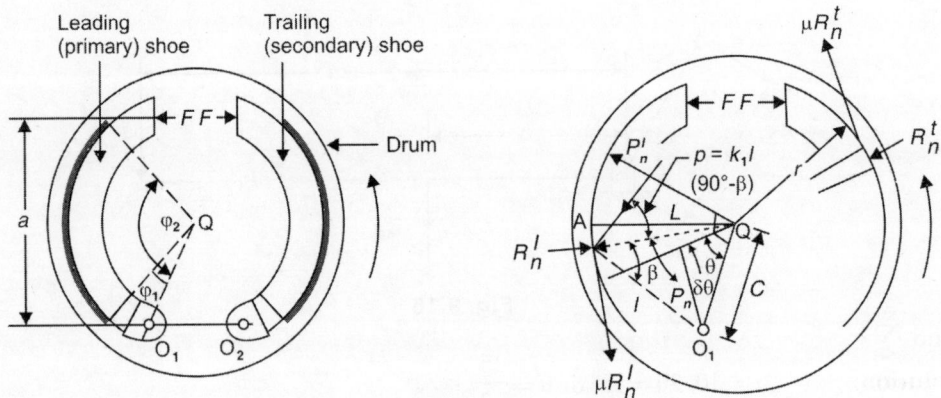

Fig. 9.17

It consists of two semicircular shoes s_1 and s_2 whose outer surfaces are lined with friction material like ferodo. Each shoe is pivoted at one end about a fixed fulcrum O_1 and O_2 and other end of each is in contact with a cam. When brakes are applied, cam rotates or two pistons in a common hydraulic cylinder are actuated. Shoes press against the inner surface of drum.

For anticlockwise rotation of the drum, the left shoe is known as the leading or primary shoe and the right shoe as the trailing or secondary shoe.

The pressure p at any point A on the contact surface will be proportional to its distance l from the pivot.

Considering the leading shoe, $p \propto l = k_1 l$, where k_1 is a constant.

The direction of p is perpendicualr to $O_1 A$.

The normal pressure, $P_n = k_1 l \cos(90° - \beta) = k_1 l \sin \beta$ $\qquad$ ...(9.17)

$\qquad\qquad\qquad = k_1 c \sin \theta \quad (O_1 L = 1 \sin \beta = c \sin \theta)$

$\qquad\qquad\qquad = k_2 \sin \theta$

$\qquad\qquad\qquad$ (where k_2 is another constant $= k_1 c$)

P_n is maximum when $\theta = 90°$.

Let P_n^l = maximum intensity of normal pressure on the leading shoe.

$$P_n \max = p_n^l = k_2 \sin 90° = k_2$$

or
$$p_n = P_n^l = \sin \theta \qquad \qquad ...(9.18)$$

Let ω = width of brake lining, μ = coefficient of friction

Consider a small element of brake lining on the leading shoe that makes an angle $\delta\theta$ at the centre.

Normal reaction on the elemental surface $= R_n^l =$ area × pressure $= (r\delta\theta\omega)p_n = r\delta\theta\omega P_n^l \sin\theta$

Taking moments about the fulcrum O_1,

$$Fa - \sum_{\varphi_1}^{\varphi_2} R_n^l c \sin\theta + \sum_{\varphi_1}^{\varphi_2} \mu R_n^l (r - \cos\theta) = 0 \qquad \qquad ...(9.19)$$

where

$$\sum_{\varphi_1}^{\varphi_2} R_n^l c \sin\theta + \int_{\varphi_1}^{\varphi_2} rcwP_n^l \sin^2\theta \, d\theta = \int_{\varphi_1}^{\varphi_2} rcwP_n^l \frac{1}{2}(1 - \cos 2\theta) \, d\theta$$

$$= rcw \, P_n^l \frac{1}{2} \left[\theta - \frac{\sin 2\theta}{2} \right]_{\varphi_1}^{\varphi_2}$$

$$= \frac{rc \, wP_n^l}{4} \left(2\varphi_2 - 2\varphi_1 - \sin 2\varphi_2 + \sin 2\varphi_1 \right)$$

and $\displaystyle \sum_{\varphi_1}^{\varphi_2} \mu R_n^l (r - c\cos\theta) = \int_{\varphi_1}^{\varphi_2} \mu r^2 wP_n^l \sin\theta \, d\theta - \int_{\varphi_1}^{\varphi_2} \mu r \, cwP_n^l \sin\theta \cos\theta \, d\theta$

$$= \mu r^2 wP_n^l (-\cos\theta)_{\varphi_1}^{\varphi_2} - \int_{\varphi_1}^{\varphi_2} \mu rcwP_n^l \frac{1}{2} \sin 2\theta \, d\theta$$

$$= \mu r^2 wP_n^l (\cos\varphi_1 - \cos\varphi_2) - \mu rcwP_n^l \frac{1}{2} \left(\frac{-\cos 2\theta}{2} \right)_{\varphi_1}^{\varphi_2}$$

$$= \frac{\mu r\omega P_n^l}{4} \left[4r(\cos\varphi_1 - \cos\varphi_2) - c(\cos 2\varphi_1 - \cos 2\varphi_2) \right]$$

Taking moments about the fulcrum O_2 for the trailing shoe,

$$Fa - \sum_{\varphi_1}^{\varphi_2} R_n^t c \sin\theta - \sum_{\varphi_1}^{\varphi_2} \mu R_n^t (r - c\cos\theta) = 0$$

where $\displaystyle \sum_{\varphi_1}^{\varphi_2} R_n^t c \sin\theta = \frac{rcwP_n^t}{4} \left[2\varphi_2 - 2\varphi_1 - \sin 2\varphi_2 + \sin 2\varphi_1 \right]$

and $\displaystyle \sum_{\varphi_1}^{\varphi_2} \mu R_n^t (r - \cos\theta) = \frac{\mu rwP_n^t}{4} \left[4r(\cos\varphi_1 - \cos\varphi_2) - c(\cos 2\varphi_1 - \cos 2\varphi_2) \right]$

Therefore, P_n^l and P_n^t, the maximum pressure intensities on the leading and the trailing shoes, can be determined.

Braking torque, $T_B = \sum_{\varphi_1}^{\varphi_2} \mu R_n^l r + \sum_{\varphi_1}^{\varphi_2} \mu R_n^t r$

$$= \int_{\varphi_1}^{\varphi_2} \mu r^2 w P_n^l \sin\theta \, d\theta + \int_{\varphi_1}^{\varphi_2} \mu r^2 w P_n^t \sin\theta \, d\theta$$

$$= r^2 \mu w \left(P_n^l + P_n^t \right) (-\cos\varphi)_{\varphi_1}^{\varphi_2}$$

$$= r^2 \mu w \left(P_n^l + P_n^t \right) (\cos\varphi_1 - \cos\varphi_2) \qquad \ldots(9.20)$$

For the same applied force F on each shoe, P_n^l is not equal to P_n^t and $P_n^l > P_n^t$. Usually, more than 50% of the total braking torque is supplied by the leading shoe.

The leading shoe is self-energizing, whereas the trailing shoe is not. This is because the friction forces acting on the leading shoe help the applied force F, and that on the trailing shoe oppose it. On reversing the direction of drum rotation, the right shoe will become self-energizing whereas the left will not be so any longer.

If the moment of frictional force is greater than the moment of normal reaction, actuating force F will be negative and the brake will become self locking.

Example 9.9: The following data refer to an internal expanding shoe brake

Force F on each shoe = 180 N

Coefficient of friction, $\mu = 0.3$

Internal radius of the brake drum, $r = 150$ mm

Width of the brake lining, $w = 40$ mm

Distances: $a = 200$ mm, $c = 120$ mm

Angles: $\varphi_1 = 30°$, $\varphi_2 = 135°$

Determine the braking torque applied when the drum rotates: (i) counterclockwise, and (ii) clockwise.

Solution:

(i) For the leading shoe,

$$Fa - \int_{\varphi_1}^{\varphi_2} R_n^l c \sin\theta + \int_{\varphi_1}^{\varphi_2} \mu R_n^l (r - \cos\theta) = 0$$

$$180 \times 0.2 - \frac{0.15 \times 0.12 \times 0.04 \times P_n^l}{4} \left(2 \times 135 \times \frac{\pi}{180} - 2 \times 30 \times \frac{\pi}{180} - \sin 270° + \sin 60° \right)$$

$$+ \frac{0.3 \times 0.15 \times 0.04 \times P_n^t}{4} \left[4 \times 0.15 (\cos 30° - \cos 135°) - 0.12 (\cos 60° - \cos 270°) \right] = 0$$

$36 - 0.000996 P_n^l + 0.000398 P_n^t = 0$, $P_n^t = 60201$ N/m^2

For the trailing shoe,

$= 36 - 0.000996 P_n^l - 0.000398 P_n^t = 0$, $P_n^l = 25825$ N/m^2

Braking torque, $T_B = r^2 \mu w \left(P_n^l + P_n^t \right) (\cos\varphi_1 - \cos\varphi_2)$

$$= (0.15)^2 \times 0.3 \times 0.04 (60201 + 25825)(\cos 30° - \cos 135°)$$

$$= 36.54 \text{ N·m}$$

(ii) When the rotation is reversed, P_n^l and P_n^t are interchanged and thus, the braking torque is the same.

9.7 EFFECT OF BRAKING

The retardation of vehicle on applying the brake can be determined by reducing the system to an equivalent static system by applying D'Alembert's principle.

Consider a vehicle moving up an inclined plane shown in Fig. 9.18.

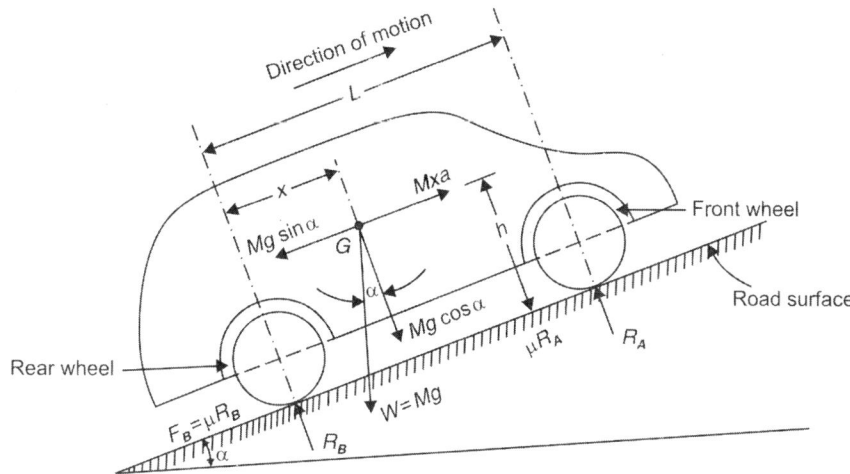

Fig. 9.18

Let α = angle of inclination of the plane with the horizontal

W = weight of the car = $M \times g$

M = mass of the vehicle

x = distance of CG of the car from the rear axle

h = height of CG of car from inclined surface.

L = wheel-base of the car (i.e. distance between the centres of the rear and front wheels)

R_A = reaction between front wheel and ground.

R_B = reaction between rear wheel and ground.

μ = coefficient of friction between the tyres and road surface.

a = retardation of vehicle.

Brakes are applied to rear wheels only

For equilibrium,

$$R_A + R_B = Mg \cos\alpha$$
$$\mu R_B + Mg \sin\alpha = Ma$$

Taking moments about G, the centre of mass of the vehicle,

$$R_B x + \mu R_B \times h - (Mg \cos\alpha - R_B) \tag{9.20a}$$

From Eq. (*i*) $R_A = Mg \cos\alpha - R_B$

∴ Eq. (9.20a) becomes

$$R_B x + \mu R_B \times h - \left(Mg\cos\alpha - R_B\right)(L-x) = 0$$

or $R_B\left(x + \mu h + L - x\right) = Mg\cos\alpha(L-x)$

or $R_B = \dfrac{Mg\cos\alpha(L-x)}{L+\mu h}$

or $\dfrac{\mu Mg\cos\alpha(L-x)}{L+\mu h} + Mg\sin\alpha = Ma$

or
$$a = g\cos\alpha\left[\frac{\mu(L-x)}{L+\mu h} + \tan\alpha\right] \qquad \dots(9.21)$$

On a level road, $\alpha = 0$, and so

$$a = g\,\frac{\mu(L-x)}{L+\mu h} \qquad \dots(9.22)$$

When the vehicle moves down a plane,

$$a = g\cos\alpha\left[\frac{\mu(L-x)}{L+\mu h} - \tan\alpha\right] \qquad \dots(9.23)$$

Brakes applied to front wheels only

$$R_A + R_B = Mg\cos\alpha \qquad \dots(9.23a)$$
$$\mu R_A + Mg\sin\alpha = Ma \qquad \dots(9.23b)$$

Taking moments about G,

$$R_B x + \mu R_A \times h - R_A\left(L-x\right) = 0 \qquad \dots(9.23c)$$

From Eqs (9.23a) and (9.23b),

$$\left(Mg\cos\alpha - R_A\right)x + \mu R_A \times h - R_A\left(L-x\right) = 0$$

or $Mgx\cos\alpha = R_A\left(x - \mu h + L - x\right),\ R_A = \dfrac{Mgx\cos\alpha}{L - \mu h}$

Therefore, Eq. (9.23b) becomes $\mu = \dfrac{Mgx\cos\alpha}{L-\mu h} + Mg\sin\alpha = Ma$

or
$$a = g\cos\alpha\left[\frac{\mu x}{L-\mu h} + \tan\alpha\right] \qquad \dots(9.24)$$

On a level road, $\alpha = 0$, and therefore,

$$a = g\,\frac{\mu x}{L-\mu h} \qquad \dots(9.25)$$

When vehicle moves down a plane,

$$a = g\cos\alpha\left(\frac{\mu x}{L-\mu h} - \tan\alpha\right) \qquad \dots(9.26)$$

Brakes applied to all four wheels

$$R_A + R_B = Mg \cos \alpha \qquad \qquad ...(9.26a)$$

$$\mu R_A + \mu R_B + Mg \sin \alpha = Ma \qquad \qquad ...(9.26b)$$

or $\quad \mu(R_A + R_B) + Mg \sin \alpha = Ma$

or $\quad \mu Mg \cos \alpha + Mg \sin \alpha = Ma$

or $\qquad \qquad a = g \cos \alpha (\mu + \tan \alpha) \qquad \qquad ...(9.27)$

On a level load, $\alpha = 0$. Therefore,

$$a = g\mu \qquad \qquad ...(9.28)$$

When vehicle moves down the plane,

$$a = g \cos \alpha (\mu - \tan \alpha) \qquad \qquad ...(9.29)$$

Example 9.10: A car moving along a level road is having the following data:

Wheel base of car = 2.85 m

Height of CG from road surface = 600 mm

Distance of CG from rear axle = 1.2 m

Speed on level road= 60 km/h

Coefficient of friction between tyres and road = 0.6

Determine the minimum distance at which the car may be stopped when brakes are applied to: (i) to the rear wheels, (ii) the front wheels, and (iii) to all the four wheels.

Solution:

Wheel base L = 2.85 m, h = 600 mm = 0.6 m, x = 1.2 m, α = 0,

$\mu = 0.6$ and initial speed $u = 60$ km/h = $\dfrac{60 \times 1000}{60 \times 60}$ m/s = 16.67 m/s, final speed, $v = 0$

Let S = distance travelled by the car before coming to rest.

(i) The value of 'a' when brakes are applied to rear wheels, is given as:

$$\text{Retardation, } a = \frac{(L - x)\,\mu \times g \times \cos \alpha}{(\mu h + L)} + g \sin \alpha$$

$$= \frac{(L - x)\,\mu \times g \times \cos \alpha}{(\mu h + L)} + g \sin 0 \qquad (\because \alpha = 0 \text{ for level road})$$

$$= \frac{(L - x)\,\mu \times g}{(\mu h + L)} \qquad (\because \cos 0 = 1 \text{ and } \sin 0 = 0)$$

$$= \frac{(2.85 - 1.2) \times 0.6 \times 9.81}{(0.6 \times 0.6 + 2.85)} = 3.025 \text{ m/s}^2$$

$\therefore \qquad \qquad \alpha = -3.025 \text{ m/s}^2$

Now using equation, $v^2 - u^2 = 2 \times a \times S$, we get

$$0^2 - 16.67^2 = 2 \times (-3.025) \times S,$$

$(\because v = 0,\ u = 16.67 \text{ and } a = -3.025 \text{ m/s}^2)$

or $\qquad\qquad 16.67^2 = 2 \times 3.025 \times S$

or $\qquad\qquad S = \dfrac{16.67^2}{2 \times 3.025} = 45.93$ m.

(ii) The retardation a when brakes are applied to the front wheels, is given by

$$a = \frac{\mu \times x \times g \cos \alpha}{(L - \mu h)} + g \sin \alpha$$

$$= \frac{0.6 \times 1.2 \times 9.81 \times \cos 0°}{(2.85 - 0.6 \times 0.6)} + 9.81 \sin 0° \qquad\qquad (\because \alpha = 0)$$

$$= \frac{0.6 \times 1.2 \times 9.81}{(2.85 - 0.36)} = 2.836 \text{ m/s}^2$$

$$a = -2.836 \text{ m/s}^2 \qquad\qquad (\text{--ve sign is due to retardation})$$

Now using, $v^2 - u^2 = 2a \times S$,

$$0^2 - 16.67^2 = 2 \times (-2.836) \times S$$

or $\qquad\qquad S = -\dfrac{16.67^2}{2 \times (-2.836)} = 48.99$ m

(iii) The retardation a when the brakes are applied to all the four wheels, is given by

$$a = g \times (\mu \cos \alpha + \sin \alpha)$$
$$= 9.81 \, (0.6 \cos 0° + \sin 0°) \qquad\qquad [\text{for a } \textit{level road}, \alpha = 0]$$
$$= 9.81 \times 0.6 = 5.886 \text{ m/s}^2$$

$\therefore \qquad\qquad a = -5.886 \text{ m/s}^2$

Now using equation $v^2 - u^2 = 2aS$,

$$0^2 - 16.67^2 = 2 \times (-5.886) \times S$$

$\therefore \qquad\qquad S = -\dfrac{16.67^2}{2(-5.886)} = 23.6$ m.

9.8 DYNAMOMETERS

Dynamometer is a device used to measure the frictional torque obtained by applying the brake. It is a brake with additional device used to determine the power developed by the machine, while maintaining its speed at certain value.

Types of Dynamometers

There are mainly two types of dynamometers:

   (i) **Absorption dynamometers:** In absorption dynamometers, work done against friction is converted into heat while being measured. They are used to measure moderate powers only. For example, prony break dynamometer and rope brake dynamometer.

   (ii) **Transmission dynamometers:** In transmission dynamometers, available power is utilised after being measured. They are suitable to measure large powers. For example, belt transmission dynamometer and the torsion dynamometer.

9.8.1 Prony Brake Dynamometer

A prony brake dynamometer consists of two wooden blocks clamped together on a revolving pulley. The pulley is fixed to a shaft of an engine. The upper block carries a lever at the end of which a weight is suspended.

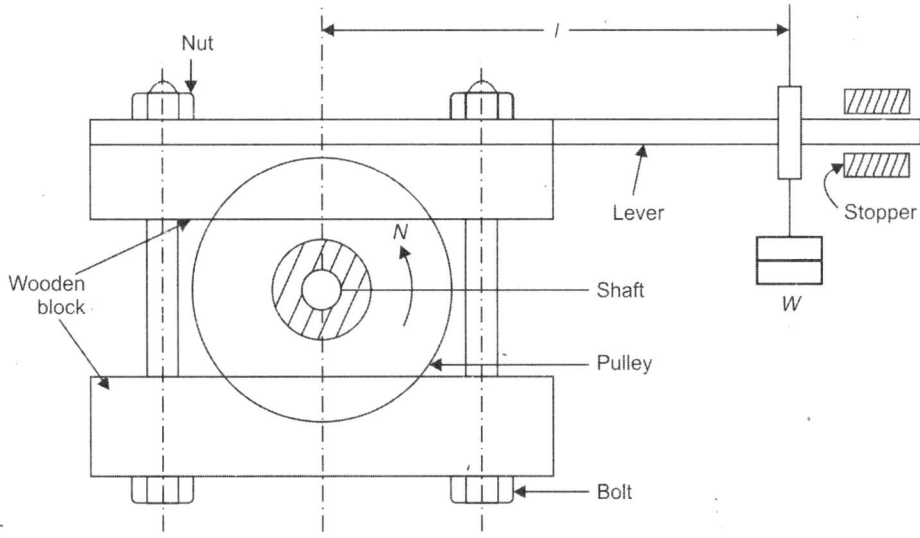

Fig. 9.19: Prony brake dynamometer

The friction between the blocks and pulley tends to rotate the blocks in the direction of rotation of the shaft. But suspended weight opposes this tendency. Each block embraces less than half of the pulley. The pressure on the pulley can be adjusted using nut and bolt until the engine runs at required speed. The arm of the lever remains horizontal in the equilibrium position for a certain weight W suspended at the end of lever.

Frictional torque $= W \times l = Mg \times l$

Power of the engine $= T \times W = Mgl \times \dfrac{2\pi N}{60} = MNK$ ...(9.30)

where $K = gl \times \dfrac{2\pi}{60} =$ constant for a particular brake.

From Eq. (9.30), it is clear that power of the engine is independent of size of the pulley and coefficient of friction.

9.8.2 Rope Brake Dynamometer

Rope brake dynamometer consists of a rope wound over the rim of a pulley keyed to the shaft of the engine whose power is to be measured. The upper end of the rope is attached to spring balance S and the lower end carries the dead weight. The ropes are spaced on the pulley across the width of the rim by three to four U-shaped wooden blocks which prevent the rope from slipping off the pulley.

To measure the power of an engine, the engine is made to run at a constant speed. For this, the torque transmitted by the engine must be equal to the frictional torque due to ropes.

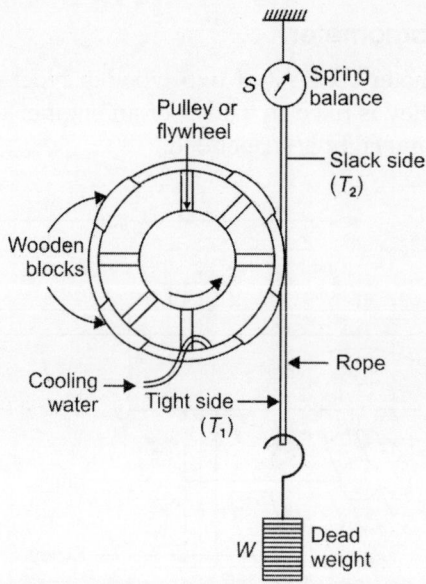

Fig. 9.20: Rope brake dynamometer

Let N = constant speed of the engine shaft

W = dead weight

S = spring balance reading

r = mean radius of the wheel

Net load on the pulley, $F_t = W - S = Mg - S$

Frictional torque $= F_t \times r = (Mg - S) r$ = torque transmitted (T)

$\therefore$ Power of the machine $= T \times w$

$$= (Mg - S)r \times \frac{2\pi N}{60} \qquad ...(9.31)$$

A cooling arrangement is required if the power of the engine is high as in that case heat produced due to friction between the rope and the pulley will be large.

A rope brake dynamometer is generally used to test the power of engine as it is easy to manufacture, inexpensive and requires no lubrication.

9.8.3 Belt Transmission Dynamometer

A belt transmission dynamometer measures the difference in tensions $(T_1 - T_2)$ on tight side (T_1) and slack side (T_2) when it is running from one pulley to another pulley. This dynamometer is also known as Tatham dynamometer.

It consists of a driving pulley A, driven pulley B and two intermediate pulleys 1 and 2 which are fixed to a lever having a fulcrum at the mid point of two pulley centres. The driving pulley A is rigidly fixed to the shaft of an engine whose power is to be measured. An endless belt runs over the driving and the driven pulleys through two intermediate pulleys. The weight of suspended mass at the end of the lever balances the difference in tensions $(T_1 - T_2)$. Balancing mass at the other end is used for initial equilibrium. Motion of the lever is restricted by two stops (s) one on each side.

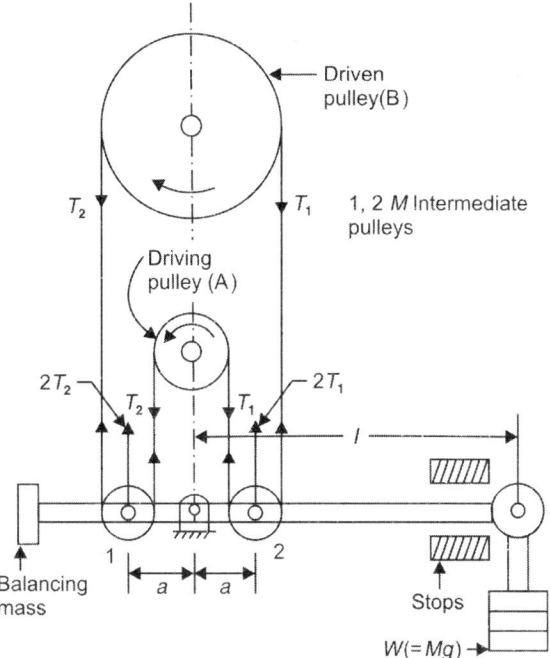

Fig. 9.21

Taking moment about fulcrum O,

$$Mgl + 2T_2 \times a - 2T_1 \times a = 0$$

or $\qquad Mgl - 2a(T_1 - T_2) = 0$

or $\qquad (T_1 - T_2) = \dfrac{Mgl}{2a}$ $\qquad\qquad\qquad\qquad\qquad$ (9.32)

If v is belt speed in m/s, D is diameter of pulley A and N is speed of engine shaft (rpm) then, Power of engine, $P = (T_1 - T_2)\, v$

or $\qquad\qquad\qquad\qquad P = (T_1 - T_2) \times \dfrac{\pi DN}{60}$ Watt

9.8.4 Torsion Dynamometer

It works on the principle that the torque transmitted is directly proportional to the angle of twist. Therefore, if angle of twist can be measured by some means, then torque transmitted can be calculated.

From torsion equation, $\dfrac{T}{J} = \dfrac{C\theta}{L}$

$$T = \dfrac{C\theta}{L} \times J \quad \left[\text{where } J = \dfrac{\pi D^4}{32} \text{ (for solid shaft)} \right]$$

$$= \dfrac{\pi(D_o^4 - D_i^4)}{32} \text{ (for hollow shaft)}$$

or $\qquad\qquad\qquad T = k\theta \quad$ where $k = \dfrac{C \times J}{L}$ = constant for a shaft

$\Rightarrow$ Torque transmitted $\propto$ angle of twist (θ)

In above expression,

    T = torque transmitted by the shaft

    C = modulus of rigidity of the shaft material (N/m²)

    θ = angle of twist (radian)

    J = polar moment of inertia (m⁴)

    L = length of shaft (m)

9.8.5 Bevis–Gibson Torsion Dynamometer

This consists of two similar discs A and B fixed on the shaft at points as far apart as possible, a lamp and a movable torque finder arrangement as shown in Fig. 9.22.

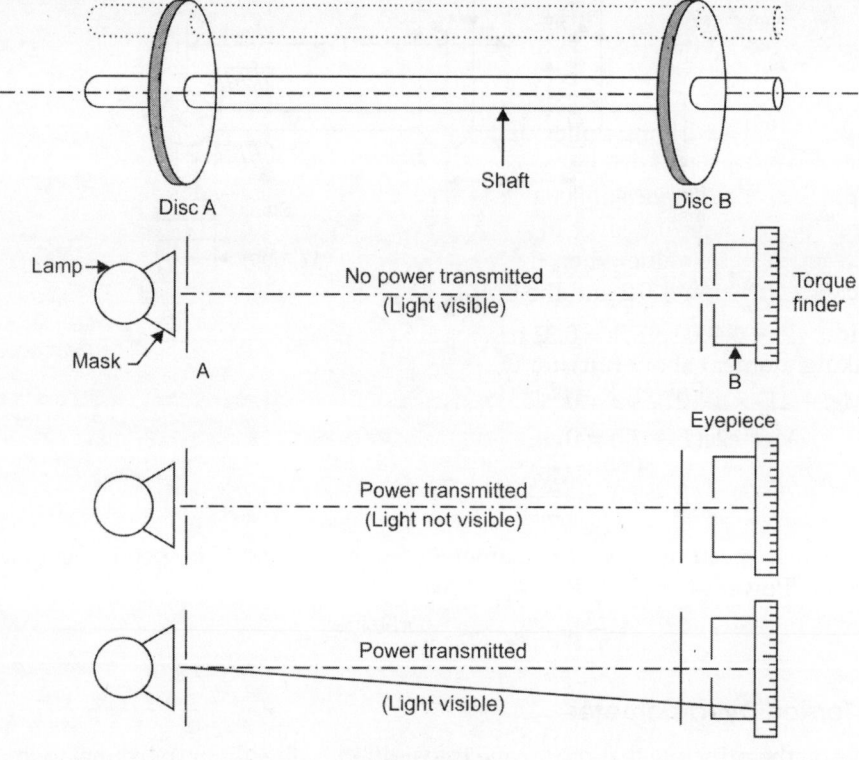

Fig. 9.22

The lamp is masked and fixed on the bearing of the shaft. The torque finder has an eyepiece capable of moving circumferentially. Each disc has a small radial slot near its periphery. Similar slots are also made at the same radius on the mask of the lamp and on the torque finder.

When the shaft rotates freely and does not transmit any torque, all the four slots are in a line and a ray of light from the lamp can be seen through the eyepiece after every revolution. When torque is transmitted, the shaft twists and the slot in the disc B shifts its position. The ray of light can no longer pass through the four slots. However, if the eyepiece is moved circumferentially by an amount equal to the displacement, the flash will again be visible once in each revolution of the shaft. The eyepiece is moved by a micrometer spindle. The angle of twist may be measured up to one hundredth of a degree.

Example 9.11: Calculate the power of an engine which is running at a constant speed of 500 rpm and carries a rope brake dynamometer. The dead weight on the engine and spring balance readings are 750 N and 150 N respectively. The diameter of flywheel is 2 m that of rope is 20 mm.

Solution: $N = 500$ rpm, $W = 750$ N, $S = 150$ N, $D = 2$ m, $d = 20$ mm $= 0.02$ m.

$$\text{Power of the engine } = (W - S)\,r \times \frac{2\pi N}{60}$$

Here, mean radius $r = \dfrac{(D + d)}{2}$

$\therefore$
$$\text{Power} = (W - S) \times \frac{(D + d)}{2} \times \frac{2\pi N}{60}$$

$$= (750 - 150) \times \frac{(2 + 0.02)}{2} \times \frac{2\pi \times 500}{60}$$

$$= 31714 \text{ Watt} = 31.714 \text{ kW.}$$

Example 9.12: The driving pulley of belt transmission dynamometer rotates at 400 rpm. The diameter of each of driving pulley and intermediate pulleys is 320 mm. The load W is suspended at a distance of 800 mm from the fulcrum O. Find the value of dead mass required to maintain the lever in a horizontal position when the power transmitted is 8 kW. Also find its value when the belt just begins to slip on the driving pulley. The coefficient of friction is 0.2 and the maximum tension in the belt is 1400 N.

Solution: (i) $N = 400$ rpm, $a = 0.32$ m, $l = 0.8$ m, $P = 8$ kW $= 800$ W

$$\text{Power, } P = (T_1 - T_2) \times v = \frac{Mgl}{2a} \times rw = \frac{Mgl}{2a} \times r \times \frac{2\pi N}{60}$$

or
$$8000 = \frac{M \times 9.81 \times 0.8}{2 \times 0.32} \times 0.16 \times \frac{2\pi \times 400}{60}$$

or
$$M = 97.39 \text{ kg}$$

Maximum tension, $T_1 = 1400$ N, $v = 0.2$, $\theta = \pi$ rad

$$\frac{T_1}{T_2} = e^{v\theta} = e^{0.20 \times \pi} = 1.87$$

$$T_2 = \frac{T_1}{1.87} = \frac{1400}{1.87} = 748.66 \text{ N}$$

$$T_1 - T_2 = \frac{Mgl}{2a}$$

or
$$(1400 - 748.66) = M \times 9.81 \times \frac{0.8}{2 \times 0.32}$$

or
$$M = 53.12 \text{ kg}$$

Example 9.13: The wheels of a bicycle are of diameter 800 mm. A rider on this bicycle is travelling at a speed of 16 km/hr on a level road. The total mass of the rider and the bicycle is 110 kg. A brake is applied to the rear wheel. The pressure applied on the brake is 100 N and coefficient of friction is 0.06. Before the cycle comes to rest, find:

(i) Distance travelled by the bicycle

(ii) Number of turns of its wheel.

Solution:

$$d = 800 \text{ mm} = 0.8 \text{ m}; v = 16 \text{ km/hr} = \frac{16 \times 100}{60 \times 60} \text{m/s} = 4.44 \text{ m/s}$$

mass $(m) = 110$ kg; pressure applied on brake, $(R_N) = 100$ N; $\mu = 0.06$

Let s = distance travelled by the bicylce before it comes to rest

n = number of turns of the wheel.

(i) When the bicycle comes to rest, the total kinetic energy of the bicycle and the rider will be absorbed by the brakes. The work will be done agaisnt friction.

∴ Work done against friction = KE of the bicycle and the rider

or   Frictional force × distance = $\frac{1}{2}mv^2$

or $(\mu R_N) \times \text{s} = \frac{1}{2}mv^2$ (∵ frictional force = μR_N)

or $0.06 \times 100 \times \text{s} = \frac{1}{2} \times 110 \times 4.44^2$

∴ $s = \dfrac{1 \times 110 \times 4.44^2}{2 \times 0.06 \times 100} = 180.7 \text{ m}$

(ii) Total distance travelled by the bicycle before it comes to rest

$$= 180.7 \text{ m}$$

Distance travelled in one turn = circumference of wheel

$$= \pi \times d$$
$$= \pi \times 0.8$$

∴   Number of turns of wheel $= \dfrac{\text{Total distance travelled}}{\text{Distance travelled in one turn}}$

$$= \frac{180.7}{\pi \times 0.8} = 71.89$$

Example 9.14: The brake drum of a single block brake of diameter 300 mm is rotating at 400 rpm in the clockwise direction as shown in Fig. 9.23. The lever OA is pivoted at O and brake shoe is rigidly fixed to the lever. The coefficient of friction for the brake lining is 0.35. The force applied at the end of lever is 800 N. Find the braking torque.

Solution:

$d = 300 \text{ mm} = 0.3 \text{ m}$, or $r = 0.15 \text{ m}$; $N = 400$ rpm; $\mu = 0.35$; force at $A = 800$ N

Let T_B = braking torque

R_n = normal reaction

μR_n = frictional force

Taking the moments of all forces about the fulcrum O, we get

$$R_n \times 300 + \mu R_n \times 100 = 800 \times (300 + 400)$$

or $300R_n \times 0.35 \times R_n \times 100 = 800 \times 700$

or $300R_n \times 350R_n = 800 \times 700$

or $750R_n = 800 \times 700$

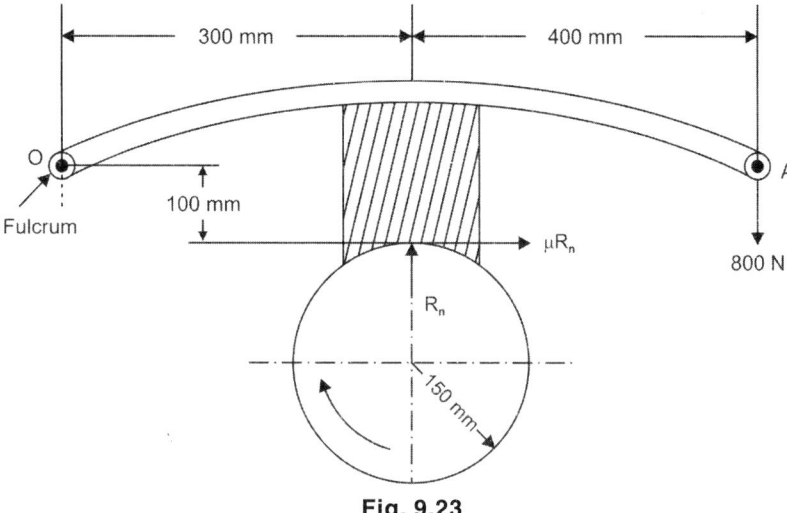

Fig. 9.23

$$R_n = 800 \times \frac{700}{750} = 746.67 \, \text{N}$$

Braking torque = Frictional force × radius

$$= (\mu \, R_n) \times r = 0.35 \times 746.67 \times 0.15$$

$$= 39.2 \, \text{N.m}$$

Example 9.15: A band brake acts on the 3/4 of circumference of a drum of 450 mm diameter which is keyed to the shaft. The band brake provides a braking torque of 225 N·m. One end of the band is attached to a fulcrum pin of the lever and the other end to a pin 100 mm from the fulcrum. If the operating force is applied at 500 mm from the fulcrum and the coefficient of friction is 0.25, find the operating force when the drum rotates in (i) the anticlockwise direction, and (ii) the clockwise direction.

Solution:

Given: Diameter of drum d = 450 mm, r = 225 mm = 0.225 m

T_B = 225 N·m, μ = 0.25, L = 500 mm

α = 100 mm, $\theta = \dfrac{3}{4} \times 2\pi = 1.5\pi$

Let T_1 = tension in band on tight side

T_2 = tension in band on slack side

F = applied force

(*i*) The drum rotates anticlockwise and one end of the band is connected to the fulcrum at O, so the applied force F acts upwards.

Tension ratio is given as

$$\frac{T_1}{T_2} = e^{\mu \theta} = e^{.25 \times 1.5\pi} = 3.248 \qquad \text{...(i)}$$

Braking torque is given as

$$T_B = (T_1 - T_2)r = (T_1 - T_2)0.225$$

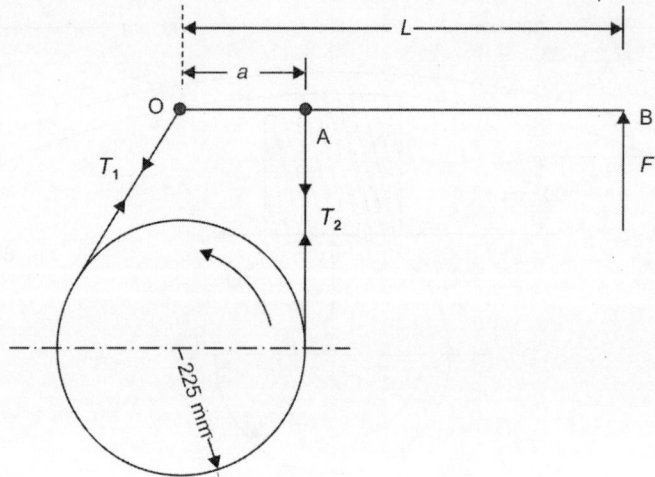

Fig. 9.24

$$T_1 - T_2 = \frac{225}{0.225} = 1000 \qquad \qquad ...(ii)$$

Solving Eqs. (i) and (ii), $T_2 = 444.84$ N

$$T_1 = 144.84 \text{ N}$$

Taking moments about fulcrum O,

$$L \times F = T_2 \times a$$

$$500 \times F = T_2 \times 100 = 444.84 \times 100$$

$$F = 88.97 \text{ N}$$

(ii) When the drum rotates in clockwise sense, tensions T_1 and T_2 will interchange their positions. Now taking moments about O,

$$F \times L = T_1 \times a$$

$$F = T_1 \times \frac{a}{L} = 1444.84 \times \frac{0.100}{0.500}$$

$$F = 288.97 \text{ N}$$

Example 9.16: A vehicle moves on a road that has a slope of 15°. The wheel base is 1.6 m and the centre of mass at 0.72 m from the rear wheels and 0.8 m above the inclined plane. The speed of the vehicle is 45 km/hr. The brakes are applied to all the four wheels and the coefficient of friction is 0.4. Determine the distance moved by the vehicle before coming to rest and the time taken to do so if it moves:

(i) Up the plane

(ii) Down the plane

Solution:

Let s the distance moved by the car before coming to rest.

$$u = 45 \text{ km/hr} = \frac{45000}{3600} = 12.5 \text{ m/s}$$

(i) When the vehicle moves up:

$a = g \cos \alpha (\mu + \tan \alpha) = 9.81 \times \cos 15° (0.4 + \tan 15°) = 6.33 \text{ m/s}^2$

If retardation is uniform,

$$v^2 - u^2 = -2as$$
$$0 - u^2 = -2as$$
$$s = \frac{u^2}{2a} = \frac{12.5^2}{2 \times 6.33} = 12.34 \text{ m}$$

Also, $v = u - at$ or $0 = 12.5 - 6.33 \times t$ or $t = 1.97$ s

(ii) When the vehicle moves down:

$a = g \cos \alpha (\mu - \tan \alpha) = 9.81 \times \cos 15° (0.4 - \tan 15°) = 1.25 \text{ m/s}^2$

$$s = \frac{u^2}{2a} = \frac{12.5^2}{2 \times 1.25} = 62.5 \text{ m}$$

Also, $0 = 12.5 - 1.25 \times t$ or $t = 10$ s

Example 9.17: A differential band brake, as shown in Fig. 9.25a, has an angle of contact of 225°. The band has a compressed woven lining and bears against a cast iron drum of 350 mm diameter. The brake is to sustain a torque of 350 N·m and the coefficient of friction between the band and the drum is 0.30. Find:

(a) The necessary force (F) for the clockwise and anticlockwise rotation of the drum
(b) The value of OA for the brake to be self-locking, when the drum rotates clockwise.

Solution: $\mu = 0.30, \theta = 225°$
        $d = 350$ mm or $r = 175$ m $= 0.175$ m
        $T_B = 350$ N·m

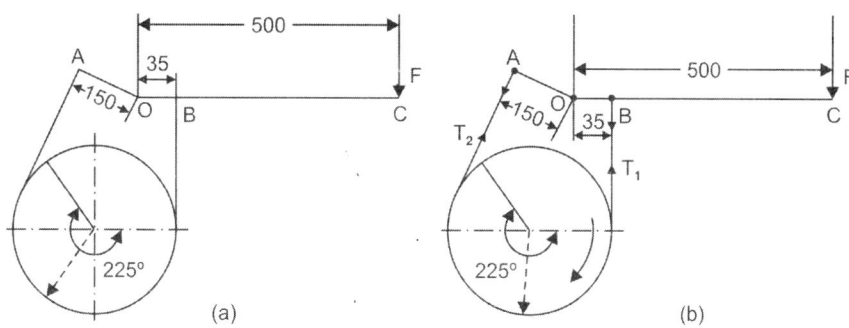

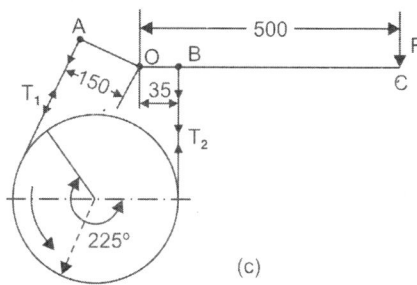

Fig. 9.25

(a) If the drum rotates in clockwise direction and $OA > OB$, the band connected to A will be having slack tension T_2 and the band connected to B experiences tight tension T_1 as shown in Fig. 5.25b.

Taking moments about fulcrum O.

$$F \times 500 + T_1 \times 35 = T_2 \times 150$$

$$100F = 30T_2 - 7T_1 \qquad \qquad ...(i)$$

Tension ratio is

$$\frac{T_1}{T_2} = e^{\mu\theta} = e^{0.30 \times 225} \times \frac{\pi}{180} = 3.2482 \qquad \qquad ...(ii)$$

Braking torque $\qquad \qquad T_B = (T_1 - T_2)$

$$350 = (T_1 - T_2) \times 0.175$$

$$T_1 - T_2 = 2000 \qquad \qquad ...(iii)$$

Solving Eqs (ii) and (iii),

$$T_1 = 2889.6 \text{ N} \text{ and } T_2 = 889.6 \text{ N}$$

Substituting the values of T_1 and T_2 in Eq. (i)

$$100F = 30T_2 - 7T_1$$

$$100F = 30 \times 889.6 - 7 \times 2889.6$$

$$F = 64.6 \text{ N}$$

If the drum rotates in anticlockwise direction and $OA > OB$, the band connected to A will be having tight tension T_1 and that connected to B experiences slack tension T_2 as shown in Fig. 9.25c.

Taking moments about fulcrum O,

$$F \times 500 = T_1 \times 150 - T_2 \times 35$$

$$F = 0.3T_1 - 0.07T_2 \qquad \qquad ...(iv)$$

Substituting the values of T_1 and T_2 in euqation (iv).

$$F = 0.3 \times 2889.6 - 0.07 \times 889.6$$

$$= 804.61 \text{ N}$$

(b) When the externally applied force F is zero, it is known as self-locking. Consider Fig. 9.25b and taking moments about fulcrum O.

$$F \times 500 = T_2 \times OA - T_1 \times OB$$

For self-locking $F = 0$, then

$$T_2 \cdot OA = T_1 \cdot OB$$

$$OA = \frac{T_1 \cdot OB}{T_2} = \frac{2889.6 \times 35}{889.6}$$

$$OA = 113.7 \text{ mm}$$

EXERCISE

9.1 What is a brake? What is the difference between a brake and a clutch?

9.2 What is the difference between brakes and dynamometers?

9.3 Differentiate between the functions of: (i) a clutch and a brake (ii) a brake and a dynamometer. Also discuss briefly the classification of friction clutches.

9.4 Describe with the help of a neat sketch the working of an external shoe brake.

9.5 Describe various types of brakes.

9.6 Explain the working of a block or shoe brake.

9.7 Show that in band and block brake, the ratio of the maximum and minimum tensions in brake straps is

$$\frac{T_o}{T_r} = \left[\frac{1+\mu\tan\theta}{1-\mu\tan\theta}\right]^n$$

where, T_o = maximum tension,
$\quad\quad T_r$ = minimum tension
$\quad\quad \mu$ = coefficient of friction between the blocks and drum.
$\quad\quad \theta$ = angle subtended by each block at the centre of the drum.

9.8 What do you understand by a self-locking and a self-energising brake?

9.9 What is the difference between a simple band brake and a differential band brake?

9.10 Describe a single shoe brake with the help of a neat sketch. What is advantage of double shoe brake over it?

9.11 Discuss the effectiveness of a band brake under various conditions.

9.12 Describe the working of a band and block brake with the help of a neat sketch.

9.13 Describe with the help of a neat sketch, the principles of operation of an internal expanding shoe. Derive the expression for the braking torque.

9.14 What is the advantage of self expanding shoe brake?

9.15 Discuss the effect of applying the brakes to a vehicle when:
   (i) Brakes are applied to the rear wheels only
   (ii) Brakes are applied to the front wheels only
   (iii) Brakes are applied to all the four wheels.

9.16 How are the dynamometers classified? What is the difference between absorption and transmission type of dynamometers? Explain with the help of diagram, any one absorption type of dynamometer.

9.17 Explain the working of any transmission dynamometer with the help of a neat sketch.

9.18 Describe the construction and operation of a prony brake or rope brake absorption dynamometer. Expain its practical examples.

9.19 Describe with the help of a neat sketch a torsion dynamometer.

9.20 The brake drum of a single block brake of diameter 300 mm is rotating at 400 rpm as shown in Fig. 9.26. The force required at the end of the lever to apply the brake is 600 N. If angle of contact is 90° and coefficient of friction between the drum and brake block is 0.3, find the braking torque. (*Ans.* 68.98 N·m)

9.21 A simple band brake is applied to a drum of 560 mm diameter with rotates at 240 rpm. The angle of contact of the band is 270°. One end of the band is fastened to a fixed pin and the other end to the brake lever 140 mm from the fixed pin. The brake lever is 800 mm long and is placed perpendicular to the diameter that bisects the angle of contact. Assuming the coefficient of friction as 0.3, determine the necessary pull at the end of the lever to stop the drum if 40 kW of power is being absorbed. Also, find the width of the band if its thickness is 3 mm and the maximum tensile stress is limted to 40 N/mm². (*Ans.* 226. 2N, 62.6 m)

9.22 A band and blocke brake having 12 blocks, each of which subtends an angle of 16° at the centre, is applied to a rotating drum of diameter 600 mm. The blocks are 75 mm thick. The drum and the flywheel mounted on the same shaft have a mass of 1800 kg

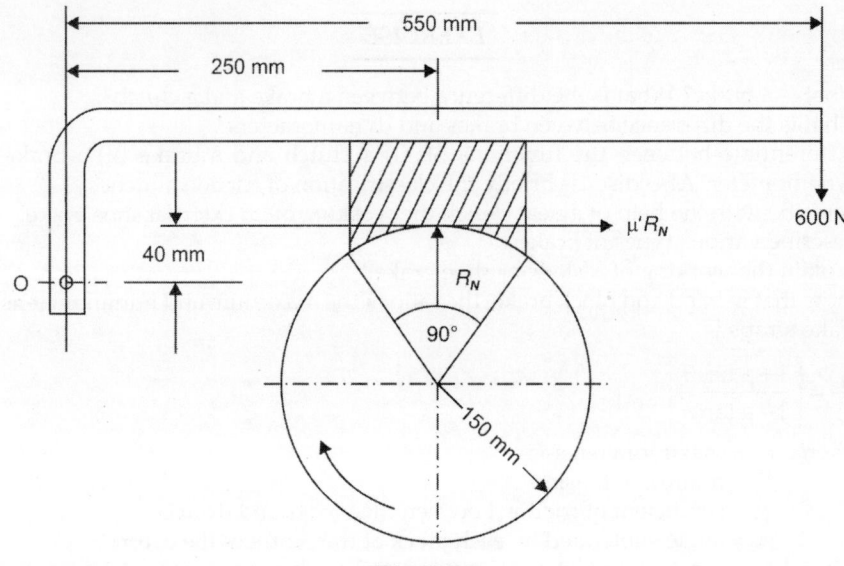

Fig. 9.26

and a combined radius of gyration of 600 mm. The two ends of tha band are attached to pins on the opposite sides of the brake fulcrum at distances of 40 mm and 150 mm from the fulcrum. If a force of 250 N is applied at a distance of 900 mm from the fulcrum, find:

(i) The maximum braking torque

(ii) The angular retardation of the drum

(iii) The time taken by the system to be stationary from the rated speed of 300 rpm. Take coefficient of friction betwee the blocks and the drum as 0.3.

(***Ans.*** 3703 N·m, 5.71 rad/s^2, 5.5 s)

9.23 An internal shoe brake has a diameter of 320 mm and width of 30 mm. The cam forces are equal. Maximum pressure is not to exceed 80 kN/m^2. $\varphi_1 = 15°$, $\varphi_2 = 145°$, $a = 220$ mm, $c = 125$ mm and $\mu = 0.32$. Determine the actuating force and the brake torque. (***Ans.*** 175.7 N, 48.04 N·m)

9.24 A vehicle having a wheel base 3.2 m has its centre of mass 1.4 m from the rear wheels and 55 mm from the ground level. It moves on a level ground at a speed of 54 km/hr. Determine the distance moved by the car before coming to rest on applying the brakes to

(i) the rear wheels

(ii) the front wheels

(iii) all the four wheels.

The coefficient of friction between the tyres and the road is 0.5.

(***Ans.*** 44.3 m, 47.9 m, 22.9 m)

9.25 In a belt transmission dynamometer, the diameters of the driving and driven pulleys are 0.36 m and 0.8 m respectively. The power transmitted from the driving to the driven shaft is 20 kW. The speed of the driving shaft is 500 rpm. If $L = 1.2$ m and $a = 400$ mm, determine the weight on the lever. (***Ans.*** 144.2 kg)

9.26 A simple band is operated by a lever 80 cm long. The brake drum is 40 cm diameter and the brake band embraces 3/4 of its circumference. One end of the band is attached to the fulcrum of the lever and the other is attached to a pin on the lever 5 cm from the fulcrum. If the brake balances a torque 3000 kg·cm on the drum shaft and the coefficient of friction between the band and the brake is 0.3, determine the minimum operating force applied at the end of the brake lever. (***Ans.*** 2.8 kg)

9.27 In a belt transmission dynamometer, the driving pulley rotates at 300 rpm. The distance between the centre of the driving pulley and the dead mass is 800 mm. The diameter of each of the driving as well as the intermediate pulleys is equal to 360 mm. Find the value of the dead mass required to maintain the lever in a horizontal position when the power transmitted is 3 kW. Also, find its value when the belt just begins to slip on the driving pulley, μ being 0.25 and the maximum tension in the belt 1200 N.

(*Ans.* 48.7 kg, 59.8 kg)

OBJECTIVE TYPE QUESTIONS

9.1 Which brake is commonly used in motor cars?
(a) band brake
(b) shoe brake
(c) band and block brake
(d) internal expanding shoe brake

9.2 The brakes commonly used in railway trains are
(a) shoe brake
(b) band brake
(c) band and block
(d) internal expanding shoe brake

9.3 In a self-locking brake, the force required to apply the brake is
(a) minimum
(b) zero
(c) maximum

9.4 When the frictional force helps the applied force in applying the brake, it is known as
(a) self-locking brake
(b) automatic brake
(c) self-energising brake
(d) none

9.5 In an internal expading shoe brake, more than 50% of total braking torque is supplied by
(a) leading shoe
(b) trailing shoe
(c) any of the two
(d) none

9.6 When brakes are applied to all the four wheels of a moving car, the distance travelled by the car before it is brought to rest, will be
(a) maximum
(b) minimum
(c) same

9.7 The ratio of tensions on the tight and slack sides in a band and block brake is given by

(a) $\dfrac{T_n}{T_0} = \left[\dfrac{1-\mu\tan\theta}{1+\mu\tan\theta}\right]^n$

(b) $\dfrac{T_n}{T_0} = \left[\dfrac{1+\mu\tan\theta}{1-\mu\tan\theta}\right]^n$

(c) $\dfrac{T_n}{T_0} = \left[\dfrac{1+\mu\tan\theta}{1-\mu\tan\theta}\right]^{1/n}$

(d) $\dfrac{T_n}{T_0} = \left[\dfrac{1-\mu\tan\theta}{1+\mu\tan\theta}\right]^{1/n}$

9.8 Tractive resistance during the propulsion of a wheeled vehicle depends on
    (a) road resistance
    (b) aerodynamic resistance
    (c) gradient resistance
    (d) all of the above
9.9 Which of the following is an absorption type dynamometer?
    (a) prony brake dynamometer
    (b) rope brake dynamometer
    (c) epicyclic train dynamometer
    (d) torsion dynamometer

ANSWERS

9.1 (d)	9.2 (a)	9.3 (b)	9.4 (c)	9.5 (a)	9.6 (b)
9.7 (b)	9.8 (d)	9.9 (a, b)			

10

Governors

10.1 INTRODUCTION

The function of a governor is to maintain the speed of an engine within specified limits whenever there is variation in the load. When the load on the shaft increases, the speed of the engine decreases. The governor acts in such a way that the supply of fuel is increased by opening the throttle valve. On the other hand, if the load on the shaft decreases, the speed of the engine increases. Governor decreases the fuel supply by closing the valve to slow the engine to its original speed.

The function of the flywheel is to control the variation in speed caused by the fluctuations in the output torque of the engine during a cycle. The operation of the flywheel is continuous whereas that of a governor is intermittent as it acts only when there is variation of load.

10.2 TYPES OF GOVERNOR

A governor can be broadly classified into two types: (1) centrifugal governors (2) inertia governors.

10.2.1 Centrifugal Governors

These are based on the balancing of centrifugal force on the rotating balls by an equal and opposite radial force, known as controlling force. The controlling force is provided by gravity effect of balls as in case of Watt governors or by a spring as in Hartnell governors. It consists of a pair of masses known as governor balls, which are attached to the two arms and rotate at a distance from the axis of rotation. The upper ends of the

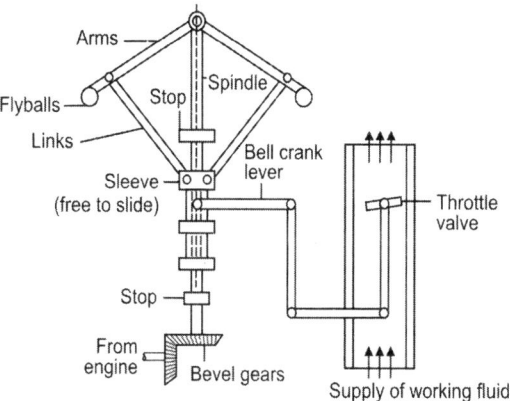

Fig. 10.1: Centrifugal governor

arms are pivoted to the spindle, which is driven by the engine through bevel gears. The arms are connected by the links to a sleeve, which is keyed to the spindle. This sleeve revolves with the spindle and can slide up and down. The sleeve is connected by a bell crank lever to a throttle valve. With the increase in the speed, the balls tend to rotate at a greater radius from the axis and sleeve rises and throttle valve is made to close to the required extent. When the speed decreases, the balls rotate at a smaller radius, sleeve falls and valve is a opened according to the requirement.

The centrifugal governors are further classified as follows:

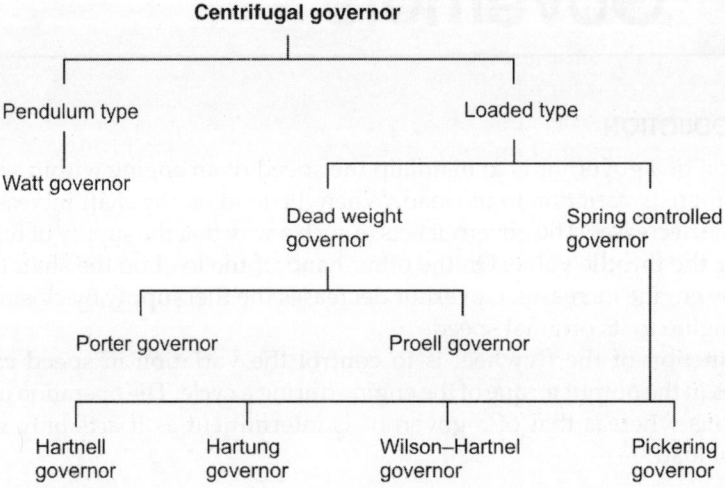

10.2.2 Inertia Governors

In these governors, the positions of the balls are affected by the forces set up by an angular acceleration/deceleration of the given spindle in addition to the centrifugal forces on the balls. Therefore, in case of centrifugal governors, the valve is operated by the change of engine speed, whereas it is operated by the rate of change of speed in case of inertia governors. Hence, the response of inertia governors is faster than that of centrifugal types.

10.3 TERMS USED IN GOVERNORS

1. **Height of governor:** It is the vertical distance from the plane of rotation of the balls to the point of intersection of the upper arms along the axis of the spindle. It is denoted by h. It decreases with the increase in speed, and increases with the decrease in speed.

2. **Equilibrium speed:** It is the speed at which the governor balls, arms, etc. are in complete equilibrium and sleeve does not tend to move upwards or downwards.

3. **Mean equilibrium speed:** It is the speed at mean position of the balls or the sleeve.

4. **Maximum and minimum equilibrium speeds:** The speeds at the maximum and minimum radius of rotation of the balls, without tending to move either way are known as maximum and minimum equilibrium speeds respectively.

5. **Sleeve lift:** It is the vertical distance travelled by sleeve on the spindle due to change in equilibrium speed.

6. **Radius of rotation:** It is the horizontal distance between the centre of the ball and the axis of rotation. It is denoted by r.

7. **Centrifugal force:** Force $F_c = mr\omega^2$ acting on the balls radially outwards, is known as centrifugal force.

8. **Controlling force:** A force equal and opposite to the centrifugal force acting radially inward is known as controlling force. This force is provided by the mass of the balls and the sleeve in porter governor and by the spring in spring controlled governors.

10.4 WATT GOVERNOR

Watt governor is the simplest form of a centrifugal governor. This governor is named after Watt who used it for steam engines. It is simple conical pendulum type governor consisting of a pair of balls attached to a spindle with the help of links. Watt governor is of three types depending upon the position of upper arms:

i. **Simple or pinned arm type Watt governor:** When the upper arms intersect on the spindle axis at point O, it is called the pinned arm type Watt governor (Fig. 10.2a).

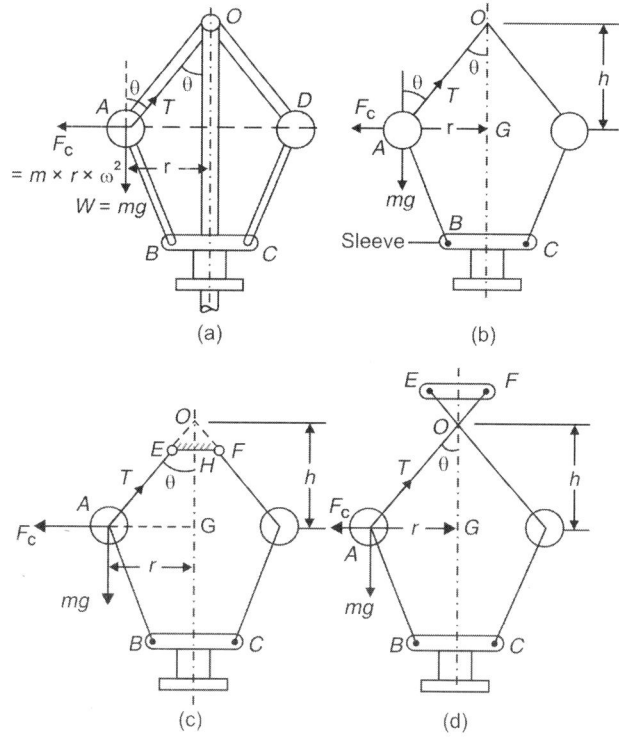

Fig 10.2

ii. **Open arm type Watt governor:** In this, upper arms are connected to a horizontal link and when produced intersect at O (Fig. 10.2b).

iii. **Cross arm type Watt governor:** When upper arms cross the spindle and are connected by a horizontal link, governor is known as a cross arm Watt governor (Fig. 10.2c).

In each case, lower links are fixed to a sleeve free to move on the spindle driven by the engine. As the spindle rotates, the balls take up a position depending upon the speed of the spindle. When the speed increases, the balls move outwards due to centrifugal force and pull the sleeve upwards, as a result the height of the governor (h) decreases, when the speed decreases, the balls move inwards, sleeve move downwards and the height of the governor (h) increases.

Let m = mass of each ball in kg

           ω = angular velocity of the arms, balls and sleeve about the axis of the spindle (rad/s)

           T = tension in the arm (N)

           r = radius of rotation of balls (m)

           h = height of the governor (m)

It is assumed that the weight of arms, links and sleeve are negligible as compared to the weight of the balls. Now, each ball is in equilibrium under the action of following forces:

    (i) Centrifugal force $F_c = mr\omega^2$

    (ii) Weight of the balls $W = mg$

    (iii) Tension T in the upper arm

There will be no tension in the lower links if the sleeve is assumed massless and also friction is neglected.

Now taking moment about O

$$F_c \times h = W \times r$$

or $$mr\omega^2 \times h = mg \times r$$

or $$h = \frac{g}{\omega^2} = \frac{g}{\left(\dfrac{2\pi N}{60}\right)^2} \quad \left(\because \omega = \frac{2\pi N}{60}\right)$$

or $$h = \frac{9.81}{N^2} \times \left(\frac{60}{2\pi}\right)^2 = \frac{895}{N^2}$$

$$h = \frac{895}{N^2} \qquad \qquad ...(10.1)$$

The height of a Watt governor is inversely proportional to the square of the speed. Therefore, at high speeds value of h is small and the movement of sleeve is very less. This sleeve movement is insufficient to effect the change in fuel supply. Hence, Watt governor is not suitable at high speeds. It can work satisfactorily from 50 to 80 rpm. This drawback is overcome by putting some load on the sleeve or by means of a spring.

Example 10.1: Calculate the change in governor height for a Watt governor when speed varies from 60 to 61 rpm.

Solution:

$$N_1 = 60 \text{ rpm}, \; N_2 = 61 \text{ rpm}$$

Initial governor height, $h_1 = \dfrac{g}{\omega_1^2} = \dfrac{895}{N_1^2} = \dfrac{895}{(60)^2}$

or $$h_1 = 0.248 \text{ m}$$

Final governor height, $h_2 = \dfrac{895}{N_2^2} = \dfrac{895}{(61)^2} = 0.24 \text{ m}$

Change in governor height = $h_1 - h_2 = 0.248 - 0.24$
$$= 0.008 \text{ m} = 8 \text{ mm}.$$

Example 10.2: Figure 10.2a shows a simple watt governor. In this, length $OA = 640$ mm and angle $\theta = 30°$. What will be the percent change in speed for a 50 mm rise in level of the balls?

Solution: $OA = 640$ mm, $\theta = 30°$

Initial height, $h = OA \cos 30° = 640 \times .866$

$$h = 554.24 \text{ mm}$$

Final height, $h' = h - 50 = 554.24 - 50 = 504.24 \text{ mm}$

Now, $h = \dfrac{g}{\omega^2}$ and $h' = \dfrac{g}{\omega'^2}$ or $\omega'^2 = \dfrac{g}{h'}$

$\therefore$

$$\frac{\omega'}{\omega} = \sqrt{\frac{g}{h^1}} \times \sqrt{\frac{h}{g}} = \sqrt{\frac{h}{h'}}$$

$$= \sqrt{\frac{554.24}{504.24}} = 1.0484$$

Percentage change in speed $= \dfrac{\omega' - \omega}{\omega} \times 100$

or

$$= \left(\frac{\omega'}{\omega} - 1\right) \times 100$$

$$= (1.0484 - 1) \times 100$$

$$= 4.84 \%$$

10.5 PORTER GOVERNOR

The porter governor is a modified Watt governor, the sleeve of which is loaded with a heavy mass. The central load with sleeve moves up and down the spindle.

Let M = mass of central load (kg)

m = mass of each ball (kg)

f = force of friction between sleeve and spindle (N)

h = height of the governor (m)

r = radius of rotation (m)

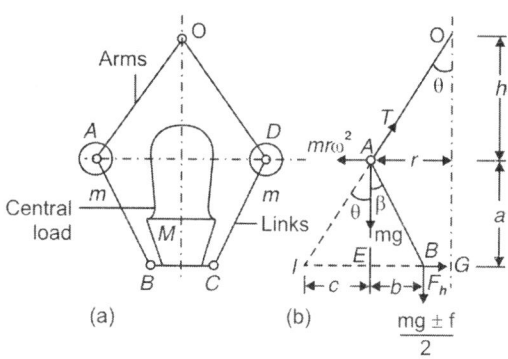

Fig. 10.3: Porter governor.

θ = angle of inclination of upper arm to the vertical

β = angle of inclination of lower link to the vertical

T = tension in upper arm (N)

The instantaneous centre of rotation of the link AB lies at the point of intersection (I) of lines perpendicular to the direction of motions of points A and B.

The force of friction (f) always acts in a direction opposite to that of the motion of sleeve. When the sleeve moves up, the force of friction acts in the downward direction and the total force in downward direction on sleeve is ($Mg + f$). When the sleeve moves down, the total force on the sleeve will be ($Mg - f$). Considering the equilibrium of lower, left half of the governor. Taking moment about I,

$$F_c \times AE = (mg \times IE) + \frac{(Mg \pm f)}{2} \times IB$$

or
$$mr\omega^2 \times a = (mg \times c) + \frac{(Mg \pm f)}{2} \times (c + b)$$

or
$$mr\omega^2 = mg \times \frac{c}{a} + \frac{(Mg \pm f)}{2} \times \left(\frac{c}{a} + \frac{b}{a}\right)$$

$$= mg \times \tan\theta + \frac{(Mg \pm f)}{2} [\tan\theta + \tan\beta]$$

$$= \tan\theta \left[mg + \frac{(Mg \pm f)}{2}\left(1 + \frac{\tan\beta}{\tan\theta}\right) \right]$$

$$= \tan\theta \left[mg + \frac{(Mg \pm f)}{2}(1 + k) \right] \qquad \left(k = \frac{\tan\beta}{\tan\theta}\right)$$

or
$$mr\omega^2 = \frac{r}{h} \left[mg + \frac{(Mg \pm f)}{2}(1 + k) \right] \qquad \qquad ...(10.2)$$

or
$$\omega^2 = \frac{1}{mh} \left[\frac{2mg + (Mg \pm f)(1 + k)}{2} \right]$$

or
$$\left(\frac{2\pi N}{60}\right)^2 = \frac{g}{h} \left[\frac{2mg + (Mg \pm f)(1 + k)}{2mg} \right]$$

or
$$N^2 = \frac{895}{h} \left[\frac{2mg + (Mg \pm f)(1 + k)}{2mg} \right] \qquad \qquad ...(10.2a)$$

Notes:

i. When the length of arms are equal to length of links and points O and B lie on the same vertical line. $\tan\theta = \tan\beta$

or
$$k = \frac{\tan\beta}{\tan\theta} = 1$$

$$N^2 = \frac{895}{h} \left[\frac{mg + (Mg \pm f)}{mg} \right] \qquad \qquad ...(10.3)$$

ii. If the friction force is absent, $f = 0$

$$N^2 = \frac{895}{h}\left[\frac{2m + M(1+k)}{2m}\right]$$ ...(10.4)

iii. If $\qquad k = 1$ and $f = 0$

$$N^2 = \frac{895}{h}\left[\frac{m+M}{m}\right]$$ ...(10.5)

The comparison of Eq. (10.5) with Eq. (10.1) for Watt governor indicates that the height of the governor in increased by $\left[\frac{m+M}{m}\right]$.

Examples 10.3: The length of upper and lower arms of a Porter governor are 20 cm and 25 cm respectively. Both the arms are pivoted on the axis of rotation. The central load is 150 N, the weight of each ball is 20 N and the friction of the sleeve together with the resistance of the operating gear is equivalent of a force of 30 N at the sleeve. If the limiting inclinations of the upper arms to the vertical are 30° and 40°, determine the range of speed of the governor.

Solution:

Central load, $\qquad W = 150$ N

weight of each ball, $\qquad \omega = 20$ N $= mg$

$\qquad F = 30$ N

Let maximum speed $\qquad = N_1$

$\qquad$ Corresponding angle $= 40° = \theta_1$

Maximum radius of rotation, $r_1 = AG = AO \sin 40°$

or $\qquad r_1 = 20 \sin 40° = 12.85$ cm $= 0.1285$ m

$\qquad h_1 = AO \cos 40° = 20 \cos 40° = 15.32$ cm $= 0.1532$ m

$$BG = \sqrt{(AB)^2 - (AG)^2} = \sqrt{(25)^2 - (12.85)^2} = 21.44 \text{ cm}$$

$$\tan \beta_1 = \frac{AG}{BG} = \frac{12.85}{21.44} = 0.6$$

$$k_1 = \frac{\tan \beta_1}{\tan \theta_1} = \frac{0.6}{0.8391} = 0.715$$

$$N_1^2 = \frac{w + \left(\dfrac{W+F}{2}\right)(1+k_1)}{w} \times \frac{895}{h_1} \qquad \text{(sleeve moves upward)}$$

$$= \frac{20 + \left(\dfrac{150+30}{2}\right)(1+0.715)}{20} \times \frac{895}{0.1532} = 50927.95$$

or $\qquad N_1 = 225.7 \approx 226$ rpm

Let minimum speed $= N_2$, corresponding angle $\theta_2 = 30°$

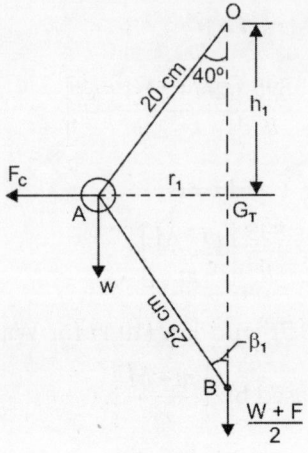

Fig. 10.4

Minimum radius of rotation, $r_2 = 20 \sin 30° = AG = 10$ cm

$$h_2 = AO \cos 30° = 20 \cos 30° = 17.32 \text{ cm} = 0.1732 \text{ m}$$

$$BG = \sqrt{(AB)^2 - (AG)^2} = \sqrt{(25)^2 - (10)^2} = 22.9 \text{ cm}$$

$$\tan \beta_2 = \frac{AG}{BG} = \frac{10}{22.9} = 0.4367$$

$$\tan \theta_2 = \tan 30° = 0.5774$$

$$k_2 = \frac{\tan \beta_2}{\tan \theta_2} = \frac{0.436}{0.5774} = 0.755$$

$$N_2^2 = \frac{w + \left(\dfrac{W - F}{2}\right)(1 + k_2)}{w}$$

$$= \frac{20 + \left(\dfrac{150 - 30}{2}\right)(1 + 0.755)}{20} \times \frac{895}{0.173} = 32373.98$$

or $\qquad N_2 = 179.9 \approx 180$ rpm

Range of speed $= N_1 - N_2 = 226 - 180 = 46$ rpm

Example 10.4. Each arm of a Porter governor is 300 mm long and is pivoted on the axis of rotation. Each ball has a mass of 6 kg and the sleeve weighs 18 kg. The radius of rotation of the ball is 200 mm when the governor begins to lift and 250 mm when the speed is maximum. Determine the maximum and the minimum speed and the range of speed of thhe governor.

Solution:

Length of each arm, $\quad l = 300$ mm

Mass of ball, $\qquad m = 6$ kg

Mass of sleeve, $\qquad M = 18$ kg

Maximum radius of rotation, $r_2 = 250$ mm
Minimum radius of rotation, $r_1 = 200$ mm
Since, all arms are of equal lengths

$\therefore$ $\qquad$ $\tan \theta = \tan \beta$ and $k = \dfrac{\tan \beta}{\tan \theta} = 1$

Friction force, $\qquad f = 0$
For minimum speed (using relation 10.2A)

$$mr_1\omega r_1^2 = \frac{r_1}{h_1}\left[mg + \frac{Mg}{2}(1+k)\right]$$

or $\qquad$ $$m\omega_1^2 = \frac{1}{h_1}\left[mg + \frac{Mg}{2}(1+1)\right] \qquad (k = 1)$$

or $\qquad$ $$\omega_1^2 = \frac{1}{mh_1}(m + M)g \qquad \text{...(i)}$$

From Fig. 10.5, $\qquad h_1 = \sqrt{l^2 - r_1^2} = \sqrt{(300)^2 - (200)^2}$

or $\qquad$ $h_1 = 100\sqrt{5}$ mm $= 0.1\sqrt{5}$ m

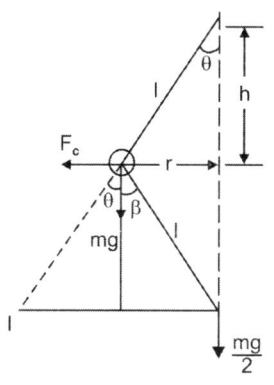

Fig. 10.5

Substituting values in Eq. (i) above,

$$\omega_1^2 = \frac{1}{6 \times 0.1\sqrt{5}}(6 + 18) \times 9.81 = 175.49$$

or $\qquad$ $\omega_1 = 13.24$ rad/s

$\therefore$ $\qquad$ $$N_1 = \frac{\omega_1 \times 60}{2\pi} = \frac{13.24 \times 60}{24} = 126.5 \text{ rpm}$$

For maximum speed (at r_2),

$$h_2 = \sqrt{l^2 - r_2^2} = \sqrt{(300)^2 - (250)^2} = 165.83 \text{ mm}$$

$$\omega_2^2 = \frac{1}{mh_2}(m + M)g$$

$$= \frac{1}{6\times.16583} (6 + 18) \times 9.81 = 236.6$$

or $\qquad \omega_2 = 15.38 \text{ rad/s}$

$\therefore \qquad N_2 = \dfrac{\omega_2 \times 60}{2\pi} = \dfrac{15.38 \times 60}{2\pi} = 146.9 \text{ rpm}$

Range of speed $\qquad = N_2 - N_1 = 146.9 - 126.5$
$$= 20.4 \text{ rpm.}$$

10.6 PROELL GOVERNOR

In Proell governor, two balls are attached on the upward extensions of lower links.

Considering the equilibrium of left half of the governor.

Taking moment about instantaneous centre I,

$$F_e \times EJ = mg \times IJ + \frac{(Mg \pm f)}{2} \times IB$$

Substituting dimensions from Fig. 10.6,

$$mr'\omega^2 \times e = mg \times (c + r - r') + \frac{(Mg \pm f)}{2}(c + b)$$

$$mr'\omega^2 = \frac{1}{e}\left[mg(c + r - r') + \frac{(Mg \pm f)}{2}(c + b)\right] \qquad ...(10.6)$$

When extension link AE is vertical, its obliquity effect is neglected, i.e. $r = r'$

$$mr'\omega^2 = \frac{1}{e}\left[mgc + \frac{(Mg \pm f)}{2}(c + b)\right]$$

or $\qquad mr'\omega^2 = \dfrac{a}{e}\left[mg \cdot \dfrac{c}{a} + \dfrac{(Mg \pm f)}{2}\left\{\dfrac{c}{a} + \dfrac{b}{a}\right\}\right]$

$$= \frac{a}{e}\left[mg\tan\theta + \frac{(Mg \pm f)}{2}(\tan\theta + \tan\beta)\right]$$

(From geometry, $\qquad \dfrac{c}{a} = \tan\theta, \dfrac{b}{a} = \tan\beta$)

or $\qquad mr'\omega^2 = \dfrac{a}{e}\tan\theta\left[mg + \dfrac{(Mg \pm f)}{2}\left(1 + \dfrac{\tan\beta}{\tan\theta}\right)\right] \qquad ...(10.6B)$

But $\qquad \tan\theta = \dfrac{r}{h} = \dfrac{r'}{h}$ and $\dfrac{\tan\beta}{\tan\theta} = k$

$\therefore \qquad mr'\omega^2 = \dfrac{a}{e} \times \dfrac{r'}{h}\left[mg + \dfrac{(Mg \pm f)}{2}(1 + k)\right]$

or $\qquad \omega^2 = \dfrac{a}{e} \times \dfrac{g}{h}\left[\dfrac{2mg + (Mg \pm f)(1 + k)}{2mg}\right]$

(multiplying and dividing by g)

or
$$\left(\frac{2\pi N}{60}\right)^2 = \frac{a}{e} \times \frac{g}{h}\left[\frac{2mg + (Mg \pm f)(1+k)}{2mg}\right]$$

or
$$N^2 = \frac{895}{h} \times \frac{a}{e}\left[\frac{2mg + (Mg \pm f)(1+k)}{2mg}\right] \qquad \qquad ...(10.7)$$

(i) When
$$\tan \beta = \tan \theta, \ k = 1$$

$$N^2 = \frac{895}{h} \times \frac{a}{e}\left[\frac{mg + (Mg \pm f)}{mg}\right] \qquad \qquad ...(10.8)$$

(ii) If friction is absent, i.e. $f = 0$

$$N^2 = \frac{895}{h} \times \frac{a}{e}\left[\frac{2m + M(1+k)}{2m}\right] \qquad \qquad ...(10.9)$$

(iii) If
$$k = 1 \text{ and } f = 0, \text{ then}$$

$$N^2 = \frac{895}{h} \times \frac{a}{e}\left[\frac{m+M}{m}\right] \qquad \qquad ... (10.10)$$

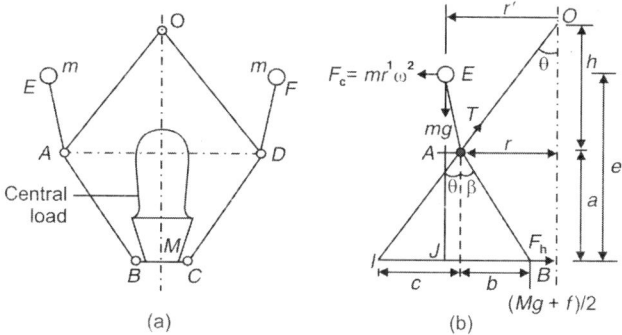

Fig. 10.6

Example 10.5: All the links of a Proell governor are 25.5 cm long and are pivoted on the axis of rotation. The mass of each ball is 4.6 kg and the balls are attached to the extensions of the lower links, which are 6.4 cm long. These extensions are parallel to the governor axis when the radius of rotation of the balls is 15.2 cm. Determine the equilibrium speed at this configuration with a central mass of 36.4 kg on the sleeve.

Solution: $l = 25.5$ cm, $\tan \theta = \tan \beta$, hence $k = 1$

$m = 4.6$ kg, $AE = 6.4$ cm (Refer to Fig. 10.6)

$r = 15.2$ cm, $M = 36.4$ kg, $N = ?$

Using Eq. (10.6b) for $f = 0$ and $k = 1$

$$mr\omega^2 = \frac{h}{e}\tan\theta\left[mg + \frac{Mg}{2}(1+k)\right]$$

or
$$\omega^2 = \frac{h\tan\theta}{emr}(m+M)\,g \qquad \qquad (k=1) \qquad ...(i)$$

$$h = \sqrt{l^2 - r^2} = \sqrt{(25.5)^2 - (15.2)^2} = 20.47 \text{ cm}$$

$$e = h + AE = 20.47 + 6.4 = 26.87 \text{ cm}$$

$$\tan \theta = r/h$$

Substituting in Eq. (i), we have

$$\omega^2 = \frac{h \times r}{emr \times h}(m + M)g = \frac{1}{em}(m + M)g$$

$$= \frac{1}{\left(\dfrac{26.87}{100}\right) \times 4.6}(4.6 + 36.4) \times 9.81 = 325.4$$

or

$$\omega = 18.03 \text{ rad/s}$$

and

$$N = \frac{18.03 \times 60}{2\pi} = 172.26 \text{ rpm.}$$

10.7 HARTNELL GOVERNOR

A Hartnell governor is a spring controlled governor. Two bell-crank levers, each carrying a ball at one end and a roller at the other, are pivoted to the frame at O and O'. The rollers fit into a groove in the sleeve. The frame is attached to the governor spindle and hence rotates with it. A helical spring in compression provides equal downward forces on the two rollers through a collar on the sleeve.

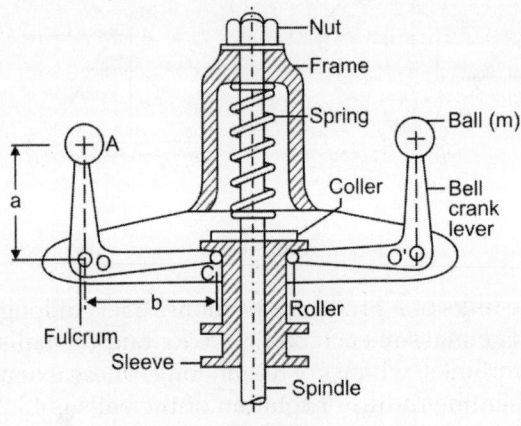

Fig 10.7

When the speed increases and the balls move away from the spindle axis, the bell-crank levers move on the pivot and lift the sleeve against the spring force. If the speed decreases, the sleeve moves downwards. The movement of the sleeve is transferred to the throttle valve of the engine to control the amount of fuel supplied to the engine. The spring force may be adjusted with the help of a nut by a screwing it up or down on the frame.

Let r_1 = minimum radius of rotation at ω_1 speed

r_2 = maximum radius of rotation at ω_2 speed

S_1 = spring force exerted on the sleeve at ω_1

S_2 = spring force exerted on the sleeve at ω_2

F_{c_1} = $mr_1\omega_1^2$ = centrifugal force at ω_1 speed

F_{c_2} = $mr_2\omega_2^2$ = centrifugal force at ω_2 speed

s = stiffness of spring

a = length of vertical arm of lever

b = length of horizontal arm of lever

h = sleeve lift or compression of spring when radius of rotation changes from r_1 to r_2.

Figure 10.8 shows the forces acting on the bell crank lever at minimum and maximum radius positions.

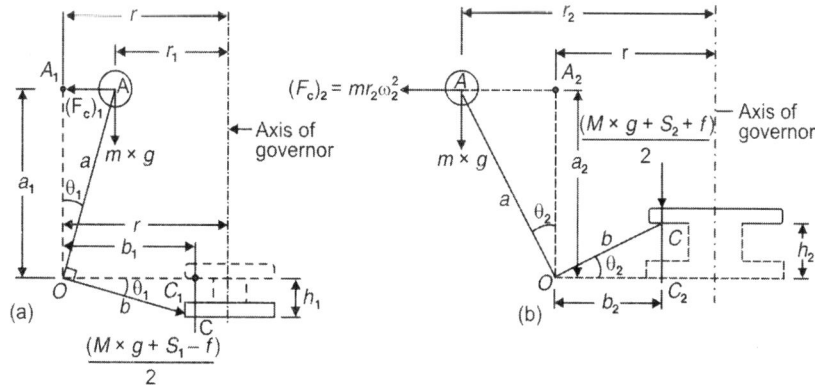

Fig. 10.8: (a) Position of minimum radius (b) Position of maximum radius

(i) Minimum radius position: From Fig. 10.8a,

$$\theta_1 = \frac{CC_1}{OC} = \frac{AA_1}{OA} \text{ or } \frac{h_1}{b} = \frac{r - r_1}{a}$$

or

$$h_1 = \frac{b}{a}(r - r_1) \qquad \text{...(10.10a)}$$

Taking moment about fulcrum O,

$$F_{c_1} \times a_1 = \frac{(Mg + S_1 - f)}{2} b_1 + mg\,(r - r_1) \qquad AA_1 = (r - r_1) \quad \text{...(10.10b)}$$

(ii) Maximum radius position: From Fig. 10.8b,

$$\theta_2 = \frac{CC_2}{OC} = \frac{AA_2}{OA} \text{ or } \frac{h_2}{b} = \frac{r_2 - r}{a}$$

∴

$$h_2 = \frac{b}{a}(r_2 - r) \qquad \text{...(10.10c)}$$

Adding Eqs (10.10a) and (10.10c), total sleeve lift, $h = h_1 + h_2$

or

$$h = \frac{b}{a}(r - r_1) + \frac{b}{a}(r_2 - r) = \frac{b}{a}(r - r_1 + r_2 - r)$$

or

$$h = \frac{b}{a}(r_2 - r_1) \qquad \text{...(10.10d)}$$

Taking moment about fulcrum O',

$$F_{c_2} \times a_2 = \left(\frac{Mg + S_2 + f}{2}\right) b_2 - mg \, (r_2 - r) \qquad \text{...(10.10e)}$$

Since θ is generally small, obliquity effects of arms of the bell-crank levers are neglected, i.e. $a_1 = a_2 = a$, $b_1 = b_2 = b$ and $r - r_1 = 0$, $r_2 - r = 0$

Then Eqs (10.10b) and (10.10e) reduces to

$$F_{c_1} \times a = \left(\frac{Mg + S_1 - f}{2}\right) b \qquad \text{...(10.10f)}$$

and

$$F_{c_2} \times a = \left(\frac{Mg + S_2 + f}{2}\right) b \qquad \text{...(10.10g)}$$

Subtracting Eq. (10.10f) from Eq. (10.10g) and neglecting friction, i.e. $f = 0$

$$(F_{c_2} - F_{c_1}) \, a = \frac{(S_2 - S_1)}{2} b$$

or

$$(S_2 - S_1) = \frac{2a}{b} (F_{c_2} - F_{c_1}) \qquad \text{...(10.10h)}$$

But spring force = spring stiffness × sleeve lift

$$(S_2 - S_1) = s \times h$$

$$= s \times \frac{b}{a} (r_2 - r_1) \qquad \text{[from Eq. (iv)] \quad ...(10.10i)}$$

From Eqs (10.10h) and (10.10i),

$$s \times \frac{b}{a} (r_2 - r_1) = \frac{2a}{b} (F_{c_2} - F_{c_1})$$

or

$$s = 2\frac{a^2}{b^2} \times \left[\frac{F_{c_2} - F_{c_1}}{r_2 - r_1}\right] \qquad \text{...(10.11)}$$

Since stiffness of the spring is constant for all positions,

$$\left[\frac{F_{c_2} - F_{c_1}}{r_2 - r_1}\right] = \frac{F_c - F_{c_1}}{r - r_1} = \frac{F_{c_2} - F_c}{r_2 - r}$$

Where F_c is centrifugal force at intermediate position 'r'.

$$\therefore \qquad F_c = F_{c_1} + (F_{c_2} - F_{c_1})\left[\frac{r - r_1}{r_2 - r_1}\right] \qquad \text{...(10.12)}$$

and

$$F_c = F_{c_2} + (F_{c_2} - F_{c_1})\left[\frac{r_2 - r}{r_2 - r_1}\right] \qquad \text{...(10.13)}$$

Example 10.6: In a spring governor of Hartnell type, the mass of each ball is 1 kg, length of vertical arm of the bell crank lever is 100 mm and that of horizontal arm is 50 mm. The distance of fulcrum of each bell crank lever is 80 mm from the axis of rotation of the governor. The extreme radii of rotation of the balls are 75 mm and 112.5 mm. The maximum equilibrium speed is 5% greater than the minimum equilibrium speed, which is 360 rpm. Find, neglecting obliquity of arms, initial compression of the spring and equilibrium speed corresponding to the radius of 100 mm.

Solution: $m = 1$ kg, $a = 100$ mm $= 0.1$ m, $b = 50$ mm $= 0.05$ m, $r = 80$ mm $= 0.08$ m, $r_1 = 75$ mm $= 0.075$ m, $r_2 = 112.5$ mm $= 0.1125$ m, $N_1 = 360$ rpm or

$$\omega_1 = \frac{2\pi N_1}{60} = \frac{2\pi \times 360}{60} = 37.68 \text{ rad/s}$$

$$\omega_2 = 105\,\omega_1 = 1.05 \times 37.68 = 39.56 \text{ rad/s}$$

Centrifugal force at minimum speed, $F_{c_1} = mr_1\,\omega_1^2$

$$F_{c_1} = 1 \times 0.075 \times (37.68)^2 = 106.6 \text{ N}$$

Centrifugal force at maximum speed, $F_{c_2} = mr^2\omega^2_2$

$$F_{c_2} = 1 \times 0.1125 \times (39.56)^2 = 176.4 \text{ N}$$

Neglecting friction and obliquity of arms, at minimum speed,

$$F_{c_1} \times a = \left(\frac{Mg + S_1}{2}\right)b \Rightarrow S_1 = 2F_{c_1}\frac{a}{b} - Mg$$

or
$$S_1 = 2 \times 106.6 \times \frac{100}{50} - 0 = 426.4 \text{ N}$$

Similarly,
$$S_2 = 2F_{c_2} \times \frac{a}{b} - Mg = 2 \times 176.4 \times \frac{100}{50} - 0 = 705.6 \text{ N}$$

Sleeve lift,
$$h = (r_2 - r_1)\frac{a}{b} = (112.5 - 75) \times \frac{50}{100}$$

$$= 18.75 \text{ mm} = 0.01875 \text{ m}$$

Stiffness of the spring, $s = \dfrac{S_2 - S_1}{h} = \dfrac{705.6 - 426.4}{0.01875} = 14890 \text{ N/m} = 14.89 \text{ N/mm}$

Initial compression of spring $= \dfrac{S_1}{s} = \dfrac{426.4}{14.89} = 28.6 \text{ mm}$

Centrifugal force at any instant,

$$F_c = F_{c_1} + (F_{c_2} - F_{c_1})\frac{r - r_1}{r_2 - r_1} \qquad \text{(at } r = 100 \text{ mm} = 0.1 \text{ m)}$$

$$= 106.6 + (176.4 - 106.6) \times \frac{(0.1 - 0.075)}{(0.1125 - 0.075)}$$

$$= 153 \text{ N}$$

But
$$F_c = mr\omega^2$$

or
$$153 = 1 \times 0.1\left(\frac{2\pi N}{60}\right)^2 = 0.0011N^2$$

$\Rightarrow$
$$N = 373 \text{ rpm.}$$

10.8 HARTUNG GOVERNOR

Hartung governor is a spring controlled governor in which the vertical arms of bell-crank lever are fitted with the spring balls. The spring balls press against the frame of the governor when the rollers at the horizontal arm press against the sleeve.

Let m = mass of each ball (kg)

M = mass of sleeve (kg)

S = spring force (N)

F_c = centrifugal force (N) = $mr\omega^2$

s = stiffness of the spring

a = length of the vertical arm of bell crank lever

b = length of the horizontal arm of bell crank lever

r = radial distance of the masses

ω = angular velocity of balls at radius r

r_0 = radius at which spring force is zero.

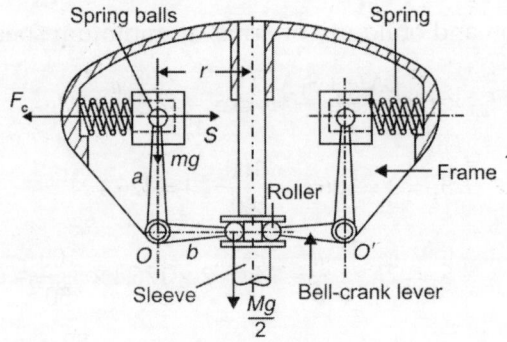

Fig. 10.9

Neglecting the obliquity effect of the arm and taking moment about fulcrum O,

$$F_c \times a = S \times a + \frac{Mg}{2} \times b$$

or

$$mr\omega^2 \times a = s\,(r - r_0) \times a + \frac{Mg}{2} \times b \qquad\qquad [S = s \times (r - r_0)]$$

Example 10.7: In a spring controlled Hartung type of governor, the length of the ball arm is 84 mm and the sleeve arm 126 mm. When in the mid position, each spring is compressed by 60 mm and the radius of rotation of the mass centres is 160 mm. The mass of the sleeve is 18 kg and of each ball 4 kg. Spring stiffness is 12 kN/m of compression and total lift of the sleeve 24 mm. Determine the ratio of the range of speed to the mean speed of the governor. Also, find the speed in the mid position. Neglect the moment due to the revolving masses when the arms are inclined.

Solution: m = 4 kg, r = 160 mm, M = 18 kg, s = 12 kN/m, a = 84 mm, r_0 = 160 – 60 = 100 mm, b = 126 mm, h = 24 mm

In the mid-pisition (Fig. 10.10a),

$$mr\omega^2 \cdot a = s(r - r_0) \cdot a + \frac{Mg}{2} \cdot b$$

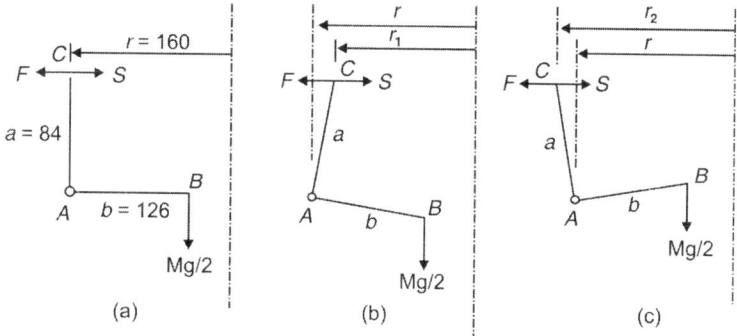

Fig. 10.10

$$4 \times 0.16\omega^2 \times 0.084 = 12000 \times 0.06 \times 0.084 + \frac{18 \times 9.81}{2} \times 0.126$$

$$= 60.48 + 11.125$$

$$\omega^2 = 1332$$

$$\omega = 36.5 \text{ rad/s}$$

$$N = \frac{36.5 \times 60}{2\pi} \text{ or } 348.5 \text{ rpm}$$

so, mean speed = 348.5 rpm

For the minimum speed, from the Fig. 10.10b, (neglecting obliquity of arms)

$$\frac{r - r_1}{a} = \frac{h}{b}$$

or

$$r_1 = r - h \cdot \frac{a}{b} = 0.16 - \frac{0.24}{2} \cdot \frac{0.084}{0.126} = 0.152 \text{ mm}$$

$$4 \times 0.152\omega_1^2 \times 0.084 = 12000 \times (0.152 - 0.1) \times 0.084 + \frac{18 \times 9.81}{2} \times 0.126$$

$$= 52.416 + 11.125$$

$$\omega_1^2 = 1244, \ \omega_1 = 35.27 \text{ rad/s}$$

$$N_1 = \frac{35.27 \times 60}{2\pi} \text{ or } 336.8 \text{ rpm}$$

∴ Minimum speed = 336.8 rpm

For the maximum speed, from Fig. 10.10c, (neglecting obliquity of arms)

$$\frac{r_2 - r}{a} = \frac{h}{b}$$

or

$$r_2 = r + h \cdot \frac{a}{b} = 0.16 + \frac{0.24}{2} \cdot \frac{0.084}{0.126} = 0.168 \text{ mm}$$

$$4 \times 0.168\omega_2^2 \times 0.084 = 12000 \times (0.168 - 0.1) \times 0.084 + \frac{18 \times 9.81}{2} \times 0.126$$

$$= 68.544 + 11.125$$

$$\omega_2^2 = 1411.4, \ \omega_2 = 37.57 \ \text{rad/s}$$

$$N_2 = \frac{37.57 \times 60}{2\pi} \ \text{or} \ 358.75 \ \text{rpm}$$

∴   Maximum speed = 358.75 rpm

Range of speed = 358.75 − 336.8 = 21.95 rpm

Ratio of range of speed to mean speed = $\dfrac{21.95}{348.5}$ = 0.063

10.9 WILSON–HARTNELL GOVERNOR

A Wilson–Hartnell governor is a spring loaded governor. The vertical arms of the bell crank lever support the balls and the horizontal arms carry two rollers at their ends. The balls are connected by two springs known as main springs which are arranged symmetrically on either side of the sleeve. When the speed of the engine increases, the ball radius increases, the springs exert an inward pull S on the balls and the rollers press against the sleeve, which is raised and closes the throttle valve.

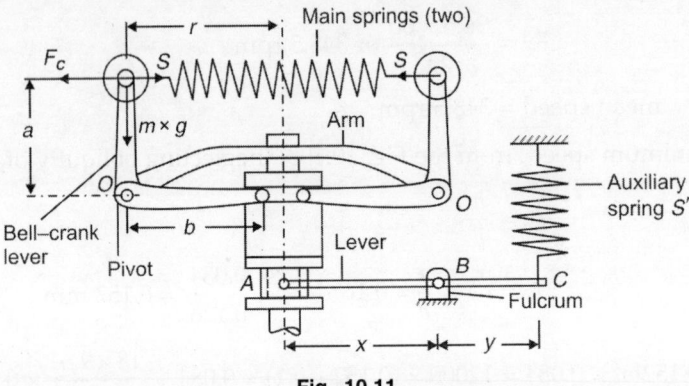

Fig. 10.11

An adjustable auxiliary spring is attached to the sleeve through a lever. The other end of the lever fits into the groove in the sleeve. The lever is pivoted at the fulcrum B. The auxiliary spring tends to keep the sleeve down and helps main springs (two are simultaneously in tension).

Let m = mass of each ball

M = mass of sleeve

s_m = stiffness of each of the main spring

s_a = stiffness of the auxiliary spring

S' = force applied by the auxiliary spring.

The total downward force on the sleeve = weight of sleeve + force due to auxiliary spring S' ± friction force = $Mg + S' \times \dfrac{y}{x} \pm f$

For minimum speed, taking moment about O,

$$(F_{c_1} - S_1)a_1 = \frac{1}{2}\left(Mg + S_1' \times \frac{y}{x} - f \right)b_1 + mgc_1 \qquad \text{...(10.14)}$$

For maximum speed, taking moment about O,

$$(F_{c_2} - S_2)a_2 = \frac{1}{2}\left(Mg + S_2' \times \frac{y}{x} - f\right)b_2 + mgc_2 \qquad ...(10.15)$$

Neglecting obliquity effect and friction, i.e.

$$a_1 = a_2 = a,\ b_1 = b_2 = b \text{ and } c_1 = c_2 = 0 \text{ and } f = 0$$

$$(F_{c_2} - S_2)a = \frac{1}{2}\left(Mg + S_1'\frac{y}{x}\right)b \qquad ...(10.15a)$$

$$(F_{c_2} - S_2)a = \frac{1}{2}\left(Mg + S_2'\frac{y}{x}\right)b \qquad ...(10.15b)$$

Subtracting Eq. (10.15a) from Eq. (10.15b)

$$(F_{c_2} - F_{c_1})a - (S_2 - S_1)a = (S_2' - S_1')\frac{yb}{2x} \qquad ...(10.15c)$$

There are two main springs. Extension of each spring is $2(r_2 - r_1)$. Therefore, force exerted by main springs,

$$(S_2 - S_1) = 2 \times \text{force exerted by each spring}$$
$$= 2 \times \text{stiffness of each spring} \times \text{extension of each spring}$$
$$= 2 \times s_m \times 2\,(r_2 - r_1)$$
$$= 4\,s_m(r_2 - r_1) \qquad ...(10.15d)$$

Let $\qquad h_1 =$ total sleeve lift from r_1 to r_2

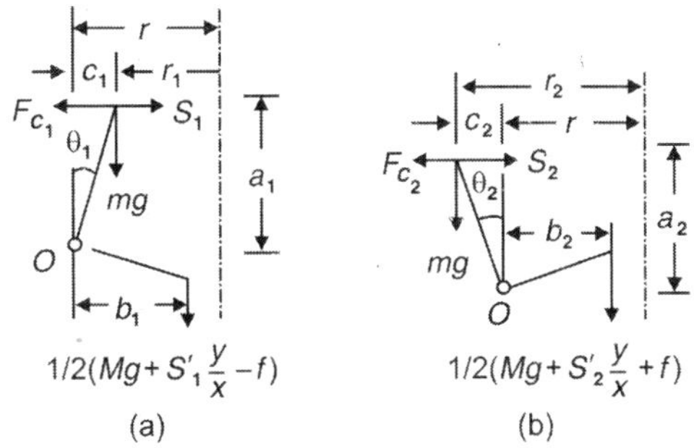

Fig. 10.12

Angle turned by ball arm and sleeve arm are same, hence,

$$\frac{h_1}{b} = \frac{r_2 - r_1}{a} \text{ or } h_1 = \frac{b}{a}\,(r_2 - r_1)$$

Let $\qquad h_2 =$ deflection of auxiliary spring in downward direction,

then (from Fig. 10.13)

$$\frac{h_1}{x} = \frac{h_2}{y} \text{ or } h_2 = h_1 \times \frac{y}{x}$$

or

$$h_2 = \frac{b}{a}\,(r_2 - r_1) \times \frac{y}{x}$$

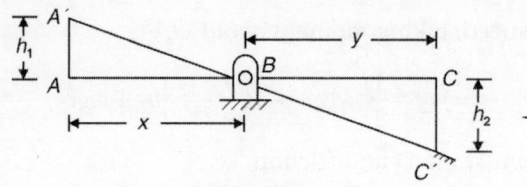

Fig. 10.13

Net force of auxiliary spring, $S_2' - S_1' = s_a h_2$

or $$S_2' - S_1' = s_a \times \frac{b}{a}(r_2 - r_1)\frac{y}{x}$$

substituting values of $(S_2 - S_1)$ and $(S_2' - S_1')$ in Eq. ...(iii)

$$(F_{c_2} - F_{c_1})a - 4s_m a(r_2 - r_1) = s_a \times \frac{b}{a}(r_2 - r_1)\frac{y}{x} \times \frac{yb}{2x}$$

or $$(F_{c_2} - F_{c_1}) = 4s_m(r_2 - r_1) - \frac{s_a}{2}\left(\frac{b}{a}\frac{y}{x}\right)^2$$

Dividing both sides by $(r_2 - r_1)$,

$$\frac{\left(F_{c_2} - F_{c_1}\right)}{\left(r_2 - r_1\right)} = 4s_m + \frac{s_a}{2}\left(\frac{b}{a}\cdot\frac{y}{x}\right)^2 \qquad \qquad ...(10.16)$$

If auxiliary spring is not used, i.e. $s_a = o$, then

$$\frac{\left(F_{c_2} - F_{c_1}\right)}{\left(r_2 - r_1\right)} = 4s_m \qquad \qquad ...(10.17)$$

Example 10.8: The following particulars refer to Wilson–Hartnell governor: mass of each ball = 2 kg, minimum radius = 125 mm, maximum radius = 175 mm, minimum speed = 240 rpm, maximum speed = 250 rpm, length of the ball arm of each bell crank lever = 150 mm, length of sleeve arm of each bell crank lever = 100 mm, stiffness of the each ball springs = 0.2 kN/m. Find the equivalent stifffness of the auxiliary spring referred to the sleeve.

Solution: $m = 2$ kg, $r_2 = 175$ mm, $r_1 = 125$ mm, $N_1 = 240$ rpm, $\omega_1 = \dfrac{2\pi N_1}{60} = \dfrac{2\pi \times 240}{60} = 8\pi = 25.14$ rad/s

$$N_2 = 250 \text{ rpm}, \omega_2 = \frac{2\pi N_2}{60} = \frac{2\pi \times 250}{60} = 26.18 \text{ rad/s}$$

$$a = 150 \text{ mm},$$
$$b = 100 \text{ mm} = .1 \text{ m}$$
$$s_m = 0.2 \text{ kN/m} = 200 \text{ N/m}$$
$$F_{c_1} = mr_1 w_1^2 = 2 \times 0.125 \times (25.14)^2 = 158 \text{ N}$$
$$F_{c_2} = mr_1 w_2^2 = 2 \times 0.175 \times (26.18)^2 = 240 \text{ N}$$

Now, using relation

$$\frac{F_{c_2} - F_{c_1}}{r_2 - r_1} = 4s_m + \frac{s_a}{2}\left(\frac{b}{a} \times \frac{y}{x}\right)^2$$

$$\frac{(240 - 158)}{(0.175 - 0.125)} = 4 \times 200 + \frac{s_a}{2} \times \left(\frac{0.1}{0.15} \times 1\right) \qquad \left(\frac{y}{x} \text{ taken as } 1\right)$$

$$s_a = 3780 \text{ N/m}$$

or $$s_a = 3.78 \text{ kN/m.}$$

10.10 PICKERING GOVERNOR

Pickering governor is used in clock and gramophone. It consists of a central spindle and three straight leaf springs symmetrically placed around it. The upper end of each spring is fixed with spindle through a nut and lower end is fastened to the sleeve. Each spring carries a mass m at its centre.

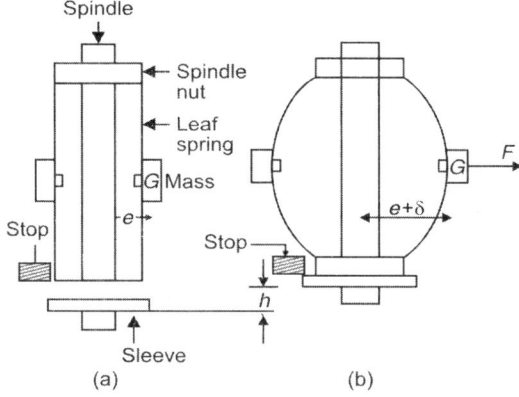

Fig. 10.14

As the speed of the spindle increases, increasing centrifugal force at the centre of the leaf spring causes it to bend. This makes the sleeve to move up. The movement of sleeve is limited by a stop.

Let m = mass attached at the centre of each spring
  e = distance between centre of mass and spindle axis, when governor is at rest
  ω = angular speed of the spindle (rad/s)
  δ = deflection of the centre of leaf spring at angular speed ω
  E = modulus of elasticity of spring material
  I = moment of inertia of the cross-section of spring about

  neutral axis $= \dfrac{bt^3}{12}$ (where b and t are width and thickness of spring)

h = lift of the sleeve at deflection $\delta' = 2.4\dfrac{\delta^2}{l}$

Centrifugal force, $F_c = m(e + \delta)\omega^2$

Deflection of leaf spring with both ends fixed and carrying a load at the centre,

$$\delta' = \frac{F_c l^3}{192EI} = \frac{m(e + \delta)\omega^2 l^3}{192EI} \qquad \text{...(10.18)}$$

Example 10.9: A gramophone is driven by a pickering governor. The mass of each disc attached to the centre of a leaf spring is 20 g. Each spring is 5 mm wide and 0.125 mm thick. The effective length of each spring is 40 mm. The distance from the spindle axis to the centre of gravity of the mass when the governor is at rest is 10 mm. Find the speed of the turntable when the sleeve has risen 0.8 mm and the ratio of the governor speed to the turntable speed is 10.5. Take $E = 210$ kN/mm².

Solution: $m = 20$ g $= 0.02$ kg, $e = 10$ mm $= 0.01$ m, $E = 210$ kN/mm² $= 210 \times 10^3$ N/mm², $b = 5$ mm, $t = 0.125$ mm.

The moment of inertia of the spring about its neutral axis,

$$I = \frac{bt^3}{12} = \frac{5(0.125)^3}{12} = 0.8 \times 10^{-3} \text{ mm}^4$$

Since the effective length of each spring is 40 mm and lift of sleeve $(h) = 0.8$ mm, therefore, length of spring between fixed ends,

$$l = 40 - 0.8 = 39.2 \text{ mm}$$

Sleeve lift, $\qquad h = \dfrac{2.4\delta^2}{l}$

or $\qquad 0.8 = \dfrac{2.4\delta^2}{l} = \dfrac{2.4\delta^2}{39.2} = 0.06\delta^2$

∴ $\qquad \delta^2 = 0.8/0.06 = 13.3$ or $\delta = 3.65$ mm

Let $\qquad N = $ speed of the governor

$\qquad N_1 = $ speed of the turntable

then $\qquad \dfrac{N}{N_1} = 10.5$

Now, $\qquad \delta = \dfrac{m(e+\delta)\omega^2 l^3}{192EI}$

$$3.65 = \frac{0.02(10+3.65)\omega^2 (39.2)^3}{192 \times 210 \times 10^3 \times 0.8 \times 10^{-3}}$$

$$= \frac{16445\omega^2}{32256} = 0.51\omega^2$$

or $\qquad \omega^2 = \dfrac{3.65}{0.51} = 7.156$

or $\qquad \omega = 2.675$ rad/s

$$N = \omega \times \frac{60}{2\pi} = 2.675 \times \frac{60}{2\pi}$$

$$N = 25.5 \text{ rpm}$$

∴ $\qquad N_1 = \dfrac{N}{10.5} = \dfrac{25.5}{10.5} = 2.43$ rpm.

10.11 SPRING CONTROLLED GRAVITY GOVERNOR

Figure 10.15a shows a spring controlled governor which consists of two bell crank levers pivoted on the moving sleeve. The rollers at the ends of the horizontal arms of levers press against a cap fixed to the governor spindle. The spring is compressed between the cap and the sleeve. As the speed increases, the balls move outwards and the cap is pressed downwards by the rollers.

Let S_1 = spring force at minimum radius position

S_2 = spring force at maximum radius position

For minimum radius position, taking moment about B,

$$\left[mg + \frac{(Mg + S_1)}{2} \right] \times b_1 = F_{c_1} \times a_1 \qquad \qquad ...(10.19)$$

For the minimum radius position shown in Fig. 10.15b, the instantaneous centre lies at B whereas for the maximum radius position shown in Fig. 10.15c, the instantaneous centre lies at I.

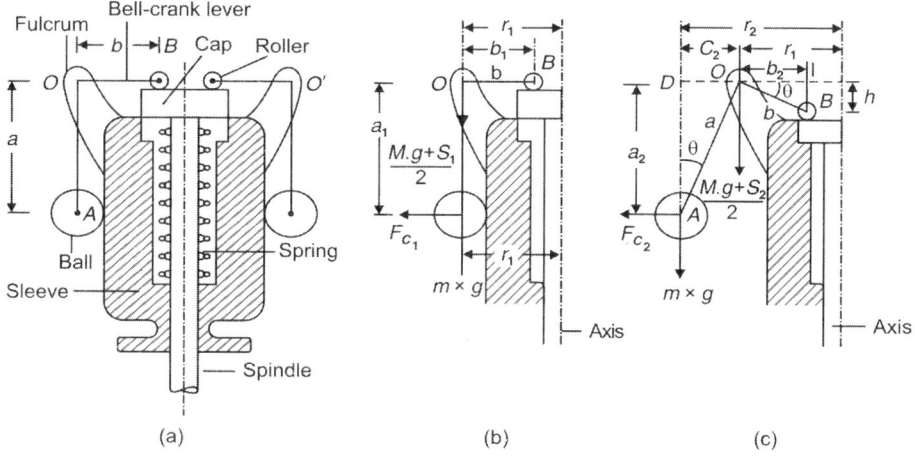

Fig. 10.15

For maximum radius position, taking moment about I,

$$mg + ID + \frac{Mg + S_2}{2} \times I_o = F_{c_2} \times a_2 \qquad \qquad ...(10.20)$$

Here, $\qquad\qquad F_{c_2} = mr_1\omega_1^2$ and $F_{c_2} = mr_2\omega_2^2$

Example 10.10: In a spring controlled gravity governor, the mass of each ball is 1.6 kg. Distance of fulcrum from the axis of rotation is 60 mm. The bell-crank lever has a vertical arm 120 mm long and a horizontal arm 50 mm long. The mass of the sleeve is 6.5 kg. The sleeve begins to rise at 200 rpm and the rise of sleeve for 5% increase is 9 mm. Determine the initial thrust in the spring and its stiffness.

Solution: m = 1.6 kg, M = 6.5 kg, r_1 = 60 mm, N_1 = 200 rpm, $a = a_1$ = 200 mm, $b = b_1$ = 50 mm, h = 9 mm

$$\omega_1 = \frac{2\pi \times 200}{60} = 20.94 \text{ rad/s}$$

(i) For initial position, taking moment abolut B,

$$mr_1^2\omega_1^2 a_1 = mgb_1 + \frac{Mg + S_1}{2}b_1$$

or $\quad [1.60 \times 0.06 \times (20.94)^2] \times 0.12 = \left[1.6 \times 9.891 + \frac{6.5 \times 9.81 + S_1}{2}\right] \times 0.05$

∴ $\qquad\qquad\qquad\qquad S_1 = 107$ N

(ii) When sleeve rises through 9 mm, radius is increased by c_2

$$c_2 = 9 \times \frac{120}{50} = 21.6 \text{ mm}$$

or $\qquad\qquad r_2 = 60 + 21.6 = 81.6$ mm

$$\omega_2 = 20.94 \times 1.05 = 21.99 \text{ rad/s}$$

$$a_2 = \sqrt{a^2 - c_2^2} = \sqrt{(120)^2 - (21.6)^2} = 118 \text{ mm}$$

$$b_2 = \sqrt{(50)^2 - (9)^2} = 49.2 \text{ mm}$$

Taking moments about I,

$$(mr_2\omega_2^2)a_2 = mg(b_2 + c_2) + \frac{Mg + S_2}{2}b_2$$

$$1.6 \times 0.816 \times (21.99)^2 \times 0.118 = 1.6 \times 9.81(0.0492 + 0.0216)$$

$$+ \frac{6.5 \times 9.81 + S_2}{2} \times 0.0492$$

$$S_2 = 193.8 \text{ N}$$

Stiffness of spring $\qquad = \dfrac{S_2 - S_1}{h} = \dfrac{193.8 - 107}{9} = 9.6$ N/mm.

10.12 SENSITIVENESS OF A GOVERNOR

A governor is said to be sensitive if it quickly responds to a small change in speed.

The sensitiveness of a governor is defined as the ratio of the difference between the maximum and the minimum speeds (i.e. range of speed) to the mean equilibrium speed.

$$\text{Sensitiveness} = \frac{\text{range of speed}}{\text{mean speed}}$$

If $\qquad N_1$ = minimum equilibrium speed

$\qquad\qquad N_2$ = maximum equilibrium speed

$\qquad N$ = mean equilibrium speed = $\dfrac{N_1 + N_2}{2}$

then,

$$\text{sensitiveness} = \frac{N_2 - N_1}{N} = \frac{(N_2 - N_1)}{\left(\dfrac{N_1 + N_2}{2}\right)}$$

$$= \frac{2(N_2 - N_1)}{(N_1 + N_2)} \qquad\qquad\qquad\text{...(10.21)}$$

10.13 HUNTING

If the speed of the engine controlled by the governor fluctuates continuously above and below the mean speed, the governor is said to be hunting. This is caused by a too sensitive governor that changes the fuel supply by a large amount when a small change in the speed of rotation takes place.

When the load on engine increases, the sleeve falls to its lowest position and opens fuel supply to the engine in excess of its requirerment so that the speed increases rapidly. As a rersult, now sleeve rises to its highest position and closes the fuel supply to the engine so that the speed falls below the mean value. This process of fluctuation in speed continues.

10.14 ISOCHRONISM

A governor is said to be isochronous if the equilibrium speed is constant for all radii of rotation of the balls within the working range, or if its range of speed is zero, i.e. $N_1 - N_2 = 0$.

For a Porter governor,

$$N_1^2 = \frac{m + \dfrac{M}{2}(1+k)}{m} \times \frac{895}{h_1}$$

$$N_2^2 = \frac{m + \dfrac{M}{2}(1+k)}{m} \times \frac{895}{h_2}$$

For isochronous governor, $N_1 = N_2$.

∴ $h_1 = h_2$, which is not possible.

Therefore, a Porter governor cannot be isochronous.

For a Hartnell governor (neglecting friction),

$$m r_1 \omega_1^2 \times a = \frac{1}{2}(Mg + S_1) \times b$$

and $$m r_2 \omega_2^2 \times a = \frac{1}{2}(Mg + S_2) \times b$$

For isochronism, $\omega_1 = \omega_2$

∴ $$\frac{Mg + S_1}{Mg + S_2} = \frac{r_1}{r_2}$$ ...(10.22)

10.15 STABILITY

A governor is said to be stable when for every speed within working range, there is only one radius of rotation of the governor balls at which the governor is in equilibrium.

A governor is said to be unstable, if the radius of rotation decreases as the speed increases.

10.16 EFFORT OF A GOVERNOR

Effort of a governor is the mean force that acts on the sleeve for 1 % change in speed.

Effort of a governor is the mean force acting on the sleeve to raise or lower it for a given change of speed. When the speed is constant, the force acting on the sleeve is zero. But when the speed changes, a force (F) is exerted on the sleeve which tends to move it.

Mean force $\left(\dfrac{0+F}{2} = \dfrac{F}{2}\right)$ is the effort of the governor.

For a Porter governor,

$$h = \frac{mg + \dfrac{Mg}{2}(1+k)}{mg} \times \frac{895}{N^2} \qquad \qquad ...(10.22a)$$

If speed is increased from N to $(1 + c) N$ and F is the force applied on the sleeve to prevent it from moving,

then $$h = \frac{mg + \dfrac{(Mg + F)}{2}(1+k)}{mg} \times \frac{895}{(1+c)^2 N^2} \qquad \qquad ...(10.22b)$$

Dividing Eq. (10.22b) by Eq. (10.22a) and simplifying gives

$$F = \frac{2c}{(1+k)}\,[2mg + Mg\,(1+k)] \qquad \qquad ...(10.23)$$

Effort, $$\frac{F}{2} = \frac{cg}{(1+k)}\,[2m + M(1+k)] \qquad \qquad ...(10.24)$$

1. When $\tan\theta = \tan\beta$ or $k = 1$,

effort, $$\frac{F}{2} = cg\,(m + M) \qquad \qquad ...(10.25)$$

2. If friction of sleeve is also considered,

effort, $$\frac{F}{2} = c\,(mg + Mg + f) \qquad \qquad ...(10.26)$$

3. For a Watt governor, $M = 0$

effort, $$\frac{F}{2} = cmg \qquad \qquad ...(10.27)$$

For a Hartnell governor,

$$mr\omega^2\,a = \frac{1}{2}(Mg + S)\,b \qquad \qquad ...(10.27a)$$

If speed is increased from ω to $(1 + c)\omega$ and F is the force applied on the sleeve to prevent its movement,

$$mr\,(1 + c)^2\omega^2 a = \frac{1}{2}(Mg + S + F)b \qquad \qquad ...(10.27b)$$

Dividing Eq. (10.27b) by Eq. (10.27a)

$$\frac{1}{(1+c)^2} = \frac{Mg+S}{Mg+S+F}$$

or $$\frac{Mg+S+F}{Mg+S} = (1+c)^2 = 1 + c^2 + 2c$$

or $$\frac{F}{Mg+S} = 1 + c^2 + 2c - 1 = 2c \qquad \text{(by } C - D \text{ and neglecting } c^2)$$

Effort, $$\frac{F}{2} = c(Mg + S) \qquad \qquad ...(10.28)$$

10.17 POWER OF A GOVERNOR

The power of a governor is the work done on the sleeve for a given percentage change of speed.

$$\text{Power} = \text{effort} \times \text{displacement of sleeve}$$

For Porter governor:

When speed changes from N to $(1 + c)N$, height of governor changes from h to h_1.

Now, $$h = \frac{mg + \dfrac{Mg}{2}(1+k)}{mg} \times \frac{895}{N^2} \qquad \qquad ...(10.28a)$$

and, $$h_1 = \frac{mg + \dfrac{Mg}{2}(1+k)}{mg} \times \frac{895}{(1+c)^2 N^2} \qquad \qquad ...(10.28b)$$

∴ $$\frac{h_1}{h} = \frac{1}{(1+c)^2} \qquad \qquad ...(10.28c)$$

Displacement of sleeve, $x = 2(h - h_1)$

or $$x = 2h\left(1 - \frac{h_1}{h}\right)$$

$$= 2h\left[1 - \frac{1}{(1+c)^2}\right] \qquad \qquad [\textit{from Eq. (10.28c)}]$$

or $$x = 2h\left[\frac{(1+c)^2 - 1}{(1+c)^2}\right] = 2h\left[\frac{1+c^2+2c-1}{1+c^2+2c}\right]$$

$$x = 2h\left[\frac{2c}{1+2c}\right] \qquad \qquad \text{(neglecting } c^2) \quad ...(10.28d)$$

Power of governor = effort × displacement

$$= cg(m + M) \times 2h\left[\frac{2c}{1+2c}\right] \qquad [\textit{from Eqs (10.25) and (10.28d)}]$$

$$= \frac{4c^2}{(1+2c)}(m + M)\, gh \qquad \qquad ...(10.29)$$

If $k \neq 1$, displacement of sleeve $\approx (1 + k)(h - h_1)$

or $\qquad x = (1 + k)h \times \left[\dfrac{2c}{1+2c}\right]$ ...(10.29a)

$\therefore$ Power $= \dfrac{cg}{(1+k)} \; [2m + M(1 + k)] \times (1 + k) \, h \times \left[\dfrac{2c}{1+2c}\right]$ [*from* Eqs (10.24) and (10.29a)]

$\qquad = \dfrac{4c^2}{(1+2c)} \times \left[m + \dfrac{M}{2}(1+k)\right] gh$ ...(10.30)

Example 10.11: The mass of each ball of a Porter governor is 4 kg and mass of sleeve is 20 kg. It has equal arms each 200 mm in length and pivoted on the axis of rotation. The radius of rotation of the ball is 100 mm when the governor begins to lift. If the fractional increase of speed is 1%, then determine the governor effort and power.

Solution: $m = 4$ kg, $M = 20$ kg, $r = 100$ mm $= 0.1$ m, fractional increase of speed $= 1\% = 0.1 \Rightarrow C = 0.01$

$$l = 200 \text{ mm}$$

$$h = \sqrt{(200)^2 - (100)^2} = 173.2 \text{ mm} = 0.1732 \text{ m}$$

As arms are of equal length and pivoted on the axis, $\tan \theta = \tan \beta$ or $k = 1$

Effort $\qquad = cg\,(m + M)$

$\qquad\qquad = 0.01 \times 9.81 \times (4 + 20) = 2.354$ N

$\qquad$ Power of governor $= \dfrac{4c^2}{(1+2c)}\,(m + M)\,gh$

$$= \dfrac{4 \times (0.01)^2}{1 + 2 \times 0.01}\,(4 + 20) \times 9.81 \times 0.1732$$

$$= 0.016 \text{ N m}$$

$$= 16 \text{ N mm}$$

Example 10.12: Each ball of a Porter governor has a mass of 3 kg and the mass of the sleeve is 15 kg. The governor has equal arms each 200 mm long and pivoted on the axis of rotation. When the radius of rotation of the balls is 120 mm, the sleeve begins to rise and 160 mm at the maximum speed. Determine:

   (i) The range of speed
   (ii) The lift of the sleeve
   (iii) The effort of the governor
   (iv) The power of the governor.

What will be the effect of friction at the sleeve if it is equivalent to 8 N?

Solution:

   Refer Fig. 10.16,

$$h_1 = \sqrt{0.2^2 - 0.12^2} = 0.16 \text{ m}$$

$$h_2 = \sqrt{0.2^2 - 0.16^2} = 120 \text{ mm} = 0.12 \text{ m}$$

$$N_1^2 = \dfrac{895}{h_1}\left(\dfrac{m+M}{m}\right) = \dfrac{895}{0.16}\left(\dfrac{3+15}{3}\right) = 33563$$

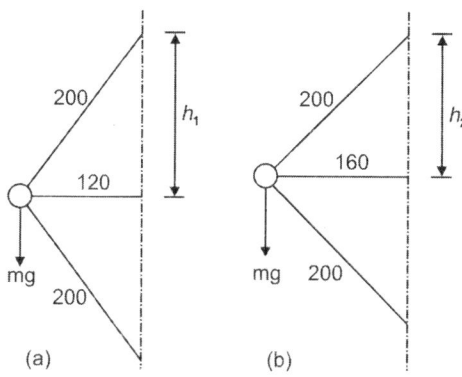

Fig. 10.16

$$N_1 = 183.2 \text{ rpm}$$

And
$$N_2^2 = \frac{895}{0.12}\left(\frac{3+15}{3}\right) = 44750 \text{ or } N_2 = 212.5 \text{ rpm}$$

(i) Range of speed $= 212.5 - 183.2 = 29.3$ rpm

(ii) Lift of sleeve $= 2(h_1 - h_2) = 2(0.16 - 0.12) = 0.08$ m

(iii) Effort $= (m + M)\, cg$

where $cN = (212.5 - 183.2) = 29.2$ or $c = 29.2/183.2 = 0.16$

or

effort $= (3 + 15) \times 0.16 \times 9.81 = 28.3$ N

(iv)
$$\text{Power} = (m + M)gh\left(\frac{4c^2}{1+2c}\right)$$

$$= (3 + 15) \times 9.81 \times 0.16 \left(\frac{4 \times 0.16^2}{1 + 2 \times 0.16}\right) = 2.26 \text{ N·m}$$

or

power $=$ effort $\times$ displacement $= 28.3 \times 0.08 \approx 2.2 \approx$ N·m

When friction is considered

$$N_1^2 = \frac{895}{h_1}\left(\frac{mg + (Mg - f)}{mg}\right)$$

$$= \frac{895}{0.16}\left(\frac{3 \times 9.81 + (15 \times 9.81 - 8)}{3 \times 9.81}\right) = 32042$$

$$N_1 = 179 \text{ rpm}$$

$$N_2^2 = \frac{895}{h_2}\left(\frac{mg + (Mg + f)}{mg}\right)$$

$$= \frac{895}{0.12}\left(\frac{3 \times 9.81 \times (15 \times 9.81 + 8)}{3 \times 9.81}\right) = 46777$$

$$N_2 = 216.3 \text{ rpm}$$

(i) Range of speed $= 216.3 - 179 = 37.3$ rpm

(ii) Lift of sleeve $=$ same as before $= 0.08$ m

(iii) Effort = $(mg + Mg + f)c$

where $\qquad\qquad c = 37.3/179 = 0.208$ or

∴ Effort $\qquad\qquad = (3 \times 9.81 + 15 \times 9.81 + 8) \times 0.208 = 38.4$ N

(iv) $\qquad$ Power = effort × displacement = $38.4 \times 0.08 = 3.07$ N·m

Example 10.13: In a Hartnell governor, the radius of rotation of the balls is 60 mm at the minimum speed of 240 rpm. The length of the ball arm is 130 mm and of the sleeve arm 80 mm. The mass of each ball is 3 kg and of the sleeve 4 kg. The stiffness of the spring is 20 N/mm. Determine:

(i) The speed when the sleeve is lifted by 50 mm
(ii) The initial compression of the spring
(iii) The governor effort
(iv) The power.

Solution:

$$a = 130 \text{ mm}, b = 80 \text{ mm}, h = 50 \text{ mm}$$

$$r_1 = 60 \text{ mm}, N_1 = 240 \text{ rpm}, s = 20\,000 \text{ N/m}$$

$$m = 3 \text{ kg}, M = 4 \text{ kg}$$

$$\omega = \frac{2\pi \times 240}{60} = 8\pi$$

$$\frac{r_2 - r_1}{a} = \theta = \frac{h}{b} \Rightarrow r_2 = r_1 + \frac{ah}{b} = 60 + \frac{130 \times 50}{80} = 141 \text{ mm}$$

(i) $$s = 2 \times \frac{a^2}{b^2}\left(\frac{F_2 - F_1}{r_2 - r_1}\right)$$

Now, $$F_1 = mr_1\omega_1^2 = 3 \times 0.06 \times (8\pi)^2 = 113.7 \text{ N}$$

or $$20000 = 2 \times \frac{0.13^2}{0.08^2}\left(\frac{F_2 - 113.7}{0.141 - 0.06}\right)$$

$$F_2 = 113.7 + 306.7 = 420.4 \text{ N}$$

$$3 \times 0.141 \times \left(\frac{2\pi \times N_2}{60}\right)^2 = 420.4$$

$$N_2^2 = 90638 \Rightarrow N_2 = 302 \text{ rpm}$$

(ii) $$mr_1\omega_1^2 a = \frac{1}{2}(Mg + F_{s_1})b$$

$$3 \times 0.06 \times (8\pi)^2 \times 0.13 = \frac{1}{2}(4 \times 9.81 + F_{s_1}) \times 0.8$$

$$F_{s_1} = 330.3 \text{ N}$$

$$\text{Initial compression} = \frac{330.3}{20000} = 0.0165 \text{ m} = 16.5 \text{ m}$$

(iii) Governor effort is also the average force applied on the spring.

$$\text{Effort} = \frac{20000 \times 0.05}{2} = 500 \text{ N}$$

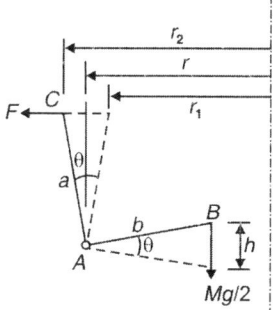

Fig. 10.17

(iv)   Power = effort × displacement

$$= 500 \times 0.05 = 25 \text{ N·m}$$

10.18 CONTROLLING FORCE

The governor balls rotating in a circular path experience a force which acts radially outwards. This force is known as centrifugal force. A force equal and opposite to the centrifugal force, acting radially inward is known as controlling force.

Controlling force is provided by the rotating masses in a Watt governor, the weight of the balls and that of the sleeve in a Porter governor and by the compressed spring in case of a Hartnell governor.

Controlling force curve: A graph that shows the variation of the controlling force with the radius of rotation is called controlling force curve or diagram. The radius of rotation is taken along X-axis and controlling force (F_c) is taken along Y-axis. This curve is useful in finding the stability and sensitiveness of the governor.

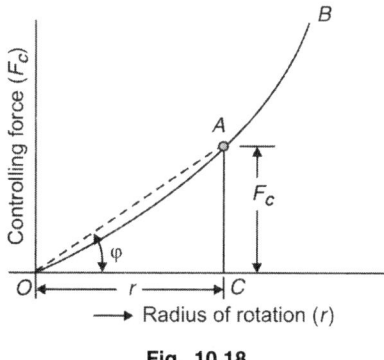

Fig. 10.18

In Fig. 10.18 *OAB* is the controlling force curve.

Controlling force is given by,

$$F_c = mr\omega^2 = mr\left(\frac{2\pi N}{60}\right)^2$$

or

$$\frac{F_c}{r} = m \times \left(\frac{2\pi}{60}\right)^2 N^2$$

or
$$\tan \varphi = KN^2 \qquad \qquad ...(10.30a)$$

$$\left(\text{here } \frac{F_c}{r} = \tan \varphi \text{ and } m \times \left(\frac{2\pi}{60}\right)^2 = \text{constant } K\right)$$

Using this relation Eq. (10.30a), the value of angle φ is determined for various values of speed (N) and a number of lines OC, OC_1, OC_2 etc. known as speed curves, are drawn.

For Porter governor, $\qquad F_c = \tan \theta \left[mg + \frac{Mg \pm f}{2}(1+k) \right]$ $\qquad ...(10.30b)$

For Hartnell governor, $F_c = \frac{(Mg + S \pm f)}{2} \times \frac{b}{a}$ $\qquad ...(10.30c)$

Controlling force curves are drawn using above relations Eq. (10.30b or 10.30c).

The intersection of the speed curves with the controlling force curve provides the speeds of the governor corresponding to a particular radius.

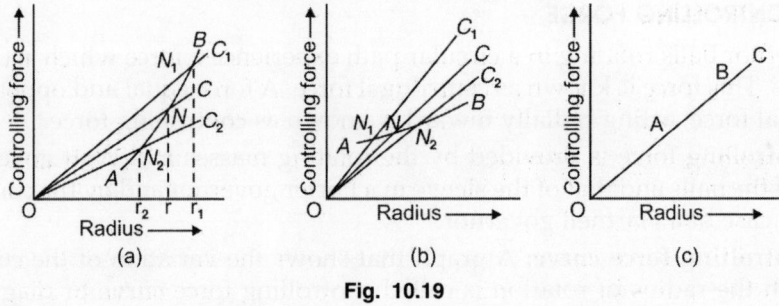

Fig. 10.19

In Fig. 10.19a radius of balls increases from mean radius r to r_1 when speed increases from mean speed N to N_1 and it decreases from r to r_2 when speed decreases from N to N_2. Such a governor is said to be a stable governor.

If the radius of balls decreases with increase in speed or vice versa, it is said to be an unstable governor (Fig. 10.19b).

Therefore, for a governor to be stable, the slope of the controlling force curve must be greater than that of the speed curve.

In Fig. 10.19c, controlling force curve is a straight line passing through the origin. In this, balls can take up any radius. Such a governor is an isochronous governor.

When friction is taken into account, two controlling force curves for ascending and descending sleeve are obtained (Fig. 10.20).

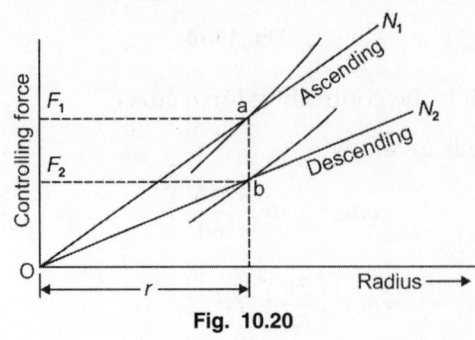

Fig. 10.20

Friction force f is positive when the sleeve moves up and negative when the sleeve moves down.

The radius of balls does not change between the speeds N_1 and N_2 corresponding to vertical portion ab while the direction of movement of sleeve changes from ascending to descending, i.e. governor becomes insensitive between speeds N_1 and N_2.

Coefficient of Insensitiveness

The ratio $\dfrac{N_1 - N_2}{N}$ is known as the coefficient of insensitiveness of the governor.

$\therefore$ Coefficient of insensitiveness $= \dfrac{N_1 - N_2}{N}$

where N is speed when friction is neglected.

10.19 STABILITY OF SPRING CONTROLLED GOVERNORS

For spring controlled governors, the relation of F_c is of the form, $F_c = ar \pm b$, where a and b are constants and value of b can be +ve, −ve or zero. Slope of the curve, $\tan \varphi = \dfrac{F_c}{r}$

or $$\tan \varphi = \frac{ar \pm b}{r} = a \pm \frac{b}{r}$$

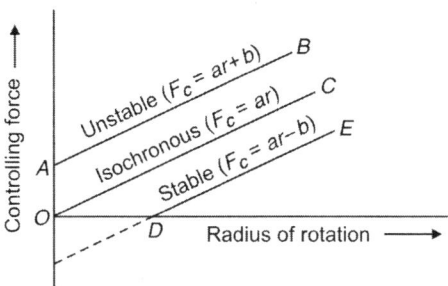

Fig 10.21

(i) If $b = 0$, $\tan \varphi = \dfrac{mr\omega^2}{r} = m\omega^2$ as m is constant, ω will be constant. That means governor runs at same speed for different radii of rotation. Governor will be isochronous

(ii) If b is +ve, $F_c = ar + b$

Slope ($\tan \varphi$) decreases with increase in r. Governor is unstable.

(iii) If b is −ve, $F_c = ar - b$.

Slope ($\tan \varphi$) increases with increase in r. Governor is stable.

Example 10.14: A Porter governor has equal arms. Each arm is 240 mm long and is pivoted on the axis os rotation. Each ball has a mass of 5 kg and the load on the sleeve is 18 kg. The ball radius is 150 mm when the sleeve begins to rise and 200 mm at the maximum speed. Find the range of speed. Also determine the coefficient of insensitiveness if the friction at the sleeve is equivalent to a force of 10 N.

Solution: $m = 5$ kg, $M = 18$ kg, $r_1 = 150$ mm, $r_2 = 200$ mm.

$$h_1 = \sqrt{l^2 - r_1^2} = \sqrt{(0.24)^2 - (0.15)^2} = \sqrt{0.0576 - 0.0225} = 0.1873 \text{ m}$$

$$N_1^2 = \frac{895}{h_1}\left(\frac{m+M}{m}\right) = \frac{895}{0.1873} \times \left(\frac{5+18}{5}\right) = 21980.78$$

or $\qquad N_1 = 148.26$ rpm

Now, $\qquad h_2 = \sqrt{l^2 - r_2^2} = \sqrt{(.24)^2 - (.2)^2} = \sqrt{.0576 - .04} = 0.1327 \text{ m}$

and $\qquad N_2^2 = \frac{895}{h_2}\left(\frac{m+M}{m}\right) = \frac{895}{0.1327} \times \left(\frac{5+18}{5}\right) = 31024.9$

or $\qquad N_2 = 176.14$ rpm

Range of speed $= N_2 - N_1 = 176.14 - 148.26 = 27.88$ rpm

Coefficient of insensitiveness $= \dfrac{N_1 - N_2}{N}$

$$= \frac{N_1 - N_2}{\dfrac{N}{}} \frac{(N_1 + N_2)}{2(N_1 + N_2)} = \frac{N_1^2 - N_2^2}{2N^2}$$

$$= \frac{\dfrac{895}{h}\left[\dfrac{mg + (Mg + f)}{mg}\right] - \dfrac{895}{h}\left[\dfrac{mg + (Mg - f)}{mg}\right]}{2 \times \dfrac{895}{h}\left[\dfrac{m+M}{m}\right]}$$

$$= \frac{f}{(m+M)g} = \frac{10}{(5+18) \times 9.81}$$

$$= 0.04432 \text{ or } 4.432\%$$

Example 10.15: The controlling force in a spring controlled governor is 1500 N when radius of rotation is 200 mm and 887.5 N when radius of rotation is 130 mm. The mass of each ball is 8 kg. If the controlling force curve is a straight line, then find:
  (i)   Controlling force when radius of rotation is 150 mm
  (ii) The speed of the governor when radius of rotation is 150 mm
  (iii) Increase in initial tension so that governor is isochronous
  (iv) Isochronous speed.

Solution: $\quad F = 1500$ when $r = 200$ mm $= 0.2$ m; $F = 887.5$ N
when $\qquad r = 130$ mm $= 0.13$ m; $m = 8$ kg

The controlling force curve is a straight line curve. As controlling force curve gives the variation of controlling force (F_c) with radius of rotation (r), hence, the controlling force can be expressed as,

$$F_c = ar + b \qquad \qquad \text{...(i)}$$

where a and b are constant.

$$F_c = 1500 \text{ N when } r = 0.2 \text{ m}$$

∴ $\qquad 1500 = a \times 0.2 + b \qquad \qquad \text{...(ii)}$

$$F = 887.5 \text{ N when } r = 0.13 \text{ m}, \qquad \qquad \text{...(iii)}$$

∴ $\qquad 887.5 = a \times 0.13 + b$

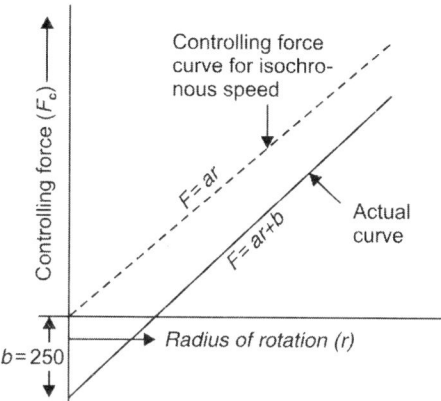

Fig 10.22

Subtracting Eq. (iii) from Eq. (ii), we get

$$1500 - 887.5 = 0.2a - 0.13a$$

or $$612.5 = 0.07a$$

∴ $$a = \frac{612.5}{0.07} = 8750$$

Substituting the value of a in Eq. (ii), we get

$$1500 = 8750 \times 0.2 + b$$

or $$b = 1500 - 8750 \times 0.2 = 1500 - 1750 = -250$$

Substituting the values of a and b in Eq. (i),

$$F = 8750 \times r - 250 \qquad \text{...(iv)}$$

(i) Controlling force when $r = 150$ mm $= 0.15$ m

Substituting $r = 0.15$ m in Eq. (iv), controlling force

$$F = 8750 \times 0.15 - 250 = 1062.5 \text{ N}$$

(ii) Speed of governor when $r = 150$ mm $= 0.15$ m

Let N = speed of governor

F_c = controlling force = 1062.5 N

$$F_c = m \times \omega^2 \times r$$

or $$1062.5 = 8 \times \omega^2 \times 0.15$$

∴ $$\omega^2 = \frac{1062.5}{8 \times 0.15} = 885.4$$

or $$\omega = \sqrt{885.4} = 29.75 \text{ rad/s}$$

But $$\omega = \frac{2\pi N}{60} \text{ or } N = \frac{60 \times \omega}{2\pi} = \frac{60 \times 29.75}{2\pi}$$

$$= 284.1 \text{ rpm}$$

$$N = 284.1 \text{ rpm}$$

(iii) The governor will be isochronous if the controlling force curve passes through the origin. The controlling force curve will pass through the origin if b is zero. The value of b will be zero if initial tension is increased to 250 N.

(iv) Let N = isochronous speed

For isochronous speed, b is zero.

$\therefore$ Controlling force, $F_c = ar + b = ar + 0 = a \times r = 8750 \times r$ N

But controlling force $F_c = m \times \omega^2 \times r = 8 \times \omega^2 \times r$

Equating the two values of F_c,

$$8 \times \omega^2 \times r = 8750 \times r$$

or
$$\omega^2 = \frac{8750}{8} = 1093.75$$

$\therefore$
$$\omega = \sqrt{1093.75} = 33.07 \text{ rad/s}$$

$\therefore$ Isochronous speed, $N = \dfrac{60 \times \omega}{2} \pi = 315.8$ rpm.

Example 10.16: The arms of a porter governor are 30 cm long. The upper arms are pivoted on the axis of rotation. The lower arms are attached to the sleeve at a distance of 4 cm from the axis of rotation. The mass of load on the sleeve is 70 kg and the mass of each ball is 10 kg. Determine the equilibrium speed with radius of rotation of balls as 20 cm. If friction is equivalent to 19.62 N at sleeve, what will be the range of speed for this position?

Solution: $m = 10$ kg, $M = 70$ kg, $F = 19.62$ N

$r = AD = 20$ cm $= 200$ mm $= 0.2$ m

$AC = AB = 30$ cm $= 300$ mm $= 0.30$ m

The configuration of the governor is shown in Fig. 10.23.

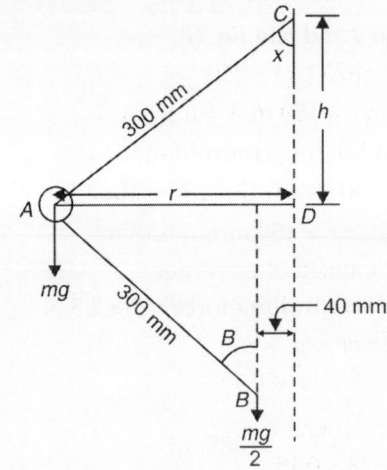

Fig. 10.23

Height of governor $\quad h = \sqrt{AC^2 - AD^2} = \sqrt{(0.30)^2 - (0.2)^2} = 0.2236$ m

$$AH = AD - HD = 0.20 - 0.04 = 0.16 \text{ m}$$

$$HB = \sqrt{AB^2 - AH^2} = \sqrt{(0.3)^2 - (0.16)^2} = 0.2537 \text{ m}$$

From triangle ADC,

$$\tan \theta = \frac{AD}{DC} = \frac{0.20}{0.2236} = 0.8944$$

From triangle *AHB*,

$$\tan \beta = \frac{AH}{HB} = \frac{0.16}{0.2537} = 0.6306; \text{ and } k = \frac{\tan \beta}{\tan \theta} = \frac{0.6306}{0.8944} = 0.7050$$

Equilibrium speed *N* is given by

$$N^2 = \frac{m + \dfrac{M}{2}(1 + k)}{m} \cdot \frac{895}{h} = \frac{10 + \dfrac{70}{2}(1 + 0.7050)}{10} \cdot \frac{895}{0.2236}$$

$$N = 166.99 \text{ rpm}$$

Let N_1, N_2 = minimum and maximum equilibrium speed.
For minimum speed,

$$N_1^2 = \frac{mg + \left(\dfrac{Mg - F}{2}\right)(1 + k)}{mg} \cdot \frac{895}{h}$$

$$= \frac{10 \times 9.81 + \left(70 \times 9.81 - \dfrac{19.62}{2}\right)(1 + 0.7050)}{10 \times 9.81} \cdot \frac{895}{0.2236}$$

$$N_1 = 164.94 \text{ rpm}$$

and for maximum speed,

$$N_2^2 = \frac{mg + \left(\dfrac{Mg + F}{2}\right)(1 + k)}{mg} \cdot \frac{895}{h}$$

$$N_2^2 = \frac{10 \times 9.81 + \left(70 \times 9.81 + \dfrac{19.62}{2}\right)(1 + 0.7050)}{10 \times 9.81} \cdot \frac{895}{0.2236}$$

$$N_2 = 169.03 \text{ rpm}$$

Range of speed = $N_2 - N_1$ = 169.03 – 164.94 = 4.09 rpm

Example 10.17: In a Proell governor, the mass of each ball is 8 kg and the mass of sleeve 120 kg. Each arm is 180 mm long. The length of extension of lower arms to which the balls are attached is 80 mm. The distance of pivots of arms from axis of rotation is 30 mm and the radius of rotation of the balls is 160 mm when the arms are inclined at 40° to the axis of rotation. Determine:

(i) The equilibrium speed
(ii) The coefficient of insensitiveness if the friction of the mechanism is equivalent to 30 N
(iii) The range of speed when the governor is inoperative.

Solution: $m = 8$ kg, $r' = 160$ mm

$$b = c = 180 \sin 40° = 115.7 \text{ mm}$$
$$r = b + 30 = 115.7 + 30 = 145.7 \text{ mm}$$
$$a = 180 \cos 40° = 137.9 \text{ mm}$$
$$AD = r' - r = 160 - 145.7 = 14.3 \text{ mm}$$
$$DE = \sqrt{80^2 - 14.3^2} = 78.7 \text{ mm}$$
$$e = a + ED = 137.9 + 78.7 = 216.6 \text{ mm}$$

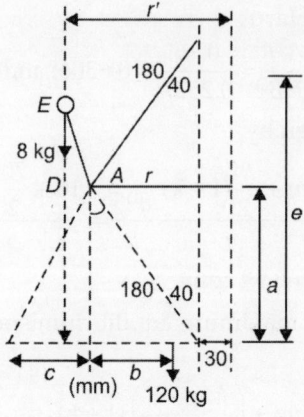

Fig. 10.24

(i) Taking moments about I,

$$mr\omega^2 \times e = mg \times (c + r - r') + \frac{Mg}{2} \times (b + c)$$

$$8 \times 0.16 \times \omega^2 \times 0.2166 = 8 \times 9.81 \times (0.1157 + 0.1457 - 0.16)$$

$$+ \frac{120 \times 9.81}{2} \times (0.1157 + 0.1157)$$

$$0.2773\omega^2 = 7.958 + 136.2$$

$$\omega^2 = 519.9, \ \omega = 22.8 \text{ or } N = \frac{22.8 \times 30}{\pi} = 217.7 \text{ rpm.}$$

(ii) Considering the friction, let ω_1 and ω_2 be the maximum and minimum speeds respectively.

$$8 \times 0.16 \times \omega_2^2 \times 0.2166 = 8 \times 9.81 \times (0.1157 + 0.1457 - 0.16)$$

$$+ \frac{120 \times 9.81 - 30}{2} \times (0.1157 + 0.1157)$$

$$0.2773\,\omega_2^2 = 7.958 + 132.7$$

$$\omega_2^2 = 507.4$$

$$\omega_2 = 22.52 \text{ or } N_2 = \frac{22.52 \times 30}{\pi} = 215.1 \text{ rpm}$$

$$0.2773\,\omega_1^2 = 7.958 + \frac{120 \times 9.81 + 30}{2} \times (0.1157 + 0.1157)$$

$$\omega_1^2 = 532.4$$

$$\omega_1^2 = 23.07 \text{ or } N_1 = \frac{23.07 \times 30}{\pi} = 220.3 \text{ rpm}$$

Coefficient of insensitiveness $= \dfrac{N_1 - N_2}{N} = \dfrac{220.3 - 215.1}{217.7} = 0.0239 \text{ or } 2.39\%$

(iii) Range of speed = 220.3 – 215.1 = 5.2 rpm.

Example 10.18: The arms of a Hartnell governor are of equal length. When the sleeve is in the mid-position, the masses rotate in a circle of diameter 150 mm (the arms are vertical in the mid-position). Neglecting friction, the equilibrium speed for this position is 360 rpm. Maximum variation of speed, taking friction into account, is to be $\pm 6\%$ of the mid-position speed for a maximum sleeve movement of 30 mm. The sleeve mass is 5 kg and the friction at the sleeve 35 N.

Assuming that the power of the governor is sufficient to overcome the friction by 1% change of speed on each side of the mid-position, find (neglecting obliquity effect of arms):

(i) the mass of each rotating ball
(ii) the spring stiffness
(iii) the initial compression of the spring.

Solution:
$$\omega = \frac{2\pi \times 360}{60} = 37.7 \text{ rad/s}$$

(i) Considering the friction at the mid-position,

$$mr\omega_1^2 a = \frac{1}{2}(Mg + F_s + f)b$$

$$m \times \left(\frac{0.150}{2}\right) \times (37.7 \times 1.01)^2 = \frac{1}{2}(5 \times 9.81 + F_s + 35) \quad (a = b) \qquad \text{...(i)}$$

and $$mr\omega_2^2 a = \frac{1}{2}(Mg + F_s - f)b$$

$$m \times \left(\frac{0.150}{2}\right) \times (37.7 \times 0.99)^2 = \frac{1}{2}(5 \times 9.81 + F_s - 35) \qquad \text{...(ii)}$$

Subtracting Eq. (ii) from Eq. (i),

$$m \times 0.075 \times (37.7)^2 [(1.01)^2 - (0.99)^2] = \frac{1}{2}(35 + 35)$$

or $$m = 8.21 \text{ kg}$$

(ii) In the extreme positions,

$$mr_2\omega_2^2 a = \frac{1}{2}(Mg + F_{s_2} + f)b$$

$$8.21 \times \left(0.075 + \frac{0.03}{2}\right) \times (37.7 \times 1.06)^2 = \frac{1}{2}(5 \times 9.81 + F_{s_2} + 35) \quad (a = b)$$

$$F_{s_2} = 2275.8$$

$$mr_1\omega_1^2 a = \frac{1}{2}(Mg + F_{s_2} - f)b$$

$$8.21 \times \left(0.075 - \frac{0.03}{2}\right) \times (37.7 \times 0.94)^2 = \frac{1}{2}(5 \times 9.81 + F_{s_2} - 35)$$

$$F_{s_1} = 1223.2 \text{ N}$$

$$h_1 s = F_{s_2} - F_{s_1}$$

$$0.03 \times s = 2275.8 - 1223.2$$

$$s = 35088 \text{ N/m or } 35.088 \text{ N/mm}$$

(iii) Initial compression $= \dfrac{F_{s_1}}{s} = \dfrac{1223.2}{35.088} = 34.86 \text{ mm}$

EXERCISE

10.1 Compare the functions of a flywheel and a governor.

10.2 Describe different types of governor.

10.3 What are centrifugal governors? How do they differ from inertia governors? Why is centrifugal governor preferred over the inertia governor?

10.4 Explain the function of a Watt governor. Derive an expression for the height in case of a Watt governor. What are its limitations?

10.5 What are the effects of friction and of adding a central weight to the sleeve of a Watt governor?

10.6 What is the effect of friction on the functioning of a Porter governor? Derive its governing equation take into account the friction at the sleeve.

10.7 Explain the term, height of the governor. Derive an expression for the height in the case of Proell governor, and also write its limitations.

10.8 What are spring controlled governors? Describe the function of a Hartnell governor and derive a relation to find the stiffness of the spring.

10.9 Explain the working of a Hartung governor with a neat sketch.

10.10 Why an auxiliary spring is used along with main springs in a Wilson-Hartnell governor? Derive a relationship for stiffness of these springs.

10.11 Describe the function of a Pickering governor or a spring controlled gravity governor.

10.12 Describe the effect of friction on the sensitiveness of a governor.

10.13 Define the terms: Stability, effort and power of a governor.

10.14 Define the following terms relating to governors:
    (i) Stability
    (ii) Sensitiveness
    (iii) Isochronism
    (iv) Hunting

10.15 What is stability of a governor? Derive the condition for stability.

10.16 Explain the term sensitiveness and isochronism in connection with governors.

10.17 What do you mean by the term hunting of centrifugal governors?

10.18 Describe Hartnell type governor with the help of a neat sketch. Derive an expression for equilibrium speed.

10.19 What is isochronism? In what type of governors can it be achieved? Find the condition of isochronism in case of a Hartnell governor.

10.20 What is the controlling force of a governor? How do the controlling force curves indicate the stability or instability of a governor? What is the shape of this curve for an isochronous governor.

10.21 Define 'coefficient of insensitiveness of governors?

10.22 Prove that the sensitiveness of a Proell governor is greater than that of a Porter governor.

10.23 In an open arm type Watt governor, $AE = 400$ mm, $EF = 50$ mm and angle $\theta = 35°$ (Refer to Fig. 10.2b). Determine the percentage change in speed when θ decreases to 30°.   (*Ans.* 3.44%)

10.24 In a Porter governor, each of the four arms is 400 mm long. The upper arms are pivoted on the axis of the sleeve, whereas the lower arms are attached to the sleeve at a distance of 45 mm from the axis of rotation. Each ball has a mass of 8 kg and the load on the sleeve is 60 kg. What will be the equilibrium speeds for the two extreme radii of 250 mm and 300 mm of rotation of the governor balls.

        (*Ans.* 147 rpm, 159.1 rpm, 12.1 rpm)

10.25 Each arm of a Proell governor is 240 mm long and each rotating ball has a mass of 3 kg. The central load acting on the sleeve is 30 kg. The pivots of all the arms are 30 mm from the axis of rotation. The vertical height of the governor is 190 mm. The extension links of the lower arms are vertical and the governor speed is 180 rpm when the sleeve is in the midposition. Determine the lengths of the extension links and the tension in the upper arms. (*Ans.* 104 mm, 223 N)

10.26 The mass of each ball of a Proell governor is 7.5 kg, and the load on the sleeve is 80 kg. Each of the arms is 300 mm long. The upper arms are pivoted on the axis of rotation, whereas the lower arms are pivoted to links of 40 mm from the axis of rotation. The extensions of the lower arms to which the balls are attached are 100 mm long and are parallel to the governor axis at the minimum radius. Determine the equilibrium speeds corresponding to extreme radii of 180 mm and 240 mm. (*Ans.* 165.3 rpm, 175.9 rpm)

10.27 In a spring loaded Hartnell type of governor, the mass of each ball is 4 kg and the lift of the sleeve is 40 mm. The governor begins to float at 200 rpm when the radius of the ball path is 90 mm. The mean working speed of the governor is 16 times the range of speed when friction is neglected. The lengths of the ball and roller arms of the bell-crank lever are 100 mm and 80 mm respectively. The pivot centre and the axis of governor are 115 mm apart. Determine the initial compression of the spring, taking into account the obliquity of arms.

Assuming the friction at the sleeve to be equivalent to a force of 15 N, determine the total alteration in speed before the sleeve begins to move from the mid-position. (*Ans.* 51.37 mm, 5.7 rpm)

10.28 In a Wilson–Hartnell type of governor, the mass of each ball is 5 kg. The lengths of the ball arm and the sleeve arm of each bell–crank lever are 100 mm and 80 mm respectively. The stiffness of each of the two springs attached directly to the balls is 0.4 N/mm. The lever for the auxiliary spring is pivoted at its mid point. When the radius of rotation is 100 mm, the equilibrium speed is 200 rpm. If the sleeve is lifted by 8 mm for an increase in speed of 6%, find the required stiffness of the auxiliary spring. (*Ans.* 11219 N/m)

10.29 Each spring of a Pickering governor of a gramophone is 6 mm wide and 0.12 mm thick with a length of 48 mm. A mass of 25 g is attached to each leaf spring at the centre. Distance between spindle axis and centre of mass when the governor is at rest is 8 mm. The ratio of the governor speed to the turn table speed is 10. Determine the speed of the turn table for a sleeve lift of 0.6 mm. Take $E = 200$ GN/m^2. (*Ans.* 575 rpm)

10.30 The following particulars refer to a spring controlled gravity governor:

The mass of each ball = 0.75 kg; mass of sleeve = 5 kg; the lengths of the arms a and b of the bell-crank lever are 110 mm and 44 mm respectively; the distance of the fulcrum O of each bell-crank lever from the axis of rotation = 55 mm; minimum radius of rotation of ball = 55 mm; equilibrium speed at minimum radius = 240 rpm; sleeve lift is 10 mm and corresponding equilibrium speed is 252 rpm. Find the stiffness of the spring. (*Ans.* 7.185 N/mm)

10.31 Each ball of a Porter governor has a mass of 6 kg and the mass of the sleeve is 40 kg. The upper arms are 300 mm long and are pivoted in the axis of rotation whereas the lower arms are 250 mm long and are attached to the sleeve at a distance of 40 mm from the axis. Determine the equilibrium speed of the governor for a radius of rotation of 150 mm for 1% change in speed. Also find the effort and the power for the same speed change.

10.32 The lengtyhs of the ball and sleeve arms of the bell-crank lever of a Hartnell governor are 140 and 120 mm respectively. The mass of each governor ball is 5 kg. The fulcrum of the bell-crank lever is at a distance of 160 mm. At the mean speed of the governor which is 250 rpm, the ball arms are vertical and the sleeve arms are horizontal. The sleeve moves up by 12 mm for an increase in speed of 4%. Neglecting friction, determine:

   (i) The spring stiffness

   (ii) The minimum equilibrium speed when the sleeve moves by 24 mm

   (iii) The sensitiveness of the governor

   (iv) The spring stiffness for the governor to be isochronous at the mean speed.

10.33 In a Hartung governor, each of rotating balls has a mass of 9 kg and each spring has a stiffness of 270 N/cm. The length of each spring is 11.4 cm when the radius of rotation of the balls is 6.9 cm and the equilibrium speed is 360 rpm. Neglecting the mass of the sleeve, determine the free length of each spring and also investigate whether the governor is isochronous and stable or not.　　(*Ans.* 14.67 cm, not isochronous, stable.)

10.34 The sleeve arm and the ball arm of a Hartnell governor are 9 cm and 10 cm long respectively. At the mean position of the sleeve, the radius at which the balls rotate is 12 cm when the equilibrium speed is 300 rpm. Each rotating ball has a mass of 2 kg and the sleeve movement is + 2 cm from the mean position. The minimum speed is 96% of the mean speed. Determine (i) the stiffness of the spring and (ii) the maximum speed of the governor. Neglect the frictional resistance.　　(*Ans.* 65.5 N/cm, 308 rpm)

10.35 The following particulars refer to a Porter governor: All the arms of governor are 178 mm long and hinged at a distance of 38 mm from the axis of rotation. The mass of each ball is 1.15 kg and mass of sleeve is 20 kg. The governor sleeve begins to rise at 280 rpm when the links are at an angle of 30° to the vertical. Assuming the frictional force to be constant, determine the minimum and maximum speeds of rotation when the inclination of arm to the vertical is 45°.

10.36 Each arm of a Porter governor is 180 mm long and is pivoted on the axis of rotation. The mass of each ball is 4 kg and of the sleeve 18 kg. The radius of rotation of the balls is 100 mm when the sleeve begins to rise and 140 mm when at the top. Determine the range of speed. Aslo, find the coefficient of insensitiveness if the friction at the sleeve is 15 N.

10.37 A Porter governor has all four arms 25 cm long. The upper arms are attached on the axis of rotation and the lower arms are attached on a sleeve at a distance of 3 cm from the axis. The mass of each ball is 5 kg and mass of the sleeve 50 kg. The extreme radii of rotation are 15 cm and 20 cm. Determine the range of speed of governor.

                                                             (*Ans.* 30.12 rpm)

10.38 In a spring loaded Hartnell type governor, the extreme radii of rotation of the balls are 8 cm and 12 cm. The ball arm and sleeve arm of the bell crank are equal in length. Mass of each ball is 2 kg. If the speeds at the two extreme positions are 400 and 420 rpm, find (i) the initial compression of the central spring and (ii) the spring constant.

                                           (*Ans.* 9179 N/m, 0.061 m)

10.39 In a spring-controlled governor, the controlling force curve is a straight line. The balls are 400 mm apart when the controlling force is 1500 N and 240 mm when it is 800 N. The mass of each ball is 10 kg. Determine the speed at which the governor runs when the balls are 300 mm apart. By how much should the initial tension be increased to make the governor isochronous? Also, find the isochronous speed.

                        (*Ans.* 254.2 rpm, 250 N, 282.5 rpm)

OBJECTIVE TYPE QUESTIONS

10.1 The height of a Watt governor is expressed as

   (a) $h = \omega^2/g$

   (b) $h = g \cdot \omega^2$

   (c) $h = g \cdot \omega$

   (d) $h = g/\omega^2$

10.2 Tick the correct statement:
   (a) Hartnell governor is spring loaded type governor
   (b) Watt governor is spring loaded type governor
   (c) Porter governor is spring loaded type governor
   (d) Proell governor is spring loaded type governor

10.3 An isochronous governor is
   (a) very stable
   (b) less sensitive
   (c) infinitely sensitive
   (d) none of these

10.4 For the same lift of sleeve, the range of speed of Proell governor as compared to Porter governor is
   (a) less
   (b) same
   (c) more
   (d) may be more or less

10.5 Governor effort is defined as
   (a) force applied for 0% change in speed
   (b) force applied for 10% change in speed
   (c) force applied for 1% change in speed
   (d) none of these

10.6 Tick the correct statement:
   (a) The governor does not have any control on the varying load on the engine
   (b) The governor reduces the speed fluctuation during a cycle of engine
   (c) The governor does not have any control on the speed of engine
   (d) The governor maintains the speed of the engine within prescribed limits for varying torque output conditions

10.7 With the increase of governor speed,
   (a) radius of rotation and height of governor increases
   (b) radius of rotation and height of governor decreases
   (c) radius of rotation decreases but height of governor increases
   (d) radius of rotation increases but height of governor decreases

10.8 The frictional resistance at the sleeve of a governor
   (a) increases sensitivity of governor
   (b) decrease sensitivity of governor
   (c) does not affect the sensitivity of governor
   (d) may increase or decrease sensitivity depending upon when speed is increasing or decreasing

10.9 If the ball masses of a governor occupy a definite specified position for each speed in the working, it is said to be
   (a) stable
   (b) hunting
   (c) isochronous
   (d) sensitive

10.10 Which one of the following is the gravity controlled type governor
   (a) Proell
   (b) Hartung
   (c) Pickering
   (d) Hartnell

10.11 In a Hartnell governor, if a spring of lower stiffness is used, then the governor will be
    (a) isochronous
    (b) more sensitive
    (c) less sensitive
    (d) none of these

10.12 A hunting governor is
    (a) more stable
    (b) less sensitive
    (c) more sensitive
    (d) none of these

10.13 When a governor is infinitely sensitive, it is called
    (a) isochronous
    (b) Hartung
    (c) sensitive
    (d) none of these

10.14 For spring controlled governors, the controlling force curve will generally bed
    (a) circle
    (b) parabola
    (c) hyperbola
    (d) none of these

10.15 Which governor cannot be isochronous?
    (a) Watt
    (b) Porter
    (c) Hartnell
    (d) Hartung

10.16 Governor power is defined as
    (a) only effort
    (b) product of effort and sleeve fit
    (c) not defined
    (d) none of these

10.17 The controlling force curve is a relationship between
    (a) controlling force and speed of rotation
    (b) controlling force and radius of rotation
    (c) controlling force and lift of governor
    (d) controlling force and height of governor

10.18 The relation between the controlling force F_c and radius of rotation (r) of a governor is given by $F_c = ar - b$, the governor will be
    (a) isochronous
    (b) stable
    (c) unstable
    (d) hunt

10.19 The main objective of controlling force diagram is to
    (a) determine the sensitiveness of the governor
    (b) determine the stability of governor
    (c) determine both (a) and (b)
    (d) none of these

10.20 The centrifugal governors as compared to inertia governors are
    (a) less sensitive and more simple
    (b) more sensitive and more simple
    (c) more sensitive and less simple
    (d) less sensitive and less simple

10.21 The speed of Watt governor for a height of 9.81 cm is equal to
    (a) 9.81 rad/s
    (b) 0.981 rad/s
    (c) 14 rad/s
    (d) 10 rad/s

10.22 The height of Porter governor with equal arms pivoted at equal distance from the axis of rotation is expressed as

    (a) $h = \left(\dfrac{m+M}{m} \right) \dfrac{895}{N}$

    (b) $h = \left(\dfrac{m+M/2}{m} \right) \dfrac{895}{N^2}$

    (c) $h = \left(\dfrac{m+M}{m} \right) \dfrac{895}{N^2}$

    (d) $h = \left(\dfrac{m/2+M}{m} \right) \dfrac{895}{N^2}$

10.23 With the decrease of governor speed,
    (a) radius of rotation decreases but height of governor increases
    (b) radius of rotation and height of governor decrease
    (c) radius of rotation and height of governor increase
    (d) radius of rotation increases but height of governor decreases

10.24 In Porter governor with arms equal and pivoted at equal distance from axis of rotation and considering friction F, the height of the governor when speed is decreasing is given by

    (a) $h = \left(\dfrac{mg+Mg-F}{mg} \right) \dfrac{895}{N^2}$

    (b) $h = \left(\dfrac{mg+Mg+F}{mg} \right) \dfrac{895}{N^2}$

    (c) $h = \left(\dfrac{mg+Mg-F}{mg} \right) \dfrac{895}{N}$

    (d) $h = \left(\dfrac{mg+Mg+F}{mg} \right) \dfrac{895}{N}$

10.25 The spring rate of each ball spring of Wilson–Hartnell governor having no auxiliary spring, is given by

    (a) $s_b = \dfrac{F_{c_2} - F_{c_1}}{r_2 - r_1}$

    (b) $s_b = \dfrac{F_{c_2} - F_{c_1}}{4(r_2 - r_1)}$

    (c) $s_b = \dfrac{F_{c_2} - F_{c_1}}{2(r_2 - r_1)}$

    (d) $s_b = \dfrac{4(F_{c_2} - F_{c_1})}{r_2 - r_1}$

10.26 For Hartnell governor with equal arms the lift of the sleeve is given by

    (a) $h = r_2 + r_1$

    (b) $h = \dfrac{r_2 - r_1}{2}$

    (c) $h = r_2 - r_1$

    (d) $h = \dfrac{r_2 + r_1}{2}$

10.27 Sensitivity of governor is expressed as

    (a) $\dfrac{N_1 - N_2}{N_1 + N_2}$

    (b) $\dfrac{2(N_1 - N_2)}{N_1 + N_2}$

    (c) $\dfrac{2(N_1 + N_2)}{N_1 - N_2}$

    (d) $\dfrac{N_1 - N_2}{2(N_1 - N_2)}$

10.28 A ...... governor is a spring loaded governor.

    (a) Watt

    (b) Hartnell

    (c) Porter

    (d) Proell

10.29 The ratio of height of a Porter governor to that of a Watt governor when the lengths of the links and the arms are the same is

    (a) $\dfrac{M + m}{M}$

    (b) $\dfrac{M + m}{m}$

    (c) $\dfrac{M}{M + m}$

    (d) $\dfrac{m}{M + m}$

10.30 A Hartnell governor is a/an .... governor.

    (a) dead weight

    (b) pendulum type

    (c) inertia

    (d) spring-loaded

10.31 The governor is said to be .... when the speed of the engine fluctuates continuously above and below the mean speed.

    (a) isochronous

    (b) hunting

    (c) insensitive

    (d) stable

10.32 The condition of isochronism can be realised in a .... governor.
  (a) Watt
  (b) Porter
  (c) Proell
  (d) Hartnell

10.33 Effort of a governor is the force exerted by it on the
  (a) balls
  (b) sleeve
  (c) upper links
  (d) lower links

10.34 In a Wilson–Hartnell governor the balls are connected by
  (a) one spring
  (b) two spring in series
  (c) two parallel springs
  (d) four springs

10.35 The force resisting the outward movement of balls is known as .... of the governor.
  (a) effort
  (b) centripetal force
  (c) controlling force
  (d) inertia force

10.36 In a governor, if the equilibrium speed is constant for all radii of rotation of balls, the governor is said to be
  (a) stable
  (b) unstable
  (c) inertia
  (d) isochronous

10.37 If the controlling force of a spring controlled governor is expressed as $F_c = ar + b$, where r is the radius of rotation and a and b are constants, it is a/an .... governor.
  (a) isochronous
  (b) centrifugal
  (c) dead-weight
  (d) unstable

10.38 When the sleeve of a Porter governor moves upwards, the governor speed
  (a) increases
  (b) decreases
  (c) remains unaffected

10.39 Which of the following governor is used to drive a gramophone?
  (a) Watt governor
  (b) Porter governor
  (c) Pickering governor
  (d) Hartnell governor

10.40 For two governors A and B, the lift of the sleeve of governor A is more than that of governor B, for a given fractional change in speed. It indicates that
  (a) governor A is more sensitive than governor B
  (b) governor B is more sensitive than governor A
  (c) both governors A and B are equally sensitive
  (d) none of the above

10.41 In a Hartnell governor, if a spring of greater stiffness is used, then the governor will be
    (a) more sensitive
    (b) less sensitive
    (c) isochronous

10.42 A hunting governor is
    (a) more stable
    (b) less sensitive
    (c) more sensitive
    (d) none of these

10.43 Isochronism in a governor is desirable when
    (a) the engine operates at low speeds
    (b) the engine operates at high speeds
    (c) the engine operates at variable speeds
    (d) one speed is desired under one load

10.44 The sensitiveness of a governor is given by

    (a) $\dfrac{\omega_{mean}}{\omega_2 - \omega_1}$

    (b) $\dfrac{\omega_2 - \omega_1}{\omega_{mean}}$

    (c) $\dfrac{\omega_2 - \omega_1}{2\omega_{mean}}$

    (d) None

10.45 Power of a governor is equal to

    (a) $\dfrac{c^2}{1+2c}(m+M)h$

    (b) $\dfrac{2c^2}{1+2c}(m+M)h$

    (c) $\dfrac{3c^2}{1+2c}(m+M)h$

    (d) $\dfrac{4c^2}{1+2c}(m+M)h$

ANSWERS

10.1 (d)	10.2 (a)	10.3 (c)	10.4 (d)	10.5 (c)	10.6 (a)
10.7 (d)	10.8 (b)	10.9 (a)	10.10 (c)	10.11 (b)	10.12 (c)
10.13 (a)	10.14 (d)	10.15 (d)	10.16 (b)	10.17 (b)	10.18 (b)
10.19 (c)	10.20 (a)	10.21 (d)	10.22 (c)	10.23 (a)	10.24 (a)
10.25 (b)	10.26 (c)	10.27 (b)	10.28 (b)	10.29 (b)	10.30 (d)
10.31 (b)	10.32 (d)	10.33 (b)	10.34 (c)	10.35 (c)	10.36 (d)
10.37 (d)	10.38 (a)	10.39 (c)	10.40 (a)	10.41 (b)	10.42 (c)
10.43 (d)	10.44 (b)	10.45 (d)			

11

Gyroscopic Motion

11.1 INTRODUCTION

Whenever a couple is applied about Z-axis on a body spinning about X-axis, its axis of spin rotates about Y-axis. This couple applied on the body is known as gyroscopic couple and the effect produced by the gyroscopic couple on the body is known as gyroscopic effect. The spinning body is known as gyroscope. Earlier, gyroscopes were used for stabilising seaborne ships and for gyroscopic compasses. During world war II, gyroscopes were used for applications like bomb sights, control of aeroplanes and guided missiles. The gyroscopic effect is observed in aeroplanes while taking turn, in steering and pitching of ships and in automobiles while negotiating a curve. Gyroscopic couple produces some undersirable effects also in a shaft roating in bearings. It increases bearing reactions in a crank shaft of automobile as it negotiates a curve and in ship propellers as the ship pitches and rolls in a rough sea, etc.

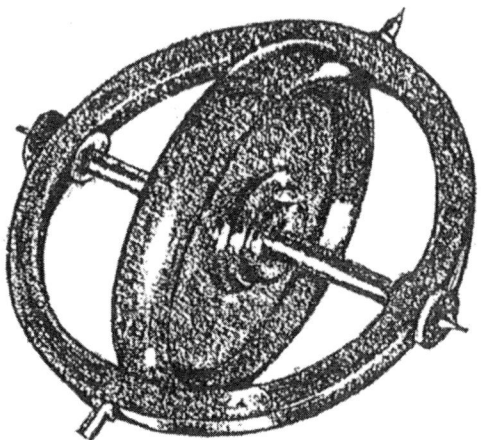

Fig. 11.1: Gyroscope

11.2 ANGULAR VELOCITY

The rate of change of angular displacement with time is known as angular velocity. It is denoted by ω (omega). It is a vector quantity.

$$\omega = \lim_{\delta t \to 0} \left(\frac{\delta \theta}{\delta t} \right) = \frac{d\theta}{dt}$$

The angular velocity of a rotating body is specified by: (i) the magnitude of velocity (ii) the direction of the axis of shaft and (iii) sense of rotation of the shaft (clockwise or anticlockwise).

The magnitude of angular velocity vector is respresented by the length of the vector. Its direction is represented by drawing the vector parallel to the axis of shaft.

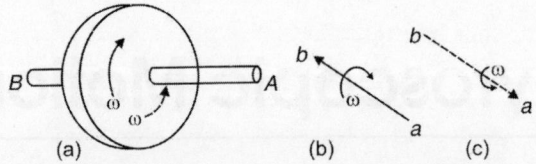

Fig. 11.2: Representation of angular velocity

Sense of rotation of shaft is represented by using right-handed screw rule, i.e. if a screw is rotated in clockwise direction, it goes away from the viewer Fig. (11.2) and vice-versa.

11.3 ANGULAR ACCELERATION

The rate of change of angular velocity with respect to time is known as angular acceleration. It is denoted by α (alpha). It is a vector quantity. Consider a shaft mounted with a disc spinning about the horizontal axis OX (known as axis of spin) with an angular velocity ω (rad/s) in anticlockwise direction when viewed from the front.

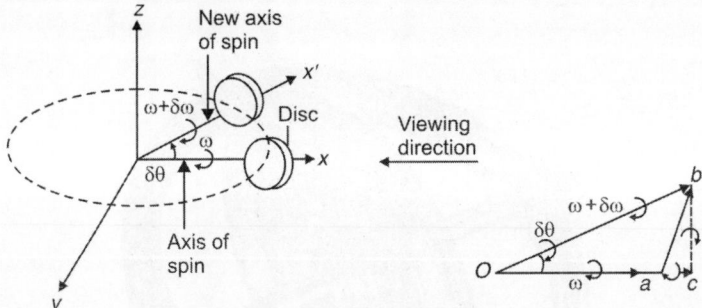

Fig. 11.3: Angular acceleration

In time δt, the magnitude of the angular velocity changes to $(\omega + \delta\omega)$ and axis of spin turns through an angle $\delta\theta$ in the direction OX'. Initial angular velocity ω is respresented by vector $\overrightarrow{oa}$ and final angular velocity $(\omega + \delta\omega)$ is represented by vector $\overrightarrow{ob}$. Vector $\overrightarrow{ab}$ represents the changes in the angular velocity of shaft in time δt (or angular acceleration). The vector $\overrightarrow{ab}$ is resolved into two components, one parallel to oa (ac) and the other perpendicular to oa (cb).

Change in angular velocity in the direction of oa,

$$ac = (\omega + \delta\omega) \cos \delta\theta - \omega$$

Rate of change of angular velocity $= \dfrac{(\omega + \delta\omega)\cos\delta\theta - \omega}{\delta t}$

Angular acceleration, $\qquad \alpha_t = \lim\limits_{\delta t \to 0} \dfrac{(\omega + \delta\omega)\cos(\delta\theta - \omega)}{\delta t}$

As $\delta t \to 0$, $\delta\theta \to 0$ and $\cos\delta\theta \to 1$

$\therefore \qquad\qquad\qquad\qquad \alpha_t = \lim\limits_{\delta t \to 0} \dfrac{(\omega + \delta\omega - \omega)}{\delta t} = \lim\limits_{\delta t \to 0} \dfrac{\delta\omega}{\delta t} = \dfrac{d\omega}{dt}$

Change in angular velocity in the direction of cb (perpendicular to oa)

$$cb = (\omega + \delta\omega)\sin\delta\theta$$

Rate of change of angular velocity $= \dfrac{(\omega + \delta\omega)\sin\delta\theta}{\delta t}$

Angular acceleration, $\qquad \alpha_c = \lim\limits_{\delta t \to 0} \dfrac{(\omega + \delta\omega)\sin\delta\theta}{\delta t}$

Since angle $\delta\theta$ is very small, $\sin\delta\theta \to \delta\theta$

Therefore, $\qquad \alpha_c = \lim\limits_{\delta t \to 0} \dfrac{(\omega + \delta\omega)\delta\theta}{\delta t} = \lim\limits_{\delta t \to 0} \dfrac{\omega\delta\theta + \delta\omega\,\delta\theta}{\delta t}$

or $\qquad\qquad\qquad \alpha_c = \lim\limits_{\delta t \to 0} \dfrac{\omega\delta\theta}{\delta t} \qquad\qquad$ ($\delta\omega\delta\theta$ is very small, hence neglected)

$$\alpha_c = \omega \dfrac{d\theta}{dt} = \omega\omega_p \qquad \left(\text{where } \dfrac{d\theta}{dt} = \omega_p \text{ is angular velocity of precession}\right)$$

Total angular acceleration, $\alpha =$ vector sum of α_t and α_c

or $\qquad\qquad\qquad\qquad \alpha = \dfrac{d\omega}{dt} + \omega\dfrac{d\theta}{dt} = \dfrac{d\omega}{dt} + \omega\,\omega_p$

Notes:

i. $\dfrac{d\omega}{dt}$ represents the change in the magnitude of the angular velocity of the shaft.

ii. $\omega\dfrac{d\theta}{dt}$ represents the change in the direction of the axis of spin.

iii. Angular acceleration, α_c acts clockwise along Z-axis when viewed from front.

11.4 GYROSCOPIC COUPLE

Consider a disc having moment of inertia I, spinning with an angular velocity ω (rad/s) about a horizontal axis of spin OX, in anticlockwise direction when seen from the front. The plane perpendicular to the axis of spin, i.e. YOZ is the called plane of spinning. Let axis of spin rotates through a small angle $\delta\theta$ in the horizontal plane (XOZ) about an axis OY to new position OX' in time δ_t. The horizontal plane XOZ is called the plane of precession and axis OY is known as axis of precession.

Angular momentum of the disc $= I\omega$

Angular momentum (a vector quantity) is represented by the vector $\overrightarrow{oa}$ when the axis of spin is OX and by $\overrightarrow{ob}$ when the axis of spin is changed to OX'.

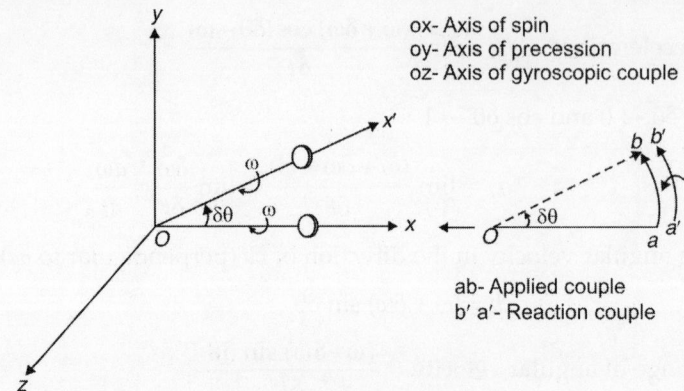

ox- Axis of spin
oy- Axis of precession
oz- Axis of gyroscopic couple

ab- Applied couple
b′ a′- Reaction couple

Fig.11.4: Gyroscopic couple

Change in angular momentum due to change in the direction of the axis of spin,

$$\overline{ob} - \overline{oa} = \overline{ab} = \overline{oa} \cdot \delta\theta = I\omega\delta\theta$$

∴   Rate of change of angular momentum,

$$= I\omega \frac{\delta\theta}{\delta t}$$

This results by the application of a couple to the disc, therefore applied couple to cause the axis of spin to precess,

$$C = \underset{\delta t \to 0}{\text{Lt}} \left(I\omega \frac{\delta\theta}{\delta t} \right) = I\omega \frac{d\theta}{dt}$$

$\dfrac{d\theta}{dt}$ is the angular velocity of the axis of spin, called as angular velocity of precession

and is denoted by ω_p (i.e. $\dfrac{d\theta}{dt} = \omega_p$)

∴ $C = I\omega\omega_p$ ...(11.1)

This applied couple is known as active gyroscopic couple (represented by *ab*). The vector *ab* lies in horizontal plane *XOZ*. For very small angle of precession δθ, vector *ab* will be perpendicular to the vertical plane *XOY*. Therefore, the axis *OZ* about which the couple acts, is called the axis of active gyroscopic couple and the plane *XOY* is called the plane of gyroscopic couple. Active gyroscopic couple acts in clockwise (CW) direction.

A reactive gyroscopic couple is also applied to the axis of spin that tends to rotate it in the opposite direction (CCW). It is represented by *b′a′*.

The effect of the gyroscopic couple on a rotating body is known as gyroscopic effect. This principle is used in a spinning body known as gyroscope that is free to move in other directions under the applied forces. The gyroscopes are used in ships, aeroplanes, monorail cars, gyrocompasses, etc.

Example 11.1: A flywheel having a mass of 20 kg and radius of gyration 300 mm is given a spin of 500 rpm about its axis which is horizontal. The flywheel is suspended at a point 250 mm from the plane of rotation of the flywheel. Find the rate of precession of the wheel.

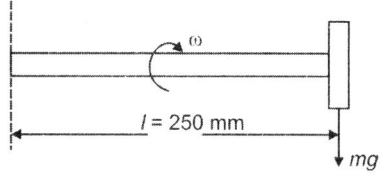

Fig. 11.5

Solution: Mass of flywheel, $m = 20$ kg

Radius of gyration, $k = 300$ mm $= 0.3$ m

$N = 500$ rpm, arm length, $l = 250$ mm $= 0.25$ m

$\therefore$ Angular velocity of flywheel, $\omega = \dfrac{2\pi N}{60}$

$$\omega = \frac{2\pi \times 500}{60} = 52.36 \text{ rad/s}$$

Moment of inertia of flywheel, $I = mk^2$

$$I = 20 \times (0.3)^2 = 1.8 \text{ kg m}^2$$

Gyroscopic couple, $C = I\,\omega\omega_p$

but gyroscopic couple is provided by moment of weight (mg) about A i.e.,

$$C = mg \times l$$

$\therefore$ $\qquad\qquad I\omega\omega_p = mg \times l$

or $\qquad 1.8 \times 52.36 \times \omega_p = 20 \times 9.8 \times 0.25$

$$\omega_p = 0.52 \text{ rad/s.}$$

11.5 EFFECT OF GYROSCOPIC COUPLE ON BEARINGS

Consider a disc of mass m and radius of gyration k mounted centrally on a horizontal axle supported in bearings A and B. The disc is spinning about the axle in anticlockwise direction when viewed from the right hand side bearing B. The axle precesses about the vertical Y-axis in the clockwise direction when viewed from top.

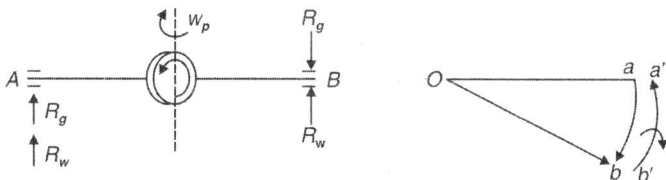

Fig. 11.6

Mass moment of Inertia of disc $I = mk^2$

The gyroscopic couple acting on the disc, $C = I\omega \cdot \omega_p$

The reactive gyroscopic couple is clockwide (Figure 11.6) when viewed from the front. It tends to raise the bearing A and lower the bearing B.

Force at bearing A due to gyroscopic couple,

$$R_{gA} = \frac{C}{l} = \frac{I\omega\omega_p}{l} \qquad\qquad \text{(upwards)}$$

Force at bearing B due to gyroscopic couple,

$$R_{gB} = \frac{C}{l} = \frac{I\omega\omega_p}{l} \qquad \text{(downwards)}$$

Force at each bearing due to weight of the disc,

$$R_{WA} = R_{WB} = \frac{mg}{2} \qquad \text{(upwards)}$$

$\therefore$ Reaction at bearing A, $R_A = \frac{C}{l} + \frac{mg}{2}$ (upwards)

Reaction at bearing B, $R_B = \frac{C}{l} - \frac{mg}{2}$ (downwards)

Example 11.2: A disc of mass 6 kg and radius of gyration 100 mm is mounted centrally on a horizontal axle of 100 mm length between the bearings. The disc spins about the axle at 1000 rpm anticlockwise when viewed from right hand side bearing. The axle precesses about the vertical axis at 60 rpm in anticlockwise direction when viewed from top. Find the resultant reaction at each bearing due to the mass and the gyroscopic effect.

Solution : Mass of disc, $m = 6$ kg

Radius of gyration, $k = 100$ mm $= 0.1$ m

Moment of inertia of disc, $I = mk^2 = 6\,(0.1)^2 = 0.06$ kg·m^2

Length of axle, $l = 100$ mm $= 0.1$ m

$$N = 1000 \text{ rpm}, N_p = 60 \text{ rpm}$$

$$\omega = \frac{2\pi N}{60} = \frac{2\pi \times 1000}{60} = 104.7 \text{ rad/s}$$

$$\omega_p = \frac{2\pi N_p}{60} = \frac{2\pi \times 60}{60} = 2\pi = 6.28 \text{ rad/s}$$

Gyroscopic couple, $C = I\omega\omega_p = .06 \times 104.7 \times 6.28 = 39.45$

$$C = 39.45 \text{ N·m}$$

Refer to Fig. 11.6 to draw the couple diagram.

The reactive gyroscopic couple is anticlockwide when viewed from front and tends to lower the bearing A and raise the bearing B.

Force at bearing A, due to gyroscopic couple,

$$(R_{gA}) = \frac{C}{l} = \frac{39.45}{0.1} = 394.5 \text{ N} \qquad \text{(downwards)}$$

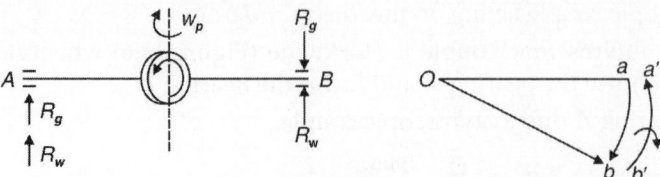

Fig. 11.7

Force at bearing B, due to gyroscopic couple $= \dfrac{C}{l} = \dfrac{39.45}{0.1}$

$$R_{gB} = 394.6 \text{ N} \qquad\qquad \text{(upwards)}$$

Force at each bearing due to weight of disc, R_W

$$R_{WA} = R_{WB} = \dfrac{m.g}{2} = \dfrac{6\times9.81}{2} = 29.43 \text{ N} \qquad\qquad \text{(upwards)}$$

∴ Total reaction at bearing A, $R_A \quad = \dfrac{C}{l} - \dfrac{mg}{2}$

$$= 394.5 - 29.43 = 365.07 \text{ N} \qquad\qquad \text{(downwards)}$$

Total reaction at bearing B, $R_B = R_{gB} + R_{wB}$

$$= 394.5 + 29.43 = 423.93 \text{ N} \qquad\qquad \text{(upwards)}$$

11.6 EFFECT OF GYROSCOPIC COUPLE ON AN AEROPLANE

The engine or propeller of an aeroplane rotates in clockwise direction when viewed from rear or tail end and the aeroplane takes a turn towards left. Let mass of the engine and propeller be m kg and its radius of gyration k m. The engine rotates with angular velocity ω (rad/s). Linear velocity of aeroplane $= v$ m/s.

Mass moment of inertia of rotating parts, I

$$I = m \, k^2 \text{ kg·m}^2$$

Aeroplane takes a turn in a curvature of radius R meter.

Angular velocity of precession of aeroplane $\omega_p = \dfrac{v}{R}$ rad/s

∴ Gyroscopic couple acting on the aeroplane,

$$C = I \, \omega \omega_p$$

Angular momentum diagram is shown in Fig. 11.8.

$\overrightarrow{oa}$ = angular momentum vector before turning

$\overrightarrow{ob}$ = angular momentum vector after turning

$\overrightarrow{ab}$ is active gyroscopic couple and $\overrightarrow{b'a'}$ is reactive gyroscopic couple. $b'a'$ is perpendicular to oa in the limit. The reactive gyroscopic couple ($b'a'$) acts in the anticlockwise direction and it tends to raise the nose and depress the tail of the aeroplane.

When the aeroplane takes a right turn, the effect of reactive gyroscopic couple will be to dip the nose and raise the tail of the aeroplane (Fig. 11.8c)

Example 11.3 The moment of inertia of an aeroplane air screw is 20 kg·m² and the speed of rotation 1000 rpm clockwise when viewed from the front. The speed of flight is 200 km per hour. Find the gyroscopic reaction of the air screw on the aeroplane when it makes a left handed turn on a path of 150 m radius.

Solution : Moment of inertia, $I = 20$ kg·m²

$$N = 1000 \text{ rpm}, \ \omega \cdot \dfrac{2\pi N}{60} = \dfrac{2\pi \times 1000}{60} = 104.72 \text{ rad/s}$$

Speed of aeroplane, $v = 200 \text{ km/hr} = \dfrac{200 \times 1000}{60} = 55.5 \text{ m/s}$

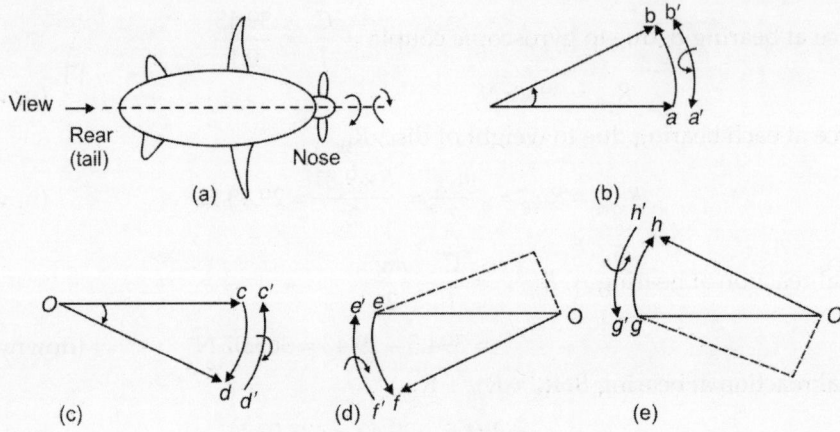

Fig. 11.8: Aeroplane taking a turn

Radius of curvature, $R = 150$ m

Angular velocity of precession $\omega_p = \dfrac{v}{R} = \dfrac{55.5}{150}$

$$\omega_p = 0.37 \text{ rad/s}$$

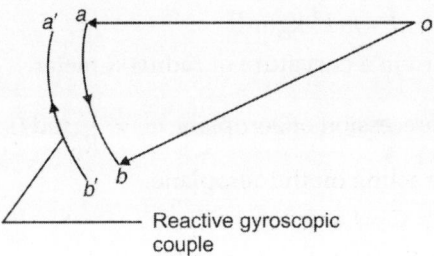

Reactive gyroscopic couple

Fig. 11.9

Gyroscopic couple on the aeroplane,

$$C = I\omega\omega_p = 20 \times 104.72 \times 0.37 = 775 \text{ N·m}$$

The reactive gyroscopic couple $b'a'$ is clockwise when looked from front. The effect of this couple will be to depress the nose and raise the tail.

11.7 GYROSCOPIC EFFECT ON NAVAL SHIPS

The terms used in connection with the motion of naval ships are described below:

  (i) The fore or the front end of the ships is known as bow.

 (ii) The rear end of ship is called as stern or aft.

(iii) The right hand side of the ship when viewed from the stern is called starboard.

 (iv) The left hand side of the ship when viewed from the stern is called port.

  (v) Taking turn to the left or right side when moving forward is known as steering.

 (vi) Pitching is up and down motion of the ship about the transverse axis.

(vii) Rolling is the angular motion of the ship about the longitudinal axis.

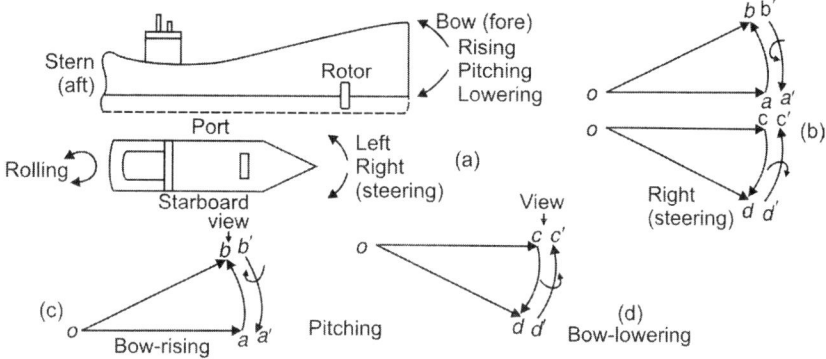

Fig. 11.10: Naval ship and couple diagram

Assume ω to be the angular velocity of the rotor in the clockwise direction when viewed from the rear end.

11.7.1 Effect of Gyroscopic Couple during Steering

Angular momentum vector of rotor before steerig is $\overrightarrow{oa}$. When the ship steers to left, the anglar momentum vector changes from $\overrightarrow{oa}$ to $\overrightarrow{ob}$. The reactive gyroscopic coupe $b\ 'a'$ is perpendicular to oa in the limit and acts in anticlockwise direction when viewed from front. This reactive gyroscopic couple tends to raise the bow and lower the stern. When the ship steers to right, the direction of reactive gyroscopic couple is reversed so that it tends to lower the bow and raise the stern.

11.7.2 Effect of Gyroscopic Couple on Ship during Pitching

The pitching of ship is assumed to take place with simple harmonic motion. The angular displacemet θ of the axis of spin from its mean position after time t,

$$\theta = \phi \sin \omega_0 t$$

where ϕ = amplitude of swing or maximum angle turned from the mean position in radians

ω_0 = angular velocity of SHM $= \dfrac{2\pi}{\text{time period of SHM}} = \dfrac{2\pi}{t}$

Angular velocity of precession, $\omega_p = \dfrac{d\theta}{dt} = \dfrac{d}{dt} (\phi \sin \omega_0 t)$

∴ $\omega_p = \phi\, \omega_0 \cos \omega_0 t$

ω_p will be maximum when $\cos \omega_0 t = 1$

∴ Maximum angular velocity of precession,

$$\omega_p(\text{max}) = \phi\, \omega_0 = \phi \times \frac{2\pi}{t}$$

Maximum gyroscopic couple, $C_{\text{max}} = I\omega\omega_p(\text{max})$

$$= I\omega \times \left[\phi \times \frac{2\pi}{t}\right]$$

When the pitching is upward, the reactive gyroscopic couple is clockwise on seeing from top and therefore, the ship would move towards starboard side (right). When the pitching is downward (bow lowered), the reactive couple is anticlockwise and the ship turns towards port side (left).

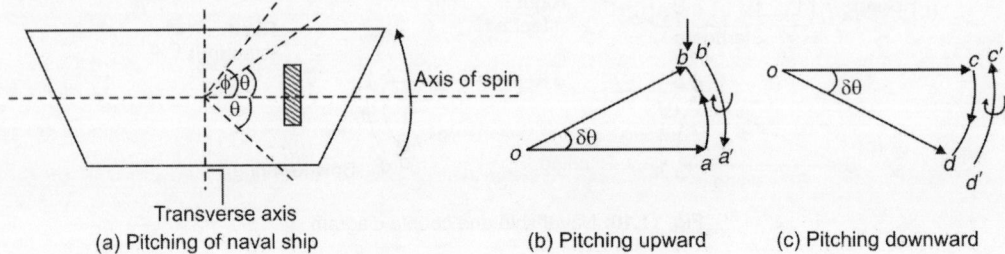

(a) Pitching of naval ship (b) Pitching upward (c) Pitching downward

Fig. 11.11: Pitching of a ship and couple diagram

Angular acceleration, $\alpha = \dfrac{d^2\theta}{dt^2} = \dfrac{d}{dt}(\omega_p) = -\phi\omega_0^2 \sin \omega_0 t$

Angular acceleration is maximum if $\sin \omega_0 t = 1$

$\therefore$ Maximum angular acceleration, $\alpha_{max} = \phi\omega_0^2$.

11.7.3 Effect of Gyroscopic Couple on Ship during Rolling

During rolling of ship, axis of spin of rotor and axis of rolling of ship are parallel for all positions. There is no precession of the axis of spin and thereofre, there is no gyroscopic effect.

Example 11.4: A turbine rotor of a sea vessel having a mass of 950 kg rotates at 1200 rpm clockwise while looking from the stern. The vessele pitches with an angular velocity of 1.2 rad/s. What will be the gyroscopic couple transmitted to the hull when the bow rises? The radius of gyration of rotor is 300 mm.

Solution: Mass of rotor, $m = 950$ kg

Radius of gyration of rotor, $k = 300$ mm $= 0.3$ m

Moment of inertia, $I = mk^2 = 950 \times (0.3)^2 = 85.5$ kg·m^2

$$N = 1200 \text{ rpm}, \omega = \frac{2\pi N}{60} = \frac{2\pi \times 1200}{60} = 125.66 \text{ rad/s}$$

$\omega_p = 1.2$ rad/s

Gyroscopic couple, $C = I\omega\omega_p = 85.5 \times 125.66 \times 1.2$

$= 12892.7$ N·m $= 12.89$ kN·m

Effect of reaction couple is to turn the ship towards starboard side (right).

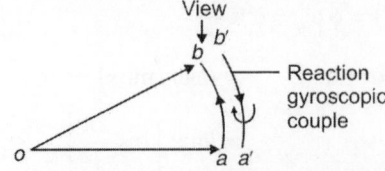

Fig. 11.12: Reaction gyroscopic couple (clockwise)

Example 11.5: The turbine rotor of a ship has a mass of 2 tonnes and rotates at 1800 rpm clockwise when viewed from stern. The radius of gyration of the rotor is 300 mm. Determine the gyroscopic couple and its effect when:

(a) The ship turns right at the radius of 200 m with a speed of 36 km/hr
(b) The ship pitches with the bow rising at an angular velocity of 1 rad/s
(c) The ship rolls at an angular velocity of 0.1 rad/s.

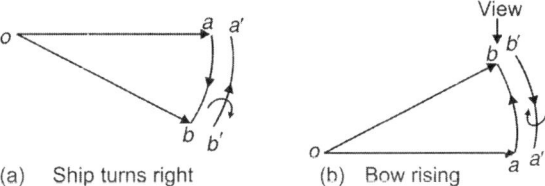

(a)   Ship turns right   (b)   Bow rising

Fig.11.13

Solution: Mass of rotor, $m = 2$ ton $= 2000$ kg

$$k = 300 \text{ mm} = 0.3 \text{ m}, I = mk^2 = 2000 \times (0.3)^2 = 180 \text{ kg m}^2$$

$$N = 1800 \text{ rpm}, \omega = \frac{2\pi N}{60} = \frac{2\pi \times 1800}{60} = 188.5 \text{ rad/s}$$

(a) $R = 200 \text{ m}, v = 36 \text{ km/hr} = \dfrac{36 \times 1000}{3600} = 10 \text{ m/s}$

$$\omega_p = \frac{v}{R} = \frac{10}{200} = 0.05 \text{ rad/s}$$

Gyroscopic couple, $c = I\omega\omega_p = 180 \times 188.5 \times 0.05$

$$= 1696.5 \text{ N·m}$$

The effect is to lower the bow and raise the stern.

(b) $\omega_p = 1 \text{ rad/s}$

Gyroscopic couple, $C = I\omega\omega_p = 180 \times 188.5 \times 1 = 33930 \text{ N·m}$

The effect of reaction couple is to move the ship towards starboard side.

(c) $\omega_p = 0.1 \text{ rad/s}$

$$C = I\omega\omega_p = 180 \times 188.5 \times 0.1$$
$$= 3393 \text{ N·m}$$

But the axis of spin is alway parallel to the axis or rolling in all positions, hence there is no gyroscopic effect on the ship.

11.8 STABILITY OF AN AUTOMOBILE WHILE MOVING IN A CURVED PATH

Consider a four-wheeled vehicle having a mass m taking a turn towards left. The wheels A and C are inner wheels and B and D are outer wheels. The centre of gravity (*CG*) of the vehicle lies vertically above the road surface.

Let, r = radius of wheels in m

R = radius of curvature in m

h = distance of centre of gravity, vertically above the road surface

x = width of track in m

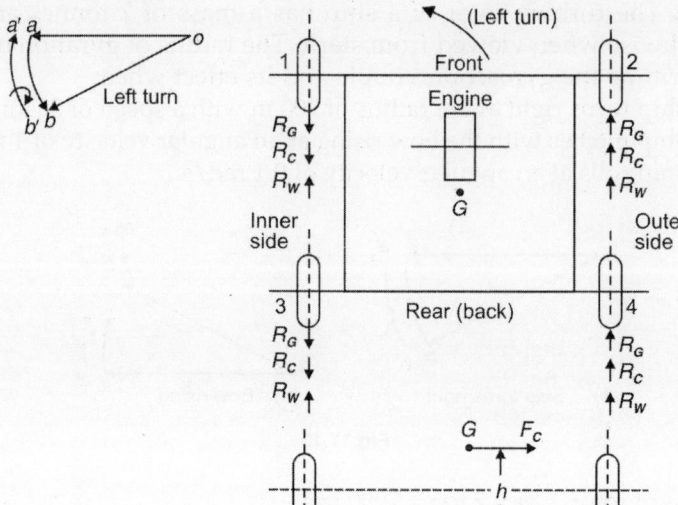

Fig. 11.14

I_w = mass moment of inertia of each wheel

ω_w = angular velocity of wheels

$\quad = \dfrac{v}{r}$ rad/s

v = linear velocity of vehicle

$\quad = r_w \omega_w$ m/s

ω_e = angular velocity of the rotating parts of the engine (rad/s)

ω_p = angular velocity of precession

$\quad = \dfrac{v}{R}$ rad/s

G = gear ratio $= \dfrac{\omega_e}{\omega_w}$

(i) *Reaction due to weight of the vehicle:*

Assuming that the weight of vehicle is equally distributed over the four wheels, weight on each wheel $= \dfrac{mg}{4}$ N (downwards). The road surface will offer an equal and opposite reaction to the wheels in upward direction. Reaction of road on each wheel $(R_w) = \dfrac{mg}{4}$ N (upwards).

(ii) *Reaction due to gyroscopic couple:*

Gyroscopic couple due to four wheels, $C_w = 4 I_w \omega_w \omega_p$

$$C_w = 4 I_w \frac{v}{r} \cdot \frac{v}{R} = 4 I_w \frac{v^2}{rR}$$

Gyroscopic couple due to rotating parts of the engine,

$$C_e = I_e \omega_e \omega_p = I_e(G\omega_w)\,\omega_p \qquad\qquad [G = \omega_e/\omega_w]$$

$$C_e = I_e\, G\frac{v}{r}\cdot\frac{v}{R} = I_e G\,\frac{v^2}{rR}$$

Total gyroscopic couple, $C_G = C_w \pm C_e$

or

$$C_G = 4I_w\frac{v^2}{rR} \pm GI_e\frac{v^2}{rR} = \frac{v^2}{rR}\,(4I_w \pm GI_e) \qquad\qquad ...(11.2)$$

Positive sign is used when the engine parts rotate in the same direction as the wheels and the negative sign when they rotate in the opposite direction.

Assuming that C_G is positive, the reactive gyroscopic couple on it is clockwise when viewed from the rear of the vehicle. The reactive couple is provided by equal and opposite forces on the outer and the inner wheels of the vehicle. Let force on each wheel be P N

$$P \times x + P \times x = C_G \text{ or } P = \frac{C_G}{2x}$$

Force P will act upwards on inner wheels and downwards on outer wheels. Therefore, reaction of ground on each outer wheel,

$$R_G = \frac{C_G}{2x}\text{ (upwards)}$$

and reaction of ground on each inner wheel,

$$R_G = \frac{C_G}{2x}\text{ (downwards)}$$

(iii) *Reaction due to centrifugal couple:*

Since the vehicle moves on a curved path, therefore a centrifugal force acts on the vehicle in the outward direction at the centre of mass (take CG) of the vehicle.

Centrifugal force $(F_c) = mR\omega_p^2$

$$= mR\left(\frac{v}{R}\right)^2$$

$$= m\frac{v^2}{R}$$

This force tends to overturn the vehicle outwards. Overturning couple,

$$C_c = F_c \times h = \frac{mv^2}{R}\times h$$

This couple is provided by equal and opposite forces acting on outer and inner wheels. Let force on each wheel be P_c N,

$$P_c \times x + P_c \times x = C_c$$

or

$$P_c = \frac{C_c}{2x}$$

Force P_c acts upwards on inner wheels and downwards on outer wheels.

∴ Reaction of ground on each outer wheel,

$$R_c = \frac{C_c}{2x}\text{ (upwards)}$$

Reaction of ground on each inner wheel,

$$R_c = \frac{C_c}{2x} \text{ (downwards)}$$

Total vertical reaction on each outer wheel,

$$R_o = \frac{mg}{4} + \frac{C_G}{2x} + \frac{C_c}{2x} \text{ (upwards)} \qquad \qquad ...(11.3)$$

Total vertical reaction on each inner wheel,

$$R_i = \frac{mg}{4} - \frac{C_G}{2x} - \frac{C_c}{2x} \text{ (upwards)} \qquad \qquad ...(11.4)$$

If vehicle is running at high speed, R_i may be zero or negative. In this case, the reaction of the ground on the inner wheels will be downward. This will cause the inner wheels to be lifted up from the gournd and overturn the vehicle. For stability of vehicle (i.e. for positive R_i), the condition is,

$$\frac{mg}{4} - \frac{C_G}{2x} - \frac{C_c}{2x} \geq 0 \qquad \qquad ...(11.5)$$

or
$$\frac{mg}{4} \geq \frac{C_G}{2x} + \frac{C_c}{2x} \text{ or } \frac{mg}{4} \geq \frac{(C_G + C_c)}{2x}$$

or
$$R_w \geq R_G + R_c.$$

Example 11.6: Each wheel of a four-wheeled, rear engine automobile has a moment of inertia of 2.4 kg·m^2 and an effective diameter of 660 mm. The rotating parts of the engine have a moment of inertia of 1.2 kg·m^2. The gear ratio of engine to the rear wheel is 3:1. The engine axis is parallel to the rear axle and the crankshaft rotates in the same sense as the road wheels. The mass of the vehicle is 2200 kg and the centre of the mass is 550 mm above the road level. The track width of the vehicle is 1.5 m. Determine the limiting speed of the vehicle around a curve with 80 m radius so that all the four wheels remain in contact with road surface.

Solution:

$$m = 2200 \text{ kg, } r = 0.33 \text{ m, } R = 80 \text{ m}$$
$$h = 0.55, x = 1.5 \text{ m}$$
$$I_w = 2.4 \text{ kg·m}^2, I_e = 1.2 \text{ kg·m}^2$$
$$G = \frac{\omega_e}{\omega_w} = 3$$

(i) *Reaction due to weight*

$$R_w = \frac{mg}{4} = \frac{2200 \times 9.81}{4} = 5395.5 \text{ N (upwards)}$$

(ii) *Reaction due to gyroscopic couple*

$$C_w = 4I_w \times \frac{v^2}{rR} = 4 \times 2.4 \times \frac{v^2}{0.33 \times 80} = 0.364v^2$$

$$C_e = I_e G \omega_w \omega_p = 1.2 \times 3 \times \frac{v^2}{0.33 \times 80} = 0.136v^2$$

∴
$$C_G = C_w + C_e = 0.364v^2 + 0.136v^2 = 0.5v^2$$

Reaction on each outer wheel,

$$R_{Go} = \frac{C_G}{2x} = \frac{0.5v^2}{2 \times 1.5} = 1.67v^2 \ \text{(upwards)}$$

Reaction on each inner wheel, $R_{Gi} = 0.167\,v^2$ (downwards)

(iii) *Reaction due to centrifugal couple*

$$C_c = \frac{mv^2}{R} h = \frac{2200 \times v^2}{80} \times 0.55 = 15.125v^2$$

Reaction on each outer wheel,

$$R_{Go} = \frac{C_c}{2x} = \frac{15.125v^2}{2 \times 1.5} = 5.042v^2 \ \text{(upwards)}$$

Reaction on each inner wheel, $R_{Gi} = 5.042\,v^2$ (downwards)

For maximum safe speed, the condition is

$$R_w = R_{Gi} + R_{Go}$$

$$5395.5 = (0.167 + 5.042)v^2$$

$$v^2 = 1035.8, \ v = 32.18 \ \text{m/s}$$

or

$$v = \frac{32.18 \times 3600}{1000} = 115.9 \ \text{km/h.}$$

11.9 STABILITY OF A TWO-WHEEL VEHICLE

Consider a two-wheel vehicle taking a left turn. The rider of the vehicle tilts the vehicle inwards to neutralise the overturning effect and to stay in equilibrium while taking a turn. The angle of inclination of the vehicle to the vertical is known as the angle of heel (θ).

Let v = linear velocity of vehicle on the track = $r_w \times \omega_w$

r = radius of the wheels

R = radius of the track

I_w = moment of inertia of each wheel

I_e = moment of inertia of rotating parts of the engine

m = total mass of the vehicle and the rider

ω_w = angular velocity of the wheels

ω_e = angular velocity of rotating parts of the engine

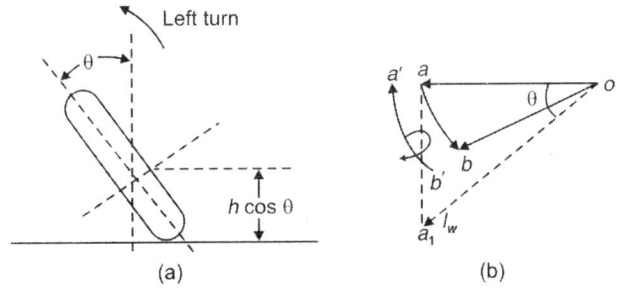

Fig.11.15: Stability of a two-wheel vehicle taking left turn

G = gear ratio = ω_e/ω_w

h = height of centre of mass of the vehicle and the rider

θ = inclination of vehicle to the vertical (angle of heel)

11.9.1 Effect of Gyroscopic Couple

When the vehicle moves over the curved path, the axis of spin is inclined to the horizontal at an angle θ. Therefore, the angular momentum vector $I\omega$ due to spin is represented by oa_1, inclined to OX (X-axis) at an angle θ. But the axis of precession is vertical. Hence, it is necessary to take horizontal component of the spin vector (oa_1).

Horizontal component of angular momentum vector,

$$= I\omega \cos\theta = (2I_w\,\omega_w \pm I_e\omega_e)\cos\theta$$

Gyroscopic couple, $\quad C_G = I\omega\cos\theta \times \omega_p$

or $\qquad\qquad\qquad C_G = (2I_w\,\omega_w \pm I_e\,\omega_e)\cos\theta \times \dfrac{v}{R}$ $\qquad\qquad \left(\omega_p = \dfrac{v}{R}\right)$

or $\qquad\qquad\qquad = (2I_w\,\omega_w \pm I_e\,G\omega_w)\dfrac{v}{R}\cos\theta$

$$= (2I_w \pm GI_e)\dfrac{v}{r}\cdot\dfrac{v}{R}\cos\theta \qquad\qquad \left(\omega_w = \dfrac{v}{r}\right)$$

$$C_G = (2I_w \pm GI_e)\dfrac{v^2}{rR}\cos\theta \qquad\qquad\qquad ...(11.6)$$

(i) When the engine is rotating in the same direction as that of wheels, then the positive sign is used and if the engine rotates in opposite direction, then negative sign is used.

(ii) The reaction couple b 'a' is clockwise when viewed from the rear of the vehicle and tends to overturn it in outward direction.

11.9.2 Effect of Centrifugal Couple

A centrifugal force $\left(F_c = m\dfrac{v^2}{R}\right)$ acts horizontally through the centre of gravity (CG) in outward direction.

$\therefore\quad$ Overturning couple due to centrifugal force,

$$C_c = F_c \times h\cos\theta = \dfrac{mv^2}{R} \times h\cos\theta \qquad\qquad ...(11.7)$$

$\therefore\quad$ Total overturning couple, $C_o = C_G + C_c$

$$C_o = (2I_w \pm GI_e)\dfrac{v^2}{rR}\cos\theta + \dfrac{mv^2}{R}h\cos\theta$$

$$C_o = \dfrac{v^2}{R}\left[\dfrac{2I_w \pm GI_e}{r} + mh\right]\cos\theta \qquad\qquad ...(11.8)$$

Balancing couple due to weight of the vehicle,

$$C_b = mg \times h\sin\theta$$

For equilibrium, overturning couple = balancing couple

$$\frac{v^2}{R}\left[\frac{2I_w \pm GI_e}{r} + mh\right]\cos\theta = mgh\sin\theta$$

From the expression, the angle of heel θ can be determined, so that the vehicle does not skid.

Example 11.7: Each wheel of a motorcycle has 600 mm diameter and moment of inertia of 1.2 kg·m². The total mass of the motorcycle and the rider is 180 kg and the combined centre of mass is 580 mm above the ground level when the motorcycle is upright. The moment of inertia of the rotating parts of the engine is 0.2 kg·m². The engine speed is 5 times the speed of the wheels and is in the same sense. Determine the angle of heel necessary when the motorcycle takes a turn of 35 m radius at a speed 54 km/h.

Solution:

$$m = 180 \text{ kg}, r = 0.3 \text{ m}, R = 35 \text{ m}, h = 0.58 \text{ m},$$
$$I_w = 1.2 \text{ kg.m}^2, I_e = 0.2 \text{ kg.m}^2,$$
$$v = \frac{54 \times 1000}{3600} = 15 \text{ m/s}, G = \frac{\omega_e}{\omega_w} = 5$$

Gyroscopic couple $C_G = (2I_w + GI_e)\dfrac{v^2}{rR}\cos\theta$

$$= (2 \times 1. + 5 \times 0.2) \times \frac{(15)^2}{0.3 \times 35} \times \cos\theta = 72.86\cos\theta$$

Centrifugal couple $C_c = m\dfrac{v^2}{R}h\cos\theta = 180 \times \dfrac{(15)^2}{35} \times 0.58 \times \cos\theta$

$$= 671.14\cos\theta$$

Total overturning couple $= (72.86 + 671.14)\cos\theta = 744\cos\theta$

Rightening couple $= mgh\sin\theta$
$$= 180 \times 9.81 \times 0.58\sin\theta = 1024\sin\theta$$

∴ $\quad\quad 1024\sin\theta = 744\cos\theta$

or $\quad\quad\quad \tan\theta = \dfrac{744}{1024} = 0.727$ or $\theta = 36°$

11.10 EFFECT OF GYROSCOPIC COUPLE ON A RIGID DISC FIXED AT AN ANGLE TO A ROTATING SHAFT

Consider a circular disc fixed rigidly to a rotating shaft such that the polar axis of the disc makes an angle θ with the axis of shaft. Let the shaft rotates with an angular velocity ω rad/s in clockwise direction when viewed from the left end of shaft.

Let *OX* be the axis of shaft,

OP be the Polar axis of disc and,

OD be the diametral axis of the disc.

m = mass, r = radius and t = thickness of the disc

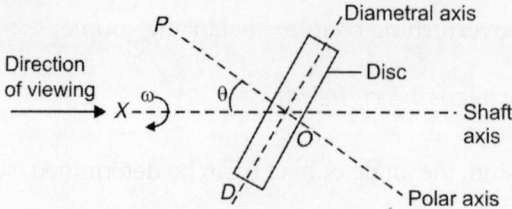

Fig.11.16: Effect of gyroscopic couple on a disc fixed rigidly at a certain angle to a rotating shaft.

Mass moment of inertia of the disc about polar axis, OP

$$I_p = \frac{mr^2}{2}$$

Mass moment of inertia of the disc about diametral axis, OD

$$I_d = m\left[\frac{t^2}{12} + \frac{r^2}{4}\right] = \frac{mr^2}{4} \text{ (t is neglected for a thin disc)}$$

(i) Angular velocity of spin (about polar axis OP) = $\omega \cos\theta$

Precession is produced about OD.

$\therefore$ Angular velocity of precession (about OD) = $\omega \sin\theta$

Gyroscopic couple acting on disc, $C_p = I_p \times \omega \cos\theta \times \omega \sin\theta$

$$C_p = \frac{1}{2} I_p\omega^2 \sin 2\theta \text{ (for spinning about polar axis)}$$

Its effect is to rotate the disc anticlockwise when viewed from the top.

(ii) For spinning about the diametral axis (OD)

angular velocity of spin (about OD) = $\omega \sin\theta$

angular velocity of precession (about OP) = $\omega \cos\theta$

gyroscopic couple acting on the disc, $C_d = I_d \times \omega \sin\theta \times \omega \cos\theta$,

$$C_d = \frac{1}{2} I_d \omega^2 \sin 2\theta$$

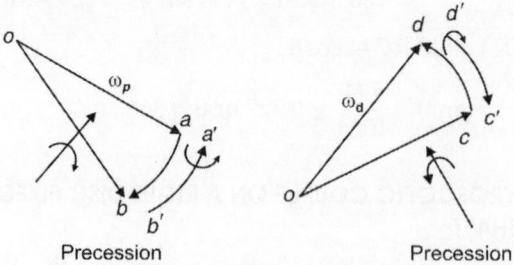

Fig. 11.17: (a) About polar axis (b) About diametral axis

Its effect is to rotate the disc clockwise when viewed from the top.

Resultant gyroscopic couple on the disc, $C = C_p - C_d$

$$C = \frac{1}{2}(I_p - I_d) \omega^2 \sin 2\theta$$

$$= \frac{1}{2}\left[\frac{mr^2}{2} - \frac{mr^2}{4}\right]\omega^2 \sin 2\theta$$

$$c = \frac{mr^2}{8}\omega^2 \sin 2\theta.$$

Example 11.8: A shaft carries a uniform thin disc of mass 40 kg and diameter 600 mm. The plane of the disc is not perfectly at right angle to the axis of the shaft but has an error of 1.5°. Determine the gyroscopic couple acting on the bearing if the shaft rotates at 840 rpm.

Solution : Mass $m = 40$ kg, $r \dfrac{600}{2} = 300$ mm $= 0.3$ m

$$\theta = 1.5°, N = 840 \text{ rpm}$$

$$\omega = \frac{2\pi N}{60} = \frac{2\pi \times 840}{60} = 88 \text{ rad/s}$$

Gyroscopic couple acting on the bearing c

$$c = \frac{mr^2}{8}\omega^2 \sin 2\theta = \frac{40 \times (0.3)^2}{8} \times (88)^2 \sin 3°$$

$$c = 182.38 \text{ N·m}$$

ADDITIONAL SOLVED EXAMPLES

Example 11.9: A uniform disc of 200 mm diameter has a mass of 10 kg. It is mounted centrally on the horizontal shaft, which runs in bearings that are 150 mm apart. The disc spins with a uniform speed of 2000 rpm in vertical plane in counter-clockwise direction looking from right hand side bearing. The shaft precesses with a uniform velocity 50 rpm in the horizontal plane in the anticlockwise direction when looking from the top. Determine the reactions at each bearing due to the mass and gyroscopic effects.

Solution: Mass of disc, $m = 10$ kg, radius, $r = \dfrac{200}{2} = 100$ mm $= 0.1$ m

Length of shaft, $l = 150$ mm $= 0.15$ m

$$N = 2000 \text{ rpm}, N_p = 50 \text{ rpm}$$

$$\omega = \frac{2\pi N}{60} = \frac{2\pi \times 2000}{60} = 209.44 \text{ rad/s}$$

$$\omega_p = \frac{2\pi N_p}{60} = \frac{2\pi \times 50}{60} = 5.24 \text{ rad/s}$$

Mass moment of inertia of disc, $I = \dfrac{mr^2}{2} = \dfrac{10 \times (0.1)^2}{2} = 0.05$ kg·m²

Gyroscopic couple, $C = I\omega\omega_p = 0.05 \times 209.44 \times 5.24$

$$= 54.87 \text{ N.m}$$

The reactive gyroscopic couple $b'a'$ is anticlockwise when viewed from front and tends to lower the bearing A and raise the bearing B.

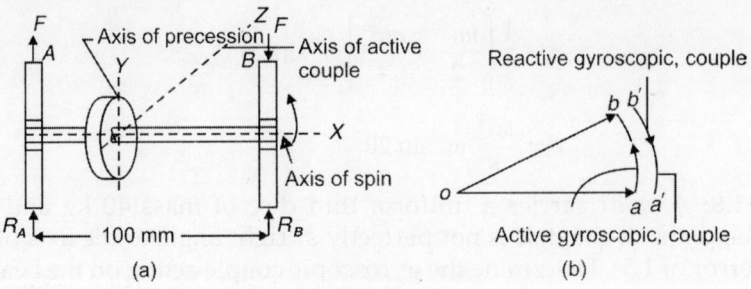

(a)

(b)

Fig. 11.18

Force at bearing A due to gyroscopic couple,

$$R_{GA} = \frac{C}{l} = \frac{54.87}{0.15} = 365.8 \text{ N (downwards)}$$

Force at bearing B due to gyroscopic couple $= \dfrac{C}{l}$

$$R_{GB} = \frac{54.87}{0.15} = 365.8 \text{ N (upwards)}$$

Force at each bearing due to weight of disc, R_w

$$R_{wA} = R_{wB} = \frac{mg}{2} = \frac{10 \times 9.8}{2} = 49 \text{ N (upwards)}$$

Total reaction at bearing A, $R_A = \dfrac{-C}{l} + \dfrac{mg}{2} = -316.8$ (downwards)

Total reaction at bearing B, $R_B = \dfrac{C}{l} + \dfrac{mg}{2} = 414.8$ (upwards)

Example 11.10: A ship is pitching through a total angle of 15°, the oscillation may be taken as simple harmonic, and the complete period is 32 seconds. The turbine rotor has a mass of 6000 kg; its radius of gyration is 450 mm and it is rotating at 2400 rpm. Calculate the maximum value of the gyroscopic couple set up by the rotor. If the rotation of the rotor is clockwise when looking from aft, in which direction will the bow tend to turn while falling. What is the maximum angular acceleration to which the ship is subjected while pitching?

Solution:
$$2\phi = 15°, \phi = \frac{15}{2} \times \frac{\pi}{180} = \frac{\pi}{24} \text{ radians}$$

$$t_p = 32 \text{ s}$$
$$m = 6000 \text{ kg}, k = 450 \text{ mm} = 0.45 \text{ m}$$

Moment of inertia, $I = mk^2 = 6000 \times (0.45)^2$
$$= 1215 \text{ kg·m}^2$$

$$\omega_1 = \frac{2\pi}{t_p} = \frac{2\pi}{32} = \frac{\pi}{16} \text{ rad/s}$$

$$N = 2400 \text{ rpm}$$

$\therefore$
$$\omega = \frac{2\pi N}{60} = \frac{2\pi \times 2400}{60} = 80\pi = 251.32 \text{ rad/s}$$

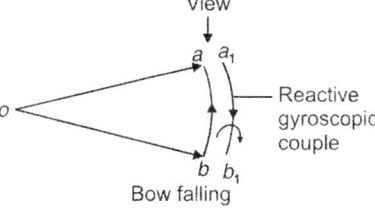

Fig. 11.19

Maximum angular velocity of precession, ω_p (max) $= \phi\,\omega_1$

$$\omega_p\,(\text{max}) = \frac{\pi}{24} \times \frac{\pi}{16} = 0.0257 \text{ rad/s}$$

Maximum gyroscopic couple, $C_{max} = I\omega\omega_p(\text{max})$

$$= 1215 \times 251.32 \times 0.0257$$
$$= 7847.8 \text{ N·m}$$

Maximum angular acceleration, $\alpha_{max} = \phi\,\omega_1^2$

$$= \frac{\pi}{24} \times \left(\frac{\pi}{16}\right)^2 = 0.005 \text{ rad/s}^2$$

Reactive gyroscpic couple is anticlockwise, therefore ship will steer towards left.

Example 11.11 A four-wheeled trolley car has a total mass of 3000 kg. Each axle with its two wheels and gears has a total moment of inertia of 32 kg·m². Each wheel is of 450 mm radius. The centre distance between two wheels on an axle is 1.4 m. Each axle is driven by a motor with a speed ratio of 1:3. Each motor along with its gear has a moment of inertia of 16 kg·m² and rotates in the opposite direction to that of the axle. The centre of mass of the car is 1 m above the rails. Calculate the limiting speed of the car when it has to travel around a curve of 250 m radius without the wheels leaving the rails.

Solution: $m = 3000$ kg, $r = 0.45$ m, $R = 80$ m, $h = 1$ m, $x = 1.4$ m

$$I_w = \frac{32}{2} = 16 \text{ kg·m}^2, I_m = 16 \text{ kg·m}^2$$

(i) Reaction due to weight

$$R_w = \frac{mg}{4} = \frac{3000 \times 9.81}{4} = 7357.5 \text{ N} \quad \text{(upwards)}$$

(ii) Reaction due to gyroscopic couple

$$C_w = 4I_w\frac{v^2}{rR} = 4 \times 16 \times \frac{v^2}{0.45 \times 250} = 0.569\,v^2$$

$$C_m = 2I_mG\omega_w\omega_p \quad \text{(as there are two motors)}$$

$$= 2 \times 16 \times 3 \times \frac{v^2}{0.45 \times 250} = 0.853v^2$$

$$C_G = C_w - C_m \quad \text{(motors rotate in opposite direction)}$$

$$= 0.569v^2 - 0.853v^2 = -0.284v^2$$

Reaction on each outer wheel,

$$R_{Go} = \frac{C_G}{2x} = \frac{0.284 v^2}{2 \times 1.4} = 0.1014 v^2 \text{ (downwards)}$$

Reaction on each inner wheel, $R_{Gi} = 0.1014 v^2$ (upwards)

(iii) *Reaction due to centrifugal couple*

$$C_c = \frac{mv^2}{R} h = 3000 \times \frac{v^2}{250} \times 1 = 12 v^2$$

$$R_{co} = \frac{C_c}{2x} = \frac{12 v^2}{2 \times 1.4} = 4.286 v^2 \text{ (upwards)}$$

$$R_{ci} = \frac{C_c}{2x} = 4.286 v^2 \text{ (downwards)}$$

Total reaction on outer wheel = $7357.5 - 0.1014 v^2 + 4.286 v^2$

$$= 7357.5 + 4.1846 \, v^2$$

Total reaction on inner wheel = $7357.5 - 0.1014 v^2 - 4.286 v^2$

$$= 7357.5 - 4.186 \, v^2$$

Thus, the reaction on the outer wheel is always positive (upwards). There are chances that the inner wheels leave the rails.

For maximum speed, $7357.5 - 4.1846 \, v^2 = 0$

$$v^2 = 1758.2, v = 41.93 \text{ m/s}$$

or

$$v = \frac{41.93 \times 3600}{1000} = 151 \text{ km/h.}$$

Example 11.12: Each road wheel of a motorcycle is of 600 mm diameter and has a moment of inertia of 1.1 kg·m². The motorcycle and the rider together weigh 220 kg and the combined centre of mass is 620 mm above the ground level when the motorcycle is upright. The moment of inertia of the rotating parts of the engine is 0.18 kg·m². The engine rotates at 4.5 times the speed of the road wheels in the same sense. Find the angle of heel necessary when the motorcycle is taking a turn of 35 m radius at a speed of 72 km/hr.

Solution: $I_w = 1.1$ kg·m², $m = 220$ kg

$\quad h = 620$ mm $= 0.62$ m, $I_e = 0.18$ kg·m²

$$G = \frac{\omega_e}{\omega_w} = 4.5, R = 35 \text{ m}, v = 72 \text{ km/hr}$$

$$v = \frac{72 \times 1000}{3600} = 20 \text{ m/s}$$

Radius of wheel, $r_w = \dfrac{600}{2} = 300$ mm $= 0.3$ m

Condition of equilibrium: Overturning couple = Balancing couple

$$\frac{v^2}{R}\left[\frac{2I_w + GI_e}{r_w} + mh\right]\cos\theta = mgh \sin\theta$$

$$\frac{(20)^2}{35}\left[\frac{2 \times 1 + 4.5 \times 0.18}{0.3} + 220 \times 0.62\right]\cos\theta = 220 \times 9.8 \times 0.62 \sin\theta$$

or $$1673.52 \cos \theta = 1336.72 \sin \theta$$

$$\tan \theta = \frac{1673.52}{1336.72} = 1.25$$

∴ $$\theta = \tan^{-1}(1.25) = 51.34°$$

Thus, the angle of heel θ is 51.34°.

Example 11.13: The turbine rotor of a ship has a mass of 2000 kg and rotates at a speed of 3000 rpm clockwise when looking from stern. The radius of gyration of the rotor is 0.5 m. Determine the gyroscopic couple and its effects upon the ship when the ship is steering to the right in a curve of 100 m radius at a speed of 16.1 knots (1 knot = 1855 m/hr). Also calculate the torque and its effect when the ship is pitching in simple harmonic motion, the bow falling with its maximum velocity. The period of pitching is 50 seconds and total angular displacement between the two extreme positions of pitching is 12°. Find the maximum acceleration during pitching motion.

Solution: $m = 2000$ kg, $N = 3000$ rpm, $k = 0.5$ m

$R = 100$ m, $v = 16.1$ knots

$$= 16.1 \times \frac{1855}{3600}$$

$$= 8.3 \text{ m/s}$$

$$\omega = \frac{2\pi N}{60}$$

$$= \frac{2\pi \times 3000}{60}$$

$$= 314.2 \text{ rad/s}$$

$$I = mk^2 = 2000 \times (0.5)^2 = 500 \text{ kg·m}^2$$

$$\omega_p = \frac{v}{R} = \frac{8.3}{100} = 0.083 \text{ rad/s}$$

Gyroscopic couple, $C = I\omega\omega_p = 500 \times 314.2 \times 0.083$

$$= 13040 \text{ N·m}$$

$$= 13.04 \text{ kN·m}$$

The reactive gyroscopic couple $b'a'$ is clockwise and it tends to raise the stern and lower the bow.

Pitching: $t_p = 50$ s, $2\phi = 12°$, $\phi = 6°$

or $$\phi = 6 \times \frac{\pi}{180}$$

$$= 0.105 \text{ rad}$$

$$\omega_1 = \frac{2\pi}{t_p}$$

$$= \frac{2\pi}{50}$$

$$= 0.1257 \text{ rad/s}$$

$$\omega p_{(max)} = \phi \omega_1$$
$$= 0.105 \times 0.1257$$
$$= 0.0132 \text{ rad/s}$$
$$C_{max} = I \omega \omega_{p(max)}$$
$$= 500 \times 314.2 \times 0.0132$$
$$= 2074 \text{ N·m}$$

The effect is that the reactive gyroscopic couple is anticlockwise (seen in top view) and it tends to turn the ship towards the left,

$$\alpha_{max} = \phi \omega_1^2$$
$$= 0.105 \times (0.1257)^2 = 0.00166 \text{ rad/s}^2.$$

Example 11.14: A gyrowheel D of mass 0.5 kg, with a radius of gyration of 20 mm, is mounted on a pivoted frame C as shown in Fig. 11.20. The axis AB of the pivots passes through the centre of rotation O of the wheel, but the centre of gravity G of the frame C is 10 mm below O. The frame has a mass of 0.30 kg and the speed of rotation of the wheel is 3000 rpm in the anticlockwise direction.

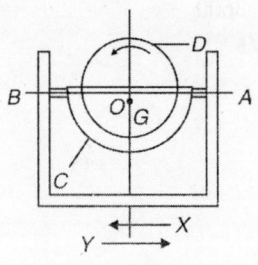

Fig. 11.20

The entire unit is mounted on a vehicle so that the axis AB is parallel to the direction of motion of the vehicle. If the vehicle travels at 15 m/s in a curve of 50 m radius, find the inclination of the gyrowheel from the vertical, when
(i) The vehicle moves in the direction of the arrow X taking a left hand turn along the curve.
(ii) The vehicle reverse at the same speed in the direction of arrow 'Y' along the same path.

Solution: $m_1 = 0.5$ kg; $k = 20$ mm $= 0.02$ m; $OG = h = 10$ mm $= 0.01$ m; $m_2 = 0.3$ kg; $N = 3000$ rpm or $\omega = 2\pi \times 3000/60 = 314.2$ rad/s, $v = 15$ m/s; $R = 50$ m.

Mass moment of inertia of the gyrowheel,
$$I = m_1.k^2 = 0.5 \, (0.02)^2 = 0.0002 \text{ kg·m}^2$$

and angular velocity of precession,
$$w_p = v/R = 15/50 = 0.3 \text{ rad/s}$$

Let $\theta =$ angle of inclination of gyrowheel from the vertical.
(i) Gyroscopic couple about O,
$$C_1 = I \omega \omega_p \cos \theta = 0.0002 \times 314.2 + 0.3 \cos \theta \text{ N·m}$$
$$= 0.019 \cos \theta \text{ N·m (anticlockwise)}$$

and centrifugal couple about O,

$$C_2 = \frac{m_2 v^2}{R} \times h \cos\theta = \frac{0.3(15)^2}{50} \times 0.01 \cos\theta \text{ N·m}$$

$$= 0.0135 \cos\theta \text{ N·m (anticlockwise)}$$

∴ Total overturning couple, $C_1 - C_2 = 0.019 \cos\theta - 0.0135 \cos\theta$

$$= 0.0055 \cos\theta \text{ N·m (anticlockwise)}$$

(–ve sign due to opposite direction

We know that balancing couple due to weight ($W_2 = m_2 g$) of the frame about O

$$= m_2 g h \sin\theta = 0.3 \times 9.81 \times 0.01 \sin\theta \text{ N·m}$$

$$= 0.029 \sin\theta \text{ N·m (clockwise)}$$

Since the overturning couple must be equal to the balancing couple for equilibrium condition, therefore,

$$0.0055 \cos\theta = 0.029 \sin\theta$$

or $\tan\theta = \sin\theta/\cos\theta = 0.0055/0.029 = 0.1896$

∴ $\theta = 10.74°$

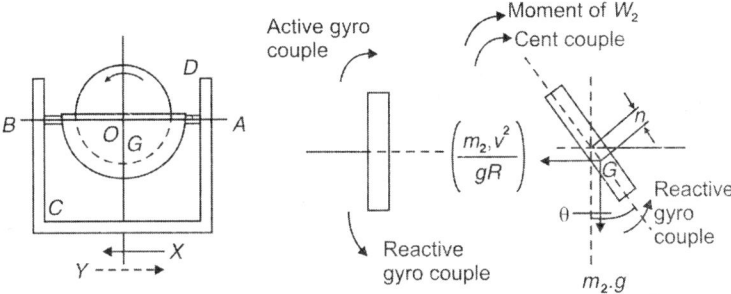

Fig. 11.21

When the vehicle reverses at the same speed in the direction of arrow Y, then the gyroscopic and centrifugal couples (C_1 and C_2) will be in clockwise direction about O and the balancing couple due to weight ($W_2 = m_2 g$) of the frame about O will be in anticlockwise direction.

∴ Total overturning couple $= C_1 + C_2 = 0.019 \cos\theta + 0.0135 \cos\theta$

$$= 0.0325 \cos\theta \text{ N·m}$$

Equating the total overturning couple to the balancing couple, we have,

$$0.0325 \cos\theta = 0.029 \sin\theta$$

or $\tan\theta = \sin\theta/\cos\theta = 0.0325/0.029 = 1.1207$

∴ $\theta = 48.26°$.

Example 11.15: An aeroplane makes a complete half circle of 50 m radius, towards left when flying at 200 km/hr. The rotary engine and the propeller of the plane have a mass 40 kg with a radius of gyration of 0.30 m. The engine runs at 2400 rpm clockwise, when viewed from the rear. Find the gyroscopic couple on the plane and state its effect on it. What will be the effect, if the aeroplane turns to its right instead of the left?

Solution: $m = 40$ kg, $R = 50$ m

$$v = 200 \text{ km/hr } = \frac{200 \times 1000}{3600} = 55.55 \text{ m/s}$$

$$\omega_p = \frac{v}{R} = \frac{55.55}{50} = 1.111 \text{ rad/s}$$

$k = 0.30$ m

Moment of inertia $I = mK^2 = 40 \times (0.30)^2 = 3.6 \text{ kg·m}^2$

$$N = 2400 \text{ rpm}$$

$$\omega = \frac{2\pi N}{60} = \frac{2\pi \times 2400}{60} = 251.327 \text{ rad/s}$$

We know that the gyroscopic couple is given by the relation,

$$C = I\omega\omega_p$$
$$= 3.6 \times 251.327 \times 1.111 = 1005.20 \text{ N-m}$$

Effect of gyroscopic couple on plane:

   (i) When turning to left, it will raise the nose and dip the tail.

   (ii) When turning to right, it will lift the tail and dip the nose.

Example 11.16: One of the driving axles of a locomotive with its two wheels has a moment of inertia of 350 kg·m . The wheels are of 1.85 m diameter. The distance between the planes of the wheels is 1.5 m. When travelling at 100 km/hr, the locomotive passes over a defective rail which causes the right hand wheel to fall 12 mm and rise again in a total time of 0.1 s, the vertical movement of the wheel being SHM. Find the maximum gyroscopic couple.

Solution: $I = 350$ kg·m^2

$D = 1.85$ m or $R = 0.925$ m

$x = 1.5$ m

$$v = \frac{100 \times 1000}{3600} = 27.777 \text{ m/s}$$

The maximum value of gyroscopic couple is given by

$$C = I\omega w_{p\max}$$

Angular velocity of locomotive, $\omega = \dfrac{v}{R} = \dfrac{27.777}{0.925} = 30.03 \text{ rad/s}$

Defective rail causes the right hand wheel to fall 12 mm in 0.1 s, we can find the maximum velocity during falling or rising,

$$v_{\max} = \text{amplitude} \times \text{angular velocity}$$

But   amplitude $= \dfrac{1}{2}$ (rise or fall)

$$= \frac{1}{2} \times 12 = 6 \text{ mm}$$

and   angular velocity $= \dfrac{2\pi}{\text{time}} = \dfrac{2\pi}{0.1} = 20\pi$

So $v_{\max} = 6 \times 20\pi = 376.99 \text{ mm/s} = 0.377 \text{ m/s}$

Angular velocity of precession,

$$\omega_{p\ max} = \frac{v_{max}}{x} = \frac{0.377}{1.5} = 0.25 \text{ rad/s}$$

Now

$$C_{max} = I\omega\omega_{p\ max}$$

$$= 350 \times 30.03 \times 0.25 = 2627.63 \text{ N·m}$$

Example 11.17: A 150 kW, 1750 rpm electric motor is installed in the cab of a large power shovel. The motor axis is horizontal and the rotor inertia is 140 kg·m². The rotor bearings are 1.25 m apart, and the motor shaft is connected to the driven member by use of a flexible coupling. If the maximum slewing speed of the shovel is 6 rad/s, what will be the additional bearing loads due to the gyroscopic effect?

Solution: $N = 1750$, $\omega = \dfrac{2\pi N}{60} = \dfrac{2\pi \times 1750}{60} = 183.25$ rad/s

$I = 140$ kg·m², $x = 1.25$ m, $\omega_p = 6$ rad/s

Gyroscopic couple $C = I\omega\omega_p = 140 \times 183.25 \times 6 = 153930$ N·m

Additional bearing loads due to gyroscopic couple

$$P = \frac{C}{x} = \frac{153930}{1.25} = 123144 \text{ N}$$

Example 11.18: The electric motor has a total mass of 10 kg and is supported by the mounting brackets A and B attached to the rotating disc. The armature of the motor has a mass of 2.5 kg and radius of gyration of 35 mm and turns counterclockwise at a speed of 1725 rpm as viewed from A to B. The turntable revolves about its vertical axis at the contant rate of 48 rpm in the direction shown. Determine the vertical components of the forces supported by mounting brackets at A and B; $AB = 0.24$ m, $OA = 0.12$ m

Solution: $m_1 = 10$ kg, $m_2 = 2.5$ kg, $k = 0.035$ m

$$N = 1725$$

$$\omega = \frac{2\pi N}{60} = \frac{2\pi \times 1725}{60} = 180.64 \text{ rad/s}$$

Angular velocity of precession,

$$N_p = 48 \text{ rpm}$$

∴

$$\omega_p = \frac{2\pi \times 48}{60} = 5.03 \text{ rad/s}$$

Moment of inertia of armature,

$$Im_2k^2 = 2.5 \times (0.035)^2$$
$$= 3.0625 \times 10^{-3} \text{ kg·m}^2$$

Gyroscopic couple,

$$C = I\omega\omega_p = 3.0625 \times 180.64 \times 5.03 = 2.78 \text{ N·m}$$

Force on bearings:

Let F_A and F_B be the forces at A and B respectively. The effect of gyroscopic couple will be to lift part A and depress part B.

Taking moment about B, we get

$$F_A \times 0.24 = 2.78$$

$$F_A = \frac{2.78}{0.24} = 11.583 \text{ N}$$

Say, $F_A = F_B = 11.6 \text{ N}$

Reaction due to weight of motor at A and B,

$$\frac{F}{2} = \frac{m_1 g}{2} = \frac{10 \times 9.81}{2}$$

$$\frac{F}{2} = 49.05 \text{ N}$$

Reaction at $A = \dfrac{F}{2} - F_A = 49.05 - 11.6 = 37.45 \text{ N}$ (part A will rise upwards)

Reaction at $B = \dfrac{F}{2} + F_B = 49.05 + 11.60 = 60.55 \text{ N}$ (Part B will depress)

EXERCISE

11.1 What do you mean by gyroscopic couple? Derive a relation for its magnitude.

11.2 In what way the angular velocity can be represented?

11.3 Write a short note on gyroscope.

11.4 What do you mean by spin, precession and gyroscopic planes?

11.5 What is active and reactive gyroscopic couple? Explain.

11.6 Explain the application of gyroscopic couple to the motion of an aircraft while taking a turn.

11.7 Describe the gyroscopic effect on sea vessels.

11.8 What is the effect of the gyroscopic couple on the stability of a four-wheeler while negotiating a curve?

11.9 What will be the effect of the gyroscopic couple on a disc fixed at a certain angle to a rotating shaft?

11.10 What makes the rider of a two-wheeler tilt to one side while taking a turn?

11.11 A uniform disc having a mass of 8 kg and a radius of gyration of 150 mm is mounted on one end of a horizontal arm of length 200 mm. The other end of arm can rotate freely in a universal bearing. The disc is given a clockwise spin of 250 rpm as seen from the disc end of the arm. Determine the motion of the disc if the arm remains horizontal.   [*Ans.* $\omega_p = 3.47$ rad/s clockwise]

11.12 A disc with radius of gyration 60 mm and a mass of 4 kg is mounted centrally on a horizontal axle of 80 mm length between the bearings. It spins about the axle at 800 rpm counterclockwise, when viewed from the right-hand side bearing. The axle precesses about a vertical axis at 50 rpm in clockwise direction when viewed from above. Determine the resultant reaction at each bearing due to mass and gyroscopic effect.   [*Ans.* 98.6 N ↑, 59.4 N ↓]

11.13 A disc supported between two bearings on a shaft of negligible weight has a mass of 80 kg and a radius of gyration of 300 mm. The distances of the disc from the bearings are 300 mm to the right from the left hand bearing and 450 mm to the left from the right hand bearing. The bearings are supported by thin vertical cords. When the disc rotates at 100 rad/s in the clockwise direction viewed from the left hand bearing, the cord supporting the left hand side bearing breaks. Find the angular velocity of precession at the instant the cord is cut and discuss the motion of the disc. [*Ans.* 0.327 rad/s]

11.14 An aeroplane flying at 250 km/hr turns towards left and completes a quarter circle of 60 m radius. The mass of rotating engine and the propeller of the plane is 450 kg with a radius of gyration of 320 mm. The engine speed is 2000 rpm clockwise when viewed from the rear. Determine the gyroscopic couple on the aircraft and state its effect.

In what way is the effect changed when:

(i) The aeroplane turns towards the right,

(ii) The engine rotates clockwise when viewed from the front (nose end) and the aeroplane turns: (a) left (b) right?

(*Ans.* 10.713 kN·m, raise the nose (i) tail raised
(ii) (a) tail raised (b) nose raised

11.15 The rotor of the turbine of ship has a mass of 2500 kg and rotates at a speed of 3200 rpm counterclockwise when viewed from stern. The rotor has radius of gyration of 0.4 m. Determine the gyroscopic couple and its effect when:

(i) The ship steers to the left in a curve of 80 m radius at a speed of 15 knots (1 knot = 1860 m/h)

(ii) The ship pitches 5 degrees above and 5 degrees below the normal position and the bow is descending with its maximum velocity. The pitching motion is simple harmonic with a periodic time of 40 seconds.

(iii) The ship rolls and at the instant, its angular velocity is 0.4 rad/s clockwise when viewed from stern. Also find the maximum angular acceleration during pitching.

(*Ans.* (i) 12981 N·m, lower the bow (ii) 1837.5 N·m starboard side (iii) No effect, α_{max} = 0.00215 rad/s^2)

11.16 A 2.2 tonne racing car has a wheel base of 2.4 m and a track of 1.4 m. The centre of mass of the car lies at 0.6 m above the ground and 1.4 m from the rear axle. Equivalent mass of engine parts is 140 kg with radius of gyration of 150 mm. The back axle ratio is 5. The engine shaft and flywheel rotate clockwise when viewed from front. Each wheel has a diameter of 0.8 m and a moment of inertia of 0.7 kg·m^2. Determine the load distribution on the wheels when the car is rounding a curve of 100 m radius at a speed of 72 km/hr to the: (i) left (ii) right.

(*Ans.* (i) 4431.8 N, 8223.8 N, 2567.2 N, 6359.2 N
(ii) 8158.2 N, 4366.2 N, 6424.8 N, 2632.8 N)

11.17 The total mass of a four-wheeled trolley car is 1800 kg. The car runs on rails of 1.6 m gauge and rounds a curve of 24 m radius at 36 km/h. The track is banked at 10°. The external diameter of the wheels is 600 mm and each pair with axle has a mass of 180 kg with radius of gyration of 240 mm. The height of centre of mass of the car above the wheel base is 950 mm. Determine the pressure on each rail allowing for centrifugal force and gyroscopic couple actions.

(*Ans.* 11999.16 N downwards, 6692.4 N upwards)

11.18 The moment of inertia of a pair of locomotive driving wheels with the axle is 200 kg·m^2. The distance between the wheel centres is 1.6 m and the diameter of the wheel treads is 1.8 m. Due to defective ballasting, one wheel falls by 5 mm and rises again in a total time of 0.12 seconds while the locomotive travels on a level track at 100 km/hr. Assuming that the displacement of the wheel takes place with simple harmonic motion, determine the gyroscopic couple produced and the reation between the wheel and rail due to this couple.

(*Ans.* 505 N.m; 315.6 N)

11.19 A flywheel of mass 10 kg and radius of gyration 200 mm is spinning about its axis, which is horizontal and is suspended at a point distant 150 mm from the plane of rotation of the flywheel. Determine the angular velocity of precession of the flywheel. The spin speed of flywheel is 900 rpm. (*Ans.* 0.39 rad/s)

11.20 A horizontal axle *AB*, 1 m long is pivoted at the mid point *C*. It carries a weight of 20 N at *A* and a wheel weighing 50 N at *B*. The wheel is made to spin at a speed of 600 rpm in a clockwise direction looking from its front. Assuming that the weight of the flywheel is uniformly distributed around the rim whose mean diameter is 0.6 m, calculate the angular velocity of precession of the system around the vertical aixs through *C*. (*Ans.* 0.52 rad/s)

11.21 An aeroplane runs at 600 km/h. The rotor of the engine weighs 4000 N with radius of gyration of 1 m. The speed of rotor is 3000 rpm in anticlockwise direction when seen from rear side of the aeroplane. If the plane takes a loop upwards in a curve of 100 m radius, find (i) gyroscopic couple developed and (ii) Effect of reaction gyroscopic couple developed on the body of the aeroplane.

(*Ans.* 213.5 kN·m)

11.22 An aeroplane makes a complete half circle of 50 m radius, towards left, when flying at 200 km per hour. The rotary engine and the propeller of the plane has a mass of 400 kg with a radius of gyration of 300 mm. The engine runs at 2400 rpm clockwise, when viewed from the rear. Find the gyroscopic couple on the aircraft and state its effect on it. What will be the effect, if the aeroplane turns to its right instead of turning to the left? (*Ans.* 10 kN·m)

11.23 Each paddle wheel of a steamer have a mass of 1600 kg and a radius of gyration of 1.2 m. The steamer turns to port in a circle of 160 m radius at 24 km/h, the speed of the paddles being 90 rpm. Find the magnitude and effect of the gyroscopic couple acting on the steamer. [*Ans.:* 905.6 N·m]

11.24 The rotor of the turbine of a yacht makes 1200 rpm clockwise when viewed from stern. The rotor has a mass of 750 kg and its radius of gyration is 250 mm. Find the maximum gyroscopic couple transmitted to the hull (body of the yacht) when yacht pitches with maximum angular velocity of 1 rad/s. What is the effect of this couple? (*Ans.* 5892 N·m)

11.25 The rotor of a turbine installed in a boat with its axis along the longitudinal axis of the boat makes 1500 rpm clockwise when viewed from the stern. The rotor has a mass of 750 kg and a radius of gyration of 300 mm. If at an instant, the boat pitches in the longitudinal vertical plane so that the bow rises from the horizontal plane with an angular velocity of 1 rad/s, determine the torque acting on the boat and the direction in which it tends to turn the boat at the instant. (*Ans.* 10.6 kN·m)

11.26 The mass of a turbine rotor of a ship is 8 tonnes and has a radius of gyration 0.6 m. It rotates at 1800 rpm clockwise when looking from the stern. Determine the gyroscopic effects in the following cases:

(i) If the ship travelling at 100 km/h steers to the left in a curve of 75 m radius.

(ii) If the ship is pitching and the bow is descending with maximum velocity. The pitching is simple harmonic, the periodic time being 20 seconds and the total angular movement between the extreme positions is 10°.

(iii) If the ship is rolling and at a certain instant has an angular velocity of 0.03 rad/s clockwise when looking from stern.

In each case, explain clearly how you determine the direction in which the ship tends to move as a result of the gyroscopic action.

(*Ans.* 201 kN·m ; 14.87 kN·m ; 16.3 kN·m)

11.27 The turbine rotor of a ship has a mass of 20 tonnes and a radius of gyration of 0.75 m. Its speed is 2000 rpm. The ship pitches 6° above and below the horizontal position. One complete oscillation takes 18 seconds and the motion is simple harmonic. Calculate:

(i) The maximum couple tending to shear the holding down bolts of the turbine

(ii) The maximum angular acceleration of the ship during pitching

(iii) The direction in which the bow will tend to turn while rising, if the rotation of the rotor is clockwise when looking from rear.

(*Ans.* 86.26 kN·m; 0.0128 rad/s², towards starboard)

11.28 A motor car takes a bend of 30 m radius at a speed of 60 km/hr. Determine the magnitudes of gyroscopic and centrifugal couples acting on the vehicle and state the effect that each of these has on the road reactions to the road wheels. Assume that each road wheel has a moment of inertia of 3 kg·m² and an effective road radius of 0.4 m. The rotating parts of the engine and transmission are equivalent to a flywheel of mass 75 kg with a radius of gyration of 100 mm. The engine turns in a clockwise directin when viewed from the front.

The back-axle ratio is 4:1, the drive through the gear box being direct. The gyroscopic effects of the half shafts at the back axle are to be ignored. The car has a mass of 1200 kg and its centre of gravity is 0.6 m above the road wheel. The turn is in a right hand direction. If the turn has been in a left hand direction, all other details being unaltered, which answers, if any, need modification.

(**Ans.** 347.5 N·m; 6670 N·m)

OBJECTIVE TYPE QUESTIONS

11.1 The axis of spin, the axis of precession and axis of active gyroscopic couple are contained in
   (a) one plane
   (b) two planes perpendicular to each other
   (c) three planes perpendicular to each other
   (d) none of the above

11.2 A disc is spinning with an angular velocity ω (rad/s) about its axis. The gyroscopic couple applied to the disc to precess it will be,

   (a) $\dfrac{1}{2} I\omega^2$ 　　　　　　　(b) $I\omega^2$

   (c) $\dfrac{1}{2} I\omega\omega_p$ 　　　　　　(d) $I\omega\omega_p$

11.3 A disc is rotating with uniform angular acceleration about X axis. The plane of disc is rotating with uniform angular velocity about Y-axis. Then, gyroscopic acceleration is
   (a) increasing
   (b) decreasing

(c) constant

(d) variable (may increase or decrease)

11.4 The planes of spin, precession and applied gyroscopic couple are

(a) same

(b) two planes perpendicular to each other

(c) three planes perpendicular to each other

(d) none of the above

11.5 The engine of a aeroplane rotates in clockwise direction when seen from the tail end and the aeroplane takes a turn to the left. The effect to the gyroscopic couple on the aeroplane will be

(a) to raise the nose and dip the tail

(b) to dip the nose and raise the tail

(c) to raise the nose and tail both

(d) to dip the nose and tail both

11.6 The gyroscopic effects due to rotating parts of a turbo jet engine of an aircraft while taking a turn depend on

(a) flight velocity

(b) flight altitude

(c) radius of the curve

(d) flight velocity and radius of the curve

11.7 The air screw of an aeroplane is rotating clockwise when looking from the front. The gyroscopic effect, when it makes a left turn, will be

(a) to depress the nose and raise the tail

(b) to raise the nose and depress the tail

(c) to tilt the aeroplane

(d) none of the above

11.8 A disc of 20 N weight is mounted in the middle of a shaft of 1 m length rotating in bearings at ends. Mass moment of inertia of disc = 0.05 N·m·sec², spin velocity = 200 rad/s anticlockwise looking from right, precessional velocity = 1 rad/s anticlockwise looking from top. The reactions at the bearings are

(a) 0, 20 N   (b) 20 N, 0

(c) 5 N, 15 N   (d) 15 N, 5 N

11.9 The rotor of a ship rotates in clockwise direction when viewed from the stern and the ship takes a left turn. The effect of the gyroscopic couple will be

(a) to raise the bow and stern

(b) to lower the bow and stern

(c) to raise the bow and lower the stern

(d) to lower the bow and raise the stern

11.10 If the angle of the cyclist with normal increases while taking a turn, the cyclist should have

(a) higher speed   (b) same speed as before

(c) lower speed   (d) none of the aove

11.11 When the pitching of the ship is upward, the effect of gyroscopic couple acting on it will be

(a) to move the ship towards port side

(b) to move the ship towards starboard side

(c) to raise the bow and lower the stern

(d) to raise the stern and lower the bow

11.12 The rotor of a ship is spinning in clockwise direction when viewed from the stern. The gyroscopic couple will be introduced
(a) when the ship is rolling
(b) when the ship is pitching
(c) when the ship is pitching or rolling
(d) in no case

11.13 A motor car moving at a certain speed takes a left turn in a curved path. If the engine rotates in the same direction as that of wheels, then due to centrifugal forces
(a) the reaction on inner wheels increases and on the outer wheels decreases
(b) the reaction on the outer wheels increases and on the inner wheels decreases
(c) the reation on front wheels increases and on the rear wheels decreases
(d) the reaction on the rear wheels increases and on the front wheels decreases

11.14 The gyroscopic acceleration is given by

(a) $\dfrac{\delta\omega}{\delta t}$

(b) $\omega\dfrac{\delta\theta}{\delta t}$

(c) $r\dfrac{\delta\theta}{\delta t}$

(d) $r.\dfrac{\delta\omega}{\delta t}$

11.15 For a gyrometer with spin vector $\overrightarrow{\omega_s}$ and a perpendicular precession vector $\overrightarrow{\omega_p}$, the external torque needed is in the direction of

(a) $\overrightarrow{\omega_p}$

(b) $\overrightarrow{\omega_s}$

(c) $\overrightarrow{\omega_p} \times \overrightarrow{\omega_s}$

(d) $\overrightarrow{\omega_s} \times \overrightarrow{\omega_p}$

11.16 During taking a turn a cyclist inclines at an angle with the normal to the road. The equilibrium is maintained due to,
(a) weight of the cyclist
(b) centrifugal force above
(c) centrifugal force and gyroscopic couple
(d) weight of cyclist, centrifugal force and gyroscopic couple

ANSWERS

11.1 (b)	11.2 (d)	11.3 (a)	11.4 (c)	11.5 (a)	11.6 (d)
11.7 (b)	11.8 (a)	11.9 (c)	11.10 (a)	11.11 (b)	11.12 (b)
11.13 (b)	11.14 (b)	11.15 (d)	11.16 (d)		

12

Mechanical Vibrations

12.1 INTRODUCTION

If a body has to and fro motion, it is said to be in vibration. The vibrations are due to internal elastic forces within the body. A pendulum swinging on either side of a mean position vibrates under the action of gravity.

Whenever a body is displaced from its equilibrium position by external force, work is done on it by external force against internal elastic forces which is stored as strain energy. When body is released, the internal elastic forces tend to restore the body to its equilibrium position. In absence of friction, the whole of the strain energy stored in the body is converted into kinetic energy when the body reaches the equilibrium position at which it has maximum kinetic energy. The body continues to move in opposite direction until the whole kinetic energy is used up to overcome the internal elastic forces and stored as strain energy.

12.2 CAUSES OF VIBRATIONS

Main causes of vibrations are:

  i. External exciting force acting on the system
  ii. Unbalanced forces and couples in the machine part
  iii. Earthquakes cause vibrations in structures like bridges, dams, highrise buildings, etc.
  iv. Winds can also cause vibrations in systems like telephone lines, electricity lines etc.

12.3 HARMFUL AND USEFUL EFFECTS OF VIBRATIONS

Vibrations are generally harmful because of the following reasons:

  i. They produce undesirable noise.
  ii. They produce unwanted stress in vibrating machine parts.
  iii. They cause discomfort to human beings.
  iv. They reduce performance of machines and human beings.
  v. They cause looseness of assembled parts and partial or complete failure of machine parts.

Inspite of these harmful effects, vibrations show some useful phenomena also, which are given below:

  i. Vibrations are useful in stress relieving equipments.
  ii. Phenomenon of vibration is useful in propagation of sound and in musical instruments.

iii. Phenomenon of vibration is used in vibrating screens, shakers, vibratory conveyors, sieves and compactors.

12.4 METHODS FOR REDUCING UNDESIRABLE VIBRATIONS

The undesirable vibrtions can be eliminated or reduced by the following methods:
i. By balancing the unbalanced forces and couples that cause vibrations.
ii. By resting the machinery on proper type of isolators.
iii. By using shock absorbers.
iv. By putting screens or glass if noise is produced by vibrations.

12.5 BASIC DEFINITIONS

Periodic motion: A motion which repeats itself after equal intervals of time is periodic motion.

Time period: Time taken to complete one cycle is known as time period (t). It is measured in seconds.

Frequency: The number of cycles of motion completed in one second is known as frequency. It is expressed in hertz (Hz) or cycles per sec.

$$f = \frac{1}{t}$$

Amplitude: It is the maximum displacement of a vibrating body from its mean position.

Damping: It is the resistance to the motion of a vibrating body.

Phase difference: It is the angle between two rotating vectors representing simple harmonic motions of same frequency.

Let $x_1 = A_1 \sin \omega t$

and $x_2 = A_2 \sin (\omega t + \phi)$ represent two simple harmonic motions. Then ϕ is the phase difference between the two.

Resonance: When the frequency of the external force is equal to the natural frequency of a vibrating body, the amplitude of vibrations becomes excessively large which may be dangerous. This phenomenon is known as resonance.

Simple harmonic motion (SHM): A periodic motion of a particle whose acceleration is always directed towards the mean position and is proportional to its distance from the mean position, is known as simple harmonic motion. The motion of a simple pendulum is simple harmonic motion. The basic equation of simple harmonic motion is

$$\ddot{x} = -\omega^2 x$$

Degree of freedom (DOF): The minimum number of independent coordinates required to specify the motion of a system at any instant is known as degrees of freedom of the system.

A mass supported by a spring, a simple pendulum and a slider crank mechanism are examples of single degree of freedom systems (Fig. 12.1).

A spring supported mass which can have angular motion also and a two mass, two spring system constrained to move in one direction without rotation are two degrees of freedom systems (Fig. 12.2).

A flexible beam between two supports and a cantilever beam have infinite number of degrees of freedom (Fig. 12.3).

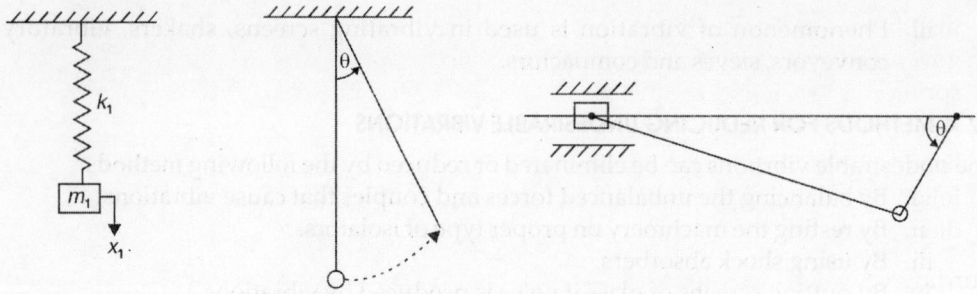

Fig. 12.1: Single degree of freedom systems

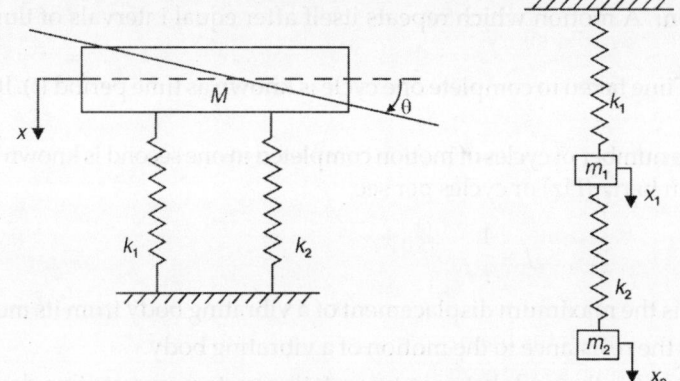

Fig. 12.2: Two degrees of freedom systems

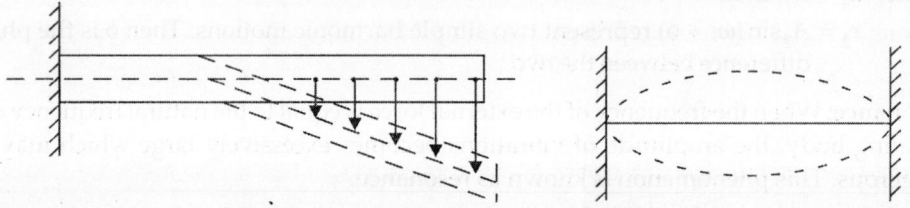

Fig. 12.3: Infinite degrees of freedom systems

12.6 TYPES OF VIBRATIONS

According to the actuating force:

i. **Free vibrations:** The vibrations of a system, when the external force is removed after giving an initial displacement to the system and then system vibrates on its own due to elastic properties, are known as free vibrations. Frequency of these vibrations is known as natural frequency (f_n). Simple pendulum is an example of free vibrations.

ii. **Forced vibrations:** The vibrations which are under the influence of external force, are called forced vibrations, e.g. electric bells, machine tools, etc. The frequency of these vibrations is that of applied force and is independent of their own natural frequency of vibrations.

According to the behaviour of vibrating system:

i. **Linear vibrations:** If in a vibratory system, mass, spring and damper behave linearly, then the resulting vibrations are known as linear vibrations. Linear vibrations are governed by linear differential equations and follow the law of superposition, e.g. spring-mass system.

ii. **Nonlinear vibrations:** If any of the basic elements of a vibratory system behaves non-linearly, vibrations are called non-linear vibrations. Non-linear vibrations does not follow the law of *super position*, e.g. vibrations of spring-mass system in transverse direction.

According to energy dissipation:

i. **Damped vibrations:** If a damper is provided in the vibratory system, the motion of the system, will be opposed by it and the energy of the system will be dissipated in friction. Such vibrations are known as damped vibrations. Amplitude of such vibrations reduces after every cycle, e.g. motion of shock absorber.

ii. **Undamped vibrations:** If no energy is dissipated in friction or other resisting force, such vibrations are called undamped vibrations, e.g. vibrations of a simple pendulum in vacuum.

According to the magnitude of external force:

i. **Deterministic vibrations:** If the magnitude of external force acting on a vibratory system is known at any instant, then the vibrations are known as deterministic vibrations.

ii. **Random vibrations:** If the magnitude of external force is nondeterministic, resulting vibrations are known as random vibrations, e.g. vibrations of builidings during an earthquake.

According to the motion with respect to axis:

i. **Longitudinal vibrations:** If a shaft is elongated and shortened so that it moves up and down along the axis of the shaft, resulting vibrations are known as longitudinal vibrations. These vibrations result in tensile and compressive stresses in the shaft (Fig. 12.4a).

ii. **Transverse vibrations:** If the particles of the body move approximately perpendicular to its axis (Fig. 12.4b) and the body/shaft is bent alternately, the resulting vibrations are known as tansverse vibrations.

iii. **Torsional vibrations:** If the shaft gets twisted and untwisted alternately, vibrations produced in the shaft are known as torsional vibrations. Torsional shear stresses are induced in the shaft (Fig. 12.4c).

12.7 ELEMENTS OF A VIBRATING SYSTEM

A vibratory system basically consists of three elements—mass, spring and damper. Energy is stored by mass in the form of kinetic energy $\left(\dfrac{1}{2}m\dot{x}^2\right)$, in the spring in the form of potential energy $\left(\dfrac{1}{2}k\dot{x}^2\right)$ and it is dissipated in the damper in the form of heat energy which opposes the motion of the system.

The equation of motion of a vibratory system shown in Fig. 12.5 can be written as

$$m\ddot{x} + c\dot{x} + kx = 0 \qquad\qquad ...(12.1)$$

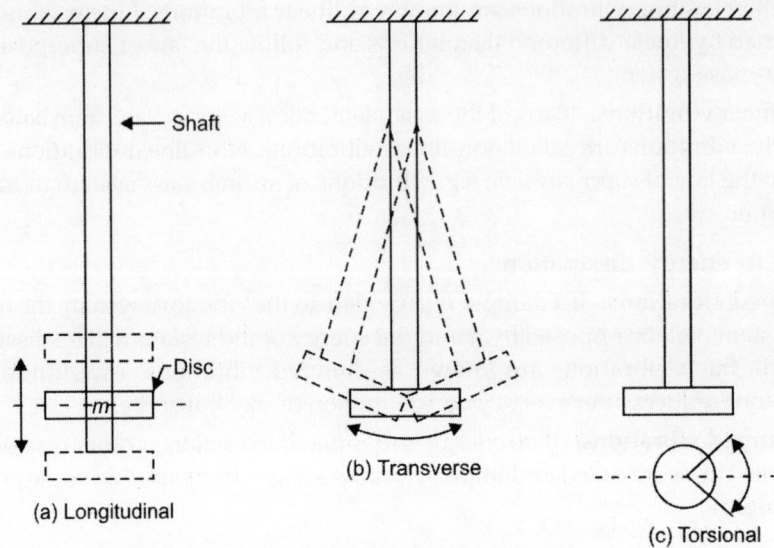

Fig. 12.4: Types of vibrations according to the motion

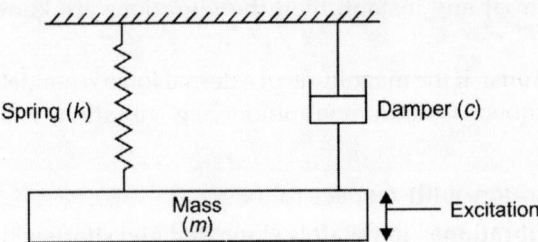

Fig. 12.5: Elements of a vibratory system

12.8 EQUIVALENT STIFFNESS OF SPRINGS

12.8.1 Springs in Series

In Fig. 12.6, two springs of stiffness k_1 and k_2 are connected in series.

Let the stiffness of equivalent spring be k_{es}. Then,

Deflection of equivalent spring = deflection of spring 1 + deflection of spring 2

or
$$\delta = \delta_1 + \delta_2$$

But
$$mg = k_{es} \cdot \delta \text{ or } \delta = \frac{mg}{k_{es}} \qquad (F = kx)$$

and
$$\delta_1 = mg/k_1, \delta_2 = mg/k_2$$

$\therefore$
$$\frac{mg}{k_{es}} = \frac{mg}{k_1} + \frac{mg}{k_2}$$

$$\frac{1}{k_{es}} = \frac{1}{k_1} + \frac{1}{k_2}$$

or
$$k_{es} = \frac{k_1 k_2}{k_1 + k_2}$$

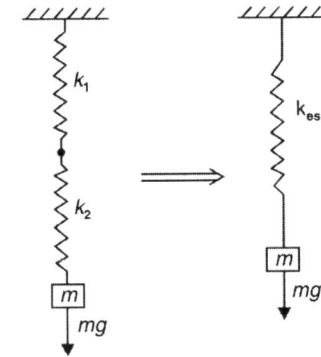

Fig. 12.6: Springs in series

12.8.2 Springs in Parallel

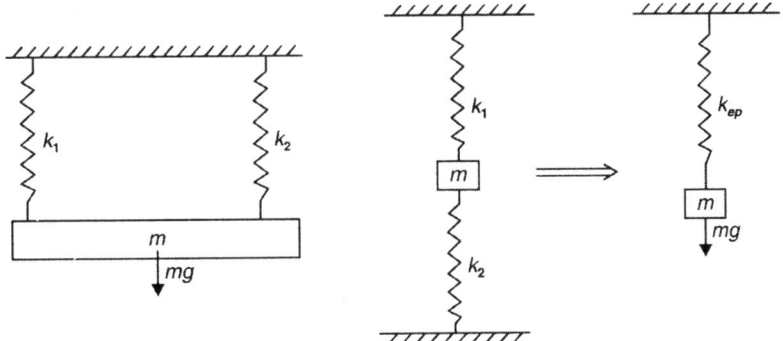

Fig. 12.7: Springs in parallel

Here,

    total load supported = load carried by spring 1 + load carried by spring 2

or $mg = m_1 g + m_2 g$

or $k_{ep}\,\delta = k_1\,\delta + k_2\,\delta$ $(F = kx)$

or $k_{ep} = k_1 + k_2$ ...(12.3)

12.9 EQUIVALENT DAMPING COEFFICIENT OF DAMPERS

12.9.1 Dampers in Series

Two dampers with damping coefficient C_1 and C_2 are in series. Their equivalent system has damping coefficient C_{es} given by

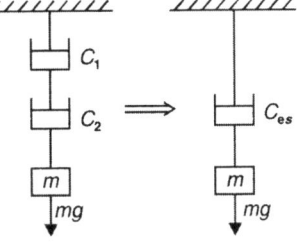

$$\frac{1}{C_{es}} = \frac{1}{C_1} + \frac{1}{C_2} \qquad ...(12.4)$$

Fig. 12.8: Dampers in series

12.9.2 Dampers in Parallel

Equivalent damping coefficient of dampers in parallel is given by,

$$C_{ep} = C_1 + C_2 \qquad\qquad ...(12.5)$$

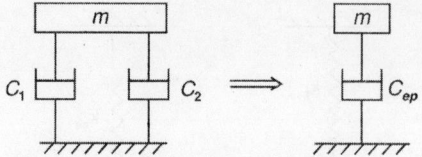

Fig. 12.9: Dampers in parallel

LONGITUDINAL VIBRATIONS

12.10 METHODS OF VIBRATION ANALYSIS

The following three methods are used to determine the natural frequency of a vibrating system.

12.10.1 Equilibrium Method

This method is based on D'Alembert's principle which states that the sum of the inertia forces and the external forces acting on a vibratory body in equilibrium must be equal to zero.

A helical spring of negligible mass is suspended vertically from a rigid support as shown in Fig. 12.10. Its free end is at A–A.

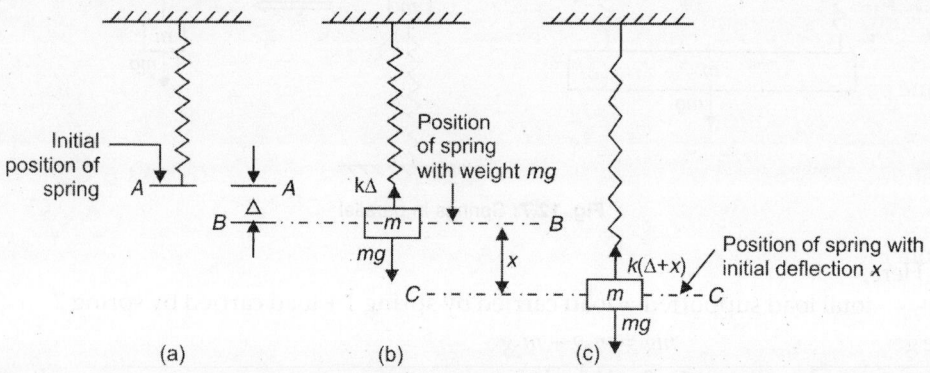

Fig. 12.10: Equilibrium method

If a mass m is suspended from the free end, the spring is stretched by a distance Δ and B-B becomes the equilibrium position (Fig. 12.10b). Thus distance Δ is known as the static deflection of the spring under the weight of the mass m.

Let k = stiffness of the spring under the weight of the mass m.

In the static equilibrium position

$$\text{upward force} = \text{downward force}$$

$$k \times \Delta = mg$$

Now, if the mass m is pulled further down through a distance x (Fig. 12.10c), the forces acting on the mass will be

inertia force = $m\ddot{x}$ (upwards)

spring force (restoring force) = $k\,(\Delta + x)$ (upwards)

weight of mass $m = mg$ (downwards)

From D'Alembert's principle, sum of inertia force and external force on the body is zero, i.e.

$$m\ddot{x} + k(\Delta + x) - mg = 0 \qquad \qquad ...(12.6)$$

or $\qquad m\ddot{x} + k\Delta + kx - mg = 0$

or $\qquad m\ddot{x} + kx = 0 \; (k\Delta = mg) \qquad \qquad ...(12.7)$

When the mass is released, it starts oscillating above and below the equilibrium position. These oscillations are longitudinal oscillations (vibrations). If there is no frictional resistance to the motion, then these oscillations will continue forever. Such vibrations are known as free longitudinal vibrations.

Equation (12.7) can be written as

$$\ddot{x} + \left(\frac{k}{m}\right)x = 0 \qquad \qquad ...(12.8)$$

Equation (12.8) is the equation of simple harmonic motion. It is analogous to

$$\ddot{x} + \omega_n^2 x = 0 \qquad \qquad ...(12.9)$$

Comparing Eqs (12.8) and (12.9), we have

$$\omega_n^2 = \frac{k}{m} \text{ or } \omega_n = \sqrt{\frac{k}{m}} \qquad \qquad ...(12.10)$$

ω_n is known as natural circular frequency of vibrations.

Time period of vibrations, $\quad T = \dfrac{2\pi}{\omega_n} = 2\pi\sqrt{\dfrac{m}{k}}$ s $\qquad \qquad ...(12.11)$

Natural linear frequency, $f_n = \dfrac{1}{T} = \dfrac{1}{2\pi}\sqrt{\dfrac{k}{m}}$ Hz $\qquad \qquad ...(12.12)$

From Eq. (12.6), $\qquad \dfrac{k}{m} = \dfrac{g}{\Delta}$

Substituting value of k/m in Eq. (12.12), we get,

$$f_n = \frac{1}{2\pi}\sqrt{\frac{g}{\Delta}} = \frac{1}{2\pi}\sqrt{\frac{9.81}{\Delta}} \qquad \qquad ...(12.13a)$$

or $$f_n = \frac{0.4985}{\sqrt{\Delta}} \text{ Hz} \qquad \qquad ...(12.13b)$$

12.10.2 Energy Method

In free undamped vibrations, the sum of kinetic energy and potential energy remains constant.

$\therefore \qquad KE + PE = \text{constant}$

or $\qquad \dfrac{d}{dt}(KE + PE) = 0$

Here, kinetic energy, $KE = \dfrac{1}{2}m\dot{x}^2$

(where $\dot{x}$ = velocity of the mass)
and potential energy PE = mean force × displacement from mean position

or
$$PE = \frac{0 + kx}{2} \times x = \frac{1}{2}kx^2$$

$$\therefore \quad \frac{d}{dt}\left[\frac{1}{2}m\dot{x}^2 + \frac{1}{2}kx^2\right] = 0$$

or
$$\frac{1}{2}m(2\dot{x}\ddot{x}) + \frac{1}{2}k(2x\dot{x}) = 0$$

or
$$m\ddot{x} + kx = 0$$

$$\therefore \quad \ddot{x} + \left(\frac{k}{m}\right)x = 0 \qquad\qquad\qquad ...(12.13c)$$

Comparing Eq. (12.13b) with the equation of SHM,

$$\omega_n = \sqrt{\frac{k}{m}}.$$

12.10.3 Rayleigh's Method

In this method, the maximum kinetic energy at the mean position (where potential energy is zero) is equal to the maximum potential energy at the extreme position (where the kinetic energy is zero).

Let the motion be simple harmonic motion.

Therefore, $\quad x = X\sin\omega_n t$

where X = maximum displacement from the mean position to the extreme position. Differentiating the above equation

$$\therefore \qquad\qquad \dot{x} = \omega_n X \cos\omega_n t,$$

$$\dot{x}_{max} = \omega_n X \quad (\text{at } t = 0)$$

KE at mean position = PE at extreme position

i.e.
$$\frac{1}{2}m(\omega_n X)^2 = \frac{1}{2}kX^2$$

or
$$m\omega_m^2 = k \text{ or } \omega_n = \sqrt{\frac{k}{m}}.$$

12.11 INERTIA EFFECT OF THE MASS OF THE SPRING

The mass of the spring and its inertia effect have been neglected in previous sections. The same is now considered.

Let $\quad m'$ = mass of the spring wire per unit length

v = velocity of the free end of the spring at the instant under consideration.

L = total length of the spring wire

Consider an element of length dy at a length y measured round the coils from the fixed end. Velocity of small element at the given instant = $\left(\frac{y}{L}v\right)$

$$KE \text{ of the element} = \frac{1}{2} \times \text{mass of element} \times (\text{velocity of element})^2$$

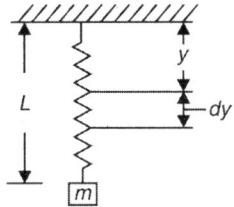

Fig. 12.11

$$= \frac{1}{2}(m' \, dy) \times \left(\frac{y}{L}v\right)^2$$

$$KE \text{ of the spring} = \int_0^L \frac{1}{2}m - v^2 \left(\frac{y}{L}\right)^2$$

$$= \frac{1}{2}\frac{m'c^2}{L^2}\int_0^L y^2 dy = \frac{1}{2}\frac{m'v^2}{L^2}\frac{L^3}{3} = \frac{1}{3}\times\frac{1}{2}(m'L)v^2$$

$$= \frac{1}{3}\times\left[\frac{1}{2}\times \text{mass of spring}\times(\text{velocity of free end})^2\right]$$

$$= \frac{1}{3}\times KE \text{ of a mass equal to that of the spring moving with}$$
the same velocity as the free end.

This shows that the inertia effect of the spring is equal to that of a mass one third of the mass of the spring, concentrated at its free end.

Thus, equivalent mass at the free end $= m + \dfrac{m_s}{3}$

where, m_s = mass of the spring = $m' \times L$

$$f_n = \frac{1}{2\pi}\sqrt{\frac{k}{m+\dfrac{m_s}{3}}} \qquad \qquad ...(12.14)$$

Consider a rod of length L suspended vertically and carrying a mass m at the free end.

Then, static deflection of rod, $\Delta = (mg/AE).L = \dfrac{mgL}{AE}$

or $$\frac{g}{\Delta} = \frac{AE}{mL}$$

Now, using Eq. (12.13a),

$$f_n = \frac{1}{2\pi}\sqrt{\frac{g}{\Delta}} = \frac{1}{2\pi}\sqrt{\frac{AE}{mL}} \qquad \qquad ...(12.15)$$

If the mass of the suspended rod is also considered,

$$f_n = \frac{1}{2\pi}\sqrt{\frac{AE}{\left(m+\dfrac{m_L}{3}\right)L}} \qquad \qquad ...(12.16)$$

12.12 DAMPED VIBRATIONS

The vibrations in which the amplitude of vibrations of a vibrating body gradually diminishes, are known as damped vibrations. These vibrations die out after sometime due to the internal molecular friction of the mass of the body and the friction of the medium in which it vibrates.

The damping basically means the diminishing of vibrations with time. The external damping can be provided by using dashpots or dampers.

Consider a helical spring suspended from a fixed support (Fig. 12.12). *A-A* is the level of the free end before the mass *m* is suspended. *B-B* is the level of static equilibrium under the weight of the mass. The mass is attached to a dashpot to dampen its motion.

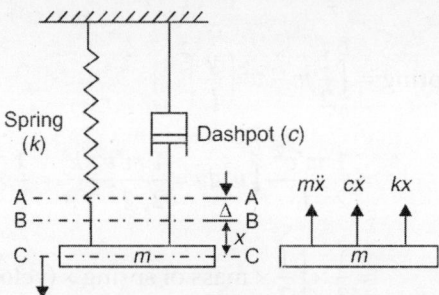

Fig. 12.12

It is assumed that the damping force is proportional to the velocity of vibration at lower values of speed.

Let the mass *m* be displaced through a distance *x* below the mean position.

Let k = stiffness of the spring

c = damping coefficient (damping force per unit velocity)

ω_n = frequency of natural undamped vibrations

x = displacement of mass from mean position at time t

$v = \dot{x}$ = velocity of the mass at time t

$a = \ddot{x}$ = acceleration of the mass at time t.

When the mass moves downwards, the frictional force of the dashpot acts in the upward direction.

Now, the forces acting on the mass are:

(i) Inertia force = $m\ddot{x}$ (upwards)

(ii) Damping force = $c\dot{x}$ (upwards)

(iii) Spring force (restoring force) = kx (upwards)

As the sum of the inertia force and the external forces on a body in any direction is to be zero, for the dynamic equilibrium of body,

$$m\ddot{x} + c\dot{x} + kx = 0$$

or $$\ddot{x} + \frac{c}{m}\dot{x} + \frac{k}{m}x = 0 \qquad \qquad ...(12.17)$$

It is a differential equation of the second order. Its solution will be of the form

$$x = Ae^{\alpha_1 t} + Be^{\alpha_2 t} \qquad \qquad ...(12.18)$$

where A and B are some constants which are determined by initial boundary conditions, α_1 and α_2 are the roots of the auxiliary equation

$$\alpha^2 + \frac{c}{m}\alpha + \frac{k}{m} = 0 \qquad \qquad ...(12.19)$$

i.e. the roots are

$$\alpha_{1,2} = -\frac{c}{2m} \pm \sqrt{\left(\frac{c}{2m}\right)^2 - \left(\frac{k}{m}\right)} \qquad \qquad ...(12.20)$$

The ratio of $\left(\frac{c}{2m}\right)^2$ to $\left(\frac{k}{m}\right)$ repesents the degree of dampness provided in the system and its square root is known as *damping factor* or *damping ratio*

$$\zeta = \sqrt{\frac{(c/2m)^2}{k/m}} = \frac{c}{2\sqrt{km}} = \frac{c}{2m\omega_n} \qquad \qquad ...(12.21)$$

or damping coefficient,

$$c = 2\zeta\sqrt{km} = 2\zeta m\omega_n = 2\zeta\frac{k}{\omega_n} \qquad \qquad ...(12.22)$$

When $\zeta = 1$, damping is known as critical damping and the corresponding value of damping coefficient c is denoted by c_c and known as critical damping coefficient.

Thus, under critical damping conditions,

$$c_c = 2\sqrt{km} = 2m\omega_n = \frac{2k}{\omega_n} \qquad \qquad ...(12.23)$$

$$\zeta = \frac{c}{c_c} = \frac{\text{actual damping coefficient}}{\text{critical damping coefficient}} \qquad \qquad ...(12.24)$$

Therefore, when

$\zeta = 1$, the damping is critical
$\zeta > 1$, the system is overdamped
$\zeta < 1$, the system is underdamped

Equation (12.17) can also be written as

$$\ddot{x} + 2\zeta\omega_n\dot{x} + \omega_n^2 x = 0 \qquad \qquad ...(12.25)$$

and roots are

$$\alpha_{1,2} = -\zeta\omega_n \pm \sqrt{\zeta^2\omega_n^2 - \omega_n^2} = \left(-\zeta \pm \sqrt{\zeta^2 - 1}\right)\omega_n$$

The exact solution of Eq. (12.25) will depend upon whether the roots $\alpha_{1,2}$ are real or imaginary.

(i) $\zeta > 1$, i.e. the system is overdamped: The roots of the auxiliary equation are real.

$$\alpha_{1,2} = \left(-\zeta \pm \sqrt{\zeta^2 - 1}\right)\omega_n$$

Therefore, the solution is

$$x = Ae^{\alpha_1 t} + Be^{\alpha_2 t}$$

$$= Ae^{\left(-\zeta + \sqrt{\zeta^2 - 1}\right)\omega_n t} + Be^{\left(-\zeta - \sqrt{\zeta^2 - 1}\right)\omega_n t} \qquad \qquad ...(12.26)$$

Constants A and B can be determined from the initial conditions. This is the equation of aperiodic motion, i.e., the system will not vibrate due to overdamping. The magnitude of the resultant displacement approaches zero with time as shown in Fig. 12.13.

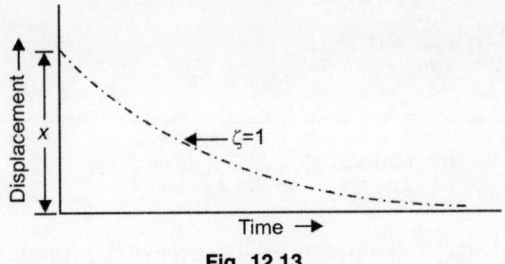

Fig. 12.13

(ii) $\zeta < 1$, **i.e. underdamped system:** The roots of the auxiliary equation are imaginary as the value of $(\zeta^2 - 1)$ will be negative. The values of roots α_1 and α_2 are,

$$\alpha_1 = \omega_n\left(-\zeta + \sqrt{\zeta^2 - 1}\right) \text{ and } \alpha_2 = \omega_n\left(-\zeta - \sqrt{\zeta^2 - 1}\right)$$

$$= \omega_n\left(-\zeta + \sqrt{(-1)(1-\zeta^2)}\right) \text{ and } \alpha_2 = \omega_n\left(-\zeta - \sqrt{(-1)(1-\zeta^2)}\right)$$

$$= \omega_n\left(-\zeta + i\sqrt{1-\zeta^2}\right) \text{ and }$$

$$\alpha_2 = \omega_n\left(-\zeta - i\sqrt{1-\zeta^2}\right) \qquad\qquad \left(\because \sqrt{(-1)} = i\,(\text{iota})\right)$$

Substituting the values of α_1 and α_2 in Eq. (12.18), we get the displacement

$$x = Ae^{\omega_n\left(-\zeta + i\sqrt{1-\zeta^2}\right)t} + Be^{\omega_n\left(-\zeta - i\sqrt{1-\zeta^2}\right)t}$$

$$= Ae^{-\omega_n\zeta t + \omega_n.i\sqrt{1-\zeta^2}\cdot t} + Be^{-\omega_n\zeta t + \omega_n.i\sqrt{1-\zeta^2}\cdot t}$$

$$= Ae^{-\omega_n\zeta t}\cdot\left(e^{\omega_n i\sqrt{1-\zeta^2}\cdot t}\right) + Be^{-\omega_n\zeta t}\cdot\left(e^{\omega_n i\sqrt{1-\zeta^2}\cdot t}\right)$$

$$= e^{-\omega_n\zeta t}\left[Ae^{\omega_n i\sqrt{1-\zeta^2}\cdot t} + Be^{-\omega_n i\sqrt{1-\zeta^2}\cdot t}\right] \qquad\qquad \text{...(12.27)}$$

Let $\qquad\qquad \omega_n\sqrt{1-\zeta^2} = \omega_d$ (frequency of underdamped vibrations) $\qquad$ ...(12.27a)

Then Eq. (12.27) becomes

$$x = e^{-\omega_n\zeta t}\left[Ae^{i\omega_d t} + Be^{-i\omega_d t}\right] \qquad\qquad \text{...(12.27b)}$$

But from the Euler's formula, we know that

$$e^{i\cdot\theta} = \cos\theta + i\sin\theta \text{ and } e^{-i\theta} = \cos\theta - i\sin\theta$$

Here $\qquad e^{i(\omega_d t)} = \cos(\omega_d t) + i\sin(\omega_d t) \text{ and } e^{-i(\omega_d t)} = \cos(\omega_d t) - i\sin(\omega_d t)$

Substituting these values in Eq. (12.27b), we get

$$x = e^{-\omega_n\zeta t}\left[A(\cos\omega_d t + i\sin\omega_d\cdot t) + B(\cos\omega_d\cdot t - i\sin\omega_d\cdot t)\right]$$

$$= e^{-\zeta\omega_n t}\left[(A+B)\cos(\omega_d\cdot t) + i(A-B)\sin(\omega_d\cdot t)\right] \qquad\qquad \text{...(12.28)}$$

$$= e^{-\zeta\omega_n t}\left[C\cos(\omega_d\cdot t) + D\sin(\omega_d\cdot t)\right] \qquad\qquad \text{...(12.29)}$$

where $\qquad\qquad C = A + B \text{ and } D = i(A - B) \qquad\qquad \text{...(12.30)}$

Hence the values of C and D can be obtained if values of A and B are known. But the values of A and B are determined from initial conditions.

Put $\quad\quad\quad\quad\quad\quad A + B = X \sin\phi$ and $i(A - B) = X\cos\phi$ in Eq. (12.28)

$$x = e^{-\zeta\omega_n \cdot t}\left[X\sin\phi\cos(\omega_d.t) + X\cos\phi\sin(\omega_d.t)\right]$$

$$= Xe^{-\zeta\omega_n \cdot t}\sin(\omega_d t + \phi) \quad\quad\quad\quad ...(12.31)$$

Constants X and ϕ are determined from initial conditions.

The solution consists of 3 terms:

- X, constant

- $e^{-\zeta\omega_n t}$, decreases with time

- $\sin(\omega_d t + \phi)$ represents repetition of motion. Therefore, resultant motion is oscillatory with decreasing amplitude.

Let $\quad X_0$ = displacement at the start of motion when $t = 0$

$\quad\quad X_1$ = displacement at the end of first oscillation when $l = T_d$

$$= Xe^{-\zeta\omega_n T_d}\sin(\omega_d T_d + \phi) \quad\quad\quad [\textit{from} \text{ Eq. (12.31)]} \ ...(12.32)$$

$$= Xe^{-\zeta\omega_n T_d}\sin\left(\omega_d\frac{2\pi}{\omega_d} + \phi\right) = Xe^{-\zeta\omega_n T_d}\sin\phi \quad\quad\quad ...(12.33)$$

$\quad\quad X_2$ = displacement at the end of second oscillation

$$= Xe^{-\zeta\omega_n \times 2T_d}\sin\phi$$

Similarly, $\quad\quad\quad\quad X_3 = Xe^{-\zeta\omega_n \times 3T_d}\sin\phi$

$$\text{..............................}$$

$$X_n = Xe^{-\zeta\omega_n \times 3T_d}\sin\phi$$

$$X_{n+1} = Xe^{-\zeta\omega_n \times (n+1)T_d}\sin\phi$$

Then $\quad\quad\quad\quad \dfrac{X_n}{X_{n+1}} = e^{\zeta\omega_n T_d} = \dfrac{X_0}{X_1} = \dfrac{X_1}{X_2} = \dfrac{X_2}{X_2} = ... \quad\quad\quad ...(12.34)$

which shows that the ratio of amplitudes of two successive oscillations is constant (Fig. 12.14).

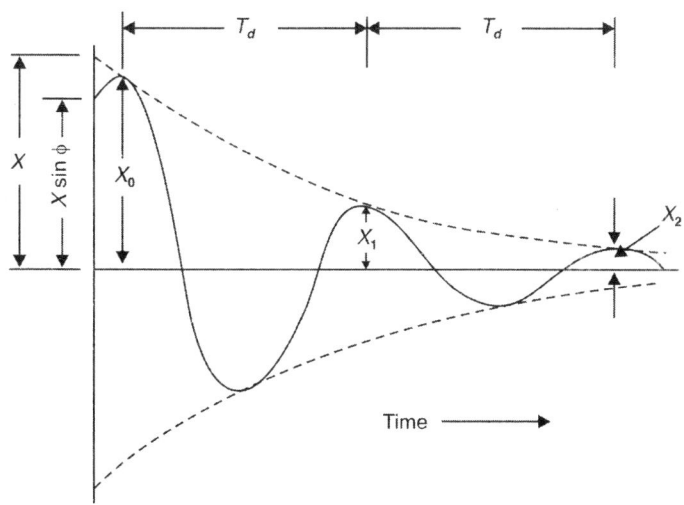

Fig. 12.14

(iii) $\zeta = 1$, **i.e. the damping is critical:** The roots of the auxiliary equation are equal, each being equal to $-\omega_n$ and the solution is

$$x = Ae^{-\omega_n t} + Be^{-\omega_n t} \cdot t$$

$$x = (A + Bt) \, e^{-\omega_n t} \qquad \qquad \qquad ...(12.35)$$

Since $e^{-\omega_n t}$ approaches zero as $t \to \infty$, the motion is aperiodic. The displacement will be approaching to zero with time.

Figure 12.15 shows the characteristics of motion for the three different cases discussed. In a critically damped system, the displaced mass returns to the position of rest in the shortest possible time without oscillation. Due to this reason, large guns are critically damped so that they return to their original position (after recoiling because of firing) in the minimum possible time. If the gun barrels are overdamped, they will take more time to return to their original positions.

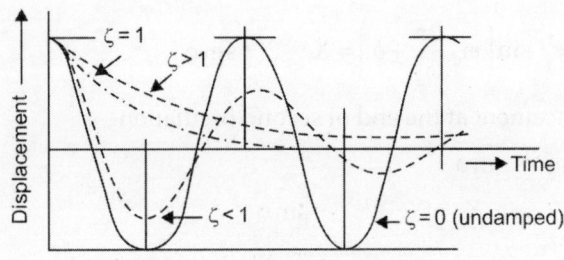

Fig. 12.15

12.13 LOGARITHMIC DECREMENT

Logarithmic decrement is defined as the natural logarithm of amplitudes of any two successive oscillations, on the same side of the mean position, in an under damped system.

Logarithmic decrement, $\delta = \log_e \left[\cdot \dfrac{X_n}{X_{n+1}} \right] = \log_e \left(e^{\zeta \omega_n T_d} \right)$ [*from* Eq. (12.34)]

or $\qquad \qquad \delta = \zeta \, \omega_n \, T_d = \zeta \, \omega_n \times \dfrac{2\pi}{\omega_d} = \zeta \ \omega_n \times \dfrac{2\pi}{\omega_n \sqrt{1 - \zeta^2}}$

$$\delta = \frac{2\pi\zeta}{\sqrt{1-\zeta^2}} \qquad \qquad \qquad ...(12.36)$$

12.14 FORCED DAMPED VIBRATIONS

A helical spring is suspended from a fixed support and a mass m is attached to its free end. A dashpot is provided in the system to dampen the vibrations.

Let B-B be the static equilibrium position under the weight of the mass. Now, if the mass is subjectd to an oscillating force $F = F_0 \sin \omega \, t$, the forces acting on the mass at any instant are

    (i) Applied oscillating force
        $F = F_0 \sin \omega t$ (downwards)
   (ii) Inertia force $= m\ddot{x}$ (upwards)
  (iii) Damping force $= c\dot{x}$ (upwards)
  (iv) Spring force (restoring force) $= kx$ (upwards)

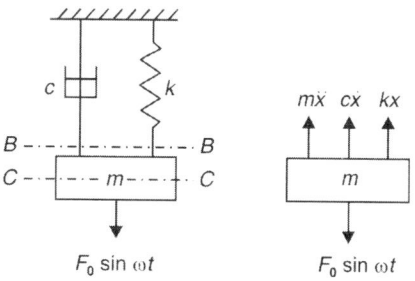

Fig. 12.16

Therefore, the equation of motion for dynamic equilibrium

$$m\ddot{x} + c\dot{x} + kx - F_0 \sin \omega t = 0$$

or
$$m\ddot{x} + c\dot{x} + kx = F_0 \sin \omega t \qquad \qquad ...(12.37)$$

Complete solution of Eq. (12.37) consists of two parts, the complementary function (*CF*) and the particular integral (*PI*).

$$CF = Xe^{-\zeta\omega_n t} \sin(\omega_d t + \phi) \qquad \qquad [from \text{ Eq. (12.31)}]$$

To obtain the *PI*, let

$$\frac{c}{m} = a, \frac{k}{m} = b, \text{ and } \frac{F_0}{m} = d$$

Then, using the operator $D = \dfrac{d}{dt}$ and $D^2 = \dfrac{d^2}{dt^2}$

∴   Eq. (12.37) becomes

$$(D^2 + aD + b)\, x = d \sin \omega t$$

$$PI = \frac{d \sin \omega t}{D^2 + aD + b}$$

$$= \frac{d \sin \omega t}{-\omega^2 + aD + b}$$

$$= \frac{1}{(b - \omega^2) + aD} \times \frac{(b - \omega^2) - aD}{(b - \omega^2) - aD}\, d \sin \omega t$$

$$= d \left[\frac{\sin \omega t (b - \omega^2) - aD \sin \omega t}{(b - \omega^2)^2 - a^2 D^2} \right]$$

$$= d \left[\frac{\sin \omega t (b - \omega^2) - a\omega \cos \omega t}{(b - \omega^2)^2 + (a\omega)^2} \right]$$

Constants R and ϕ are given by

$$R = \sqrt{(b - \omega^2)^2 + (a\omega)^2} \text{ and } \phi = \tan^{-1} \frac{a\omega}{(b - \omega^2)} \qquad \begin{bmatrix} \text{Let } (b - \omega^2) = R \cos \phi \\ \text{and } a\omega = R \cos \phi \end{bmatrix}$$

$$PI = \frac{dR\left(\sin \omega t \cos \phi - \cos \omega t \sin \phi\right)}{\left(b - \omega^2\right)^2 + (a\omega)^2}$$

$$= \frac{d\sqrt{\left(b - \omega^2\right)^2 + (a\omega)^2}}{\left(b - \omega^2\right)^2 + (a\omega)^2} \sin\left(\omega t - \phi\right)$$

$$= \frac{d}{\sqrt{\left(b - \omega^2\right)^2 + (a\omega)^2}} \sin\left(\omega t - \phi\right)$$

$$= \frac{F_0 / m}{\sqrt{\left(\dfrac{k}{m} - \omega^2\right)^2 + \left(\dfrac{c}{m}\omega\right)^2}} \sin\left(\omega t - \phi\right)$$

$$= \frac{F_0}{\sqrt{(k - m\omega^2)^2 + (c\omega)^2}} \sin\left(\omega t - \phi\right)$$

$$x = CF + PI$$

$$= Xe^{-\zeta\omega_n t} \sin\left(\omega_d t - \phi_1\right) + \frac{F_0}{\sqrt{(k - mw^2)^2 + (c\omega)^2}} \sin\left(\omega t - \phi\right) \qquad ...(12.38)$$

First part of solution (CF) represents the damped free vibrations which die out with time as $e^{-\infty} = 0$. The second part of solution (PI) represents the steady state rersponse of the system under forced vibrations condition.

The amplitude of the steady-states response is given as

$$A = \frac{F_0}{\sqrt{\left(k - m\omega^2\right)^2 + (c\omega)^2}} \qquad ...(12.39)$$

$$= \frac{F_0 / k}{\sqrt{\left(1 - \dfrac{m\omega^2}{k}\right)^2 + \left(\dfrac{c}{k}\omega\right)^2}} = \frac{F_0 / k}{\sqrt{\left(1 - \dfrac{\omega^2}{\omega_n^2}\right)^2 + \left(\dfrac{2m\omega_n\zeta\omega}{k}\right)^2}} \quad [\because C = 2m\omega_n\zeta]$$

$$A = \frac{F_0 / k}{\sqrt{\left[1 - \left(\dfrac{\omega}{\omega_n}\right)^2\right]^2 + \left(2\zeta\dfrac{\omega}{\omega_n}\right)^2}} = \left(\dfrac{k}{m} = \omega_n^2\right) \qquad ...(12.40)$$

Here, $F_0/k = \delta$, is the static deflection of the spring of stiffness k produced by a force F_0.

The frequency of the steady-state forced vibrations is the same as that of the external applied force. The phase lag (ϕ) of displacement relative to the velocity vector is obtained below.

$$\tan \phi = \frac{a\omega}{b-\omega^2} = \frac{\dfrac{c}{m}\cdot\omega}{\dfrac{k}{m}-\omega^2} = \frac{c\omega}{k-m\omega^2}$$

$$= \frac{(2m\omega_n\zeta)\omega}{k\left(1-\dfrac{m}{k}\omega^2\right)} = \frac{2\zeta\dfrac{\omega}{\omega_n}}{1-\left(\dfrac{\omega}{\omega_n}\right)^2} \quad \left(\because \frac{k}{m}=\omega_n^2\right) \qquad \text{...(12.41)}$$

Graphical Method: Assuming that the displacement of the vibrating mass under the action of the applied simple harmonic force $F_0 \sin \omega t$ is also simple harmonic and lags by an amount ϕ. It can be represented by $x = A \sin (\omega t - \phi)$ (where A = amplitude of vibrations).

Differentiating the above relation, we have

$$\dot{x} = \omega A \cos (\omega t - \phi)$$

$$= \omega A \sin\left[\frac{\pi}{2}+(\omega t - \phi)\right]$$

Again differentiating,

$$\ddot{x} = -\omega^2 A \sin (\omega t - \phi)$$

Substituting these values in the above relation, we have

$$m\ddot{x} + c\dot{x} + kx = F_0 \sin \omega t - m\omega^2 A \sin (\omega t - \phi) + c\omega A \sin\left[\frac{\pi}{2}+(\omega t - \phi)\right]$$

$$+ kA \sin (\omega t - \phi) - F_0 \sin \omega t = 0$$

$$F_0 \sin \omega t + m\omega^2 A \sin (\omega t - \phi) - c\omega A \sin\left[\frac{\pi}{2}+(\omega t - \phi)\right] - kA \sin (\omega t - \phi) = 0$$

The forces and the vector sum of the same have been shown in Fig. 12.16

In triangle *abc*,
$$F_0 = ab = \sqrt{(cb)^2 + (ca)^2}$$

or
$$\sqrt{\left(kA - m\omega^2 A\right)^2 + (c\omega A)^2} = F_0$$

or
$$A \sqrt{\left(k - m\omega^2\right) + (c\omega)^2} = F_0$$

or
$$A = \frac{F_0}{\sqrt{\left(k - m\omega^2\right)^2 + (c\omega)^2}} \qquad \text{...(12.42)}$$

and
$$\tan \phi = \frac{ca}{cb} = \frac{c\omega A}{\left(kA - m\omega^2 A\right)} \qquad \text{(*from* Fig. 12.16)}$$

$$= \frac{c\omega}{\left(k - m\omega^2\right)}$$

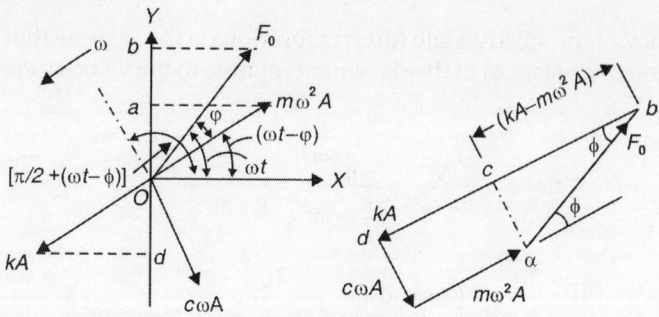

Fig. 12.17

The projected lengths of these vectors on vertical axis represents the instantaneous values of these forces at any time t.

12.15 MAGNIFICATION FACTOR

The ratio of the amplitude of the stead-state response to the static deflection under the action of force F_0 is known as *magnification factor* (*MF*). $MF = \dfrac{A}{\Delta}$

$$MF = \frac{F_0 / \sqrt{\left(k - m\omega^2\right)^2 + (c\omega)^2}}{F_0 / k} \qquad \text{(value of } A \textit{ from } \text{Eq. (12.42) and } \Delta = F_0/k)$$

$$= \frac{k}{\sqrt{\left(k - m\omega^2\right)^2 + (c\omega)^2}}$$

$$= \frac{1}{\sqrt{\left(1 - \dfrac{m}{k}\omega^2\right)^2 + \left(\dfrac{c}{k}\omega\right)^2}}$$

$$MF = \frac{1}{\sqrt{\left[1 - \left(\dfrac{\omega}{\omega_n}\right)^2\right]^2 + \left(2\zeta\dfrac{\omega}{\omega_n}\right)^2}} \qquad \qquad ...(12.43)$$

Therefore, the magnification factor depends upon:

(a) The ratio of frequencies, $\dfrac{\omega}{\omega_n}$

(b) The damping factor.

Figure 12.17 shows the relationship between the magnification factor (*MF*) and ratio of frequencies $\dfrac{\omega}{\omega_n}$ for different values of damping factor ζ. From the curve, it is clear that as damping increases (i.e. ζ increases), the value of magnification factor decreases. And when ζ decreases, the value of *MF* increases. At $\zeta = 0$ (i.e. when there is no damping), the value of magnificaton factor reaches to infinity at $\dfrac{\omega}{\omega_n} = 1$. At $\dfrac{\omega}{\omega_n} = 1$, the frequency of free vibration is equal to the frequency of forced vibrations. This condition is known as resonance.

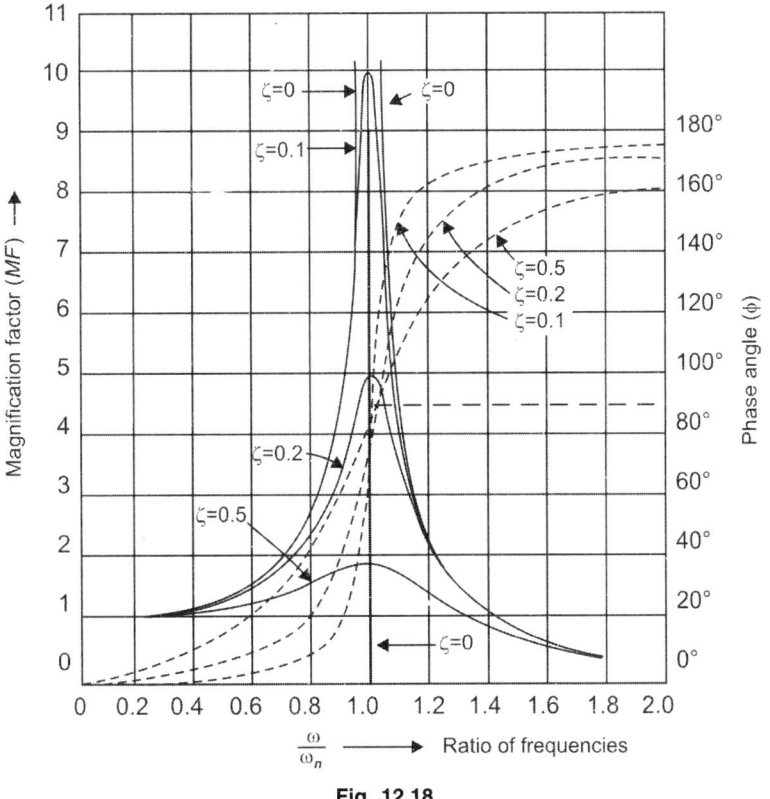

Fig. 12.18

In practice, the magnification factor cannot reach infinity due to friction which tends to dampen the vibration. However, the amplitude can reach very high values.

Following observations are made from Fig. 12.17.

(i) Irrespective of the amount of damping, the maximum amplitude of vibration occurs before the ratio ω/ω_n reaches unity or when the frequency of the forced vibration is less than that of the undamped vibrations.

(ii) Phase angle varies from zero at low frequencies to 180° at very high frequencies. It changes very rapidly near the resonance and is 90° at resonance irrespective of damping.

(iii) In the absence of any damping, phase angle suddenly changes from zero to 180° at resonance.

12.16 VIBRATION ISOLATION AND TRANSMISSIBILITY

Vibrations produced in machines due to unbalanced masses are transmitted to the foundation upon which machines are installed. To reduce these undesirable vibrations, machines are generally mounted on some vibration isolating material, like springs, dampers, etc.

Transmissibility is defined as the ratio of the force transmitted (to the foundation) to the force applied. It is measure of the effectiveness of the vibration isolating material. It is denoted by E

If the springs and damper are used as vibration isolating material then the resultant force transmitted to the foundation is the vector sum of the spring force (kA) and the damping force ($c\omega A$) which are perpendicular to each other (Fig. 12.19).

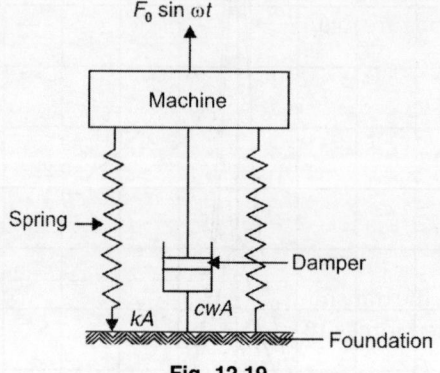

$$F_0 \sin \omega t$$

Fig. 12.19

$\therefore$ Force transmitted,
$$F_t = \sqrt{(kA)^2 + (c\omega A)^2}$$

$$= A\sqrt{(k)^2 + (c\omega)^2}$$

$$= \frac{F_0}{\sqrt{\left(k - m\omega^2\right)^2 + (c\omega)^2}}\sqrt{k^2 + (c\omega)^2}$$

$$= \frac{F_0\sqrt{1 + \left(\dfrac{c}{k}\omega\right)^2}}{\sqrt{\left(1 - \dfrac{m}{k}\omega^2\right)^2 + \left(\dfrac{c}{k}\omega\right)^2}}$$

$$F_t = \frac{F_0\sqrt{1 + \left(2\zeta\omega / \omega_n\right)^2}}{\sqrt{\left[1 - \left(\dfrac{\omega}{\omega_n}\right)^2\right]^2 + \left(2\zeta\omega / \omega_n\right)^2}} \qquad\qquad ...(12.44)$$

Transmissibility, $\qquad \varepsilon = \dfrac{\text{resultant force transmitted } (F_t)}{\text{maximum force applied}}$

$\therefore \qquad\qquad = \dfrac{F_t}{F_0} = \dfrac{\sqrt{1 + \left(2\zeta\omega / \omega_n\right)^2}}{\sqrt{\left[1 - \left(\omega / \omega_n\right)^2\right]^2 + \left(2\zeta\omega / \omega_n\right)^2}} \qquad\qquad ...(12.45)$

At resonance, $\qquad \dfrac{\omega}{\omega_n} = 1$

$\therefore \qquad\qquad \varepsilon = \dfrac{\sqrt{1 + (2\zeta)^2}}{2\zeta} \qquad\qquad ...(12.46)$

When no damper is used, $\zeta = 0$

and $\qquad\qquad \varepsilon = \dfrac{1}{\pm\left[1 - \left(\omega / \omega_n\right)^2\right]} \qquad\qquad ...(12.47)$

The variation of transmissibility (ε) with $\left(\dfrac{\omega}{\omega_n}\right)$ for different values of ζ is shown is Fig. 12.19.

Following observations are made from the Fig. 12.20.

(i) When $\omega/\omega_n < \sqrt{2}$, ε is more than 1, i.e. the transmitted force is always more than the exciting force.

(ii) When $\omega/\omega_n > \sqrt{2}$, ε is less than 1, i.e. the transmitted force is always less than the exciting force.

(iii) When $\omega/\omega_n = \sqrt{2}$, ε is 1, i.e. the transmitted force is equal to the exciting force, for all values of ζ.

(iv) When $\omega/\omega_n > 1$, the transmitted force is infinite. If damping is used, the magnitude of the transmitted force can be reduced.

(v) When $\omega/\omega_n > \sqrt{2}$, ε increases as the damping is increased.

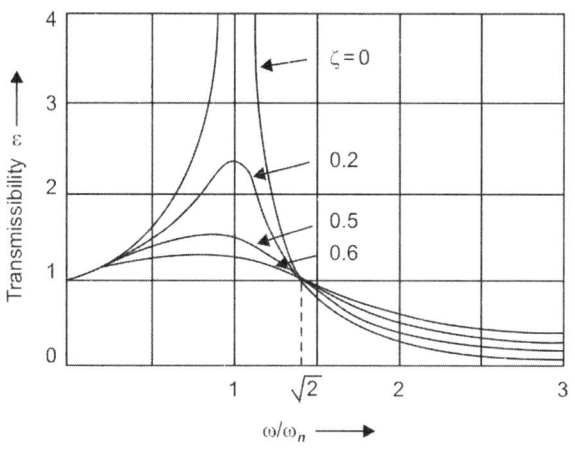

Fig. 12.20

Following observations are made from Fig. 12.20:

(i) When $\omega/\omega_n < \sqrt{2}$, ε is more than 1, i.e., the transmitted force is always more than the exciting force.

(ii) When $\omega/\omega_2 > \sqrt{2}$, ε is less than 1, i.e., the transmitted force is always less than the exciting force.

(iii) When $\omega/\omega_n = \sqrt{2}$, ε is 1, i.e., the transmitted force is equal to the exciting force, for all values of ζ.

(iv) When $\omega/\omega_n > \sqrt{2}$, ε increases as the damping is increased.

Example 12.1: In a spring mass vibrating system, the natural frequency of vibration is reduced to half the value when a second spring is added to the first spring in series. Determine the stiffness of the second spring in terms of that of the first spring.

Solution:
$$\omega_{n_2} = \frac{1}{2} \times \omega_{n_1} \qquad \text{...(i)}$$

Let the stiffness of first spring is k_1 and that of second spring is k_2.

$$\omega_{n_1} = \sqrt{\frac{k_1}{m}}, \ \omega_{n_2} = \sqrt{\frac{k_1 k_2}{(k_1 + k_2)m}}$$

Substituting in Eq. (i),

$$\sqrt{\frac{k_1 k_2}{(k_1 + k_2)m}} = \frac{1}{2}\sqrt{\frac{k_1}{m}}$$

Squaring both sides,

$$\frac{k_1 k_2}{(k_1 + k_2)\, m} = \frac{1}{4} \times \frac{k_1}{m}$$

or $\qquad\qquad\qquad 4k_2 = k_1 + k_2$

or $\qquad\qquad\qquad 4k_2 - k_2 = k_1 \ $ or $\ 3k_2 = k_1$

∴ $\qquad\qquad\qquad k_2 = \dfrac{k_1}{3}$

Example 12.2: Determine the equation of vibration of the water column in a U-tube shown in Fig. 12.21.

Solution: (a) *Newton's method*:

Let a = area of cross-section of the tube,
$\quad p$ = mass density of water,
$\quad L$ = total length of water column,

Inertia force + external force = 0

Mass × Acceleration + Weight of water column above $A - A = 0$

$$(aLp) \times \ddot{x} + (a \times 2x)pg = 0$$

or $\qquad\qquad\qquad \ddot{x} + \dfrac{2g}{L}x = 0$

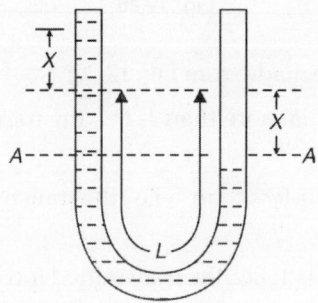

Fig. 12.21

(b) *Energy method*: At any instant,

$$\frac{d}{dt}(KE + PE) = 0, \ KE = \frac{1}{2}mv^2 = \frac{1}{2}(aLp)\dot{x}^2$$

PE = work to transfer a water column of length x from the right-hand side to the left-hand side = $mgx = (axp)\,gx = apgx^2$

$$\frac{d}{dt}\left(\frac{1}{2}aLP\dot{x}^2 + apgx^2\right) = 0$$

$$\frac{1}{2}aLp \times 2\dot{x}\ddot{x} + apg \times 2x\dot{x} = 0$$

$$\ddot{x} + \frac{2g}{L}x = 0$$

$\therefore$ Natural frequency, $\qquad \omega_n = \sqrt{\dfrac{2g}{L}}$

Example 12.3: Determine the natural frequency of a vibrating system shown in Fig. 12.22.

Solution:

Force in spring 1, $\qquad F_1 = W$

Force in spring 2, $\qquad F_2 = W\dfrac{L_1}{L_2}$

$$(F_1 \times L_1 = F_2 \times L_2)$$

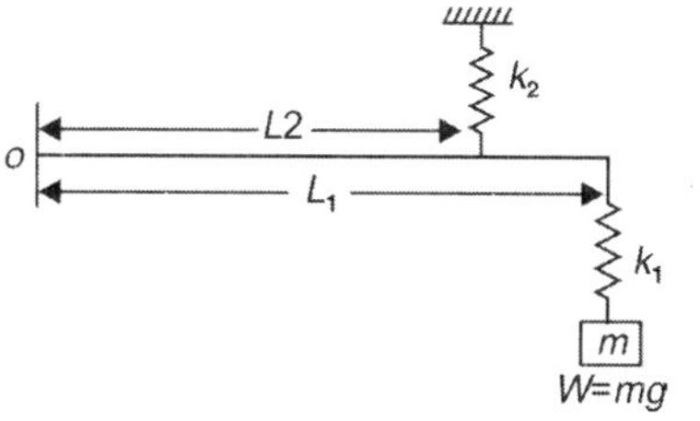

Fig. 12.22

Deflection of mass = deflection of spring

$1 + \dfrac{L_1}{L_2}$ (Deflection of spring 2)

or $\qquad \Delta = \dfrac{W}{k_1} + \dfrac{L_1}{L_2} \times \dfrac{W \times \dfrac{l_1}{l_2}}{k_2}$

$$= W\left[\frac{1}{k_1} + \frac{(L_1/L_2)^2}{k_2}\right] = mg\left[\frac{k_2 + k_1(l_1/l_2)^2}{k_1 k_2}\right]$$

$\therefore$ Natural frequency, $\qquad \omega_n = \sqrt{\dfrac{g}{\Delta}}$

or $\qquad \omega_n = \sqrt{\dfrac{k_1 k_2}{[k_1(L_1/L_2)^2 + k_2]m}}$

$$= \sqrt{\frac{k_1 k_2 (L_2/L_1)^2}{[k_1 + k_2(L_2/L_1)^2]m}}$$

Example 12.4: A coil of spring stiffness 4 N/mm supports vertically a mass of 20 kg at the free end. The motion is resisted by the oil dash-pot. It is found that the amplitude at the beginning of the fourth cycle is 0.8 times the amplitude of previous vibration. Determine the damping force per unit velocity and also find the ratio of frequency of the damped and undamped vibrations.

Solution:

$$k = 4\,N/mm, m = 20\,kg$$

$$X_3 = 0.8X_2,\, f_d = c\dot{x},\, c = ?,\, \frac{\omega_d}{\omega_n} = ?$$

$$\omega_n = \sqrt{\frac{k}{m}} = \sqrt{\frac{4000\phi}{2\phi}} = \sqrt{200} = 10\sqrt{2}\,\,rad/sec$$

$$\delta = \log_e\left(\frac{X_n}{X_{n+1}}\right) = \log_e\frac{1}{0.8} = \log_e 1.25 = 0.223$$

$$\delta = \frac{2\pi\zeta}{\sqrt{1-\zeta^2}} = 0.223 \text{ or } \frac{4\pi^2\zeta^2}{1-\zeta^2} = 0.049$$

$$4\pi^2\zeta^2 = 0.49 - 0.049\zeta^2$$

$$\zeta^2(4\pi^2 + 0.049) = 0.049$$

$$\zeta = 0.035$$

$$\frac{\omega_d}{\omega_n} = \sqrt{1-\zeta^2} = 0.999$$

$$C = 2\zeta m\omega_n = 2 \times 0.035 \times 20 \times 10\sqrt{2} = 19.7$$

Example 12.5: A vibrating system consists of a mass of 20 kg, a spring of stiffness 20 kN/m and a damper. The damping provided is only 30% of the critical value. Determine the natural frequency of the damped vibration and the ratio of two consecutive amplitudes.

Solution: $m = 20\,kg, k = 20\,kN/m = 20000\,N/m$.

Damping = 30% of critical value

i.e. $$c = 30\%\, c_c = 0.3\, c_c$$

or $$\frac{c}{c_c} = 0.3 \text{ but } \frac{c}{c_c} = \zeta$$

∴ $$\zeta = 0.3$$

$$\omega_n = \sqrt{\frac{k}{m}} = \sqrt{\frac{20 \times 1000}{20}} = 10\sqrt{10}$$

$$\omega_d = \sqrt{1-\zeta^2} \times \omega_n = \sqrt{1-(0.3)^2} \times 10\sqrt{10} = 30.17 = 30.2$$

$$\omega_d = 30.2\,\,rad/sec$$

$$\delta = \frac{2\pi\zeta}{\sqrt{1-\zeta^2}} = \frac{2\pi \times 0.3}{\sqrt{1-(0.3)^2}} = 1.9749$$

but, $$\delta = \log_e\left(\frac{X_0}{X_1}\right)$$

or $$\frac{X_0}{X_1} = e^\delta = e^{1.9749} = 7.206 = 7.21$$

∴ Ratio of two consecutive amplitude, $\dfrac{X_0}{X_1} = 7.21$.

Example 12.6: In a single-degree damped vibrating system, the suspended mass of 4 kg makes 24 oscillations in 20 seconds. The amplitude decreases to 0.3 of the initial value after 4 oscillations. Find the stiffness of the spring, the logarithmic decrement, the damping factor and damping coefficient.

Solution:

$$m = 4 \text{ kg}, n = 24, t = 20 \text{ sec}$$

$$T = \frac{t}{n} = \frac{20}{24} = 0.833 \text{ sec}, f = \frac{1}{T} = 1.2 \text{ Hz}$$

If
$$X_0 = x, X_4 = 0.3x, k = ?, \delta = ?, \zeta = ?, c = ?$$

$$\omega = \sqrt{\frac{k}{m}}, k = m\omega^2 = m(2\pi f)^2 = 4 \times (2\pi \times 1.2)^2 = 227.2 \text{ N/m}$$

$$\frac{X_0}{X_4} = \frac{x}{0.3x} = 3.33 \text{ but } \frac{X_0}{X_4} = \tau^4 e^{\zeta \omega_n T_d}$$

∴
$$e^{4\zeta \omega_n T_d} = 3.33 \text{ or } \left(\frac{X_0}{X_1}\right)^4 = 3.33$$

⇒
$$\frac{X_0}{X_1} = (3.33)^{1/4} = 1.35$$

$$\delta = \log_e \left(\frac{X_0}{X_1}\right) = \log_e (1.35) = 0.3$$

Also
$$\delta = \frac{2\pi\zeta}{\sqrt{1-\zeta^2}} \text{ or } 0.3 = \frac{2\pi\zeta}{\sqrt{1-\zeta^2}} \text{ or } 0.0022 (1-\zeta^2) = \zeta^2$$

⇒
$$\zeta = 0.04768$$
$$C_c = 2\sqrt{km} = 2\sqrt{227.2 \times 4} = 60.29$$
$$C = \zeta \times C_c = 0.04768 \times 60.98 = 2.8747$$

or
$$C = 2.88.$$

Example 12.7: The measurement on a mechanical vibrating system shows that it has a mass of 8 kg and that the springs can be combined to give an equivalent spring of stiffness 5.4 N/mm. If the vibrating system has a dashpot attached which exerts a force of 40 N, when the mass has a velocity of 1 m/sec, find critical damping coefficient, damping factor, logarithmic decrement and ratio of two consecutive amplitudes.

Solution: $m = 8 \text{ kg}, s = 5.4 \text{ N/mm} = 5400 \text{ N/m}$,

$v = 1 \text{ m/sec} = \dot{x}$

Damping force $= 40 \text{ N} = c\dot{x}, c = 40 \text{ N/m/sec}$

$$c_c = 2m\omega_n = 2m\sqrt{\frac{s}{m}} = 2 \times 8\sqrt{\frac{5400}{8}} = 416 \text{ N/m/sec}$$

$$\zeta = \frac{C}{C_C} = \frac{40}{416} = 0.096$$

$$\delta = \frac{2\pi c}{\sqrt{(c_c)^2 - c^2}} = \frac{2\pi \times 40}{\sqrt{(416)^2 - (40)^2}} = 0.6$$

or
$$\log_e \left(\frac{x_n}{x_{n+1}} \right) = 0.6 \left(\delta = \log_e \frac{x_n}{x_{n+1}} \right)$$

or
$$\frac{x_n}{x_{n+1}} = \tau^6 = 1.82$$

TRANSVERSE VIBRATIONS

The time period and natural frequency of the transverse vibrations will be same as that of the longitudinal vibrations. Transverse vibrations of shafts and beams under different types of loads and end conditions are explained below.

12.17 SINGLE CONCENTRATED LOAD

In case of shafts and beams of negligible mass carrying a concentrated mass, the force is proportional to the deflection of the mass from the equilibrium position and the relation derived for natural frequency of longitudinal vibrations is applicable here.

$$f_n = \frac{1}{2\pi} \sqrt{g/\Delta}$$

The value of Δ is different in each case,

$\Delta = \dfrac{mgL^3}{3EI}$; for cantilevers, supporting a concentrated mass at the free end. ...(12.48)

$\Delta = \dfrac{mga^2b^2}{3EIL}$; for simply supported beams. ...(12.49)

$\Delta = \dfrac{mga^3b^3}{3EIL^3}$; for beams fixed at both ends. ...(12.50)

A shaft supported in long bearings is assumed to have both ends fixed while one in short bearings is considered to be simply supported.

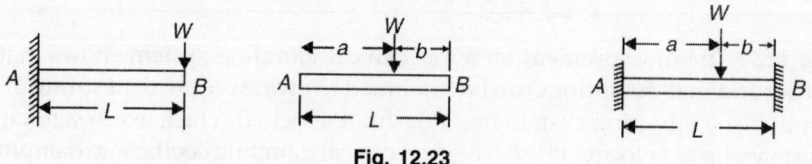

Fig. 12.23

12.18 UNIFORMLY DISTRIBUTED LOAD

Figure 12.24 shows a shaft supported at its ends and carrying a uniform mass. Let m = distributed mass per unit length, L = length of the shaft.

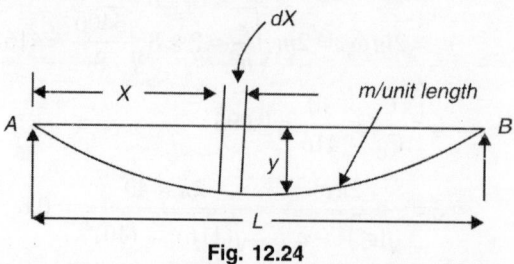

Fig. 12.24

The shaft makes transverse vibrations due to elastic forces. At any instant, let it be deflected by an amount y at a distance x from the end A. The vibrations being free and due to elastic forces, will be simple harmonic motion type.

From the theory of bending of shafts, we know,

$$EI \frac{d^4y}{dx^4} = \text{dynamic load per unit length}$$

$$= \text{centrifugal force per unit length}$$

$$= my\,\omega^2$$

or
$$\frac{d^4y}{dx^4} - \frac{my\omega^2}{EI} = 0 \qquad \qquad \text{...(12.51)}$$

or
$$\frac{d^4y}{dx^4} - \lambda^4 y = 0$$

where,
$$\lambda^4 = \frac{m\omega^2}{EI}$$

The auxiliary equation is $(D^4 - \lambda^4)\,y = 0$

This gives, $D = \pm\lambda$ and $\pm i\lambda$

The solution will be of the form,

$$y = A\sin\lambda x + B\cos\lambda x + C\sinh\lambda x + D\cosh\lambda x \qquad \text{...(12.51a)}$$

This is the general expression for the deflection in case of uniformly loaded shafts. Constants A, B, C and D are found from the end conditions.

12.18.1 Simply Supported Shaft

The boundary conditions are:

(a) $y = 0$ at $x = 0$ and $x = L$

(b) $\dfrac{d^2y}{dx^2} = 0$ at $x = 0$ and $x = L$ (bending moment is zero at ends)

When $x = 0$, $y = 0$; $B + D = 0$ \qquad ...(12.51b)

When $x = L$, $y = 0$, from equation (i),

$A\sin\lambda L + B\cos\lambda L + C\sin h\,\lambda L + D\cos h\,\lambda L = 0$ \qquad ...(12.51c)

Differentiating Eq. (12.51a) with respect to x twice, we have

$$\frac{dy}{dx} = \lambda(A\cos\lambda x - B\sin\lambda x + C\cos h\lambda x + D\sin h\lambda x)$$

$$\frac{d^2y}{dx^2} = \lambda^2(-A\sin\lambda x - B\cos\lambda x + C\sin h\lambda x + D\cos h\lambda x)$$

When $x = 0$, \qquad $\dfrac{d^2y}{dx^2} = 0$

$$\lambda^2\,(-B + D) = 0 \qquad \qquad \text{...(12.51d)}$$

When $x = L$, \qquad $\dfrac{d^2y}{dx^2} = 0$

$$\lambda^2 (-A \sin \lambda L - B \cos \lambda L + C \sin h \, \lambda L + D \cos h \, \lambda L = 0 \qquad \qquad ...(12.51e)$$

From Eqs (12.51b) and (12.51d) $B = 0$ and $D = 0$

Thus, Eqs (12.51c) and (12.51e) can be written as,

$$A \sin \lambda L + C \sin h \, \lambda L = 0$$

and $\qquad -A \sin \lambda L + C \sin h \, \lambda L = 0$

Adding these, we get, $\quad C \sin h \, \lambda L = 0$

Subtracting, $\qquad \qquad A \sin \lambda L = 0$

$\sin h \, \lambda L$ cannot be zero, because if

$$\lambda = 0, \lambda^4 = 0$$

or $\qquad \qquad \dfrac{m \omega^2}{EI} = 0$

or $\qquad \qquad \dfrac{m}{EI} (2\pi f_n)^2 = 0$

or $\qquad \qquad f_n = 0$

which means that the system does not vibrate.

$\therefore \qquad \qquad C = 0$

Thus, Eq. (12.51a) reduces to,

$$y = A \sin \lambda x \quad (B, C \text{ and } D \text{ are zero})$$

Now, when $A \sin \lambda L = 0$, A cannot be zero as B, C and D are already zero and if A is also zero, there are no vibrations.

$$\sin \lambda L = 0$$

or $\qquad \qquad \lambda L = 0, \pi, 2\pi, 3\pi, ...$

But λL cannot be equal to zero; if so, there will be no vibration.

$$\lambda = \frac{\pi}{L}, \frac{2\pi}{L}, \frac{3\pi}{L}, ...$$

or

$$\left[\frac{m \omega^2}{EI}\right]^{1/4} = \frac{\pi}{L}, \frac{2\pi}{L}, \frac{3\pi}{L}, ...$$

$$\omega^{1/2} = \frac{\pi}{L}\left(\frac{EI}{m}\right)^{1/4}, \frac{2\pi}{L}\left(\frac{EI}{m}\right)^{1/4}, \frac{3\pi}{L}\left(\frac{EI}{m}\right)^{1/4}, ...$$

$$\omega = (2\pi f_n) = \frac{\pi^2}{L^2}\sqrt{\frac{EI}{m}}, \frac{4\pi^2}{L^2}\sqrt{\frac{EI}{m}}, \frac{9\pi^2}{L^2}\sqrt{\frac{EI}{m}}$$

$$f_n = \frac{\pi}{2}\sqrt{\frac{EI}{mL^4}}, \frac{4\pi}{2}\sqrt{\frac{EI}{mL^4}}, \frac{9\pi}{2}\sqrt{\frac{EI}{mL^4}}$$

A simply supported shaft carrying a uniformly distributed mass has maximum deflection at the mid-span.

$$\Delta = \frac{5mg \, L^4}{384 \, EI} \quad \text{or} \quad \frac{EI}{mL^4} = \frac{5g}{384\Delta}$$

Then, taking the smallest value of f_n,

$$f_n = \frac{\pi}{2} \sqrt{\frac{5g}{384\Delta}} \qquad \text{...(12.52)}$$

This is the lowest frequency of transverse vibrations and is called the fundamental frequency.

As the equation for the displacement is $y = A \sin \lambda L$, and at node points, $y = 0$

$$\therefore \quad 0 = A \sin \lambda x = A \sin \frac{\pi}{L} x \quad \text{or} \quad \frac{\pi}{L} x = 0, \text{ i.e., } x = 0 \text{ and } x = L$$

This means a node at each end.

The next higher frequency is four times the fundamental frequency.

$$0 = A \sin \lambda x = A \sin \frac{2\pi}{L} x \quad \text{or} \quad \frac{2\pi}{L} x = 0, \text{ i.e., } x = 0, 1/2 \text{ and } L$$

i.e., it has three nodes, two at the ends and one at the centre.

The next higher frequency is nine times the fundamental frequency. It has four nodes dividing the shaft into three equal parts, and so on (Fig. 12.25).

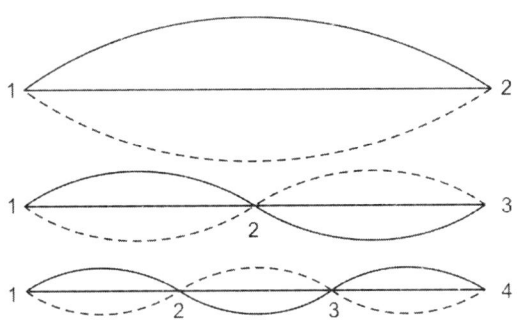

Fig. 12.25

Thus, a simply supported shaft will have an infinite number of frequencies under uniformly distributed load.

12.18.2 Cantilevers

The end conditions are:

(a) $y = 0$ at $x = 0$ (zero deflection)

(b) $\dfrac{dy}{dx} = 0$ at $x = 0$ (zero slope)

(c) $\dfrac{d^2y}{dx^2} = 0$ at $x = L$ (zero bending moment)

(d) $\dfrac{d^3y}{dx^3} = 0$ at $x = L$ (zero shear force)

Static deflection at free end, $\Delta = \dfrac{mgL^3}{8EI}$ ...(12.53)

12.18.3 Both Ends Fixed

The end conditions are:

(a) $y = 0$ and $x = 0$ and $x = L$

(b) $\dfrac{dy}{dx} = 0$ at $x = 0$ and $x = L$

Static deflection at the centre, $\Delta = \dfrac{mgL^4}{384EI}$ ...(12.54)

12.19 TRANSVERSE VIBRATIONS OF A SHAFT CARRYING SEVERAL LOADS

There are two methods to find the natural frequency of the system:

(i) Dunkerley's method, which is semiempirical. This gives approximate results but is simple. This is used when diameter of shaft is uniform.

(ii) The energy method, which gives accurate results but involves heavy calculations in case there are many loads. This is also known as Rayleigh's method.

12.19.1 Dunkerley's Method

Let $W_1, W_2, W_3, \ldots$ be the concentrated loads on the shaft due to masses $m_1, m_2, m_3, \ldots$ and $\Delta_1, \Delta_2, \Delta_3, \ldots$ the static deflections of this shaft under each load when that load acts alone on the shaft. Let the shaft carry uniformly distributed mass of m per unit length over its whole span and the static deflection at mid-span due to the load of this mass be Δ_s.

Let, f_n = frequency of transverse vibration of the whole system,

f_{ns} = frequency with the distributed load acting alone,

f_{n1}, f_{n2}, f_{n3} = frequency of transverse vibrations when each of load $W_1, W_2, W_3, \ldots$ acts alone.

Then, according to Dunkerley's empirical formula,

$$\frac{1}{f_n^2} = \frac{1}{f_{n1}^2} + \frac{1}{f_{n2}^2} + \frac{1}{f_{n3}^2} + \ldots + \frac{1}{f_{ns}^2}$$ ...(12.55)

where, $f_{n1} = \dfrac{1}{2\pi}\sqrt{\dfrac{g}{\Delta_1}} = \dfrac{\sqrt{9.81}}{2\pi}\dfrac{1}{\sqrt{\Delta_1}} = \dfrac{0.4985}{\sqrt{\Delta_1}}$

Similarly, $f_{n2} = \dfrac{0.4985}{\sqrt{\Delta_2}}$; $f_{n3} = \dfrac{0.4985}{\sqrt{\Delta_3}}$, and so on.

$$f_{ns} = \frac{\pi}{2}\sqrt{\frac{5g}{384\Delta_s}} = \frac{\pi}{2}\sqrt{\frac{5 \times 9.81}{384}} \times \frac{1}{\sqrt{\Delta_s}} = \frac{0.5614}{\sqrt{\Delta_s}}$$

$$\frac{1}{f_n^2} = \frac{1}{(0.4985)^2}(\Delta_1 + \Delta_2 + \Delta_3 + \ldots) + \frac{1}{(0.5614)^2}\Delta_s$$

$$= \frac{1}{(0.4985)^2}\left(\Delta_1 + \Delta_2 + \Delta_3 + \ldots + \frac{\Delta_s}{1.27}\right)$$

$$f_n = \frac{0.4985}{\sqrt{\Delta_1 + \Delta_2 + \Delta_3 + \ldots + \dfrac{\delta_s}{1.27}}}$$ ...(12.56)

12.19.2 Energy Method

Consider a shaft with negligible mass, carrying point loads $W_1, W_2, W_3, ...$ due to masses $m_1, m_2, m_3, ...$ Let $y_1, y_2, y_3, ...$ be total deflections under these loads.

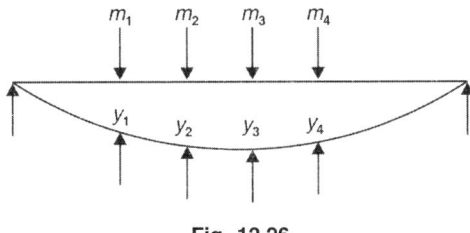

Fig. 12.26

In the extreme positions of the shaft, it possesses maximum potential energy and no kinetic energy, whereas in the mean position, it possesses maximum kinetic energy and no potential energy. Thus, the maximum potential energy of the shaft can be made equal to its maximum kinetic energy.

Maximum
$$PE = \frac{1}{2}W_{1y1} + \frac{1}{2}W_{2y2} + \frac{1}{2}W_{3y3} + ...$$

$$= \frac{g}{2}(m_1y_1 + m_2y_2 + m_3y_3 + ...)$$

$$= \frac{g}{2}\Sigma my$$

Maximum
$$KE = \frac{1}{2}m_1v_1^2 + \frac{1}{2}m_2v_2^2 + \frac{1}{2}m_3v_3^2 + ...$$

$$= \frac{1}{2}m_1(\omega y_1)^2 + \frac{1}{2}m_2(\omega y_2)^2 + \frac{1}{2}m_3(\omega y_3)^2 + ...$$

$$= \frac{\omega^2}{2}(m_1y_1^2 + m_2y_2^2 + m_3y_3^2 + ...)$$

$$= \frac{\omega^2}{2}\Sigma my^2$$

where ω is the circular frequency of vibration. Equating maximum PE and maximum KE,

$$\frac{g}{2}\Sigma my = \frac{\omega^2}{2}\Sigma my^2$$

$$\omega = \sqrt{\frac{g\Sigma my}{\Sigma my^2}}$$

$$f_n = \frac{\omega}{2\pi} = \frac{1}{2\pi}\sqrt{\frac{g\Sigma my}{\Sigma my^2}} \qquad \qquad ...(12.57)$$

12.20 WHIRLING OF SHAFTS

The centre of mass of a shaft mounted with a rotor does not generally coincides with its axis of rotation. Thus, when the shaft begins to rotate, the centre of mass of the shaft is subjected to radially outward centrifugal force. This force bends the shaft in the direction of initial eccentricity of centre of mass which further increases the eccentricity and hence, the centrifugal force. The effect is therefore, cumulative and ultimately the shaft may even fail. The bending of shaft depends upon:

(i) initial displacement of centre of mass, and

(ii) speed of rotation of shaft.

The speed at which the shaft rotates so that the deflection of the shaft from the axis of rotation becomes infinite, is known as whirling speed or critical speed. At critical speed, the shaft tends to vibrate violently in transverse direction.

Let, m = mass of the rotor

k = stiffness of the shaft

ω = angular velocity of the shaft

e = initial eccentricity of centre of mass of rotor

y = additional deflection of rotor due to centrifugal force

Now, centrifugal force acting radially outward,

$$= m(y + e)\omega^2$$

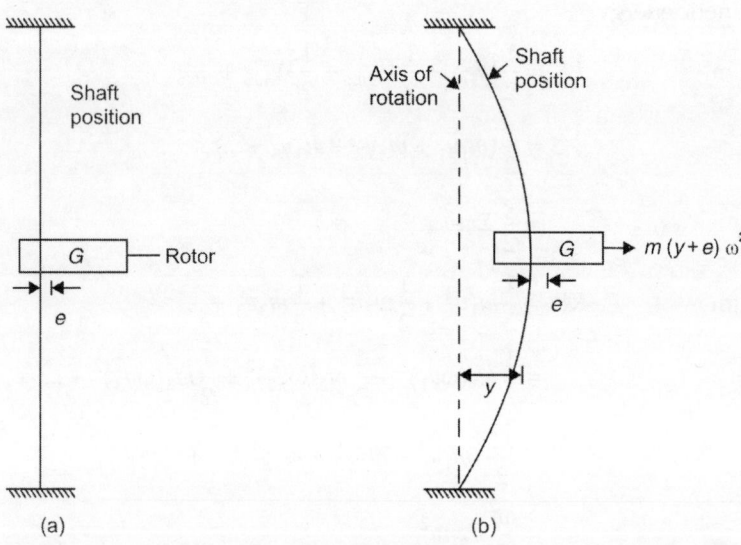

Fig. 12.27

This force is balanced by inward elastic resisting force by the shaft.

∴ Force resisting the deflection = ky

For equilibrium

$$ky = m(y + e)\,\omega^2 = my\omega^2 + me\omega^2$$

or $$ky - my\omega^2 = me\omega^2$$

or $$y(k - m\omega^2) = me\omega^2$$

$$y = \frac{me\omega^2}{(k - m\omega^2)} = \frac{e}{\dfrac{k}{m\omega^2} - 1}$$

$$y = \frac{e}{\left(\dfrac{\omega_n}{\omega}\right)^2 - 1} \qquad \left(\frac{k}{m} = \omega_n^2\right) \quad ...(12.58)$$

When $\omega = \omega_n$, the deflection y becomes infinitely large. The value of 'ω' is known as critical speed or whirling speed.

$\therefore$ Critical speed $\qquad \omega_c = \omega_n = \sqrt{\dfrac{k}{m}} = \sqrt{\dfrac{g}{\Delta}}$ $\qquad\qquad$ ...(12.59)

or $\qquad\qquad \dfrac{2\pi N_c}{60} = \sqrt{\dfrac{g}{\Delta}}$ or $N_c = \dfrac{60}{2\pi}\sqrt{\dfrac{g}{\Delta}}$

$$N_c = \dfrac{0.4985}{\sqrt{\Delta}} \qquad\qquad\qquad ...(12.60)$$

If the speed of the shaft is increased rapidly above the critical speed, i.e. $\omega > \omega_n$ or $\left(\dfrac{\omega_n}{\omega}\right)^2 < 1$, deflection of shaft ($y$) becomes negative.

At a particular value of speed, deflection of shaft (y) becomes equal to $-e$. That means centre of mass of rotor will lie on the axis of rotation and shaft will run steadily. The principal is used in running high speed turbines.

Example 12.8: A vertical shaft of diameter 60 mm and length 1.5 m is fixed at its upper end. The other end carries a disc of weight 200 N. The modulus of elasticity of the material of the shaft is $2 \times 10^5 \dfrac{N}{mm^2}$. Neglecting the weight of the shaft, determine the frequency of transverse vibrations.

Solution: $d = 60$ mm $= 0.06$ m, $L = 1.5$ m, $W = 200$ N

$E = 2 \times 10^5$ N/mm^2 = $2 \times 10^5 \times 10^6$ N/m^2 = 2×10^{11} N/m^2.

The static deflection of the shaft (D) under the load W when the shaft is placed horizontally as a cantilever, is given as

$$\Delta = \dfrac{WL^3}{3EI} = \dfrac{WL^3}{3E \times \left(\dfrac{\pi}{64}d^4\right)} \qquad\qquad \left(I = \dfrac{\pi}{64}d^4\right)$$

or $\qquad\qquad \Delta = \dfrac{200 \times (1.3)^3}{3 \times 2 \times 10^{11} \times \dfrac{\pi}{64} \times 0.06)^4} = 1.769 \times 10^{-3}$ m

Frequency of transverse vibrations, $f_n = \dfrac{1}{2\pi}\sqrt{\dfrac{g}{\Delta}}$

or $\qquad\qquad f_n = \dfrac{1}{2\pi}\sqrt{\dfrac{9.81}{1.769 \times 10^{-3}}} = 11.857 = 12$ Hz.

Example 12.9: A simply supported shaft of length 1 m and diameter 30 mm carries a mass of 70 kg placed 400 mm from one end. Determine the frequency of natural transverse vibrations, if $E = 200$ G N/m^2.

Solution: $L = 1$ m, $d = 30$ mm $= 0.03$ m, $m = 70$ kg,

$E = 200 \times 10^9$ N/m^2, $W = mg = 70 \times 9.81$ N

$a = 400$ mm $= 0.4$ m, $b = 1 - 0.4 = 0.6$ m

$$I = \frac{\pi}{64} d^4 = \frac{\pi}{64} \times (0.03)^4 = 3.97 \times 10^{-8} \text{ m}^4$$

Deflection of simply supported beam,

$$\Delta = \frac{Wa^2b^2}{3EIL}$$

or

$$\Delta = \frac{70 \times 9.81 \times (0.4)^2 \times (0.6)^2}{3 \times 200 \times 10^9 \times 3.97 \times 10^{-8} \times 1} = 1.659 \times 10^{-3} \text{ m}$$

Frequency of transverse vibrations,

$$f_n = \frac{1}{2\pi}\sqrt{\frac{g}{\Delta}} = \frac{1}{2\pi}\sqrt{\frac{9.81}{1.659 \times 10^{-3}}} = 12.245 \text{ Hz}$$

Example 12.10: A shaft 40 mm diameter and 2.5 m long has a mass of 15 kg per metre length. It is simply supported at the ends and carries three masses 90 kg, 140 kg and 60 kg at 0.8 m, 1.5 m and 2 m respectively from the left support. If $E = 200 \text{ GN/m}^2$, find the frequency of the transverse vibrations.

Solution: $d = 40$ mm $= 0.04$ m, $L = 2.5$ m

$$I = \frac{\pi}{64} \times d^4 = \frac{\pi}{64} \times (0.04)^4 = 0.1256 \times 10^{-6} \text{ m}^4$$

Frequency of transverse vibrations of shaft is given as,

$$f_n = \frac{0.4985}{\sqrt{\Delta_1 + \Delta_2 + \Delta_3 + ... + \dfrac{\Delta_s}{1.27}}}$$

Deflection,

$$\Delta = \frac{mga^2b^2}{3EIL}$$

Here, $m_1 = 90$ kg, $a_1 = 0.8$ m and $b_1 = 1.7$ m.

Hence, deflection under load W_1,

$$\Delta_1 = \frac{90 \times 9.81 \times (0.8)^2 \times (1.7)^2}{3 \times 200 \times 10^9 \times 0.1257 \times 10^{-6} \times 2.5} = 0.00866 \text{ m}$$

For Δ_2, $m_2 = 140$ kg, $a_2 = 1.5$ m, $b_2 = 1$ m

∴

$$\Delta_2 = \frac{140 \times 9.81 \times (1.5)^2 \times (1)^2}{3 \times 200 \times 10^9 \times 0.1257 \times 10^{-6} \times 2.5} = 0.1639 \text{ m}$$

For Δ_3, $m_3 = 60$ kg, $a_3 = 2$ m, $b_3 = 0.5$ m

∴

$$\Delta_3 = \frac{60 \times 9.81 \times (2)^2 \times (0.5)^2}{3 \times 200 \times 10^9 \times 0.1257 \times 10^{-6} \times 2.5} = 0.00312 \text{ m}$$

Deflection of shaft,

$$\Delta_s = \frac{5mgl^4}{384EI} = \frac{5 \times 15 \times 9.81 \times (2.5)^4}{384 \times 200 \times 10^9 \times 0.1257 \times 10^{-6}} = 0.00298 \text{ m}$$

Hence

$$f_n = \frac{0.4985}{\sqrt{0.00866 + 0.01639 + 0.00312 + \dfrac{0.00298}{127}}} = 2.85 \text{ Hz}$$

Example 12.11: A shaft is simply supported at the ends and is of 20 mm in diameter and 600 mm in length. The shaft carries a load of 19.62 N at its centre. The weight of shaft per metre length is 248.2 N. Find the critical speed of the shaft. Take Young's modulus = 200 GN/m².

Solution: $d = 20$ mm $= 0.02$ m; $L = 600$ mm $= 0.6$ m; load at centre, $W = 19.62$ N, the weight of shaft per metre length, $w = 248.2$ N; $E = 200$ GN/m² $= 200 \times 10^9$ N/m².

We know that the critical speed of a shaft which carries point loads or uniformly distributed load or combination of both, is equal to frequency of transverse vibration. The frequency of transverse vibration of the shaft may be obtained by using Dunkerley's method.

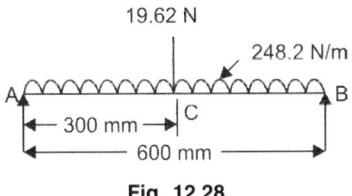

Fig. 12.28

The frequency of transverse vibration is given as,

$$f_n = \frac{0.4985}{\sqrt{\Delta_1 + \Delta_2 + \ldots\ldots + \dfrac{\Delta_s}{1.27}}} \qquad \text{...(i)}$$

For the given problem,

$$f_n = \frac{0.4985}{\sqrt{\Delta_1 + \dfrac{\Delta_s}{1.27}}} \qquad \text{(ii)}$$

Δ_1 = static deflection due to point load at the centre

$$= \frac{WL^3}{48EI} \quad \text{(For a simply supported shaft)}$$

$$= \frac{19.62 \times 0.6^3}{48 \times (200 \times 10^9) \times 7.855 \times 10^{-9}}$$

($\because W = 19.62$ N; $L = 0.6$ m; $E = 200 \times 10^9$ N/m², and

I = moment of inertia of shaft

$$= \frac{\pi}{64} \times d^4 = \frac{\pi}{64} \times (0.02)^4 = 7.855 \times 10^{-9} \text{ m}^4)$$

$$= 0.000056 \text{ m}$$

and Δ_s = static deflection at the centre due to weight of the shaft

$$= \frac{5}{384} \times \frac{w \times L^4}{EI} \quad \text{(Simply supported shaft with uniformly distributed load)}$$

$$= \frac{5}{384} \times \frac{248.2 \times 0.6^4}{(200 \times 10^9) \times (7.855 \times 10^{-9})} \quad (\because w = 248.2 \text{ N/m}; I = 7.855 \times 10^{-9} \text{ m}^4)$$

$$= 0.000266 \text{ m}$$

Substituting the values of Δ_1 and Δ_s in Eq. (ii), we get,

$$f_n = \frac{0.4985}{\sqrt{0.000056 + \dfrac{0.000266}{1.27}}} = 30.6 \text{ Hz}$$

The critical speed in rps of a shaft is equal to frequency of transverse vibration in Hz.

∴ Critical speed of shaft, $N_c = f_n = 30.6$ rps $= 30.6 \times 60$ rpm $= 1836$ rpm.

TORSIONAL VIBRATIONS

12.21 FREE TORSIONAL VIBRATIONS OF A SINGLE ROTOR

Assume a shaft of length L and of negligible mass fixed at its upper end carrying a disc of moment of inertia I at the lower end. If the disc is given a twist about its vertical axis and then released, it starts oscillating about the axis. These oscillations are known as torsional vibrations.

Let,　m = mass of the disc (kg)

　　　k = radius of gyration (m)

　　　I = mass of inertia of disc = mk^2

　　　θ = angular displacement of the disc from its mean position at any instant

　　　q = torsional stiffness of the shaft (torque required to twist the shaft per radian

　　　within elastic limit $= \left(\dfrac{GJ}{L}\right)\left(q = \dfrac{T}{\theta} = \dfrac{GJ}{L}\right)$

Here,　G = modulus of rigidity of the shaft material

　　　J = polar moment of inertia of cross-section of shaft

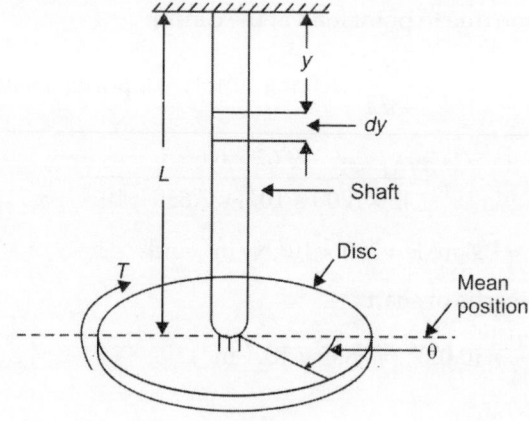

Fig. 12.29

Torques acting on disc at any instant are,

(a) Inertia torque $= -I \times a = -I\dfrac{d^2\theta}{dt^2}$

(b) Restoring torque $= -q \times \theta$

(negative sign is used as restoring torque acts opposite to the angular displacement)

For equilibrium, the sum of all torques acting on the disc must be zero.

$$\therefore \qquad -I\frac{d^2\theta}{dt^2} - q\theta = 0$$

or

$$\frac{d^2\theta}{dt^2} + \frac{q}{I}\theta = 0 \qquad \qquad ...(12.60a)$$

The fundamental equation of the simple harmonic motion is,

$$\frac{d^2\theta}{dt^2} + \omega_n^2\theta = 0 \qquad \qquad ...(12.60b)$$

Comparing Eqs (12.60a) and (12.60b),

$$\omega_n = \sqrt{\frac{q}{I}} \qquad \qquad ...(12.61)$$

The natural frequency of torsional vibrations, $f_n = \dfrac{\omega_n}{2\pi} = \dfrac{1}{2\pi}\sqrt{\dfrac{q}{I}}$ $\qquad ...(12.62)$

The time period, $T = \dfrac{1}{f_n} = 2\pi\sqrt{\dfrac{I}{q}}.$ $\qquad ...(12.63)$

12.22 INERTIA EFFECT OF MASS OF SHAFT ON TORSIONAL VIBRATIONS

Let I_1 = moment of inertia of the shaft

ω = angular velocity of free end of the shaft, consider an element of length d_y at a distance y from the fixed end.

Angular velocity of the element $= \dfrac{\omega}{L}y$

Mass moment of inertia of element $= \dfrac{I_1}{L}d_y$

KE of element $= \dfrac{1}{2} \times (MOI \text{ of element}) \times (\text{angular velocity})^2$

$$= \frac{1}{2} \times \left(I_1\frac{dy}{L}\right)\left(\frac{\omega y}{L}\right)^2$$

KE of shaft, $\quad = \displaystyle\int_0^L \frac{1}{2} \times \frac{I_1}{L}\left(\frac{y}{l}\omega\right)^2 dy$

$$= \frac{I_1\omega^2}{2L^2}\int_0^L y^2\,dy$$

$$= \frac{I_1\omega^2}{2L^3} \times \frac{L^3}{3}$$

$$= \frac{1}{3} \times \frac{1}{2}I_1\omega^2$$

$$= \frac{1}{3} \times \left[\frac{1}{2}(MOI \text{ of shaft}) \times (\text{angular velocity of free end})^2\right]$$

$= \dfrac{1}{3} \times KE$ of a disc of MOI equal to that of the shaft attached to the free end of the shaft. $\qquad ...(12.64)$

Therefore, to consider the effect of inertia of the shaft, the mass moment of inertia of the disc is increased by one-third of the mass moment of inertia of the shaft. The natural frequency of the torsional variations is then given as,

$$f_n = \frac{1}{2\pi}\sqrt{\frac{q}{I + \dfrac{I_1}{3}}} \qquad\qquad ...(12.65)$$

12.23 FREE TORSIONAL VIBRATIONS OF A TWO ROTOR SYSTEM

A shaft to which two rotors are attached at its free ends, is known as two rotor system. If the two rotors are twisted in opposite direction, the system will vibrate torsionally. Hence, some length of the shaft is twisted in one direction while the rest is twisted in the other. The section which remains unaffected is known as node (N). The shaft behaves as if clamped at the node and the two sections vibrate as two separate shafts with the same frequency but opposite in direction.

Let L_a and L_b = lengths of two portions of the shaft

I_a and I_b = MOI of rotors A and B respectively

q_a and q_b = torsional stiffness of length L_a and L_b of the shaft respectively

f_{na} and f_{nb} = natural frequencies of torsional vibrations of rotors A and B respectively

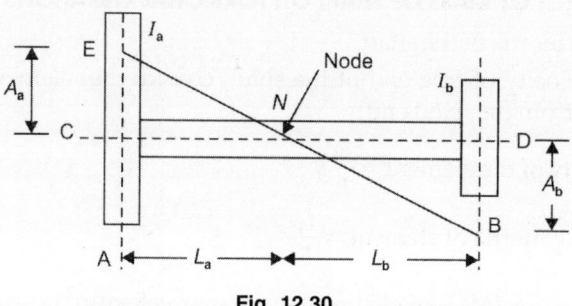

Fig. 12.30

Then $f_{na} = f_{nb}$

∴ $$\frac{1}{2\pi}\sqrt{\frac{q_a}{I_a}} = \frac{1}{2\pi}\sqrt{\frac{q_b}{I_b}}$$

or $$\frac{q_a}{I_a} = \frac{q_b}{I_b}$$

or $$\frac{GJ}{L_a I_a} = \frac{GJ}{L_b I_b}$$

Here, J = polar moment of inertia of shaft $= \dfrac{\pi d^4}{32}$

or $$\frac{L_a}{L_b} = \frac{I_b}{I_a} \qquad\qquad ...(12.66)$$

and $$L_a I_a = L_b I_b \qquad\qquad ...(12.67)$$

Hence, the node divides the length of the shaft in the inverse ratio of moment of inertia of the two rotors,

Also $\dfrac{\text{Amplitude of rotor A}}{\text{Amplitude of rotor B}} = \dfrac{A_a}{A_b} = \dfrac{L_a}{L_b}$ ...(12.68)

(In Fig. 12.30, ΔECN and ΔNDB are similar triangles)
From Eq. (12.67),

$$L_A = \frac{I_B}{I_A} \times L_B$$ ...(12.68a)

We also know that, $\quad L = L_A + L_B$
or $\quad\quad\quad\quad\quad L_B = (L - L_A)$

Substituting value of L_B in Eq. (12.68a), we have

$$L_A = \frac{I_B}{I_A} \times (L - L_A) = \frac{I_B}{I_A} \times L - \frac{I_B}{I_A} \times L_A$$

or $\quad\quad L_A\left(1 + \dfrac{I_B}{I_A}\right) = \dfrac{I_B}{I_A} \times L$

or $\quad\quad L_A\left(\dfrac{I_A + I_B}{I_A}\right) = \dfrac{I_B}{I_A} \times L$

or $\quad\quad\quad\quad L_A = \dfrac{I_B}{(I_A + I_B)} \times L$ ...(12.69)

12.24 FREE TORSIONAL VIBRATIONS OF A THREE ROTOR SYSTEM

Consider a shaft to which two rotors A and B are attached at the ends and the rotor C in between. This system is known as three rotor system (Fig. 12.31). There are two possible modes of vibrations:

(i) The outside rotors A and B rotate in the same direction and rotor C rotates in the opposite direction. The nodes lie at N_1 and N_2.

(ii) The outside rotors A and B rotate in opposite direction. There is a single node, either between A and C or between B and C.

Let us consider the case when the rotors A and B rotate in the same direction and rotor C rotates in opposite direction (Fig. 12.31b).

Let I_A, I_B and I_C = mass moment of inertia of rotors A, B and C respectively,

$\quad L_1$ = distance between rotor A and C
$\quad L_2$ = distance between rotor B and C
$\quad L_a$ = distance of node N_1 from rotor A
$\quad L_b$ = distance of node N_2 from rotor B
$\quad L_{c1}$ = distance of node N_1 from rotor $c = (L_1 - L_a)$
$\quad L_{c2}$ = distance of node N_2 from rotor $c = (L_2 - L_b)$

Now, $\quad\quad\quad\quad f_{na} = f_{nb} = f_{nc}$

or $\quad\quad \dfrac{1}{2\pi}\sqrt{\dfrac{q_a}{I_a}} = \dfrac{1}{2\pi}\sqrt{\dfrac{q_b}{I_b}} = \dfrac{1}{2\pi}\sqrt{\dfrac{q_c}{I_c}}$

or
$$\frac{q_a}{I_a} = \frac{q_b}{I_b} = \frac{q_c}{I_c}$$

The torque required to produce unit twist of C is the sum of torques required to produce a unit twist in each of the lengths L_{c1} and L_{c2}, i.e. $q_c = q_{c1} + q_{c2}$

$\therefore$
$$\frac{GJ}{L_a I_a} = \left(\frac{GJ}{L_{c1}} + \frac{GJ}{L_{c2}}\right)\frac{1}{I_c} = \frac{GJ}{L_b I_b}$$

or
$$\frac{1}{L_a I_a} = \left(\frac{1}{L_1 - L_a} + \frac{1}{L_2 - L_b}\right)\frac{1}{I_c} = \frac{1}{L_b I_b}$$

As $L_a I_a = L_b I_b$ length L_a can be expressed in terms of L_b and a quadratic equation in L_b can be obtained. There will be two values of L_b and correspondingly two values of L_a.

   (i) One set of values given by the quadratic equation gives the position of two nodes and the frequency thus obtained is known as *two-node frequency*.
   (ii) In the other set of values, one gives the position of a single node and the other is beyond the physical limits of the equation. The frequency so obtained, is known as the *single-node frequency* or *fundamental frequency*.

If A and C rotate in the same direction and B rotates in the opposite direction, a single node is obtained between B and C (Fig. 12.31c). (L_a does not give the actual node point.) And $L_a > L_1$.

Similarly, if B and C rotate in the same direction and A in the opposite direction, single node is obtained between A and C (Fig. 12.31d). (L_b does not give the actual node point). In this case, $L_b > L_2$.

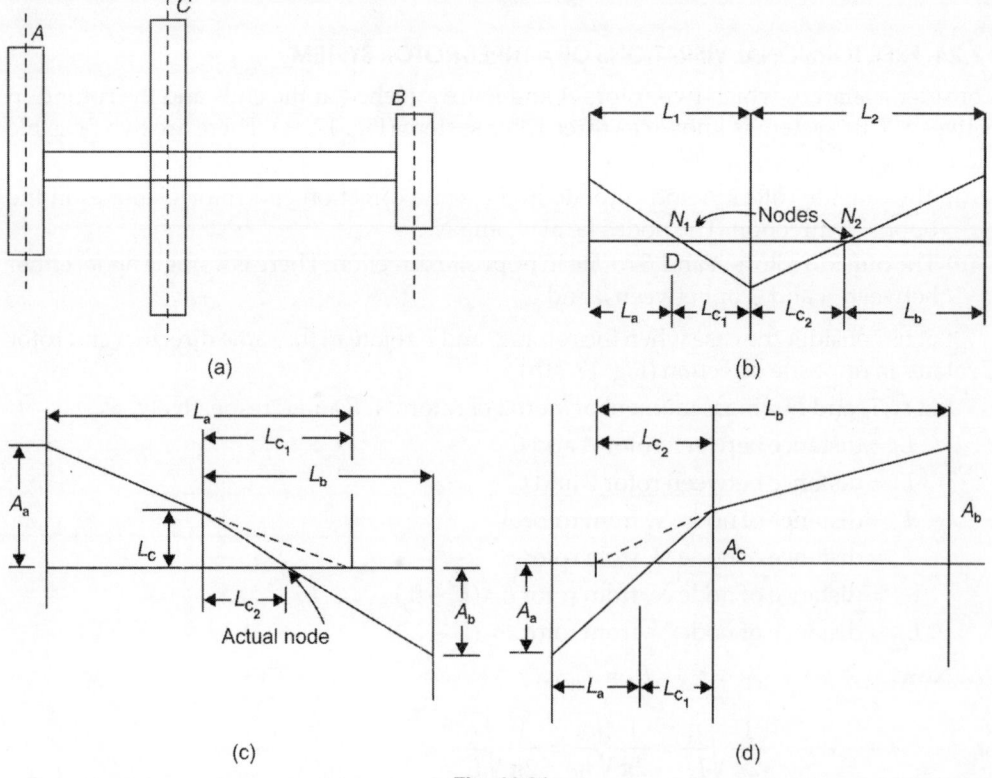

(a) (b)

(c) (d)

Fig. 12.31

Let A_a, A_b and A_c be the amplitudes of rotors A, B and C respectively.

Then, $\dfrac{A_a}{A_c} = \dfrac{L_a}{L_{c1}}$ and $\dfrac{A_b}{A_c} = \dfrac{L_b}{L_{c2}}$ (*from* Figs. 12.31b and c)

In general, possible number of node points and frequencies is one less than the number of rotors in torsional vibrating system.

12.25 TORSIONALLY EQUIVALENT SHAFT

In practice, the rotors are fixed to a shaft of different diameters at different sections. To find the frequency of such a system, the stepped shaft is replaced by a torsionally equivalent shaft of suitable diameter.

A torsionally equivalent shaft is a shaft of uniform diameter which has the same torsional stiffness as that of the stepped shaft so that it twists through the same angle under a given torque as the actual stepped shaft.

Figure 12.32a shows a shaft of different diameters and different lengths whereas Fig. 12.32b shows a shaft of uniform diameter 'd' and of length L. The shaft shown in Fig. 12.32b will become torsionally equivalent shaft if angle of twist is same in both the shafts when a torque T is applied at their opposite ends.

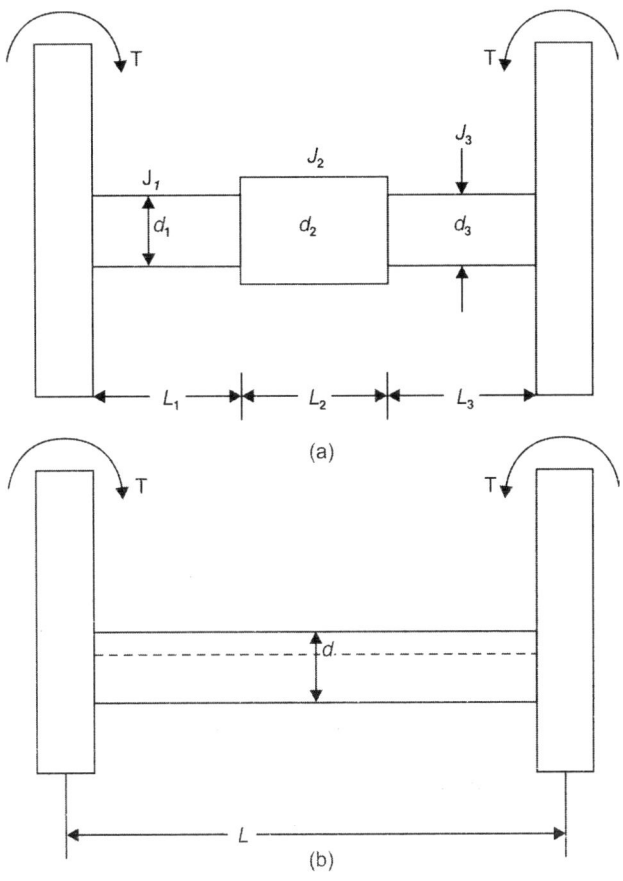

(a)

(b)

Fig. 12.32

Let d_1, d_2 and d_3 = diameters of lengths L_1, L_2 and L_3 respectively.

θ_1, θ_2 and θ_3 = angle of twist for lengths L_1, L_2 and L_3 respectively.

θ = angle of twist for a shaft of uniform dia. 'd' and length 'L'.

We know the torsional equation,

$$\frac{T}{J} = \frac{G + \theta}{L} \quad \text{or} \quad \theta = \frac{T \times L}{J \times G}$$

For a shaft of diameter 'd_1' and length 'L_1', the angle of twist is θ_1.

$$\theta_1 = \frac{T \times L_1}{J_1 \times G} \quad \text{where} \quad J_1 = \frac{\pi}{32} d_1^4$$

Similarly, the values of θ_2 and θ_3 are,

$$\theta_2 = \frac{T \times L_2}{J_2 \times G} \quad \text{where} \quad \theta_3 = \frac{T \times L_3}{J_3 \times G}$$

$\therefore$ Total angle of twist = $\theta_1 + \theta_2 + \theta_3$,

$$\theta = \frac{T \times L_1}{J_1 \times G} + \frac{T \times L_2}{J_2 \times G} + \frac{T \times L_3}{J_3 \times G}$$

For a shaft of uniform diameter 'd' and length 'L' when same torque is applied, the angle of twist is given as,

$$\theta = \frac{T \times L}{J \times G} \quad \text{where} \quad J = \frac{\pi}{32} d^4$$

For torsionally equivalent shaft,

Angle of twist in a shaft of uniform diameter = total angle of twist in a shaft of different lengths,

or
$$\theta = \theta_1 + \theta_2 + \theta_3$$

or
$$\frac{T \times L}{J \times G} = \frac{T \times L_1}{J_1 \times G} + \frac{T \times L_2}{J_2 \times G} + \frac{T \times L_3}{J_3 \times G}$$

or
$$\frac{L}{J} = \frac{L_1}{J_1} + \frac{L_2}{J_2} + \frac{L_3}{J_3}$$

or
$$\frac{L}{\dfrac{\pi}{32} d^4} = \frac{L_1}{\dfrac{\pi}{32} d_1^4} + \frac{L_2}{\dfrac{\pi}{32} d_2^4} + \frac{L_3}{\dfrac{\pi}{32} d_3^4}$$

or
$$\frac{L}{d^4} = \frac{L_1}{d_1^4} + \frac{L_2}{d_2^4} + \frac{L_3}{d_3^4}$$

In practice, to save the labour, it is assumed that the diameter of the equivalent shaft is equal to that of one of the section of the actual shaft. Let us assume that $d = d_1$, then the above equation becomes

$$\frac{L}{d_1^4} = \frac{L_1}{d_1^4} + \frac{L_2}{d_2^4} + \frac{L_3}{d_3^4}$$

or
$$L = L_1 \times \frac{d_1^4}{d_1^4} + L_2 \times \frac{d_1^4}{d_2^4} + L_3 \times \frac{d_1^4}{d_3^4}$$

or
$$L = L_1 + L_2 \left(\frac{d_1}{d_2}\right)^4 + L_3 \left(\frac{d_1}{d_3}\right)^4 + \dots\dots \qquad \dots(12.70)$$

where L = length of torsionally equivalent shaft.

12.26 FREE TORSIONAL VIBRATIONS OF GEARED SYSTEM

A geared system is shown in Fig. 12.33. It has two shafts of diameters d_a and d_b and of lengths L_1 and L_2 respectively. Shaft 1 carries a rotor A on one end and a pinion on the other end (c). Shaft 2 carries a gear D meshing with the pinion at one end and a rotor B on the other end. This system is replaced by an equivalent single shaft system with following assumptions:

    i. There is no blacklash or slip in the gear drive,

    ii. The inertia of shafts and gears is negligible, and

    iii. The load is within the elastic limits of gear teeth i.e., the gear teeth are rigid and do not distort under load.

With these assumptions, the system reduces to a two-rotor system as shown in Fig. 12.33b.

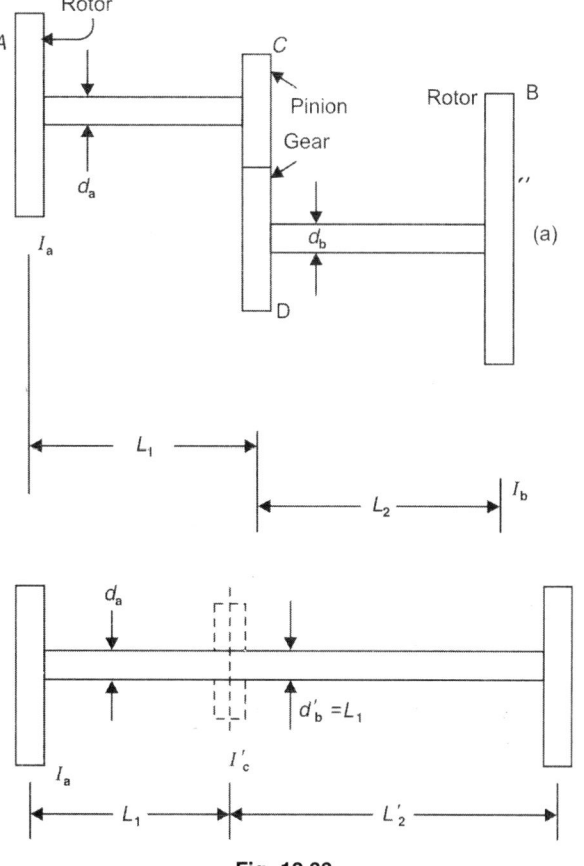

Fig. 12.33

The system 'b' will be equivalent to the system 'a' if,

    i. the kinetic energy of system 'b' is equal to that of 'a', and

    ii. the strain energy of system 'b' is equal to that of 'a'.

Equating the kinetic energies,

KE of original system = KE of section L_1 + KE of section L_2

KE of equivalent system = KE of section L_1 + KE of section L_2'

$\therefore KE$ of section L_2' = KE of section L_2

$$\frac{1}{2}I_b'L\omega_b'^2 = \frac{1}{2}I_b\omega_b^2$$

$$\frac{1}{2}I_b'\omega_a^2 = \frac{1}{2}I_b\omega_b^2 \qquad (\omega_b' = \omega_a)$$

or $\qquad\qquad I_b' = I_b\left(\frac{\omega_b}{\omega_a}\right)^2 = \frac{I_b}{G_r^2} \qquad \left(G_r = \frac{\omega_a}{\omega_b}\right)$

Equating the strain energies,

SE of section L_2' = SE of section L_2

$$\frac{1}{2}T_b'\theta_b' = \frac{1}{2}T_b\theta_b$$

$$\frac{J_b'G}{L_2'}\theta_b'\theta_b' = \frac{J_bG}{L_2}\theta_b\theta_b \qquad \left(\theta = \frac{TL}{JG}\right)$$

$$L_2' = \left(\frac{\theta_b'}{\theta_b}\right)^2 L_2\left(\frac{J_b'}{J_b}\right) \qquad (\theta = \omega t)$$

$$= \left(\frac{\omega_b'}{\omega_b}\right)^2 L_2\left(\frac{d_b'}{d_b}\right)^4 \qquad (\omega_b' = \omega_a)$$

$$= \left(\frac{\omega_a}{\omega_b}\right)^2 L_2\left(\frac{d_b'}{d_b}\right)^4$$

$$= G_r^2 L_2\left(\frac{d_b'}{d_b}\right)^4$$

Assuming the diameter of the equivalent shaft of to be equal to that of shaft 1, $d_b' = d_a$,

$$L_2' = G_r^2 L_2\left(\frac{d_a}{d_b}\right)^4$$

Length of equivalent shaft,

$$= L_1 + G_r^2 L_2\left(\frac{d_a}{d_b}\right)^4$$

In case the inertia of the gearing is not negligible, an additional rotor has to be considered at a distance l_1 from the rotor A. The mass moment of inertia of the rotor is thus given by,

$$I_c' = I_{ca} + \frac{I_{cb}}{G_r^2}$$

where I_{ca} = MOI of gear, I_{cb} = MOI of pinion.

This way, the system will act as a three-rotor system.

Example 12.12: A vertical shaft is fixed at upper end and carries a disc of weight 2000 N at its lower end. The length of shaft is 1 m and its diameter is 200 mm. The modulus of rigidity of the shaft is $8.16 \times 10^4 \ \dfrac{\text{N}}{\text{mm}^2}$. If the radius of gyration of flywheel is 275 mm, determine the frequency of torsional vibrations.

Solution: $W = 2000$ N, $L = 1$ m, $d = 200$ mm $= 0.2$ m

$$G = 8.16 \times 10^4 \ \text{N/mm}^2 = 8.16 \times 10^{10} \ \text{N/m}^2$$

$$k = 275 \ \text{mm} = 0.275 \ \text{m}, \ m = \frac{W}{g} = \frac{2000}{9.81} \ \text{kg} = 203.9 \ \text{kg}$$

$$J = \frac{\pi}{32} d^4 = \frac{\pi}{32} \times (0.2)^4 = 0.000157 = 1.57 \times 10^{-4} \ \text{m}^4$$

Torsional stiffness,

$$q = \frac{GJ}{L} = \frac{8.16 \times 10^{10} \times 1.57 \times 10^{-4}}{1}$$

$$= 1.28 \times 10^7$$

$$q = 1.28 \times 10^7 \ \text{N·m}$$

$$I = mk^2 = 230.9 \times (0.275)^2 = 15.42 \ \text{kg·m}^2$$

Frequency of torsional vibrations,

$$f_n = \frac{1}{2\pi} \sqrt{\frac{q}{I}} = \frac{1}{2\pi} \sqrt{\frac{1.28 \times 10^7}{15.42}} = 145.14 \ \text{Hz}$$

Example 12.13: Determine the frequency of torsional vibrations of the disc attached to the shaft, if both the ends of the shaft are fixed and diameter of the shaft is 40 mm. The disc has a mass of 96 kg and a radius of gyration of 0.4 m. Take modulus of rigidity for the shaft material as 85 GN/m². $L_1 = 1$ m and $L_2 = 0.8$ m.

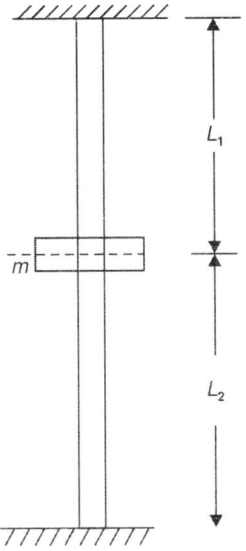

Fig. 12.34

Solution: $m = 96$ kg

$G = 85 \times 10^9$ N/m²

$k = 0.4$ m

$d = 0.04$ m

$I = mk^2$

$= 96 \times (0.4)^2$

$= 15.36$ kg·m²

$J = \dfrac{\pi}{32} d^4$

$= \dfrac{\pi}{32} \times (0.04)^4$

$= 0.251 \times 10^{-6}$ m⁴

Total torsional stiffness of shaft,

$$q = q_1 + q_2 = \dfrac{GJ}{L_1} + \dfrac{GJ}{L_2}$$

$$= 85 \times 10^9 \times 0.251 \times 10^{-6} \left(\dfrac{1}{1} + \dfrac{1}{0.8} \right)$$

$$= 48004 \text{ N·m}$$

$$f_n = \dfrac{1}{2\pi} \sqrt{\dfrac{q}{I}}$$

$$= \dfrac{1}{2\pi} \sqrt{\dfrac{48004}{15.36}}$$

$$= 8.9 \text{ Hz}$$

Example 12.14: The shaft shown in Fig. 12.35 carries two masses. The mass A is 300 kg with radius of gyration of 0.75 m and the mass B is 500 kg with radius of gyration of 0.9 m. Determine the frequency of the torsional vibrations. It is desired to have the node at the mid-section of the shaft of 120 mm diameter by changing the diameter of the section having a 90 mm diameter. What will be the new diameter?

Solution:

(i) Reducing the shaft of torsionally equivalent shaft of 100 mm diameter.

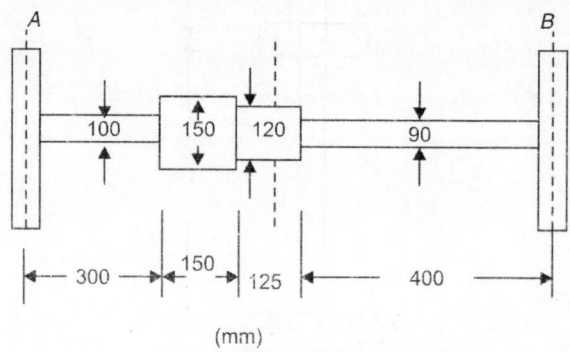

(mm)

Fig. 12.35

Length of equivalent shaft,

$$L = 300\left(\frac{100}{100}\right)^4 + 150\left(\frac{100}{150}\right)^4 + 125\left(\frac{100}{120}\right)^4 + 400\left(\frac{100}{90}\right)^4$$

$$= 300 + 31.6 + 60.2 + 609.7$$

$$= 1001.5 \text{ mm or } 1.0015 \text{ m}$$

To locate the node point, $I_a L_a = I_b L_b$

or $\quad\quad m_a k_a^2 L_a = m_b k_b^2 L_b$

or $\quad\quad 300 \times (0.75)^2 L_a = 500 \times (0.9)^2 (1.0015 - L_a)$

or $\quad\quad L_a = 0.707 \text{ m}$

$$f_n = \frac{1}{2\pi}\sqrt{\frac{GJ}{I_a L_a}} = \frac{1}{2\pi}\sqrt{\frac{84 \times 10^9 \times \frac{\pi}{32}(0.1)^4}{300 \times (0.75)^2 \times 0.707}} = 13.2 \text{ Hz}$$

(ii) Now, if the node point is to be at the mid-section of the shaft with diameter 120 mm.

$$L_a = 300 + 31.6 + \frac{60.2}{2} = 361.7 \text{ mm or } 0.3617 \text{ m}$$

Again, $\quad\quad I_a L_a = I_b L_b \text{ or } m_a k_a^2 L_a = m_b k_b^2 L_b$

$$300 \times (0.75)^2 \times 0.3617 = 500(0.9)^2 \times L_b$$

or $\quad\quad L_b = 0.1507 \text{ m}$

Let d be the new diameter of the last section of the shaft.

Then, in this case, $\quad L_b = \frac{1}{2} \times \left[125\left(\frac{100}{120}\right)^4\right] + 400\left(\frac{100}{d}\right)^4$

or $\quad\quad 150.7 = \frac{60.2}{2} + 400\left(\frac{100}{d}\right)^4, d = 135 \text{ mm}.$

Example 12.15: A shaft carries a motor at one end and a pinion on the other end. The length and diameter of this shaft are 500 mm and 60 mm respectively. There is another shaft of length 900 mm and of diameter 100 mm. This shaft carries a gear wheel at one end and a centrifugal pump on another end. The gear wheel and pinion are meshing together so that the centrifugal pump is driven by the motor. The mass moment of inertia of motor and centrifugal pump are 200 kg·m^2 and 750 kg·m^2 respectively. If the inertia of gears and shaft is neglected and pump speed is one-third of the motor then find the frequency of torsional vibrations of the system. Take the value of modulus of rigidity as 80 kN/mm^2.

Solution: $L_1 = 500 \text{ mm} = 0.5 \text{ m}; d_1 = 60 \text{ mm} = 0.06 \text{ m}; L_2 = 900 \text{ mm} = 0.9 \text{ m}; d_2 = 100 \text{ mm} = 0.1$ m; Mass moment of inertia of motor, $I_A = 200 \text{ kgm}^2$; mass moment of inertia of pump,

$I_D = 750 \text{ kgm}^2$; pump speed $= \frac{1}{3}$ of motor speed. Hence, speed of motor $= 3 \times$ pump speed;

$G = 80 \text{ kN/mm}^2 = 80 \times 10^3 \text{ N/mm}^2 = 80 \times 10^3 \times 10^6 \text{ N/m}^2 = 80 \times 10^9 \text{ N/m}^2.$

The inertia of gears and shaft is neglected. Then this becomes the two-rotor system.

The gear ratio, $G = \dfrac{\text{Speed of pinion } B}{\text{Speed of gear wheel } C} = \dfrac{\text{Speed of motor}}{\text{Speed of pump}} = 3$

The mass moment of inertia of the equivalent rotor D' is given as:

$$I'_D = \frac{I_D}{G^2} = \frac{750}{3^2} = \frac{750}{9} = 83.33 \text{ kg·m}^2$$

Additional length of the equivalent shaft, assuming its diameter as $d_1 = 60$ mm, is given as,

$$L_e = G^2 \times \left(\frac{d_1}{d_2}\right)^4 \times L_2$$

$$= 3^2 \times \left(\frac{60}{100}\right)^4 \times 900 = 1049.76 \text{ mm} = 1050 \text{ mm}$$

∴ Total length of equivalent shaft,

$$L = L_1 + L_e = 500 + 1050 = 1550 \text{ mm} = 1.55 \text{ m}$$

Figure 12.36 shows the position of node N on the equivalent system.

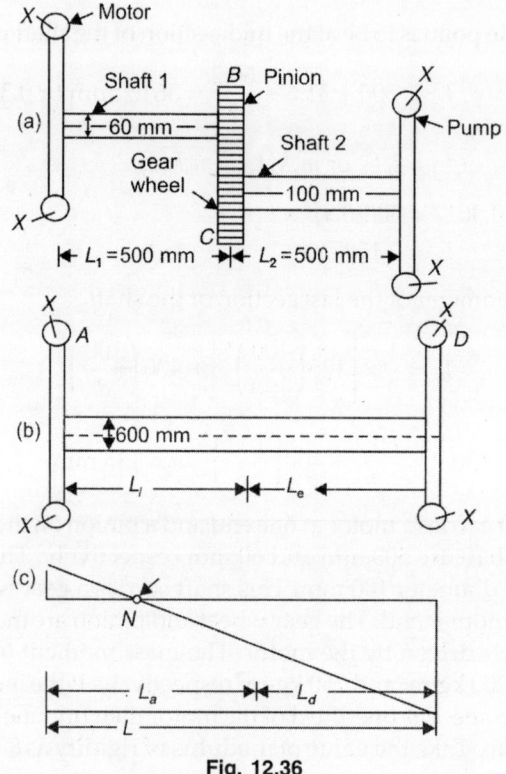

Fig. 12.36

Let L_A = distance of node N from rotor A

$L_{D'}$ = distance of node N from rotor D.

We know that,

$$L_A \times I_A = L_{D'} \times I_{D'}$$

or

$$L_A = 200 = L_{D'} \times 83.33$$

$$= (L - L_A) \times 83.33 \qquad (\because L_{D'} = L - L_A)$$

$$= (1.55 - L_A) \times 83.33$$

or
$$\frac{L_A \times 200}{83.33} = 1.55 - L_A$$

or
$$2.4 L_A = 1.55 - L_A$$

or
$$3.4 L_A = 1.55$$

or
$$L_A = \frac{1.55}{3.4} = 0.4558 \approx 0.456 \text{ m} = 456 \text{ mm}$$

The frequency of torsional vibration is given by,

$$f_{nA} = \frac{1}{2}\sqrt{\frac{G \times J}{I_A \times L_A}}$$

$$J = \frac{\pi}{34} d_1^4 = \frac{\pi}{32} \times (0.06)^4 = 1.27 \times 10^{-6} \text{ m}^4$$

$$f_{nA} = \frac{1}{2\pi}\sqrt{\frac{(80 \times 10^9) \times (1.27 \times 10^{-6})}{200 \times 0.456}} \text{ Hz} = 5.31 \text{ Hz}$$

Example 12.16: Determine the time in which the mass in a damped vibrating system would settle down to 1/50th of its initial deflection for the following data:

$$m = 200 \text{ kg}, \zeta = 0.22, k = 40 \text{ N/mm}$$

Also, find the number of oscillations completed to reach this value of deflection.

Solution: We know,
$$\frac{X_O}{X_N} = e^{\zeta \omega_n N T_d}$$

where
$$\omega_n = \sqrt{\frac{k}{m}} = \sqrt{\frac{40 \times 10^3}{200}} = 14.14 \text{ rad/s}$$

$\therefore$
$$50 = e^{0.22 \times 14.14 N T_d}$$

or total time,
$$N T_d = 126 \text{ s}$$

$$T_d = \frac{2\pi}{\sqrt{1 - \zeta^2}\omega_n} = \frac{2\pi}{\left(\sqrt{1 - (0.22)^2}\right) \times 14.14} = 0.455 \text{ s}$$

Number of oscillations completed $= \dfrac{1.26}{0.455} = 2.76.$

Example 12.17: A machine mounted on springs and fitted with a dashpot has a mass of 60 kg. There are three springs, each of stiffness 12 N/mm. The amplitude of vibrations reduces from 45 to 8 mm in two complete oscillations. Assuming that the damping force varies as the velocity, determine: (i) the damping coefficient, (ii) the ratio of frequencies of damped and undamped vibrations and (iii) the periodic time of damped vibrations.

Solution: $m = 60$, Stiffness of each spring, $k_1 = 12$ N/mm

Total stiffness, $k = 3k_1 = 3 \times 12 = 36$ N/mm $= 36 \times 10^3 \dfrac{\text{N}}{\text{m}}$

Natural frequency, $\omega_n = \sqrt{\dfrac{k}{m}} = \sqrt{\dfrac{36 \times 10^3}{60}} = 24.49$ rad/sec

(i)
$$\frac{X_0}{X_2} = \frac{X_0}{X_1} \times \frac{X_1}{X_2} = \left(\frac{X_0}{X_1}\right)^2$$

or
$$\left(\frac{X_0}{X_1}\right) = \left(\frac{X_0}{X_2}\right)^{1/2} = \left(\frac{45}{8}\right)^{1/2} = 2.37$$

$$\frac{2\pi\zeta}{\sqrt{1-\zeta^2}} = \log_e 2.37 = 0.864$$

$$1 - \zeta^2 = 52.88\zeta^2, \text{ or } \zeta^2 = 0.0185$$

or
$$\zeta = 0.136$$

damping coefficient,
$$c = 2\,m,\, \omega_n \zeta = 2 \times 60 \times 24.49 \times 0.136$$
$$= 400\,\text{N/m/s} = 0.4\,\text{N/mm/s}$$

(ii) $\dfrac{\text{Damped frequency}}{\text{Undamped frequency}} = \dfrac{\omega_d}{\omega_n} = \dfrac{\sqrt{1-\zeta^2}\,\omega_n}{\omega_n} = \sqrt{1-\zeta^2}$

$$= \sqrt{1 - (0.136)^2} = 0.99$$

(iii) $T_d = \dfrac{2\pi}{\omega_d} = \dfrac{2\pi}{\sqrt{1-\zeta^2}\,\omega_n} = \dfrac{2\pi}{\left(\sqrt{1-(0.136)^2}\right) \times 24.49} = 0.259\,\text{s}$

Example 12.18: A machine weighing 3.5 kg, vibrates in a viscous medium. A harmonic exciting force of 40 N acts on the machine and produces resonant amplitude of 18 mm with a period of 0.2 second. Determine the damping coefficient.

Solution: $m = 3.5$ kg, $F_0 = 40$ N, $A = 18$ mm

$$T = 0.2\,\text{sec}$$

$$\omega = \frac{2\pi}{T} = \frac{2\pi}{0.2} = 31.4\,\text{rad/sec}$$

At resonance, $\omega = \omega_n$

or
$$\omega_n = \sqrt{\frac{k}{m}} = 31.4 \text{ or } \frac{k}{m} = (31.4)^2 = 985.96$$

$\Rightarrow$
$$k = 985.96 \times 3.5 = 3450.86\,\text{N/m} = 3.45086\,\text{N/mm}$$

Amplitude,
$$A = \frac{F_0/k}{\sqrt{\left[1 - \left(\dfrac{\omega}{\omega_n}\right)^2\right]^2 + \left(\dfrac{2\zeta\omega}{\omega_n}\right)^2}}$$

or
$$A = \frac{F_0/k}{2\zeta} \qquad \left(\frac{\omega}{\omega_n} = 1\right)$$

or
$$0.0/8 = \frac{40/3450.86}{2\zeta}$$

or
$$\zeta = 0.3212$$

Damping coefficient, $c = 2\,m\,\omega_n\zeta$
$$= 2 \times 3.5 \times 31.4 \times 0.3212$$
$$c = 70.6\,\text{N/m/s}$$

Example 12.19: A body having a mass of 15 kg is suspended from a spring which deflects 12 mm under weight of the mass. Determine the frequency of the free vibrations. What is the viscous damping force needed to make the motion aperiodic at a speed of 1 mm/s? If, when damped to this extent, a disturbing force having a maximum value of 100 N and vibrating at 6 Hz is made to act on the body, determine the amplitude of the ultimate motion.

Solution: $m = 15$ kg, $\Delta = 12$ mm, $F_0 = 100$ N, $f = 6$ Hz

$$f_n = \frac{1}{2\pi}\sqrt{\frac{g}{\Delta}} = \frac{1}{2\pi}\sqrt{\frac{9.81}{0.012}} = 4.55 \text{ Hz}$$

The motion becomes aperiodic when the damped frequency is zero or when it is critically damped ($z = 1$), and,

$$\omega = \omega_n = \sqrt{\frac{g}{\Delta}} = \sqrt{\frac{9.81}{0.012}} = 28.59 \text{ rad/s}$$

$$c = 2m\omega_n = 2 \times 15 \times 28.59 = 857 \text{ N/m/s}$$
$$= 0.857 \text{ N/mm/s}$$

Thus, the force needed is 0.857 N at a speed of 1 mm/s

Amplitude, $$A = \frac{F_0}{\sqrt{(k - m\omega^2)^2 + (c\omega)^2}}$$

But, $$\omega = 2\pi \times f = 2\pi \times 6 = 37.7 \text{ rad/s}$$
and k can be found from

$$f_n = \frac{1}{2\pi}\sqrt{k/m}$$

or $$4.55 = \frac{1}{2\pi}\sqrt{k/15}$$

or $$k = 12260 \text{ N/m}$$

∴ $$A = \frac{100}{\sqrt{[12260 - 15 \times (37.7)^2]^2 + (857 \times 37.7)^2}}$$

$$= 0.00298 \text{ m} = 2.98 \text{ mm}$$

Example 12.20: A refrigerator unit having a mass of 35 kg is to be supported on three springs, each having a spring stiffness. The unit operates at 480 rpm. Find the value of stiffness k if only 10% of the shaking force is allowed to be transmitted to the supporting structure.

Solution: As no damper is used,

transmissibility, $$\varepsilon = \frac{1}{\pm\left[1 - \left(\dfrac{\omega}{\omega_n}\right)^2\right]}$$

$$\omega = \frac{2\pi \times 480}{60} = 16\pi \text{ and } \varepsilon = 0.1$$

$\therefore \qquad 0.1 = \dfrac{1}{\pm\left[1 - \left(\dfrac{16\pi}{\omega_n}\right)^2\right]}$

or $\qquad \pm\left[0.1 - 0.1\left(\dfrac{16\pi}{\omega_n}\right)^2\right] = 1$

If the positive sign is taken, $\dfrac{16\pi}{\omega_n} = \sqrt{-9}$ which is not possible.

Therefore, taking the negative sign, $\dfrac{16\pi}{\omega_n} = \sqrt{11}$

or $\qquad\qquad\qquad \omega_n = 15.15 \text{ rad/s}$

or $\qquad\qquad\qquad \sqrt{\dfrac{k}{m}} = \sqrt{\dfrac{k}{35}} = 15.15$

Equivalent stiffness, $\quad k = 8037 \text{ N/m} = 8.037 \text{ N/mm}$

Stiffness of each spring $= \dfrac{8.037}{3} = 2.679 \text{ N/mm}.$

Example 12.21: A spring mass system is excited by a force $F \sin \omega t$. On measuring, the amplitude of vibration is found to be 12 mm at resonance. However, at a frequency 0.8 times the resonant frequency, the amplitude reduces to 8 mm. Determine the damping ratio of the system.

Solution: At resonance ($\omega = \omega_n$), Amplitude, $A_1 = 12$ mm at $\omega = 0.8\omega_n$, $A_2 = 8$ mm

We know that the amplitude of steady state response is given as,

$$A = \dfrac{F_0/k}{\sqrt{\left[1 - \left(\dfrac{\omega}{\omega_n}\right)^2\right]^2 + \left(2\zeta\dfrac{\omega}{\omega_n}\right)^2}} \qquad \text{...(i)}$$

At resonance i.e., $\qquad \omega = \omega_n \text{ or } \dfrac{\omega}{\omega_n} = 1$

Amplitude, $\qquad\qquad A_1 = \dfrac{F_0/k}{2\zeta}$

Substituting the values, $\quad 12 = \dfrac{F_0/k}{2\zeta} \text{ or } \dfrac{F_0}{k} = 24\zeta \qquad \text{...(ii)}$

at $\omega = 0.8\omega_n$, $A_2 = 8$ mm.

Again substituting values in Eq. (i)

$$8 = \dfrac{F_0/k}{\sqrt{\left[1 - (0.8)^2\right]^2 + (2\zeta \times 0.8)^2}}$$

or $\qquad\qquad 8 = \dfrac{24\zeta}{\sqrt{0.1296 + 2.56\zeta^2}} \qquad \left(\text{from Eq. (ii), } \dfrac{F_0}{k} = 24\zeta\right)$

or
$$\left(\frac{8}{24}\right)^2 = \frac{\zeta^2}{0.1296 + 2.56\zeta^2}$$

or
$$0.0144 + 0.2844\zeta^2 = \zeta^2$$

or
$$0.7155\zeta^2 = 0.0144$$

or
$$\zeta^2 = 0.02013$$

∴
$$\zeta = 0.142$$

EXERCISE

12.1 What are vibrations?

12.2 What are the causes and effects of vibrations?

12.3 Define free vibrations, forced vibrations and damped vibrations.

12.4 Describe with the help of neat sketches the longitudinal, transverse and torsional vibrations.

12.5 What are the basic elements of a vibratory system? What is the degree of freedom?

12.6 Write short notes on the following:
(i) Critical speed   (ii) Damped vibrations

12.7 Explain the term "logarithmic decrement" as applied to damped vibrations.

12.8 Explain with the help of graphs, the variation in amplitude of forced undamped vibrations, with change in angular velocity when periodic force is constant in magnitude.

12.9 Determine the natural frequency of a vibratory system having a mass suspended from the free end of a massless spring. Discuss the effect of the inertia of the spring mass.

12.10 Describe different methods of finding the natural frequency of free longitudinal vibrations.

12.11 Define 'damped vibrations'. Discuss the effect of damping on vibratory systems. What do you mean by underdamping, overdamping and critical damping.

12.12 Define the terms: damping factor, damping coefficient and critical damping coefficient.

12.13 Describe whirling of shafts. Define whirling or critical speed of shaft.

12.14 Derive an expression for logarithmic decrement in terms of damping factor.

12.15 Derive an expression for the natural frequency of free transverse and lontidudinal vibrations by equilibrium method.

12.16 Prove that the whirling speed for a rotating shaft is the same as the frequency of natural transverse vibrations.

12.17 Discuss the effect of inertia of the shaft in longitudinal and transverse vibrations.

12.18 Derive an expression for the frequency of free torsional vibrations for a shaft fixed at one end and carrying a load on the free end.

12.19 Establish an expression for the amplitude of forced vibrations.

12.20 Discuss the effect of inertia of a shaft on the free torsional vibrations.

12.21 Explain the terms vibration isolation, transmissibility and the force transmitted to the foundation of the machine.

12.22 Prove that the ratio of two successive amplitudes in case of under damped system is constant.

12.23 What is magnification factor? Derive the relation for transmissibility when no damper is used in the system.

12.24 What is steady state response of the system in case of forced vibrations?

12.25 Find the natural frequency of a uniformly loaded simply supported shaft, making transverse vibrations due to elastic forces.

12.26 What do you mean by Torsionally equivalent shaft? Derive an expression for the same.

12.27 Establish the relation to determine the frequency of torsional vibrations of a geared system.

12.28 A car having a mass of 1500 kg deflects its spring 3 cm under its own load. Find the natural frequency of car in vertical direction.

12.29 The natural frequency of a spring-mass system is 15 Hz. An extra 3 kg mass is coupled to its mass and natural frequency reduces by 3 Hz. Find the mass and stiffness of the system.

12.30 A spring mass system has a period 0.25 seconds. What will be the new period if the spring constant is increased by 50%.

12.31 A spring mass system has a natural frequency of 12 Hz. When the spring constant is reduced by 800 N/m, the frequency is changed by 50%. Determine the mass and spring constant of the original system.

12.32 A vertical shaft 100 mm in diameter and 1 m in length has its upper end fixed at the top. At the other end, it carries a disc of weight 20 kN. The Young's modulus of the material of the shaft is 2×10^5 N/mm^2. Neglecting the weight of the shaft, determine the frequency of longitudinal vibrations and transverse vibrations.

(*Ans.* (i) 139.75 Hz, (ii) 6 Hz)

12.33 A simply supported shaft of length 1.6 m carries a mass of 120 kg placed 500 mm from one end. If E = 200 GN/m^2 and diameter of shaft is 50 mm, then find the natural frequency of transverse vibrations. (*Ans.* 14.34 Hz)

12.34 Find the frequency of transverse vibrations of a shaft which is simply supported at the ends and is of 40 mm in diameter. The length of the shaft is 5 m. The shaft carries three point loads of masses 15 kg, 35 kg and 22.5 kg at 1 m, 2 m and 3.4 m respectively from the left support. The Young's modulus for the material of the shaft is 200 GN/m^2. Neglect the weight of the shaft. (*Ans.* 2.1 Hz)

12.35 In a spring-mass vibrating system, the natural frequency of vibration is 3.56 Hz. When the amount of the suspended mass is increased by 5 kg, the natural frequency is lowered to 2.9 Hz. Determine the original known mass and the spring constant.

(*Ans.* 10 kg; 5 N/mm)

12.36 A vibrating system consists of a mass of 20 kg, a spring of stiffness 20 kN/m and a damper. The damping provided is only 30% of the critical value. Determine the natural frequency of the damped vibration and the ratio of two consecutive amplitudes. (*Ans.* 30.2 rad/s; 7.21)

12.37 The following data relate to a damped vibrating system:

$$m = 140 \text{ kg}, s = 50 \text{ N/mm and } z = 0.25$$

Determine the time in which the mass would settle down to 1/80th of its initial deflection. Also, what will be the number of oscillations completed to reach this value. (*Ans.* 0.927 s; 2.7)

12.38 A shaft is simply supported at the ends and is of 20 mm in diameter and 600 mm in length. The shaft carries a load of 9.81 N at its centre. The weight of shaft per metre length is 124.1 N. Find the critical speed of the shaft. Take Young's modulus = 200 GN/m^2. (*Ans.* 2596.5 rpm)

12.39 A shaft is simply supported at its end and is of 40 mm in diameter and of length 5 m. The shaft carries three point loads of masses 15 kg, 35 kg and 22.5 kg at 1 m, 2 m, and 3.4 m respectively from the left support. The weight of the shaft per metre length is given as 18.394 N. The Young's modulus for the material of the shaft is 200 GN/m^2. Find the critical speed of the shaft. (*Ans.* 120 rpm)

12.40 A steel bar 22 mm wide and 45 mm deep is freely supported at two points 800 mm apart and carries a load of 180 kg midway between them. Determine the natural frequency of the transverse vibration, neglecting the weight of the bar. Also find the frequency of vibration if an additional load of 180 kg is distributed uniformly along the length of the shaft. Take E = 250 GN/m^2. (*Ans.* 23.5 Hz; 19.2 Hz)

12.41 A shaft 1.2 m long has diameter of 45 mm for half the length and 60 mm for the remaining length. One end of the shaft is fixed and the other carries a rotor of mass 200 kg with a radius of gyration of 45 mm. Find the frequency of free torsional vibration neglecting the inertia of the shaft. Take G = 84 GN/m^2. (*Ans.* 3.88 Hz)

12.42 In a single-degree damped vibrating system, the suspended mass of 4 kg makes 24 oscillations in 20 seconds. The amplitude decreases to 0.3 of the initial value after 4 oscillations. Find the stiffness of the spring, the logarithmic decrement, the damping factor and damping coefficient. (*Ans.* 227 N/m; 0.3; 0.0478; 2.88 N/m/s)

12.43 A gun barrel weighs 300 kg and has a recoil spring of stiffness 250 N/mm. The barrel recoils 0.8 m on firing. Determine:
  (i) the critical recoil velocity of the gun, and
  (ii) the critical damping coefficient of the dashpot engaged at the end of the recoil stroke. (*Ans.* 23.1 m/s; 17.322 N/mm/s)

12.44 A machine weighs 18 kg and is supported on springs and dashpots. The total stiffness of the springs is 12 N/mm and damping is 0.2 N/mm/s. The system is initially at rest and a velocity of 120 mm/s is imparted to the mass. Determine:
  (i) the displacement and velocity of mass as a function of time, and
  (ii) the displacement and velocity after 0.4 s.

12.45 A rotor has a mass of 12 kg and is mounted midway on a 24 mm diameter horizontal shaft supported at the ends by two bearings. The bearings are 1 m apart. The shaft rotates at 2400 rpm. If the centre of mass of the rotor is 0.11 mm away from the geometric centre of the rotor due to a certain manufacturing defect, find the amplitude of the steady-state vibration and the dynamic force transmitted to the bearing. E = 200 GN/m^2.

Index